Hans Primas

Chemistry, Quantum Mechanics and Reductionism

Perspectives in Theoretical Chemistry

Springer-Verlag
Berlin Heidelberg New York Tokyo 1983

Author

Hans Primas
Laboratory for Physical Chemistry
Swiss Federal Institute of Technology (ETH)
ETH-Zentrum, CH-8092 Zürich

Second corrected edition 1983.
First edition (1981) published in the series:
Lecture Notes in Chemistry, Vol. 24
Springer-Verlag Berlin Heidelberg New York Tokyo

ISBN-13: 978-3-642-69367-0 e-ISBN-13: 978-3-642-69365-6
DOI: 10.1007/978-3-642-69365-6

Softcover reprint of the hardcover 2nd edition 1983

2152/3140-543210

CONTENTS

Chapter 4: Beyond Pioneer Quantum Mechanics

Chapter 5: A Framework for Theoretical Chemistry

Chapter 6: Reductionism, Holism and Complementarity

F O R E W O R D

Considering the great variety of facts, theories, forms of life that has existed at any time in the history of Western thought and that has become quite overwhelming today, scientists and philosophers have adopted the one or the other of the following three points of view:

1. *Positivism*: things are what they are - all we can do is to collect what exists and to order it.

2. *Realism*: things are not what they seem to be but there is an underlying reality. It is the task of the scientist (philosopher) to discover this reality and to explain everything on its basis.

Positivism and realism are well known in the 20th century - but they have great ancestors. Thus some early Greek thinkers assembled 'wonderful facts' such as earthquakes, solar eclipses, the periodic rising of the Nile and tried to explain them, using different explanations for different facts. The procedure is quite old; it can still be found in Herodotus and in part of the Hippocratic corpus. Today we have the tendency to collect facts and increase the precision of measurements without any theoretical rationale. Facts, it is assumed by those who follow the trend, have value in themselves no matter how well they fit into or clash with abstract conceptions and no matter what their (theoretical, or 'social') significance.

Positivism was criticised already in antiquity. "To know a lot" writes Heraclitus "does not teach reason". Anaximander developed a comprehensive point of view that explained the origin and structure of the world: the world is an ordered arrangement, it is a cosmos and it is uniform - the same laws are valid throughout. Dreams, gods, Tartarus have no room in this world, they become homeless and the question arises how belief in them can be accounted for. The question becomes urgent in the case of Parmenides who showed, by an intricate and forceful argument that there can be neither motion nor difference. For now not only gods and dreams but the entire world of commonsense turns out to be unreal.

It is interesting to see how certain features of this extreme view survive even today. Thus the theory of relativity gives what one might call a static account of motion and leaves essential properties to the consciousness of the observer while there are many physicists who claim that important differences (quantum theory vs. the laws and facts of classical mechanics, chemistry, biology) are in fact non existent. It is also well known how often philosophers and physicists have tried to understand quantum theory in classical terms. Modern realism differs from Parmenides in trying to reduce appearance to the underlying reality instead of throwing it out. It agrees with him in assuming that real is what can be so reduced.

In antiquity the two traditions just described were soon joined by a third which, for want of a better term, I shall call

3. *the structural approach*. According to this approach positivism and realism neglect phenomena which, far from being passing fancies of nimble minds are relied upon and used by them even when they propose radical views such as those of Parmenides. Positivism assumes that facts have no structure - hence all we can do is to collect them and parcel them off in convenient ways. Realism assumes that there is only one kind of structure to be considered. Commonsense and the work of scientists and philosophers are built in a different way; they are split into domains each domain being held together by certain principles. If we want to maintain the forms of life on which our science and our existence as rational human beings depends then we must take these principles into account and use them as boundary conditions of research.

The foremost ancient exponent of the structural view

was Aristotle. Aristotle rejected positivism on the grounds that we know not only facts but also principles. The principles may not cover all there is but there are sizeable domains held together by the same set of principles. It is the task of the philosopher-scientist to identify the domains and to formulate the principles appropriate to them. Aristotle rejected realism by simply pointing to the results of a research of this kind: *there are* different domains guided by different principles none of which can be pushed aside or 'reduced' to other principles. However, there may exist a general theory of being which permits us to bring order into the principles discovered.

Realism and especially the mechanical philosophy played an important role in the rise and the development of modern science. Helmholtz still maintained that an explanation was satisfactory only if it used mechanical models. Mach, Duhem and, following them, Einstein then looked for theories of a different kind (called "theories of principle" by Einstein) which covered large areas but without giving a detailed account of the systems concerned while Einstein and especially Bohr introduced the idea that such theories may be context-dependent, different theories being valid in different domains. Combining these ideas with abstract mathematics such as various algebras, lattice theory, logics then led to a powerful revival of the structural approach. Thus the search for a generalized quantum theory is exactly in Aristotle's spirit: we do not take it for granted that the quantum theories we have are the best way of dealing with everything, looking either for new interpretations or suitable approximation methods to solve hairy cases; we rather try to identify domains and theories suited for them and then look for ways of relating these theories to each other.

Professor Primas explains the structural approach and demonstrates its power in the domain of chemistry. In doing so he reviews major standpoints and problems of modern science and philosophy of science; he shows how some of the problems can be solved by the structural approach. Freeing researchers from the agnosticism of the positivists and the dogmatism of the realists he liberates their imagination and prepares a fruitful collaboration of science, philosophy and abstract (non numerical) mathematics. In the past these subjects were often separated, or combined in a one sided manner. Professor Primas rejoins them and points the way to a new philosophy of nature that is not only useful and exact but also contains a metaphysics, i.e. a comprehensive picture of the world and man's place in it.

Berkeley

Paul Feyerabend

P R E F A C E

The purpose of this book is to provide a deeper insight into the modern theories of molecular matter. It incorporates the most important developments which have taken place during the last decades and reflects the modern trend to abstraction. At the present state of the art we have acquired a fairly good knowledge of "how to compute" small molecules using the methods of quantum chemistry. Yet, in spite of many statements to the contrary and many superficial discussions, the theoretical basis of chemistry and biology is not safely in our hands.

It is all but impossible to summarize the modern developments of the theory of matter in nontechnical language. But I hope that I can give some feeling for the problems, the intellectual excitements and the worries of some theoreticians. I know very well that such an enterprise is a dangerous adventure and that one says that a clever scientist should take care of his reputation by barricading himself behind the safe wall of his speciality.

This volume is not meant to be a textbook; in many respects it has complementary goals. For good and bad reasons, most textbooks ignore the historical and philosophical aspects and go ahead on the basis of crude simplifications; many even lie like the devil and do not shrink from naive indoctrination. Some sections of this book can be read as commentaries on our standard texts, they are intended to stir the waters with controversy. These parts certainly reflect the tastes, the inclination and the prejudices of the author.

Since all our textbooks are ultraconservative and strictly censor new developments, these notes are in part also a progress report, intended to increase the chemist's information about what the theoreticians have been up to in the last decades. Modern theoretical developments like quantum logics and algebraic quantum mechanics are now sufficiently mature to demand the interest of more than a few specialists. I tried to inform rather than to instruct, to give the gist of arguments rather than routine details, and I hope the reader will obtain an overall view of the subject. One or the other reader may deplore the lack of detail or think that I make excessive demands on his previous knowledge. In that case, he is urged to consult the references where he is likely to find more complete information. The references to the bibliography - which is extensive but far from complete - are by name of the authors and date. A few sections are on a technically more difficult level and are marked by the symbol #, and some sections I wrote for my own benefit. My feelings will not be hurt if the reader chooses to skip them. I shall be delighted if there are at least a few sections in which he finds some inspiration.

The modern theory of matter reveals an amazing continuity of philosophical and scientific problems. There is no novel idea in contemporary research which could not be ultimately be traced to the ancients. Our dominant mode of explanation is still Pythagorean mathematization; like Plato we try to understand reality in terms of some underlying mathematical structure. Most words in the scientist's and philosopher's vocabulary have changed their meaning over the past few hundred years. Yet the question raised in Platon's dialogue *Parmenides* two thousand years ago are still our questions, all that has changed is our approach. Our mathematical tools are much sharper and our empirical knowledge has vastly increased while the style of modern research has become distinctly unbalanced and irresponsible.

One of the worst features of modern science is the high degree of specialization and the exclusion of all historical and philosophical aspects. It is bad that the contemporary research programs force so many researchers in one area to be totally ignorant of most other areas. The ways of thinking, experiencing and behaving, the mode of activity exhibited by contemporary science strikingly reminds one of what Shapiro (1965) has called the paranoid style. It is characteristic of this style that bridges between related problems are broken down so that things remain neatly and rigidly separated. Scientists who cultivate a paranoid research style are usually extremely acute and intense, show an exorbitant respect for compartmentalizations and computers, and firmly demand complete autonomy for their narrowly fixed ideas. They like over-precise and rigid formulations but are not able to see the associated narrowing of interests. The separation of philosophy and science has led to the so-called realistic world view and to the blindness of many experts who are entirely unaware of the abstracting and isolating nature of modern science.

Basically, experiment and theory work together beautifully and complement each other perfectly. If the reader is not convinced of the paranoid tendencies of modern science, he should ponder on the wisdom of separating experiment and theory. This dissociation poses most severe sociological and psychological problems and may easily have disastrous consequences for the development of science. The most striking characteristic of modern theoretical science is the tendency to go to higher and higher levels of abstraction. Much to the dismay of the experimentalists, the theoreticians move away from specific problems and turn to comparative studies of theoretical structures. In the words of Marshall Stone (1966), the "abstract" of today becomes the "concrete" of tomorrow. Nowadays, mathematics is the most important language of theoretical science but few scientists can indulge in the luxury of keeping up with the increasing mathematical character of our most fundamental theories. This deplorable separation of experiment and theory forces the experimentalists to discuss so-called "models" of the phenomenon under investigation, using rather superficial and unreflected ideas about abstraction, idealization and approximation. Since the new developments in theoretical chemistry require a broader mathematical training than is customary in our present-day chemical education, most chemists do not realize that new well-founded concepts and powerful mathematical techniques are available for the design and interpretation of experiments. Here, I do *not* refer to the useful bust vastly overrated methods of numerical quantum chemistry. Quantum chemistry is but a narrow subfield of theoretical chemistry and numerical quantum chemistry is nothing but a powerful tool.

Much of the material covered in this volume has been presented in lectures that I gave at the Swiss Federal Institute of Technology (ETH) in Zürich during the last ten years. I owe a special obligation to my students who criticized the lectures as well as the notes. Some of the conceptually difficult material was distributed privately several years ago in form of lecture notes. In turn I received encouragement and constructive critique from many people, yet I find it impossible to list them all in a fair way. However, I would like to single out the influence of the late Joseph Maria Jauch (1914-1974). I had the benefit of many most stimulating conversations and correspondences with Professor Jauch on interpretation problems of quantum mechanics which have been essential in shaping my views. In particular, Jauch convinced me in a long and tough discussion in May 1970 that the concept of classical observables (or "essential observables", as he called them) was the missing link between physics and chemistry.

Without the action of a friendly pressure group the various lecture notes and manuscripts would never have been transformed into the present shape. There existed a variety of preliminary versions, but again it is not possible to enumerate all the people who read and criticized them. I am greatly indebted to all of them. I would like to mention at least my former and present coworkers Dr. Peter Brand, Dr. Peter Pfeifer, Wolfgang Gasche, Dr. Ulrich Müller-Herold, Werner Gans, Guido Raggio, Eberhard Müller and Anton Amann. Their feedback was always stimulating and very essential, their original contributions have been indispensable.

A special word of thanks is due to Miss H.Rohrer who not only did an impressive and beautiful job with the final version but also drew up and typed many drafts with skill and humor and has devoted enormous energy to improve presentation and style.

Finally, I wish to express my sincere thanks to Paul K. Feyerabend who managed to find time to read the final manuscript. He even took the pains of going meticulously through the script, correcting a staggering number of linguistic errors, making many stylistic improvements, and therewith teaching me that "anything goes" is not a rule that can be applied to the English language. Any remaining errors are of course my responsibility alone.

Zürich Hans Primas

1. OPEN PROBLEMS OF PRESENT-DAY THEORETICAL CHEMISTRY

"Some like to understand what they believe in.
Others like to believe in what they understand."
Stanislaw Lec, 1959

1.1 THE OVERPRODUCTION OF TRUTH

"Knowledge of any kind is a thing to be honored and prized", says Aristotle (384-322 B.C.) in the opening sentence of his *On the Soul*. According to the Auger report *"the number of scientific journals and periodicals, which was about 100 at the beginning of the nineteenth century, reached 1'000 in 1850, more than 10'000 in 1900, approaches 100'000 in 1960 and - if this rate of growth remains constant - should be in the neighbourhood of a million at the end of the century"* (Auger, 1961, p.15). Linus Pauling (1958) guessed *"that about 100'000 new chemical facts are being discovered each year, at present"*. Stanislaw Ulam (1976, p. 288) estimated that contemporary mathematicians produce one or two hundred thousand theorems a year. Due to this explosive development of research, science has split into many different sciences. Chemistry has split into many disciplines only tenuously connected. No researcher can keep abreast with the work in his own narrow subfield. Nobody can digest even the most outstanding results of science, to say nothing of integrating them into our culture. *"The man of knowledge in our time is bowed under a burden he never imagined he would ever have: the overproduction of truth that cannot be consumed. For centuries men lived in the belief that truth was slim and elusive and that once he found it the troubles of mankind would be over. And here we are in the closing of the 20th century, choking on truth"* (Ernest Becker, 1973). This is something to worry about. The widely accepted tale *"knowledge is good for mankind"* has become suspicious.

Due to the overproduction of truth we are actually losing ground in the more fundamental aspects of science. Our preoccupation with trivial research and shallow truths make the public disinterest in scientific questions understandable. Unintegrated knowledge is boring for the nonspecialist. What matters is insight. It is not enough to have knowledge, the knowledge has to become an inner experience with each of us, it has to become part of our culture. We have an enormous amount of scientific data but we have not yet found simple words for great ideas.

1.2 CHEMICAL THEORIES

The ambitious goal of any chemical theory is to understand the underlying order in the bewildering variety of chemical phenomena. Since Robert Boyle (1627-1691) in his famous work "The Sceptical Chymist" (first published in 1661) established chemistry as a separate science founded on unbiased experimental observations, the philosophically naive view of chemistry as a purely empirical science has become increasingly popular. On the other hand, the forefather of modern chemistry is famous for his creative speculative reasoning, and since his time chemistry belongs to the heavily theory-loaded natural sciences. Chemistry like any science is the result of intuition and imagination, not of empirical facts and logic alone. Every creative scientist forms in his mind pictures outlining phenonema *by omitting everything he considers as accidental or irrelevant*. This freedom of thought implies a freedom in constructing phenomena and is characteristic for *any* theory.

There is a curious perversion associated with the "overproduction of truth" in chemistry: nowadays the specialization goes so far that we have a branch called "theoretical chemistry". Moreover, there are indications that we are approaching a situation where the representatives of the other branches of chemistry scarcely know what theoretical chemists are doing. Does it matter? Yes, since there is no chemistry without theory. The vitality of chemistry as a science is conditioned by the connections of its branches. The progressive compartmentalization leads to a badly fragmented state of chemistry. This situation calls for high-level abstractions of great simplifying, unifying and predictive power. *The most important task of contemporary theoretical chemistry is to stimulate the mutual understanding of the various branches of chemistry and its neighboring sciences.*

Ideally, theoretical reflexions should also strengthen the embedding of science in our own history and culture. For example, up to the present time, a fair and competent appraisal of alchemy and ancient chemistry is missing. Folkloristic history often refers with sardonic humour to ancient chemical theories, even though many of these theories were profound, imaginative and fruitful. It is easy to make alchemy appear ridiculous if one lives in the splendid isolation of a modern scientist. However, it would be both fair and scholarly not to forget that Robert Boyle was deeply influenced by alchemical thoughts, that Isaac Newton was passionately concerned with alchemical transmutations, that the psychological and religious implications of alchemy were Carl Gustav Jung's major preoccupations during the last thirty years of his life, and that Wolfgang Pauli proposed to revive the alchemical idea of a unitary psycho-physical language within the framework of modern physics (compare Heisenberg, 1959, p.663). The ideas of alchemy and ancient chemistry were neither vague nor crazy but they have been overshadowed by deductive Newtonian mechanics. Only after the modern theory of matter had exposed the monism of the mechanical view as philosophically naive and empirically untenable, did the way become free for a new appreciation of ancient theories.

The philosophical revolution triggered by the advent of quantum mechanics is a singular event in the history of science. The classical way of looking at phenomena is no longer defensible, the subject of science can never again be the same. Nevertheless, present-day theoretical chemistry is still under the spell of the Newtonian approach to understanding. Of course, Newton's equations of motion have been replaced by quantum mechanical ones. Yet, the monism of Newtonian physics simply has been replaced by a new monism governed by the Schrödinger equation. Many contemporary quantum chemists see no reason to revise their philosophical views but still adhere to the old dream of an objective single frame of reference that permits us to eliminate the pluralism of chemical theories and to deduce (at least in principle) all chemical phenomena from a single unifying principle.

1.3 "WE CAN CALCULATE EVERYTHING"

This dogma of modern numerical quantum chemistry (Clementi, 1973) has its own origin in the following famous statement by P.A.M. Dirac (1929a) dating from the pioneer time of quantum mechanics: "*The underlying physical laws necessary for the mathematical theory of a larger part of physics and the whole of chemistry are thus completely known, and the difficulty is only that the application of these laws leads to equations much too complicated to be soluble.*" As we know today, this claim is not correct in at least two respects. Firstly, the formalism of pioneer quantum mechanics of 1929 is a very special case of modern (generalized) quantum mechanics which we now regard as fundamental for every theory of molecular matter (compare fig.1). Secondly, the interpretative problems had not been solved in a satisfactory way in 1929 (compare fig.2). At that time, the conceptual problems of reducing chemical theories to the epistemologically very differently structured quantum mechanics were not discussed, and Dirac viewed the problem of reduction only as a very complicated mathematical problem.

The discovery of the electric nature of chemical forces by quantum mechanics took the valence problem out of its chemical isolation and made it accessible to an exact mathematical treatment. The last fifty years taught us a lot about the relation between chemistry and the Schrödinger equation, and have led to a penetrating understanding of chemical structures. Since the advent of electronic computers, numerical quantum chemistry has been a tremendous succes. Nowadays we have a number of masterful analyses of electronic wavefunctions. Many calculations have been extremely sophisticated, designed by some of the foremost researchers in this field to extract a maximum of insight from quantum theory. For simple molecules, outstanding agreement between calculated and measured data has been obtained. Yet, the concept of a chemical bond could not be found anywhere in these calculations. We can calculate bonding energies without even knowing what a chemical bond is!

Remark: In fact, even today many thoughtful theoreticians struggle for a more profound understanding of the phenomenon of chemical bonding. Compare e.g. Platt, (1961), Ruedenberg (1962, 1975), Bader and Preston (1961), England et al. (1971), Bader (1975), Woolley (1976a), Runtz et al. (1977), Kutzelnigg (1978), Bader et al. (1979a,b)

There is a danger to be sidetracked into purely numerical problems and to forget the original impetus of our enterprise: understanding the

FORMALISM	QUANTUM MECHANICS OF PIONEER TIME (1925-1935)	MODERN NONRELATIVISTIC QUANTUM MECHANICS
CONCEPT	generalization of Newtonian mechanics	logic of temporal propositions
RANGE OF APPLICATION	only for micro systems	universal
KINEMATICS		
space-time-structure	correspondence principle	Galilei group
commutation relations	Heisenberg's commutation relations	Weyl's commutation relations as projective representations of the Galilei group
algebra and observables	irreducible only type I	arbitrary W*-algebras of type I, II or III
number of degrees of freedom	finite	arbitrary
superselection rules	none	possible, defined by the kinematic symmetry
DYNAMICS		
time evolution	dualistic	undivided
unitary time evolution	time-dependent Schrödinger equation	unitary representation of time translation group
stochastic time evolution	ad-hoc via projection postulate	can be derived

Fig.1 Comparison of the formalism in pioneer quantum mechanics with the Hilbert space model of modern nonrelativistic quantum mechanics

INTERPRETATION	PIONEER QUANTUM MECHANICS (1925-1935)	MODERN NONRELATIVISTIC QUANTUM MECHANICS
REFERENT	single system (Bohr, Fock) Ensemble (von Neumann)	single system
STATUS	epistemic interpretation	ontic interpretation
OBJECTIFIABILITY	Heisenberg's uncertainty principle	existence of incompatible properties
DIALECTICS	Bohr's complementarity	non-Boolean logic of the temporal ontic propositions
STATES	$\Psi(t)$ is merely an auxiliary mathematical function to calculate expectation values	$\Psi(t)$ represents the set of all temporal propositions which are true at time t
STATISTICAL INTERPRETATION	Born statistical interpretation of $\|\Psi(t)\|^2$	the statistical interpretation can be derived
RELATION TO EMPIRICAL FACTS	correspondence principle projection postulate measurement process	the classical observables have at all times a sharp value and are directly observable without perturbing the state of the system

Fig. 2 Comparison of the interpretation in pioneer quantum mechanics with the individual and ontic interpretation of modern nonrelativistic quantum mechanics

behavior of matter. We should not confuse a useful theoretical tool with theory. Numerical quantum mechanics is a most important tool for chemistry but it cannot replace thinking. The final explanation of empirical facts is not achieved by merely calculating measurable quantities from first principles. The ultimate objective of a theory is not to determine numbers but to create a large, consistent abstract structure that mirrors the observable phenomena. We have understood a phenomenon only when we can show that it is an immediately obvious and necessary part of a larger context. The vision of some theoreticians has been narrowed down to problems that can be formulated numerically. Some even take no notice of genuine chemical and biological patterns or deny them scientific significance since they are not computable by their methods. Such a one-sided view of numerical quantum chemistry is by no means the inevitable penalty for the attempt to reduce chemistry to physics. Rather, it is the result of the utilitarian character of contemporary research which lost all philosophical ambitions and has only a very restricted insight into its own limitations.

The important concepts of chemistry have never been well-treated by ab-initio quantum chemistry so that quantum mechanics has *not* become the primary tool in the chemist's understanding of matter. Brute-force numerical quantum chemistry can hardly do justice to the qualitative features of chemistry. But without insight into the qualitative concepts we are losing chemistry. The allegedly basic methods often fail to illuminate the essential function of a molecule or a reaction that is evident to the experimental scientists. As a result practical chemists had to look for inspiration elsewhere and generated the ad-hoc semiempirical methods of quantum chemistry. This approach "*has become a part of the chemical structure theory, which has an essentially empirical (inductive) basis; it was suggested by quantum mechanics, but it is no longer just a branch of quantum mechanics*" (Linus Pauling, 1959, p.388). Despite the erudition, imagination and common sense used to create the semiempirical methods of quantum chemistry, the success of this craft remains a central enigma for the theoreticians. The models of semiempirical quantum chemistry are built upon an inadequate conceptual basis, and their mathematical foundation is so wobbly that they are a source of frustration. Moreover, they give us an image of matter that does not conform to what we are led to expect from the first principles of quantum mechanics. But experimentalists are not at all impressed by such scruples. *And properly so.*

Some contemporary theoreticians have attempted to narrow the scope of scientific inquiry by requiring operational definitions and first-principle underpinnings for all concepts. In theoretical chemistry, there is a distinct tendency to throw out typically chemical variables, admitting that they have served a noble purpose in the past but asserting that they are now obsolete. The premise underlying such a view is that the only valid meaning of any chemical concept is one which can be unequivocally defined in terms of present-day ab-initio quantum chemistry. This method of *problem solving by rejection* has been proposed for such concepts as valence, bond, structure, localized orbitals, aromaticity, acidity, color, smell, water repellence etc. A particularly powerful variant is the method of *problem solving by dissolving definitions*. Using this procedure we can easily solve even the "problem of life" by claiming that the distinction between living and non-living entities is a pre-scientific idea, obliterated by modern research. But some people consider such a line of action as unfair and shallow. We need a creative approach to the rich field of chemistry, not just a critical one that bids us to dismiss most problems of chemistry as meaningless. *The task of theoretical chemistry is to sharpen and to explain chemical concepts and not to reject a whole area of inquiry.*

1.4 SOME PUZZLES OF MOLECULAR QUANTUM MECHANICS

Viewed from a classical point of view, quantum mechanics is a strange and counterintuitive theory. Probably not all scientists using quantum mechanics as a tool have been aware of all its logical and philosophical consequences. It is both worthwhile and amusing to review briefly some old puzzles of molecular quantum mechanics. The practitioners of quantum chemistry are only rarely concerned with these problems for they know how to avoid a paradoxical situation. This know-how belongs to our background knowledge but lacks the benefit of a theoretical justification: it is distressingly easy to find what one wants by ad-hoc methods, and remarkably difficult to derive it from truly first principles. To *explain* and to *understand* necessitates the use of consistent theoretical concepts and a proper interpretation of the mathematical formalism. Therefore we must try to resolve also those paradoxes of quantum mechanics that raise philosophical questions even though such questions do not trouble quantum chemists engaged in *describing* and *computing*.

Being accustomed to the classical mode of describing natural phenomena and to the use of dialectics, not all readers may realize the theoretical significance of the following examples. We claim that these examples are not simply irrelevant riddles but bewildering enigmas that defy straightforward explanations. Paradoxes are often a source of new insights, and it is therefore not wise to push them aside as minor difficulties. Contrary to what some experts assume we are convinced that these paradoxes *can* be resolved in terms of modern quantum mechanics based on quantum logic. If we intend to attack *new* problems in the theory of molecular matter, such as theoretical molecular biology, molecular evolution, or a theory of the hierachical structures in the molecular world, then we must replace much of the background knowledge used in working quantum chemistry by a unified research program based *strictly* on quantum logic.

QUESTION 1: Do isolated quantal systems exist at all?

Quantum mechanics casts severe doubt on the existence of *isolated* systems. In contrast to classical theories, quantum mechanics predicts an *entanglement* of a system with its surroundings under the influence of even extremely weak interactions. That is, an interacting system is never represented by a product state but by a non-trivial linear combination of product states. If this is true, then it becomes very difficult to explain what we mean by a "system". This holistic character of quantum mechanics is well established; it is experimentally confirmed that even noninteracting systems can be entangled in the sense of Einstein, Podolsky and Rosen (1935) and Schrödinger (1935a,b; 1936) (compare also the detailed discussion in sect.3.7).

QUESTION 2: Is the Pauli principle a universal and inviolable fact of nature?

This question was posed by Margenau (1966). Although the Pauli principle is usually claimed to be an inviolable law, we know very well that we never need to consider all the electrons of the universe. In the face of the experimentally established correlations in noninteracting systems, the usual "explanation" by reference to vanishing exchange integral loses much of its weight.

QUESTION 3: Does quantum mechanics apply to large molecular systems?

Many scientists have suggested that quantum mechanics is not sufficient for describing large or complex molecular systems. Ludwig (1955, 1961a) proposed the existence of correlations in macromolecules that are stronger than those required by quantum mechanics. Elsasser (1961b, 1962, 1969) claimed that the laws of organisms have components (so-called "biotonic laws") that are not deducible from the principles of quantum mechanics. Wigner (1961, 1962a) thought that it is "*likely that the present laws and concepts of quantum mechanics will have to undergo modifications before they can be applied to the problem of life*". Jørgensen (1974) suggested that "*quantum mechanics is only applicable to reproducible small systems with assembly properties*". On the other hand there is no empirical evidence for a limited validity of quantum mechanics, no clear-cut failures are known.

QUESTION 4: Is the superposition principle universally valid?

Accepting the unrestricted validity of the superposition principle, one is led to grotesque situations, as exemplified by Schrödinger's (1935b) cat paradox. A coherent superposition of the states of a living and a dead cat must obviously be excluded from the theory. This paradox is closely connected with the so-called *"measurement problem of quantum mechanics"*. Replacing the theoretical discussion of the measurement process by the pragmatic *reduction postulate* requires at least a reinterpretation of the superposition principle of quantum mechanics.

QUESTION 5: Why do so many stationary states not exist?

This puzzle is a reinforced molecular version of question 4. In fact, already relatively small molecular systems do not exhibit stationary states allowed by traditional quantum chemistry.

Example: What is the reason that we can buy in the drug store D-alanine (state vector $|\Psi_D\rangle$), L-alanine (state vector $|\Psi_L\rangle$) and the racemate (density operator $|\Psi_D\rangle\langle\Psi_D| + |\Psi_L\rangle\langle\Psi_L|$), but *not* the coherent superpositions $(|\Psi_D\rangle + |\Psi_L\rangle)$ and $(|\Psi_D\rangle - |\Psi_L\rangle)$?
According to the traditional view, the last mentioned states are supposed to represent the ground state and an excited state, respectively.

QUESTION 6: Why are macroscopic bodies localized?

According to quantum mechanics localized states do exist as quasistationary states. However, traditional quantum mechanics also shows that after a sufficiently long time, the localized character of a state vector is completely lost. Hence: the question why do billard balls exist? This puzzle was posed by Einstein (1948; 1949, p.282; 1953b). In the famous Born-Einstein controversy about this problem, Born (1969, letters 80-89 and 105-116) completely missed Einstein's point; the question is therefore still open.

QUESTION 7: Why does quantum mechanics fail to account for chemical systematics?

Ab-initio quantum chemistry correctly describes *individual* small molecules; nevertheless it fails to describe the *classes* of structurally diverse but functionally related molecules which are omnipresent in chemistry (compare e.g. Hartmann, 1964; Jørgensen, 1974). We claim that this puzzle is the molecular version of Einstein's problem of question 6.

Preliminary example: Localized orbitals are related to the concepts of the working chemist but, natural orbitals, as a rule, are not. *Dilemma for the theoretician:* natural orbitals are elegant and mathematically well-defined; all current definitions of localized orbitals are ad hoc and graceless. *Related puzzle:* It seems to be an empirical rule that - without external influence by an experimenter - simple quantal systems are in eigenstates of the Hamiltonian but complex quantal systems are in coherent or localized states.

QUESTION 8: Why can approximations be better than the exact solutions?

It is well known that the lowest energy solution of the Hartree-Fock approximation does not necessarily have the symmetry of the exact ground state of the corresponding Schrödinger equation. However, in some most interesting cases, the "wrong" symmetry type SCF-solution *does* represent a physically relevant situation that cannot be encompassed by the original Schrödinger equation. What is the proper explanation for this unexpected success of the Hartree-type method?

QUESTION 9: Why is the Born-Oppenheimer picture so successful?

Deep conceptual problems are hidden in the seemingly simple separation of the electronic and nuclear motion in a molecule. Using modern singular perturbation theory, the mathematical methods used in the Born-Oppenheimer approximation can be put on a sound footing. That is, we understand the computational methods used in the adiabatic approximation. Nevertheless, we hardly understand why the Born-Oppenheimer *picture* is compatible with the *concepts* of quantum mechanics.

First example: We describe the six degrees of freedom of the ground state of the helium atom (considered as 3-particle problem with the center-of-mass motion separated) as a problem of two interacting particles in an external Coulomb potential. However, in the case of the molecule H_2^+ we discuss the very same type of differential equation in an entirely different way, and split the 6 degrees of freedom into 1 vibrational mode, 2 rotational modes, and 3 electronic type degrees of freedom. This qualitatively different description does by no means follow from a purely mathematical discussion. *Second example:* The statement "the three nuclei in the molecule C_3 form an equilateral triangle" is considered to be a meaningful proposition (which may prove to be true or false). The similar statement "the three electrons of the lithium atom form an equilateral triangle" is neither a true nor a false proposition but meaningless. Question: Which fundamental principles are responsible for this qualitatively different behavior of electrons and nuclei?

QUESTION 10: Is temperature an observable?

There are equilibrium systems having a well-defined temperature. The temperature of such systems can be measured, using a single system only with an arbitrarily small perturbation of the thermodynamic state of the system. Is there a quantum mechanical description of a *single* system such that the temperature is represented by an *observable* having a dispersion-free value in an equilibrium state? Traditional quantum statistical mechanics does not fulfill this desideratum, firstly because it refers to an ensemble (e.g. it does not refer to a steam engine but to an infinite ensemble of noninteracting steam engines), and secondly, because temperature is not represented by an observable (i.e. a self-adjoint operator acting on a Hilbert space of states).

All the ten puzzles mentioned are connected with the incapability of traditional quantum chemistry to deal with the classical properties of molecular matter. However, the experts in axiomatic quantum mechanics have known for many years, that there are no reasons to insist on the traditional formulation of quantum mechanics which excludes classical observables by an ad-hoc irreducibility postulate.

Hints: Later in this essay we will return to these questions so that some short hints should suffice for the moment. Question 1 is philosophically the deepest problem, it can be solved by reflecting about the way in which we recognize patterns in our world. Disentangling a system from its surroundings creates new phenomena which can be discovered by a pattern cognition method that rejects the Einstein-Podolsky-Rosen correlations between the object and its surroundings as irrelevant. An appropriate generalization of quantum mechanics settles questions 2,3, and 4. Quantum logic provides a natural way for introducing classical observables. Classically inequivalent states obey neither the superposition principle nor the Pauli principle. The existence of classical observables implies the existence of superselection rules (i.e. a restricted validity of the superposition principle) which explains the phenomena of question 5. Questions 5 or 6 are related to the emergence of classical observables via the interaction of the system with its environment. When we are dealing with interactions, we are always interested in stability properties. We call those quantities *robust* that persist under slight perturbations of the system. In the alanine example, the vectors $|\Psi_D\rangle$ and $|\Psi_L\rangle$ represent robust quantal states, while the eigenstates are dynamically unstable so that a superselection rule between $|\Psi_D\rangle$ and $|\Psi_L\rangle$ evolves dynamically. Similarly, in the Einstein example only the localized states are robust. Questions 5, 6, and 7 refer to mathematically ill-conditioned problems which can be discussed only with an appropriate regularization method. In general, a natural orbital analysis is a mathematically ill-conditioned problem while the calculation of localized orbitals is a robust procedure. Evidently, chemical systematics, referred to in question 7, deals only with robust properties. The regularization methods of ill-conditioned problems are related to the methods of determining classically inequivalent states. Interestingly, a modified Hartree approximation is a good method for calculating classically inequivalent states, or in other words, the Hartree (or Hartree-Fock) approximation can regularize an ill-conditioned Schrödinger problem. In this sense, the Hartree method is indeed better than the exact method (question 8). The Born-Oppenheimer picture mentioned in question 9 is also related to the emergence of new classical properties (the classical concept of "molecular structure"); it is more complicated because it involves a "weak superselection rule" which is not strictly valid. Thermodynamics (question 10) refers to a description which is independent of the existence of elementary objects so that the appropriate algebra of observables should be continuous. In fact, modern formulations of thermodynamics use a global type III_1-W*-algebra of observables which has no atoms but where the temperature as well as the chemical potential are indeed classical observables.

1.5 NEW POINTS OF VIEW ARE NEEDED

Contemporary quantum chemistry has had a remarkable historical success in the description of molecular phenomena but cannot be the basis of chemistry because it is an *incomplete theory*. It is based on the Copenhagen interpretation of quantum mechanics which suggests a *dialectical view, presupposing* the classical level of description and assuming an irreducible *duality* between the observed micro-system and the macroscopic measurement apparatus (Bohr, 1958; Fock, 1957). This is not a good starting point for the discussion of chemical problems. Every chemical and molecular-biological system is characterized by the fact that the very same object simultaneously involves both quantal *and* classical properties in an essential way. A paradigmatic example is a biomolecule with its molecular stability, its photochemical properties, its primary, secondary and tertiary structure.

The most serious deficiency of pioneer quantum mechanics is its inability to deal in a candid way with classical systems, and that it does not allow a logically consistent and useful theory of laboratory measurements. The orthodox interpretation of quantum mechanics (von Neumann, 1932; London and Bauer, 1939; Wigner, 1963) is troubled by the so-called *measurement problem of quantum mechanics*, that is by the proper way of including an objective macroscopic level into quantum mechanics. We cannot claim to have a valid theory if the interpretative problems of quantum mechanics are not resolved. These aspects have to be accounted for in a fundamental theory of matter worthy of the name. What we need is not a radical departure from the quantum mechanics of the pioneer time 1925-1935 but a generalization that *encompasses* classical theories (like Newtonian mechanics, electrodynamics, thermodynamics, chemical kinetics) as well as the traditional quantum mechanics of atoms and small molecules.

Current quantum chemistry tends to give the impression that the major difficulties in applying quantum mechanics to complex molecular systems are computational. However, *the main stumbling block for the development of a theory of large and complex molecular systems is not computational but conceptual*. Fundamental difficulties arise already at the stage where a model of a complex system has to be chosen. The concept of an *isolated system* is fundamental for every scientific inquiry but it is by no means unproblematic. For example, if we want to discuss the metabolism of a cell from a fundamental point of view, the main problem is not computation but the theoretical characterization of a cell. To start with, we have to regard a cell as an isolated object. Isolating a cell is a bold idea yet indispensable if a cell is to become the referent of a theory. Our ability to describe objects cannot go farther than our theoretical ability to isolate them. Theoretical isolation means abstraction, and abstraction generates the relevant patterns for a scientific inquiry. The creation of physical reality by abstractions is a deep conceptual problem that cannot be discussed independently of more general philosophical problems. We have to understand what a phenomenon is, how we recognize it, and how we communicate about it.

Today, philosophical discussions are banned from our top scientific journals. According to Albert Einstein (1944, p.289) "*a fateful 'fear of metaphysics' arose which has come to be a malady of contemporary empiristic philosophizing; this malady is the counterpart to that earlier philosophizing in the clouds, which thought it could neglect and dispense with what was given by the senses*". If we want to start new things rather than trying to elaborate and improve old ones, then we cannot escape reflecting on our basic conceptions.

2. ON THE STRUCTURE OF SCIENTIFIC THEORIES

"Science never makes an advance until philosophy authorizes and encourages it to do so"

Thomas Mann, 1936

2.1 A GOOD THEORY SHOULD BE CONSISTENT, CONFIRMED, AND INTUITABLE

Scientific method is one of many possible approaches to understanding the world. Science does not deal with every aspect of reality but only with a carefully selected part of reality. Science does not provide any premises for aesthetical, moral or political conclusions. There are complementary approaches such as the fine arts, poetry or religious experiences which are no less important than science.
The aim of science is to tell the truth, even if it is not the whole truth. The scientific approach depends in a complex way on culturally determined tacit presuppositions specifying what is essential and what is not. The most fundamental presupposition of science is that instead of considering the whole universe at once we may, as a reasonable approximation, regard it as consisting of parts on which we can do experiments. The basic structural pattern of scientific experience is the subject-object relation which in turn is based on the assumption of the separability of subject and object.

Every scientific theory mediates between the external world and a contemplating subject. We therefore have to consider two mappings:

(i) a mapping of the external world into the formal framework of the theory

(ii) a mapping of the formal framework of the theory into certain psychic structures of the subject.

Accordingly, every theory involves three classes of referents: *objects*, *abstractions*, and *minds*. Knowledge would be impossible without them.

Philosophers of science often reduce the role of scientific theories either to hypothetico-deductive systems (Duhem, Quine) or to an operationalistic viewpoint (Bridgeman). These aspects are important but one-sided and cannot do justice to the structure of scientific theories. There is much more to be said about scientific explanation and the creation and change of theories. Every theory is supposed to offer a systematic conceptual pattern for the observed data. Nevertheless, the final aim of a scientific theory is not to summarize data but *to understand reality*. Neither deductivism nor operationalism cares whether a theory is intelligible, intuitable, fertile - although the beauty of a theory has always been of overwhelming importance for the theoreticians. To articulate these criteria we need *normative rules* which are outside the domain of deductionism and operationalism.

A more fruitful approach is the study of the *sign relations* in science. The general theory of signs was initiated by the American philosopher *Charles Sanders Peirce* (1839-1914) and is called *semiotics*. The three members of a semiotic triad are the *sign*, its *object*, and its *interpreter*. A sign is defined as "*something which stands to somebody for something in some respect or capacity*" (Peirce, Collected Papers, Vol.2, paragraph 228). The something for which the sign stands is called the *object* or the *referent*. The somebody to whom the sign stands is its *interpreter*. According to Morris (1938), the three dimensions of semiotic analysis are the semantic, the syntactic and the pragmatic dimension. *Semantics* deals with relations of signs to their objects, *syntactics* deals with the formal relations of signs to one another, and *pragmatics* deals with the relations of signs to their interpreters.

We consider a scientific theory to be a semiotic system consisting of the following three parts:

(i) *syntactics*, realized as a mathematical formalism dealing with the logico-mathematical structure of the theory;

(ii) *semantics*, realized by an interpretation that deals with the relation of the mathematical symbols to the objects which they denote;

(iii) *pragmatics*, consisting of regulative principles of a normative kind, describing the relation of the theoretical terms to their interpreters and the possible contexts of use.

Given the formalism of a theory, we are within certain limits free to choose its interpretation (that is, the semantics and the pragmatics). If the interpretative rules and the regulative principles of different theories are not the same, then these theories have no common vocabulary.

Seen in this way, every theory has a double task: it mirrors both some aspects of outer reality and some aspects of inner reality. There are basically *three* ways to judge scientific theories. First of all, a theory has to be *logically consistent*, that is, its formalism must be free from essential logical contradictions. Secondly, a theory has to be *empirically well-confirmed*, it must not contradict empirical facts in a serious way. Thirdly, a theory has to be *intuitable*, that is, its main concepts should be knowable through intuition, by insight without rational thought. To characterize this "inner perfection" (as Albert Einstein, 1949, p.23, calls it), the theoreticians often use the vague but important terms "naturalness" and "mathematical simplicity". The first point is concerned with the intrinsic consistency of the syntax of the theory (without considering the relations to the external or internal world), the second point refers to the relations of a theory with the external reality, while the third point refers to the relations of the theory with the inner reality.

A good theory should neither neglect nor favor one of the three aspects of the semiotic triad, it ought to be consistent (in the sense of syntactics), confirmed (in the sense of semantics) and intuitable (in the sense of pragmatics) (compare fig.3). Modern physical theories have a highly developed syntactical formalism and a predominantly operational semantic interpretation. In the spirit of our time, logical consistency and empirical confirmation are considered the most important virtues of a theory. But the meaning conveyed by the theory is seldom an inner experience with us. Alchemy on the other hand, failed mainly because of the dictatorship of chemical facts but the alchemist "*experienced the presence of pre-existing ideas in physical matter*" (Jung, Coll. Works, vol.12, par.346). The general intellectual trend of our scientific and technical culture does not give much weight to the role of intuition so that pragmatics is a badly neglected aspect of contemporary science. One therefore often hears the remark that modern theories cannot be described in intuitive and imaginative terms. Nevertheless, most creative scientists will probably disapprove of the following statement in the anonymous preface

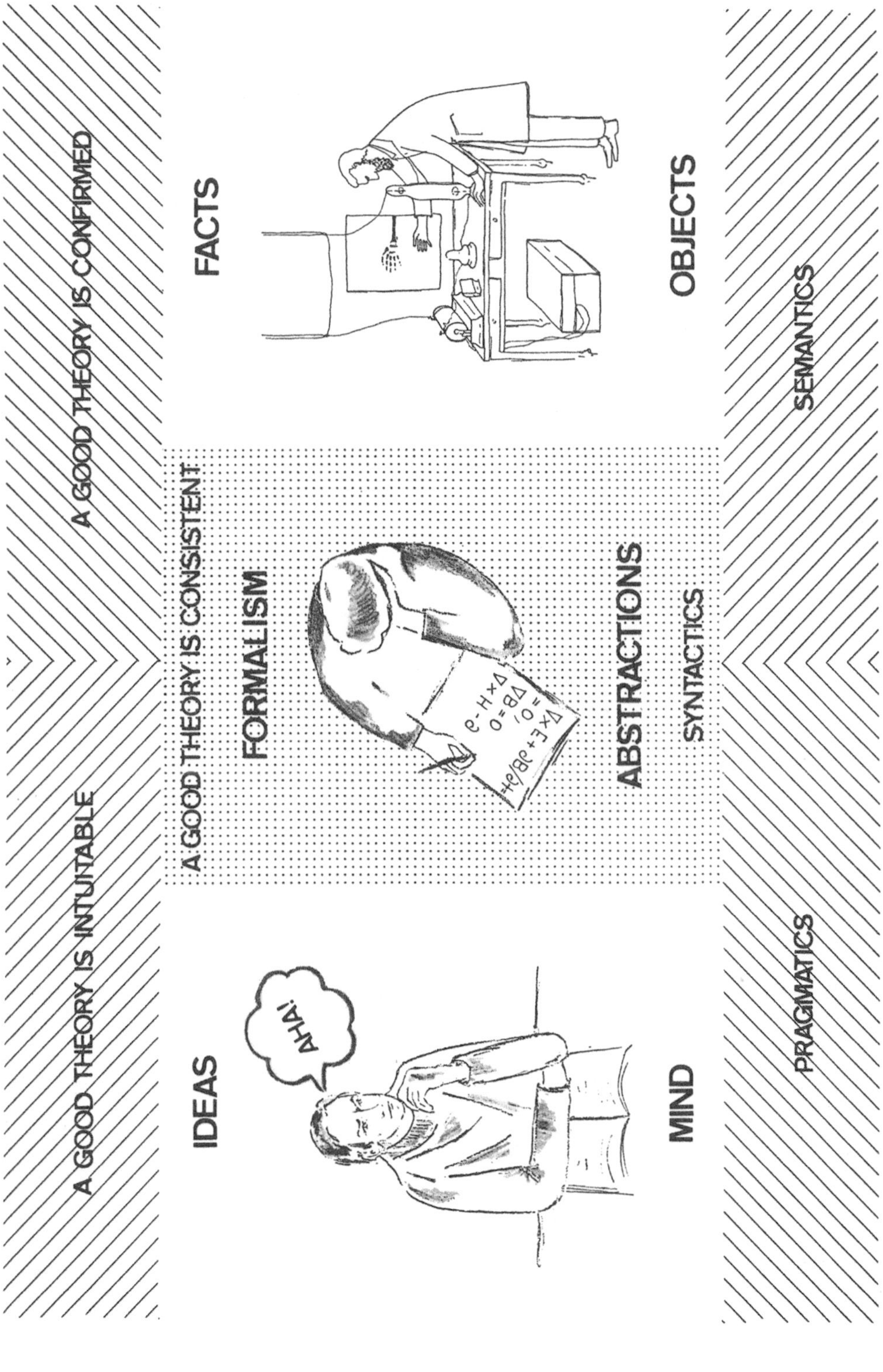

Fig.3 A theory as a semiotic system consisting of syntactics, semantics and pragmatics

(due to Andreas Osiander) to Copernicus' *"De Revolutionibus"*: *"It is not necessary that these hypotheses be true; they need not even be like the truth; it suffices when they lead to calculations which agree with observations."* (in: Copernicus, 1543): a theory that is merely logically consistent and gives accurate predictions is not yet a good theory. We want much more: *we want to see something never seen before.*

2.2 NO THEORY CAN BE PROVED TO BE FREE FROM INNER CONTRADICTIONS

Modern science is based on the platonic faith in a mathematical order in nature, or, as Galilei says "*The great book of nature is written in the mathematical language,... without whose help it is impossible to comprehend a single word of it*" (Galilei, 1623, sect.6). Since that time a most creative interaction has taken place between pure mathematics and the sciences. Contemporary science is becoming increasingly mathematized, and the language of mathematics now has a dominating influence upon modern theoretical research. The cause for this development is, to use the wording of a famous essay by Eugene Wigner, "the unreasonable effectiveness of mathematics in the natural sciences". "*The miracle of appropriateness of the language of mathematics for the formulation of the laws of physics is a wonderful gift which we neither understand nor deserve*" (Wigner, 1960).

An obvious advantage of expressing the syntax of a theory in the language of mathematics is the simplicity of checking its consistency. The syntax of a theory must be free from essential contradictions. Note that it would be too much to require that the formalism of a theory should be shown to be free from contradictions. Kurt Gödel's (1906-1978) famous incompleteness theorem says that it will never be possible to show by mathematical means that mathematics does not contain contradictions.

References: For an introduction for these advanced areas of logic one may consult the lucid text by Boolos and Jeffrey (1974). The basic papers on undecidable propositions are easily accessible in an anthology edited by Dàvies (1965).

That is, no theory that permits formalization can be shown to be free from inner contradictions with the means of this theory itself. An information-theoretic version of Gödel's incompleteness theorem, due to Chaitin (1977), says that it is possible to prove that a specific object is of computational complexity greater than n only if n is less than the computational complexity of the axioms being employed in the demonstration. Furthermore, we should not forget that the mathematical formalism of modern scientific theories uses such mysterious artifices as "real numbers", or mathematical conventions such as the "axiom of choice" that can never be corroborated. In short, freedom from logical contradictions is essential for the trivial aspects of a theory - but it should not be overstressed.

What is the axiom of choice? The axiom of choice can be formulated as follows: *"Given a collection of nonempty subsets of a set, there is a set that contains just one member from each of these"*. The axiom of choice is obviously true for finite collections. However, according to a deep result by Paul Cohen, the axiom is *independent* for infinite sets: we can adopt the axiom of choice, or replace it by the statement: the axiom of choice is false. Modern functional analysis - hence modern quantum mechanics - considers the axiom of choice to be true. That is fine, but then we accept the so-called Banach-Tarski paradox: *"Using the axiom of choice, it is theoretically possible to divide a sphere as large as the sun into little solid pieces and then without any compression or distortion of these pieces, to rearrange them, leaving none out, into a sphere of the size of a pea"*.

2.3 EXPERIMENTS CAN NEITHER PROVE A THEORY TRUE NOR FALSE

A popular fable pretends that there is a "scientific method", consisting of a set of rules which, if followed, will lead to scientific discoveries. A modern version of this fable goes as follows: *The* method of science "*consists of applying the following steps to every problem in science, formally and explicitly and regularly: 1) Devising alternative hypotheses; 2) Devising a crucial experiment (or several of them), with alternative possible outcomes, each of which will, as nearly as possible, exclude one or more of the hypotheses; 3) Carrying out the experiment so as to get a clean result; 1') Recycling the procedure, making subhypotheses or sequential hypotheses to refine the possibilities that remain; and so on*" (Platt, 1964). The fact that even successful and thoughtful scientists cannot tell what they are actually doing shows that scientific research is an art characterized by specific skills and not by a set of rules. As stressed by Michael Polanyi *there are things we know but cannot tell.* "*This is strikingly true for our knowledge of skills. I can say that I know how to ride a bicycle or how to swim, but this does not mean that I can tell how I manage to keep my balance on a bicycle or keep afloat when swimming. I may not have the slightest idea of how I do this, or even an entirely wrong or grossly imperfect idea of it, and yet I go on cycling or swimming merrily*" (Polanyi, 1962). Or, as Imre Lakatos (1922-1974) put it, a scientist knows about science just as much as a fish knows about hydrodynamics.

The logical positivists of the Vienna Circle claimed that a proposition was worthy of consideration by scientists if and only if a method for testing and verifying it could be described. The so-called *Received View* of the positivistic philosophy of science holds that theories are tested by their predictions. If sufficiently many of these predictions are true, the theory is considered as confirmed. If any of these predictions are false, the theory is considered as falsified. Even if these statements are qualified in some ways, the Received View has been shown to be defective beyond repair. A general consensus that this positivistic analysis of scientific knowledge is inadequate now seems to hold among most philosophers of science. It is not only naive and philosophically untenable but also historically false.

Reference: For a survey and critique of the mainstream of thought in the philosophy of science from 1930 to 1969, compare Suppe (1974).

The first attack against the Received View was due to Karl Popper. He used Hume's argument against the possibility of logical induction to show that scientific theories cannot be verified by any feasible empirical evidence. The critique of the verifiability theory in his now classical *Logik der Forschung* showed conclusively that there is no such thing as the verification of general propositions by empirical research. No matter how many tests a theory has passed, it is always possible to find a weakness that leads to its rejection. "*Theories are not verifiable, but they can be 'corroborated'* (Popper, 1935; on p.251 in the English edition of 1959).

Popper considered empirical falsifiability to be the conditio sine qua non of a scientific theory. However, his much discussed thesis that "*it must be possible for an empirical scientific system to be refuted by experience*" (l.c.,p.41) is not tenable. Even Popper admits that scientists do not regard a theory as falsified if it conflicts with a single experimental result. *Every falsification is fallible*, a falsification never implies the rejection of a theory.

Example: The ground state of the hydrogen molecule has been studied with great skill and accuracy by our best experimentalists. Herzberg and Monfils (1960) have determined the dissociation energy of H_2 experimentally to be D_0=36'113,6 cm^{-1} with an estimated error of less than ±0,3 cm^{-1}. At about the same time Kolos and Roothaan (1960) carried out a very precise ab-initio calculation of the ground state energy of H_2 and found an astonishingly good agreement between the calculated and observed dissociation energy, in spite of the fact that some corrections had been neglected. An improved double precision ab-initio calculation with a 100-term James-Coolidge-type expansion by Kolos and Wolniewicz (1968) gave a clamped-nuclei total energy of E = = -1,174'474'983'0178 atomic units. Correcting for all known relativistic and radiative effects, they arrived at a theoretical dissociation energy of D_0=36'117,3 cm^{-1}, i.e. 3,7 cm^{-1} lower than the experimental value, or one order of magnitude larger than the experimental error. At this time the theoretical and experimental results were considered as definitely inconsistent. This discrepancy gave rise to a furious theoretical activity, all kinds of feasible and almost impossible effects were considered, without changing the discrepancy. The situation was serious since the Helium atom (which has the same type of interactions as H_2) was in perfect agreement with the theory. Quantum mechanics was falsified, nobody was happy but probably not one of the experts troubled his head about rejecting quantum mechanics! Later experimental work showed "*that previous work was more strongly affected by overlapping lines and lack of resolution than was suspected at the time*" (Herzberg, 1970). The improved experiment gave D_0=36'118,3 cm^{-1} (Herzberg, 1970), in good agreement with theoretical value.

As clearly seen by Pierre Duhem (1861-1916), no experimentum crucis exists. "*The physicist can never subject an isolated hypothesis to experimental test but only a whole group of hypotheses; when the experiment is in disagreement with his*

predictions, what he learns is that at least one of the hypotheses constituting this group is unacceptable and ought to be modified, but the experiment does not designate which one should be changed" (Duhem, 1908; on p.187 in the English edition of 1954). Empirical data can never disprove a theory. "*Any statement can be held true come what may, if we make drastic enough adjustments elsewhere in the system*" (Quine, 1953, p.43). Of course, an attempt to maintain a view in the face of recalcitrant experience may run the risk of making the research program look "epicyclic", piling one complication upon another.
On the other hand, falsifications are an important source for theoretical and experimental innovations. "*Nature may shout NO, but human ingenuity - contrary to Weyl and Popper - may always be able to shout louder*" (Lakatos, 1974, p.249).

Recent criticisms of falsificationism also deny that falsifications play the decisive role in the growth of science that Popper and his believers attribute to them. Hilary Putnam (1962) argued that scientific theories cannot be overthrown by experiments and observations alone, but only by alternative theories. Under the influence of Norwood Russell Hanson (1958), and Thomas S. Kuhn (1958), this view is widely accepted nowadays. Even modern logical positivists admit that it is "*always possible to 'save' a theory when it encounters recalcitrant data*" (Sneed, 1971, p.289) and that "*a theory is not that kind of entity of which one could reasonably say that it is falsified or refuted*" (Stegmüller, 1975).

2.4 PRECONCEPTIONS AND PRIOR CONCEPTIONS

Another popular fable is the idea that a scientific discovery begins with "hard facts" or "unprejudiced observations". Scandalously, there is no such thing as an observation unprejudiced and uncorrupted by past experience. Science certainly depends on observations in a crucial way but a scientific inquiry never starts from facts and observations alone (as claimed by the logical positivists of the Vienna Circle). The meaningful content of any theory is always larger than its empirical content. As underlined by Pierre Duhem (1908), observations are interpretations of phenomena depending upon the acceptance of a particular theory. That is, *theories are prior to experiments*, or in the new-fashioned terminology of Norwood Russell Hanson (1958), all facts are theory-laden. "*We now realize with special clarity, how much in error are those theorists who believe that theory comes inductively from experience*" (Einstein, 1936).

We cannot help being prejudiced by our ancestors, we can never avoid cultural, social and political preconceptions. Tacit presuppositions are the basis of understanding between the members of any social group. Judgements that are endorsed by every member of a group facilitate mutual understanding *within* this group but hamper the unterstanding *between* different social groups. Scientists are no exception, they share the culture of their society. A scientist who does not want to become incapable of imagining a perspective other than his own cannot dispense with philosophical reflections which purify and free his mind.

The scientific judgement of many scientists is distorted by the influence of untenable philosophies. If this is admitted, why not rid science of all metaphysical elements, why not reject all philosophical questions? Because "not to engage in philosophical contemplations" always means to unwittingly adopt an unreflected and most likely untenable philosophical position. For example, the logical positivists took over Wittgenstein's (Tractatus, 1922) rejection of metaphysics as meaningless, since there was no way of verifying it empirically. But logical positivism was still a philosophy, also not a well-grounded one. It is (or at least, it should be) the function of a philosophy of science to disclose hidden assumptions and tacit preconceptions which limit our awareness and act as barriers to our understanding. It is not possible to get rid of prior conceptions but we can try to convert tacit preconceptions into intentional prior conceptions.

As Jean Piaget has pointed out, many philosophers have assumed a definite psychology of perception in their epistemologies. It emerges from Piaget's (1970) experimental researches that even the simplest forms of perceptual invariants (like the notion of an object) require a definite learning process. When an individual receives a set of stimuli from its environment, he *assimilates* the stimuli to the *structures* of his existing knowledge. We can interpret phenomena only in terms of this existing knowledge; at the same time our ideas about the external world are in constant change.

It is mainly due to modern movement in the philosophy of science (Hanson, 1958; Feyerabend, 1958b, 1970a-c; Kuhn, 1962) that many scientists again acknowledge that "facts" are not inescapable basic data of existence but always dependent of the observer, his culture, theories and preconceptions. "*Facts contain ideological components, older views, natural interpretations, which have vanished from sight or perhaps never were formulated in an explicit manner. These components are highly suspicious, first because of their age, because of their archaic origin, and, secondly, because their very nature protects them from critical examination and always has protected them from such an examination*" (Feyerabend, 1970a, p.310). Since facts are partly created by the theories we hold, every child and every scientist requires a long time to "see" the facts. Whether a long training is more like "objective learning" (without any intervention of normative principles) or more like an indoctrination, is a rather delicate problem.

Example: Increasing emphasis is being laid on the teaching of science in elementary schools. One reason for teaching science even in the pre-elementary-school years has been summarized by Robert Karplus (1964) as follows: "*What about the often heard recommendation that science instruction be postponed until the youngsters have reached the intellectual maturity of the middle teens? At this stage, unfortunately, educational efforts reach only the fraction of the student body which is favorably disposed toward science education because of earlier favorable experience at home or at school. For the others, many of whom form a strong dislike for science, it is too late. Their spontaneous intellectual development just does not keep pace with the expectation of the school or does not proceed in the direction of modern science*". *Learning the truth or indoctrination of an ideology?*

Every social group (e.g. the scientific community) shares a particular set of tacit preconceptions and explicit prior conceptions, called a *paradigm* by Thomas S. Kuhn (1962). A paradigm is a world view, a set of beliefs and assumptions structuring recognition of problems, meaningfulness of questions and the acceptability of explanations. A paradigm firmly establishes a conceptual scheme capable of being applied in a uniform manner to most varied phenomena. The members of every particular social group not only accept a common paradigm but require all who wish to be recognized as accepted members to operate within the framework of the paradigm. Of course, there is never a unanimous agreement about "facts" (more precisely: about the paradigms generating "facts"), the traditional solution is to regard a dissenting minority as cranks. In the words of Michael Polanyi: "*There must be all times a predominantly accepted scientific view of the nature of things, in the light of which research is jointly conducted by members of the community of scientists. A strong presumption that any evidence which contradicts this view is invalid must prevail. Such evidence has to be disregarded, even if it cannot be accounted for, in the hope that it will eventually turn out to be false or irrelevant*" (Polanyi, 1963).

Example: It is a preconceived idea of experimental science that experiments are possible and that they can be repeated. That's a bold idea! For any experiment, there is a division of the world into the experimental system and its environment (including the observer!). The first tacit preconception is that such a division can be enforced and is harmless or can be corrected for (possible objection: the interdependence of all parts of the world is destroyed). The second assumption is that the required initial conditions can be brought about by an appropriate intervention of the experimenter (possible objection: the experimenter separates parts which should remain united). Thirdly, the unlimited repeatability of an experiment under identical conditions is considered to be constitutive for a scientific empiricism (possible objection: every experiment whatsoever has to be an irreversible process and therefore cannot be repeated under exactly the same conditions).
Not everybody accepts the paradigm of experimental science, many knowing and profound thinkers belong to the dissenting minority. Recommended reading: Johann Wolfgang von Goethe's essay *"Der Versuch als Vermittler von Objekt und Subjekt"*.

The upshot that there is no neutral observation language, that every observation is theory-laden is of crucial importance for the modern theory of molecular matter. What counts as a "fact" is determined by some theory or viewpoint. Adopting different prior conceptions we will in general observe different facts. *Each viewpoint creates its own reality.*

Example: As discussed by Niels Bohr (1932) in his Faraday lecture, thermodynamic concepts stand in a complementary relation to a complete and detailed mechanical description. The introduction of thermodynamical concepts, such as temperature and entropy, requires a point of view incompatible with the mechanical point of view.

Adopting an obsolete view on the reduction of physical theories, some philosophers of science fail to see that point mechanics and thermodynamics both represent legitimate but mutually incompatible viewpoints, and make the entirely unsubstantiated claim that "*the classical thermodynamics of Clausius, Kelvin and Carathéodory is simply an incorrect theory of the world*" (Sklar, 1976).

The temperature T and the energy E are incompatible physical quantities so that they cannot have simultaneously a sharp value. In fact, the theory of thermodynamical fluctuations gives the inequality

$$\Delta(\frac{1}{T})\Delta E \geq k$$

where k is Boltzmann's constant, and ΔX denotes the standard deviation of a fluctuating quantity X. As we know from the modern theory of K-flows, the striking analogy of this thermodynamic inequality with the Heisenberg-type inequalities is by no means superficial. The mechanical equivalent is the time-energy inequality, relating the deviation ΔE of an energy measurement with the measuring time Δt by

$$\Delta t \cdot \Delta E \geq \hbar/2$$

where $\hbar$ is Planck's constant divided by 2π.

The view that the objects one observes, and the properties that they possess are constituted in part by the adopted viewpoint contradicts the naive ontological vision of a world of independently existing objects but is perfectly compatible with the objectivity of scientific knowledge. This situation is well known from quantum mechanics where "*evidence obtained under different experimental conditions cannot be comprehended within a single picture, but must be regarded as complementary in the sense that only the totality of the phenomena exhausts the possible information about the objects*" (Bohr, 1949, p.210). Niels Bohr has advocated to use "*the word phenomenon to refer exclusively to observations obtained under specified circumstances, including an account of the whole experiment*" (Bohr, 1948). In a wider context, we propose to define a *phenomenon to refer to an observation under specified circumstances and prior conceptions.*

2.5 THERE IS NO INSIGHT WITHOUT INNER PICTURES

One does not attain a real understanding of a theoretical argument by merely checking its formal proof. It is necessary to see through the technicalities, to grasp the essential idea and to have intuitive insight. Intuition, as an immediate grasping of evidence, as a way of coming to knowledge independently of deduction, is of utmost importance for every scientist.

The possibility to "understand" scientific theories in an intuitive way requires the existence of appropriate mental structures which are conceptual but to which we have no conscious access (Pylyshyn, 1973). This introspectable imagery is often described in terms of "inner pictures" but is itself neither verbal nor pictorial. We say that a good theory should be intuitable and we mean that it should be knowable through intuition, by an inner experience allowing an immediate grasping. The word "intuitable" corresponds to the German expression *anschaulich* which has nothing to do with pictures or representations in the sense of mechanical models (as Margenau, 1950, p.327, erroneously assumes). Both *anschaulich* and *intuitable* refer to *inner* pictures.

Intuitive direct awareness is hardly teachable, it depends on unpredictable bursts of imagination. If we grasp or solve a problem with a single flash of insight, the psychologists speak of *aha! reactions*. "*Unsere Sprache hat die Interjektion 'aha' eigens für die Kundgabe solcher Erlebnisse geschaffen*" (Bühler, 1918). Aha-reactions owe little to logic or reason, they are overwhelming, they have a high degree of inevitability and aesthetic appeal. In his delightful book *"A Mathematical Apology"*, the mathematician Godefrey Harold Hardy stresses the importance of the aesthetic side of theories: "*The mathematician's patterns, like the painter's or poet's, must be beautiful; the ideas, like the colours or words, must fit together in a harmonious way. Beauty is the first test: there is no permanent place in the world for ugly mathematics.*" (Hardy, 1967, p.85). This criterion applies to scientific theories as well. Paul Adrien Maurice Dirac expressed it in the following way: "*We may try to make progress by following in Hamilton's footsteps, taking mathematical beauty as our guiding beacon, and setting up theories which are of interest, in the first place, only because of the beauty of their mathematics. We may then hope that such equations will ultimately prove their value in physics, basing this hope on the belief that Nature demands mathematical beauty in her laws.*" (Dirac, 1964).

There is a close connection between aha-reactions and the psychological process that creates new ideas. This creative process is quite mysterious. Carl Friedrich Gauss (1777-1855) reports in a letter to Wilhelm Olbers about the work on his famous *Disquisitiones Arithmeticae:* "*Dieser Mangel hat mir alles Uebrige, was ich fand, verleidet; und seit 4 Jahren wird selten eine Woche hingegangen sein, wo ich nicht einen oder den andern vergeblichen Versuch, diesen Knoten zu lösen, gemacht hätte - besonders lebhaft nun auch wieder in der letzten Zeit. Aber alles Brüten, alles Suchen ist umsonst gewesen, traurig habe ich jedesmal die Feder wieder niederlegen müssen. Endlich vor ein paar Tagen ist's gelungen - aber nicht meinem mühsamen Streben, sondern bloss durch die Gnade Gottes möchte ich sagen. Wie der Blitz einschlägt, hat sich das Rätsel gelöst; ich selbst wäre nicht im Stande, den leitenden Faden zwischen dem, was ich vorher wusste, dem, womit ich die letzten Versuche gemacht hatte, - und dem, wodurch es gelang, nachzuweisen.*" (Gauss, 1805). Aha-reactions are creative acts of the human mind. The spontaneous innate ability to form concepts and ideas by creative imagination is a unique feature in man, distinguishing him from the rest of nature.

The essential ingredients of a theory are not made up by our conscious mind. Ideas are produced by our unconscious, the German word *Einfälle* characterizes them as the things that fall into our consciousness. This origin of ideas explains the impression of incorrigibility they convey until it can be realized or justified by the conscious mind. Every creative process has an integrating tendency, it brings a new unity into existence but it does not seem to serve any immediate biological purpose. We know very little about the nature of the creative process; it stems from the unconscious and develops according to its own logic.

References: There are two famous and most interesting studies by the mathematicians Henri Poincaré (1913, pp.383-394) and Jacques Hadamard (1945) of their own creative activities. Collections of personal experiences of the creative process have been edited by Kankeleit (1959) and by Dalenius et al. (1970). A valuable study of the conscious and unconscious in science and art is due to Koestler (1964). An illuminating description of the role of intuition and imagination in Chinese thinking and of the oriental approach to science has been given by the eminent Japanese physicist Hideki Yukawa (1973).

Following Carl Gustav Jung (1875-1961), we use "*the term idea to express the meaning of a primordial image*", where by image we "*do not mean the psychic reflection of an external object, but a concept derived from poetic usage, namely, a figure of fancy or fantasy-image*" (Jung, Coll. Works, vol.6, par. 732, 743). The image does not correspond to something outside but is real in its own way. "*The real*", says Jung, "*is what works*" (Coll. Works, vol.7, par. 353).

But what is then the referent of imagination? Imagination is the realm of soul. For depth psychology the referent of images is unconscious but has still objective existence. The "*unconscious psyche, common to all mankind, does not consist merely of contents capable of becoming conscious, but of latent predispositions toward identical reactions*" (Jung, Coll. Works, vol.13, par.11). For the inborn mode of psychic apprehension, Jung has proposed the term *archetype* (Jung, Coll. Works, vol.8, par.270 ff, and vol.6, par.624). The word archetype comes from the Greek and means the *prime imprinter*. "*The archetypal representations (images and ideas) mediated to us by the unconscious should not be confused with the archetype as such*" (Jung, Coll. Works, vol.8, par.417). We do not experience the archetypes themselves, they remain hidden but they express themselves by archetypal motives in ideas and imaginations, and exercise a formative influence upon human thought and behavior. The archetypes are the operative agents of the soul and put us into an imaginative style of discourse. They are the ultimate psychic realities; they are metaphysically real because they are capable of producing observable effects. Archetypes are "*psychic realities, real because they work*" (Jung, Coll. Works, vol.7, par.151). However, archetypes are not necessarily psychic structures, the psyche is just the place where they become manifest. Together with the instincts, the archetypes make up that psychic stratum which Jung calls the *collective unconscious* (Coll. Works, vol.8, par.229). The collective unconscious by its very nature is the common property of all human beings, regardless of their culture.

An activated archetype exerts a fascination, it attracts to itself the contents of consciousness "*so that the subject is gripped by it as though by an instinct*". "*Its passing over into consciousness is felt as an illumination, a revelation, or a 'saving idea'*". In the long run, archetypes "*mould the destinies of individuals by unconsciously influencing their thinking, feeling and behavior, even if this influence is not recognized until long afterwards*" (Jung, Coll. Works, vol.5, par.225, 450, 467).

If archetypes appear unconsciously in projections, they contribute to myth-formations and preconceptions. *Projection* is a psychic process by which a subjective content becomes alienated from the subject and is embodied into the object (cf. Jung, Coll. Works, vol.6, par.783). A projection is never conscious, it can only be recognized if it is dissolved. The recognition of the archetypal nature of a preconception removes its magical effect and turns it into a neutral prior conception. The dissolution of a projection gives insight into the driving forces of the unconscious but does not imply that we have dissolved the forces. To discard archetypal images "*would be a distinct loss. Our task is not, therefore, to deny the archetype, but to dissolve the projections, in order to restore their contents to the individual who has involuntarily lost them by projecting them outside himself*" (Jung, Coll. Works, vol.9.1, par.160).

By absorbing more and more unconscious content, the consciousness extends its frontiers, and this process leads to a continuous development of the ego. The emancipation from the power of the unconscious leads to empirical science by creating the external objective world. In this evolution there is the danger of overemphazising consciousness, of identifying consciousness with thinking, leading to the exaggeration that the external objective world is the only reality (compare the analysis by Erich Neumann, 1954, pp.341 and 384 ff). Forgetting that constituents of reality are also to be found in the unconscious has a sterilizing effect on creativity. An overemphasis on consciousness leads to the one-sidedness characteristic of the self-confident expert, strong-willed and incapable of seeing anything that transcends his narrow sphere of competence. A typical symptom of such a one-sided overexpansion of the ego is the absence of any genuine creative activity.

Archetypal images play a crucial role as intuitive concepts for physical phenomena. For example, the archetypal notion of *magic power* is a concept found all over the world and is used as an intuitive concept for the physical notion of *energy* (compare Jung, Coll. Works, vol.7, par.151). That *numbers* have an archetypal foundation is an old conjecture of many mathematicians, so that Jung defines number psychologically as an *archetype of order* which has become conscious (Jung, Coll. Works, vol.8, par.870).

The assimilation of unconscious archetypal contents leads to ideas expressing the meaning of archetypal image in abstract form. If the transpersonal content of ideas is repressed, then the symbols degenerate into mere signs, archetypal ideas into mere formal concepts. Such a rationalization is destructive, the ideas lose their compelling significance and living force, and are emptied of meaning. An example is the *nominalism* in mathematical number theory, that is, the view that there are no abstract entities whatsoever represented by numbers. One of the greatest creative mathematicians, Kurt Gödel (1906-1978) has not accepted this formalistic view but has maintained that "*classes and concepts may ... be conceived as real objects ... existing independently of our definition and constructions*" (Gödel, 1944, p.137). Clearly, Gödel does not assume that these mathematical classes and concepts exist in outer reality but he still holds that they exist independently of our minds. Most mathematicians consider the idea of an a priori meaning of mathematics as perfectly self-evident although they admit it only reluctantly. Jean Dieudonné describes Bourbaki's attitude towards the meaning of mathematics as follows: "*On foundations we believe in the reality of mathematics, but of course when philosophers attack us with their paradoxes we rush to hide behind formalism and say: 'Mathematics is just a combination of meaningless symbols', and then we bring out Chapters 1 and 2 on set theory. Finally we are left in peace to go back to our mathematics and do it as we have always done, with the feeling each mathematician has that he is working with something real*". (Dieudonné, 1970). Without its archetypal background, mathematics would be idle chatter or mere desire for amusement.

The dynamics of scientific evolution is governed by activated archetypes determining the typical mode of action and the currently valid paradigms. Since the archetypes cannot be evoked by the ego's will, the large-scale development of science is determined by the dynamics of the *collective* unconscious, a transpersonal reality. The fact, stressed by Thomas S. Kuhn and Paul K. Feyerabend, that the development of scientific theories owes little to logic or reason is not surprising when we accept that "*the soul has a logic of its own and an experience of its own not to be seized by languages appropriate to physical phenomena on the one hand and to mental processes on the other*" (Christou, 1963). The paradigms change because modifications occur in the collective unconscious. The existence of such changes is a historical fact (compare e.g. Erich Neumann, 1954). Whether they occur spontaneously or in reaction to the evolution of consciousness or to social and political changes is largely an open question.

Every myth, every great work of literature and art, every new conception and every theory has an unconscious core of meaning expressing archetypal truths. "*All the most powerful ideas in history go back to archetypes. ... In their present form they are variants of archetypal ideas, created by consciously applying and adapting these ideas to reality. For it is the function of consciousness not only to recognize and assimilate the external world through the gateway of the senses, but to translate into visible reality the world within us.*" (Jung, Coll. Works, vol.8, par.342). The rich mathematical symbolism of modern scientific theories expresses both aspects of the outer reality and archetypal aspects of the collective unconscious. That is, the syntactics of a theory reflects both the structure of matter and the structure of psyche. An explanation of this astounding parallelism of inside and outside worlds could be "*the fact that 'psyche' and 'matter' are not basically incommensurable in nature, but may perhaps be qualities of one and the same existential being*" (Jung, 1960).

2.6 ARE THEORIES DANGEROUS?

The current revulsion from science is more than an irresponsible repetition of contemporary fashions, it may be a *reaction against theories expressing an unduly narrow world view.* Many contemporary scientists are troubled by grave doubts about the usefulness and responsibility of the scientific endeavor. Our current knowledge allows us to do miraculous things but we are quite ignorant about whether to do so is morally defensible. The problems created by the combination of science and technology are tremendous. That we are in trouble is also seen by our experts: "*Any attempt to find automatically safe channels for the present explosive variety of progress must lead to frustration. The only safety possible is relative, and lies in an intelligent exercise of day-to-day judgement. ... We can specify only the human qualities required: patience, flexibility, intelligence.*" (von Neumann, 1955a). Are our most intelligent leading experts unaware of the need for insight and wisdom? The separation of analytical intellect and synthetic imagination has disastrous effects, it leads to a dangerous remoteness of science from the mainstream of human thinking. These are the reasons why science has lost prestige and why faith and admiration for science have strongly diminished.

The legendary image of a scientist as a humble searcher for truth is more and more replaced by the image of a scientist as a well-paid brillant expert, speaking an unintelligible professional jargon, highly competent in a narrowly defined domain but arrogantly extending his competence into fields in which he knows nothing, and neglecting the fact that science is only a small subdivision of human knowledge. Johannes Kepler (1571-1630), who asked: "*God, whom I can almost touch with my hands when observing the universe, do I find him also in myself?*", was probably a humble searcher for truth. The modern expert, fascinated in creating "*neutron bombs destroying only life, not property*" cannot be said to be searching for truth about nature. Know-how has replaced wisdom which is certainly a factor contributing to the revulsion from science.

The well-known claim that "*science as long as it limits itself to the study of the laws of nature has no moral or ethical quality*" is said to be evident. But is it true? Lack of awareness of moral content is no proof of its nonexistence. The scientific revolution of the last few centuries caused enormous moral problems. The very same people who ridicule the contemporaries of Galileo Galilei (1564-1642) for not being too happy about his move, do not seem to notice that he started not only modern science but also its shortsighted arrogance. In his *Dialogue on the Two Main Systems of the World*, Galileo assumes naively that he has a fully developed self-awareness and unrestricted insight in his own activities: "*Only the blind need a guide, but anyone with two eyes in his head and with a mind should use them to guide himself.*" (Galileo, 1632). With this Galileo characterized the modern scientist as a *man without shadow*, "*one who imagines he actually is only what he cares to know about himself*" (Jung, Coll. Works, vol.8, par.409). Adopting Galileo's doctrine of free inquiry, independent of the interference of any authority other than our private means of information implies that we rely on our own form of illumination. That is, we rely on inferences drawn from some type of innate knowledge. Of course, we no longer speak of the voice of God or the temptations of the Devil but simply and uncritically rely on "natural knowledge" (such as the claim of the *a priori* necessity of the connection between cause and effect, or the claim that the *tertium non datur* is a basic principle of logic). This natural knowledge is archetypal knowledge, emerging from the collective unconscious into the individual's mind and replacing the old autorities. The typical modern scientist has lost any understanding of why in ancient times eating from the tree of knowledge was considered as a severe taboo infringement. Nowadays, we insolently reject any guide and claim, as a matter of course that "*a scientist has no responsibility for the truths he finds in his research*".

It is said that nothing could be more humanistic than the pursuit of truth. However, there are truths and truths, shallow truths, one-sided truths and some truths are more important to pursue than others. There are even truths not compatible with other truths, or, as Niels Bohr once put it: "*A great truth is a truth whose opposite is also a great truth*" (quoted in Hans Bohr, 1967, p.328). It is not at all plain that the pursuit of scientific truth is compatible with our humanistic principles. The rationalist's traditions of scientific thought are often regarded as destructive since it is felt that the scientific world view is too narrow, scorning metaphysical, religious and artistic approaches, and since we should not increase our scientific knowledge if we cannot at the same time advance our wisdom and virtue. On the other hand, the preponderant view among contemporary scientists is that it is better to have scientific knowledge than not to, and that "*there is simply no way of deciding in advance where a basic scientific exploration will come out, or what the risk and benefits will be.*" (Thomas, 1978). Many scientists find moral problems associated with purely scientific activities unintelligible, perhaps because moral considerations could seriously damage basic scientific research. "*It is not consistent with the genius of the American people to restrict the progress of scientific knowledge by legislation or otherwise*" (Packard and Cope, 1883). Nowadays this ideology is firmly established. Although there were no good indications that the new technologies would be used responsibly, John von Neumann reconfirmed this ideology in a testimony before a Committee on Atomic Energy of the United State Senate, given on January 31, 1946: "*I do not believe that the nation would act wisely in attempting to forbid to itself to do certain things*" (von Neumann, Coll. Works, Vol.6, p.50). And some years later: "*For progress there is no cure*" (von Neumann, 1955a). The old idea that the world was put at man's disposal by God degenerates into a new boldness: *what can be done must be done.*

Modern philosophers have a different view. We quote as an example Georg Picht: "*Humanität, die ihre Eingebundenheit in die Natur verleugnet, schlägt notwendig um in Bestialität und entfesselt dann ein Spiel von desorganisierten Instinkten, die wir sonst bei keinem Raubtier kennen. Die 'bestia rationalis' wird von destruktiven Impulsen getrieben, deren sie nicht Herr zu werden vermag, weil sie mit einer Geisteskrankheit geschlagen ist, die man als 'dementia rationalis' bezeichnen kann. Charakterisieren lässt sich die Krankheit dadurch, dass eine partikularisierte Rationalität in offenem Widerspruch zu allen jenen Formen der Erkenntnis getreten ist, die in unserer philosophischen Tradition mit dem Namen 'Vernunft' bezeichnet worden ist. Vernünftig denkt, wer in der Lage ist, die Konsequenzen seines Denkens und Handelns zu überblicken, und wer bereit ist, die Verantwortung für diese Konsequenzen auf sich zu nehmen. In diesem Sinne ist eine Wissenschaft, die auf eine Theorie ihrer eigenen Konsequenzen verzichtet und nicht bereit ist, die Verantwortung für ihre technische und praktische Auswirkung zu übernehmen, widervernünftig.*" (Picht, 1976). Adopting Picht's definition of rationality, we have to admit that, starting with Galileo, science has become more and more irrational. Science *has* a moral content but we scientists lost the sense of moral responsibility for our activities. Pure scientists cannot escape the problem of responsibility; at least they are guilty of not seeing that there is no sharp dichotomy between pure science and technology.

The imaginative and creative activity of scientists and artists is not very different. To be sure, artists are usually well-informed about the irrational nature of creative processes while most scientists pretend or even believe that they proceed rationally. The driving force in an artist's or scientist's work is not the desire to benefit humanity. "*Wahrlich, es ist nicht das Wissen, sondern das Lernen, nicht das Besitzen, sondern das Erwerben, nicht das Da-seyn, sondern das Hinkommen, was den grössten Genuss gewährt*" (C.F. Gauss to W. Bolyai, letter of 2.9.1808, cf. Bolyai, 1899). For a creative scientist, research is fascinating, and fascination cannot be quenched. Both artists and scientists often work with extraordinary enthusiasm and great intensity, they even devote their life to their work. Loyalty to their vision is more important than humanistic considerations. The creative scientist is in danger of thinking *that it is he, who has* the great idea, not realizing that the great idea *has him.* The idea goes back to an archetype, and as long it remains unconscious it takes possession of the ego. A scientist possessed by archtypal power has an enormous dynamism within himself but becomes utterly unfree and suffers a loss of consciousness. Such an inflation of the consciousness "*occurs whenever people are overpowered by knowledge or by some new realization. ... The inflation has nothing to do with the kind of knowledge, but simply and solely with the fact that any new knowledge can so seize hold a weak head that he no longer sees and hears anything else. He is hypnotized by it, and instantly believes he has solved the riddle of the universe*" (Jung, Coll. Works., vol.7, par.243).

A modern scientist as a rule has no idea what is going on when he has a great idea. He uses shallow razionalizations, wonders about the bad image science has, and explains it by projecting his own irrationality on an irrational behavior of the mass media. A creative possession is never harmless but our culture has lost the instinct of the older cultures for the peril of the gods. There is the danger of losing our freedom of consciousness by succumbing to the automatism of the collective unconscious. "*An inflated consciousness is always egocentric and conscious of nothing but its own existence. It is incapapble of learning from the past, incapable of understanding contemporary events, and incapable of drawing right conclusions about the future. It is hypnotized by itself and therefore cannot be argued with*". (Jung, Coll. Works, vol.12, par.563). In the perspective of our critics, a pretty good description of a scientist.

Modern science has badly neglected the forces in man that shape science, technology and human affairs. Nevertheless, a good part of the responsibility for modern technological developments is taken by scientists. Why? "... *sie haben in ihrer wissenschaftlichen Arbeit besser gelernt, objektiv, sachlich und, was das wichtigste ist, in grossen Zusammenhängen zu denken*", as Werner Heisenberg (1969, p.272) lets his friend Carl Friedrich von Weizsäcker argue. However, how can a scientist think objectively, without losing track of bigger contexts, if he is completely in the dark about the driving forces of his work? It never seems to occur to the brilliant experts that they assume a responsibility that goes beyond all reasonable bounds of their competence. It is no longer possible to close our eyes to all but the most optimistic predictions. We have to learn to escape the danger of identification with primordial ideas and the associated possibility of an inflation which threatens the consciousness. Good theories are risky since they can be ideal carriers for projections, representing an unconscious content cut off from all influence of the conscious mind. The great danger is that we are not aware of our ignorance, or, as John Steinbeck (1902-1968) put it in his *The Winter of our Discontent*: "*I guess we're all, most of us, the wards of that 19th century science which denied existence to anything it could not measure or explain. We did not see what we couldn't explain, and meanwhile a great part of the world was abandoned to children, insane people, fools and mystics*".

2.7 SUMMING UP

Science, technology, art and religion have their origin in the creative center of the psyche. Science assumes that there is *order* in nature; scientific theories try to relate the order of external objective reality to the order of a transpersonal inner reality.

Depth psychology describes the introspectively recognizable form of psychic orderedness by *archetypes*. Myths are well-known expressions of archetypes, they were not invented but experienced. Science is the myth of modern man.

Scientific theories create order in the empirical material but every theory goes far beyond the immediate knowledge, understanding and insight. Theories are the result of creative imagination, they necessarily have an intrinsic beauty.

The formalism of mathematically formulated theory allows to describe experience on a deeper than a purely descriptive level. The propositions of the syntactical formalism obtain meaning through (i) their connection with empirical data, and (ii) their connection with the conceptual order of the mind. Ideally, the structure of the formalism of a theory should mirror both the order of the external *and* of the internal reality.

Research is an intuitive artistic vocation so that pure science should be considered on a par with the arts. All central concepts of a theory are derived from archetypal structures. Ideas are not the result of a conscious effort of will but they *appear* spontaneously. The genuine creative process is always unconscious. Archetypal images decide the fate of scientific theories just as much as do the realities of the external world.

A scientific theory is never an independent body of knowledge but always relative to its historical and cultural context. The basic preconception necessary for any scientific theory is called the *paradigm*. The adopted paradigm reflects the "Zeitgeist", that is the general intellectual and moral state of the culture. A paradigm establishes the possibility of a standard of objectivity and truth.

An empirically well-confirmed theory is unfalsifiable in the context of its paradigm, it can only be overthrown by a new paradigm. The driving forces that determine the dynamics of the paradigms are the archetypes. On the other hand, a new paradigm is not accepted in empirical science unless it has substantial explanatory success.

The classical scientific ideal of unprejudiced and objectively demonstrable knowledge is not tenable. Reality has to be constructed theoretically, it is dependent on the adopted point of view.

Philosophical questions are not without consequences in the establishment of science. Some popular philosophical views, like positivism and operationalism, are not only untenable but a threat to man's creative possibilities.

Science leads to an increase of consciousness harboring the danger of inflation of the consciousness with unconscious contents. It is doubtful if a further rapid and unrestricted development of science is compatible with vital humanistic principles.

The pursuit of truth has moral content. Every scientist who rejects external authorities has the full responsibility for the truths he finds in his research. The current revulsion from science is at least in part a sane reaction against an infantile and irrational behavior on part of the scientific community.

3. PIONEER QUANTUM MECHANICS AND ITS INTERPRETATION

Believe nothing,
no matter where you read it,
or who said it,
no matter if I have said it,
unless it agrees with your own reason
and your own common sense.
Buddha

3.1 INTRODUCTORY REMARKS AND PREVIEW

Quantum mechanics was created in a unique effort of a small group of ingenious physicists during the period of 1922 to 1927. The leading pioneers were Niels Bohr, Louis de Broglie, Max Born, Werner Heisenberg, Pascual Jordan, Wolfgang Pauli, Erwin Schrödinger and Paul Adrien Maurice Dirac. By 1928 the new mechanics was sufficiently developed to be applied to the properties of atoms, molecules, solids and radiation. The first really authoritative text was Dirac's *"The Principles of Quantum Mechanics"* of 1930. Von Neumann's brilliant *"Mathematische Grundlagen der Quantenmechanik"* of 1932 contributed much to the mathematical refinement of the new theory, while Pauli's article *"Die allgemeinen Prinzipien der Wellenmechanik"* in the *"Handbuch der Physik"* of 1933 gives an encyclopedic coverage from a more physical point of view. In spite of their slightly different positions, we regard the expositions in these classical works by Dirac (1930), von Neumann (1932) and Pauli (1933) as aspects of one single theory which we call *pioneer quantum mechanics*. More recent accounts of pioneer quantum mechanics are the monograph by Bohm (1951) and by Ludwig (1954). Ludwig's text stresses the mathematical formalism and is easier to read than von Neumann's classic, while Bohm's book is one of the few texts that give a judicious and thoughtful discussion of the interpretative problem of pioneer quantum mechanics.

Pioneer quantum mechanics is important, not only because it is relevant for any theory of molecular matter but also because it implies a striking departure from philosophical tradition, initiating a new and deeper understanding of the structure of scientific theories. Clearly, quantum mechanics stands within certain historical traditions and still undergoes - as does every great theory - further evolutions. To understand quantum mechanics as a cultural enterprise in which man has reached a new level of consciousness and to recognize the continuity of contemporary achievements with the past, we have the obligation to understand the principal historical roots. This does not mean that we should try to make the visions of the pioneers our own, since, as stressed by Martin Strauss (1970, p.262), *"the history of science knows of no instance of a physical theory correctly understood by its author"*. This is an important reason why in order to understand quantum mechanics it is necessary to know its history.

The history of quantum theory shows that even the most creative scientists are often very conservative in the sense that they tend to stick stubbornly to their own ideas. The theories they have developed themselves are emotionally important to them, they are even obsessed by their initial ideas. However, the integration of an idea or discovery into the main body of science is a process of depersonalization whereby the initial idea changes its content. Without exception, every brilliant idea is also misleading and its final form emerges only in a long process of gradual clarification. As a rule, the discoverer is disappointed by the later reinterpretations; the development he started does not correspond to his own vision of the meaning of "his" idea. The development of an idea by the scientific community is independent of the conceptions of its creator. The joy and excitement of the discoverer often changes into deep frustration. The development of pioneer quantum mechanics is an outstanding example. The great pioneers Max Planck, Albert Einstein, Louis de Broglie and Erwin Schrödinger have been deeply worried about the development of quantum mechanics. In a letter of December 22, 1950, Albert Einstein wrote to Erwin Schrödinger: *"Wenn man die Quantentheorie als (im Prinzip) endgültig ansehen will, so muss man glauben, dass eine vollständigere Beschreibung zwecklos wäre, weil es für sie keine Gesetze gäbe. Wenn es so wäre, dann würde die Physik nur mehr für Krämer und Ingenieure Interesse beanspruchen können; das ganze wäre ein trauriges Pfuschwerk"* (Przibram, 1963, p.36).

The history of early quantum theory and pioneer quantum mechanics is easily accessible and warrants but brief recapitulation here. A profound exposition is due to Jammer (1966). Additional details can be found in Whittaker (1953), Klein (1962), Kuhn et al. (1967), Scott (1967), v.d. Waerden (1967), Hermann (1969) and Gerber (1969). In section 3.2 we give a deliberately narrow review of the historical development of the Hilbert space formalism of pioneer quantum mechanics, focusing on those aspects which were of importance for the further development of quantum mechanics. We also give some examples for the odd but quite characteristic feature of history of science that the pioneers of quantum theory were not fully aware of the meaning and the implications of their discoveries.

In section 3.3 we give a terse outline of the Hilbert-space formalism of pioneer quantum mechanics from a modern point of view. It is written in a more technical language and it may be skipped; the remaining sections can be read independently of this section.

It is tempting, but mistaken, to believe that we can use the formalism of quantum mechanics without entering into a discussion of its interpretation. Only interpreted mathematical formulas express laws of nature, the interpretation gives a physical meaning to the mathematical symbols used. *Interpretations never are inherent in the formalism but must be assigned to it.* Mathematical construction is such a marvelously effective creative principle that the formal structure of a new theory may be arrived at prior to the establishment of a consistent interpretation. That is, we may be confronted with the amazing situation that we know and can use the mathematical equations of a new theory without exactly knowing what they mean. Pioneer quantum mechanics is an example: it is the most controversial theory we have and even today no general agreement exists on its interpretation (compare for example the review given by Jammer, 1974).

The interpretation of quantum mechanics is a deep problem. It is surprising how often vulgarized versions of various interpretations have been presented and how uninformed some textbook authors are who pretend to have studied Bohr's, Heisenberg's, and von Neumann's fundamental contributions. A summary of the Copenhagen interpretation is given in section 3.4. Our discussion is necessarily a very condensed summary; a deeper understanding requires the study of the original papers which are quoted rather completely. It is crucial to understand the Copenhagen view, to appreciate both its wisdom and its inadequacy.

The Copenhagen view has a distinct ring of phenomenalism and instrumentalism. According to Bohr (1931, in the English translation on p.18), "*in our description of nature the purpose is not to disclose the real essence of the phenomena but only to track down, so far as it is possible, relations between the manifold aspects of our experience*". The Copenhagen view is content with a pragmatic description of laboratory experiments; what is "really" happening is not considered as a legitimate question. The first attempt to adopt a broader view is the von Neumann-London-Bauer interpretation, discussed in section 3.5. A lasting contribution is von Neumann's caricature of quantum measurements. The aim of this theory of quantum measurements is to show that the reduction postulate is compatible both with the formalism of quantum mechanics and Born's probabilistic interpretation of the state vector. In this respect, von Neumann has been successful. However, von Neumann did not solve the so-called measurement problem, that is, he did not deduce the reduction postulate from the time-dependent Schrödinger equation. At this level of the theoretical development, quantum mechanics can predict correctly and in a consistent manner with what probabilities events will occur but it does not explain at all the fact that quantum events *do* occur.

Von Neumann's rejection of Bohr's requirement that the observational means must be described by a classical theory creates great problems. The main stumbling block is the measurement problem. The solution of this enigma should provide an explanation why interactions that lead to measurements differ so fundamentally from interactions that do not involve measurements. These difficulties have been considered to be so grave that many scientists have questioned the universal validity of quantum mechanics in the molecular domain. If we grant the unrestricted validity of the formalism of pioneer quantum mechanics, then we know only three logically consistent interpretations of quantum mechanics: the exorbitantly pragmatic Copenhagen view, the outrageously subjectivistic von Neumann-London-Bauer interpretation, and the astounding Everett interpretation. The Copenhagen view has no measurement problem because - according to Bohr - "*there is no quantum world*" (quoted from Petersen, 1963). The von Neumann-London-Bauer interpretation "solves" the measurement problem by claiming that the consciousness of the observer is responsible for quantal events. These two interpretations are logically consistent but reject the view that there are things existing objectively and independently of any measurement or observer.

The belief that there exists an objective reality outside of any mind is a regulative principle accepted by almost everybody in everyday life. A logically consistent interpretation of the formalism of quantum mechanics that allows us to accept realism as a regulative principle is possible; it is the Everett interpretation, discussed in section 3.6. The Everett interpretation agrees with the empirical facts, it has the advantage of an attractive simplicity, and it frees us from all the unnecessary restrictions of the Copenhagen view and the von Neumann-London-Bauer interpretation. However, the Everett interpretation has an unambiguous meaning only if the measurement problem is solved in a technical sense. That is, we have to explain how quantum mechanics can account for the existence of classical properties of certain quantal systems.

The existence of classical properties is, however, in direct contradiction to the superposition principle of pioneer quantum mechanics. The universal validity of the superposition principle implies a wholeness of quantum mechanics quite foreign to classical physics. This fact was put in evidence in a famous paradox due to Einstein, Podolsky and Rosen; it is discussed in section 3.7. Nowadays, the existence of the Einstein-Podolsky-Rosen correlation is experimentally well established and forces us to break with an old tradition: the idea that most physical situations can be described in excellent approximations by a closed-system analysis involving only a small part of the entire universe. Because of the omnipresence of Einstein-Podolsky-Rosen correlations, a complete description of processes localized in a small part of the world seems to require the inclusion of the whole universe, or at least, the inclusion of a background state due to the rest of the world. Or in Bohr's language, the results of an experiment cannot be separated from its conditions.

In section 3.8 we resume our discussion of the status of pioneer quantum mechanics and explain why it is - in spite of its empirical success - inadequate as a fundamental theory of molecular matter.

3.2 THE HISTORICAL DEVELOPMENT OF THE HILBERT-SPACE MODEL OF PIONEER QUANTUM MECHANICS

The initial phase: the discovery of the quantum of action

The development of quantum theory is an excellent example of the fact that science only rarely progresses in direct ways but that basic problems are often solved by achievements in other fields apparently having no connection with the problem at hand. The quantum revolution began at the intersection of thermodynamics and electrodynamics. The birthday of quantum theory can be identified with the discovery of the quantum of action by Max Planck (1858-1947). On December 14, 1900, Max Planck presented his derivation of the distribution law for black-body radiation to the "Deutsche Physikalische Gesellschaft" in Berlin. Planck (1900) postulated that light could not be radiated continuously but only in whole multiples of $h\nu$, where ν is the frequency and h a new universal constant, nowadays called Planck's constant of action. Planck regarded his hypothesis as necessary for explaining the experimentally known spectral distribution of black-body radiation; it signifies a breach with classical physics and the emergence of a new paradigm.

Remark: Planck assumed as established that a classical system with a large number of coupled degrees of freedom leads to the equipartition law of classical statistical mechanics. As we know today, this assumption is *not* a logical consequence of the equations of motions of classical physics (Bocchieri and Valz-Gris, 1972, 1974), so that the Rayleigh-Jeans radiation law is not a consequence of classical physics. Moreover, modern studies suggest that classical physics is not in such violent contradiction with the experimental results on the black-body radiation as Planck believed. (Bocchieri et al., 1974; Bocchieri and Valz-Gris, 1975; Wallace, 1974). Boyer (1969) has shown that even if one assumes the equipartition law as valid, it is possible to derive Planck's radiation law for black-body radiation from *classical* statistical mechanics without any quantum assumptions by postulating the existence of a classical fluctuating electromagnetic radiation with a Lorentz-invariant spectrum (compare also Theimer, 1971). Hence, *though quantum theory flourishes, its historical foundation has been superseded to a large extent.*

The next revolutionary step was made by Albert Einstein (1879-1955). In 1905, in the same volume of the "Annalen der Physik" in which his epoch-making papers on Brownian motion and on special relativity appeared, Einstein (1905a) pointed out that the experimental data on the photoelectric effect could be naturally interpreted by assuming that light has a particle structure. The light quanta (later called *photons*) of a monochromatic electromagnetic radiation of frequency ν would then have an energy $h\nu$, a momentum $h\nu/c$ and move with the velocity c of light. While

Planck regarded the quantum of action as a property of the *interaction* between radiation and light, Einstein put forward the different and more radical hypothesis that energy quanta have an independent existence in radiation.

Remark: This idea was too radical for Planck and he did not accept the idea of light quanta. The driving force in Planck's work was the idea that thermodynamics and electrodynamics could be unified into a consistent classical theory. When the full implications of Planck's theory became apparent, Planck did not accept them. In his proposal to elect Einstein as a member of the Prussian Academy of Science, Max Planck wrote on June 12, 1913: "*Dass er in seinen Spekulationen gelegentlich auch einmal über das Ziel hinausgeschossen haben mag, wie z.B. in seiner Hypothese der Lichtquanten, wird man nicht allzuschwer anrechnen dürfen; denn ohne einmal ein Risiko zu wagen, lässt sich auch in der exaktesten Wissenschaft keine wirkliche Neuerung einführen*" (quoted from Kirsten and Körber, 1975, p.202).

The third decisive step was taken by Niels Bohr (1885-1962) in 1913 by introducing the concept of *stationary states*. Bohr (1913) postulated that electrons in stationary states do not radiate. He assumed that electrons in atoms revolve in certain discrete orbits without radiating, but when jumping from one stationary state to another they emit a quantum of light of frequency $\nu = (E_2 - E_1)/h$, where E_1 and E_2 are the energies of the first and second orbit, respectively. With this Bohr could explain the line spectrum of hydrogen.

A further crucial step in our understanding of black-body radiation was taken by Satyendra Nath Bose (1894-1974). On June 4, 1924, Bose sent a short paper in English to Albert Einstein, in which he deduced Planck's law independently of classical electrodynamics, only assuming that the ultimate elementary regions in phase space have the volume h^3. Einstein was impressed, translated the paper into German and submitted it to the "Zeitschrift für Physik" (Bose, 1924). Bose showed that photons had the amazing property of being *strictly identical*. Bose's result immediately was extended by Einstein (1924, 1925) to the study of an ideal gas with particles of nonzero rest mass.

Remark: Einstein immediately grasped the meaning of Bose's hypothesis of the indistinguishability of photons, and generalized it to particles of nonzero rest mass. However, he never adopted the straightforward conclusion that strict identity implies indistinguishability so that a reference to trajectories of identical particles makes no sense. Einstein always insisted on a classical space-time description and rejected quantum theory as an exhaustive description of nature.

The discovery of the wave nature of matter

In 1923 Louis de Broglie (born 1892) arrived at the idea that the notion of the coexistence of particles and waves, discovered by Einstein in 1905 for light, extends to all particles. De Broglie (1923) assumed that the particle is located at a well-defined point within the wave at every instant, and that in the case of a monochromatic wave of frequency ν it possesses the energy $E = h\nu$ and the momentum $P = h/\lambda$ where λ is the wavelength. The sensational discovery of electron diffraction by Davisson and Germer (1927) gave an empirical verification of the wave nature of matter, confirming the brilliant idea of de Broglie.

Remark: De Broglie tried to develop his ideas into a "theory of the double-solution", where the particle in the wave is guided by the wave propagation. Although de Broglie defended this view throughout life (compare the review by de Broglie, 1971), it was never accepted by the majority of physicists.

Heisenberg's breakthrough

Werner Heisenberg's (1901-1976) first attempts to elucidate Bohr's rules were inspired by a strictly positivistic outlook: to formulate a new mechanics which is based exclusively on *measurable* quantities such as the frequencies ν and the intensities I of spectral lines (Heisenberg, 1925). He did not fulfill his own program but he proposed a theory in terms of energies E and coordinates x, related to measurable quantities by $h\nu_{jk} = E_j - E_k$ and $I_{jk} = |x_{jk}|^2$ where ν_{jk} and x_{jk} referred to a pair (j,k) of stationary states. By an imaginative Fourier analysis of the intensity formulas of atomic spectroscopy, Heisenberg solved the following "quantization problem": given a quantity that replaces the classical dynamical variable x(t) in the new theory, which quantity replaces the classical variable $\{x(t)\}^2$? By introducing a new type of product, Heisenberg gave an explicit solution in terms of the quantities x_{jk}. Heisenberg already noted the noncommutativity of this new product as a distressing difficulty but went ahead to develop his idea, arguing that "*in speziellen Fällen ... tritt diese Schwierigkeit nicht auf*".

Remark: Heisenberg's guiding idea was to eliminate the orbits of electrons arguing that they were not accessible to observation. He proposed a new mechanics to be based only on *directly observable quantities*, such as the frequencies and intensities of spectral lines. However, the full-fledged, algebraic formalism of Born, Heisenberg and Jordan (1926) was based on the matrix calculus and did not fulfill Heisenberg's original positivistic program.

That the lack of commutativity of coordinates and momenta is the *central feature* of Heisenberg's new mechanics was realized by Max Born (1882-1970), and independently by Paul Adrien Maurice Dirac (born 1902). Born recognized Heisenberg's multiplication as matrix multiplication and together with Pascual Jordan (1902-1980) developed a matrix formulation of the new mechanics (Born and Jordan, 1925; received on September 27, 1925). Dirac's approach was more abstract but also mathematically more elegant and conceptually deeper. Dirac (1925; received on November 7, 1925) already envisaged an axiomatic approach and started with a quantum algebra of dynamical variables which satisfy all the rules of normal numbers with the exception that the product is not necessarily commutative. In later papers, Dirac (1926a,b) called his mathematical structure the *algebra of q-numbers* (where q stands for quantum); nowadays we speak of an *algebra of observables*.

In his first paper, Dirac made another essential contribution. Bohr's correspondence rule was generally considered to be a reliable guiding principle; it expressed the belief that under conditions where the effect of Planck's quantum action can be considered to be negligible, the predictions of quantum theory should coincide with those of classical mechanics. By an ingenious application of this correspondence principle, Dirac (1925) postulated that the commutator between two observables A and B corresponds to $i\hbar$ times the Poisson bracket

$$AB - BA = i\hbar \sum_j \left(\frac{\partial A}{\partial Q_j} \frac{\partial B}{\partial P_j} - \frac{\partial B}{\partial Q_j} \frac{\partial A}{\partial P_j} \right)$$

where Q_j, P_j are the canonical dynamical variables of j-th degree of freedom and the sum is over all degrees of freedom of the system. From this general postulate it follows that the canonical q-numbers fulfill the commutation relations,

$$\begin{aligned} Q_jQ_k - Q_kQ_j &= 0 \;, \\ P_jP_k - P_kP_j &= 0 \;, \\ Q_jP_k - P_kQ_j &= 0 \qquad \text{for } j \neq k \;, \\ Q_jP_j - P_jQ_j &= i\hbar \;, \end{aligned}$$

The corresponding equations were discovered independently by Born and Jordan (1925) in their matrix representation.

Remark: The term "matrix" did not appear in Dirac's first papers, he apparently did not like this special representation of his q-numbers too much. More than Born and Jordan, Dirac realized that the most important thing was the noncommutativity of the dynamical variables. As Dirac reported "*Heisenberg was led very reluctantly to noncommutative algebra, because it was so foreign to all ideas of physicists at that time, and when it first turned up he thought there must be something wrong with his theory and tried to correct it, but he was just forced to accept it*" (quoted from Mehra, 1972, footnote 37).

Quantum dynamics

One of the most powerful ideas inspired by Bohr's correspondence principle was the conviction that most features of Hamiltonian mechanics must be correct both in classical and quantum theory (compare the influential lectures Max Born gave in Göttingen in the winter 1923/24, published as "Vorlesungen über Atommechanik", Born, 1925). In his basic paper Heisenberg (1925) already postulated that the *form* of the classical equations of motion can be retained and that only the kinematical interpretation of the canonical variable q(t) as a location has to be rejected. After Heisenberg's breakthrough, Born and Jordan (1925), and independently Dirac (1925) postulated that the Hamiltonian formalism should be preserved with the only modification that the dynamical variables be represented by noncommuting quantities. With this the Hamiltonian equations of motion of quantum mechanics are given by

$$\dot{Q}_j = \frac{\partial H}{\partial P_j} \quad , \quad \dot{P}_j = - \frac{\partial H}{\partial Q_j}$$

where $H = H(P,Q)$ is the Hamiltonian of the system. More generally, Dirac's correspondence principle in terms of Poisson brackets directly gives the equation of motion for an arbitrary dynamical variable A as

$$i\hbar\dot{A} = AH - HA \quad .$$

In the following papers by Dirac (1926a,b,c) and in the famous "Dreimännerarbeit" by Born, Heisenberg and Jordan (1926) the new mechanics had already reached the status of a fully developed theory of mechanical systems of any finite number of degrees of freedom together with a theory of angular momentum and a representation theory of the three-dimensional group of rotations.

Remark: It shows the great genius of the pioneers that they were able to develop a new mechanics in spite of the lack of deeper insight into their guiding principles. For decades the unexplained success of the correspondence principle and Hamiltonian formalism remained a mystery. As we know today, one reason is that *classical* mechanics was not yet fully understood. Although invariance principles played a decisive role in Galileo Galilei's (1564-1642) thinking, and although Eugen Wigner and Hermann Weyl saw the group theoretical implications of the new quantum mechanics from its very beginning, the first full discussion of the Galilei group in quantum mechanics is due to Valentine Bargmann (born 1908), thirty years after Heisenberg's breakthrough! Bargmann's (1954) study led to an unexpected discovery: the invariance under Galilei transformations implies a superselection rule (the so-called Bargmann superselection rule) which guarantees the strict conversation of mass. The corresponding discussion for classical mechanics was initiated by Loinger (1962). The group theoretical background of the correspondence principle became clear only after these works. Both classical mechanics and pioneer quantum mechanics presuppose the Galilean spacetime geometry for the formulation of their laws. This spacetime concept is given by the Galilei group and determines to some extent the laws governing the behavior of matter. Both classical mechanics and pioneer quantum mechanics are elementary but different representations of the Galilei group. This common feature is the reason for many formal similarities between classical mechanics and pioneer quantum mechanics. Nowadays the correspondence arguments which were so important in the development of quantum mechanics are obsolete; they are replaced by the representation theory of the Galilei or the Lorentz group.

Schrödinger's wave mechanics

Inspired by the thesis of Louis de Broglie and by Hamilton's comparison of classical mechanics and geometric optics, Erwin Schrödinger (1887-1961) interpreted the classical trajectories of particles as the "rays" associated with "matter waves". In 1926, Schrödinger created a wave mechanics of matter, intended as a rival to matrix mechanics, restoring the classical description of radiation. (Schrödinger, 1926a, received on January 27, 1926). From a variation principle Schrödinger obtained a differential equation which the wave Ψ must satisfy,

$$H(q, \frac{\hbar}{i}\frac{\partial}{\partial q})\ \Psi(q) = E\ \Psi(q) \quad ,$$

where $H(q,p)$ is the classical Hamiltonian function and $2\pi\hbar$ is Planck's constant.

In his second note, Schrödinger (1926b) introduced the dynamics by a time-dependent wave function Ψ_t satisfying the equation

$$i\hbar\partial\Psi_t/\partial t = H\Psi_t \ ,$$

nowadays called the *time-dependent Schrödinger equation*.

In a stationary state, one can put

$$\Psi_t = e^{-iEt/\hbar}\Psi \quad ,$$

where Ψ satisfies the *time-independent Schrödinger equation*

$$H\Psi = E\Psi \quad ,$$

which states that Ψ must be an eigenfunction and E an eigenvalue of the Hamiltonian H.

In his first two papers, Schrödinger thought that the wave function Ψ described a matter wave in physical three-dimensional space. In his fourth communication, Schrödinger (1926d, received on June 21, 1926) interpreted Ψ as a scalar field and $|\Psi|^2$ as the electrical charge density but already realized that this interpretation breaks down for many-electron problems. Schrödinger's function Ψ of an n-electron problem refers to a 3n-dimensional configuration space, but originally Schrödinger developed his equations for one particle (n=1) so that the 3-dimensional physical space coincided with the 3n-dimensional configuration space. Realizing this situation, Schrödinger was forced to abandon de Broglie's ideas on particle localizations and - much against his own motivating ideas - the interpretation of his wave function Ψ as a matter wave in three-dimensional physical space. In contradistinction to the electromagnetic field, Schrödinger's Ψ-field is not a "real wave", being propagated in physical space, but an abstract entity in a high-dimensional configuration space.

In his paper "Ueber das Verhältnis der Heisenberg-Born-Jordanschen Quantenmechanik zu der meinen", Schrödinger (1926c; received on March 18, 1926) showed that from a formal mathematical viewpoint his wave mechanics was identical with the matrix mechanics. Independently, a much more precise equivalence proof was given by Wolfgang Pauli (1900-1958) in a letter to Pascual Jordan on April 12, 1926 (the letter is reprinted in v.d. Waerden, 1973).

Remark: Schrödinger had a deep aversion to Heisenberg's new mechanics. "*Ich hatte von seiner Theorie natürlich Kenntnis, fühlte mich aber durch die mir sehr schwierig scheinenden Methoden der transzendenten Algebra und durch den Mangel an Anschaulichkeit abgeschreckt, um nicht zu sagen abgestossen*" (footnote 2 in Schrödinger,

1926c). In private discussion, Schrödinger was more articulate. During a visit at Bohr's institute in Copenhagen in September 1926, Schrödinger reportedly exclaimed desperately: *"Wenn es doch bei dieser verdammten Quantenspringerei bleiben soll, so bedaure ich, mich mit der Quantentheorie überhaupt beschäftigt zu haben"* (quoted from Heisenberg, 1956a). Heisenberg expressed his aversion to Schrödinger's wave mechanics in a letter to Wolfgang Pauli on June 8, 1926: *"Je mehr ich über den physikalischen Teil der Schrödingerschen Theorie nachdenke, desto abscheulicher finde ich ihn"* (quoted from Hermann, 1963, p.65). Note that from a modern point of view, both Schrödinger and Heisenberg refer to one and the same physical theory: pioneer quantum mechanics.

The revelation of the meaning of Schrödinger's Ψ-function

When it became plain that Schrödinger's original idea of matter waves was untenable, the question of a correct interpretation of Schrödinger's Ψ-function became urgent. Referring to the early development of the formalisms of matrix and wave mechanics, Born (1955) reminisced in his Nobel lecture: *"Was aber dieser Formalismus eigentlich bedeutete, war keineswegs klar. Die Mathematik war, wie es öfters vorkommt, klüger als das sinngebende Denken"*. Almost simultaneously with Schrödinger's qualms about the interpretation of Ψ as a matter wave (Schrödinger, 1926d, received on June 21, 1926), Max Born (1882-1970) published his "Quantenmechanik der Stossvorgänge" (Born, 1926a, received on June 25, 1926), where he preferred Schrödinger's wave mechanics over the matrix mechanics as "die tiefste Fassung der Quantengesetze", and introduced a new probabilistic interpretation of Schrödinger's Ψ-function. According to this probabilistic interpretation of wave mechanics, Schrödinger's Ψ-function is a *probability amplitude* so that the absolute square $|\Psi|^2$ is a probability density in configuration space. When an event can occur in several alternative ways, the probability *amplitude* for the event equals the sum of the probability amplitudes for each way separately (superposition principle for probability amplitudes). Accordingly, the probability of this event is not given by $P = |\Psi_1|^2 + |\Psi_2|^2$ but by $P = |\Psi_1+\Psi_2|^2$, so that one speaks of an interference of probability amplitudes. Born's probabilistic interpretation (Born, 1926a-b, 1927) was quickly accepted by most physicists (with the notable exceptions of Erwin Schrödinger and Albert Einstein).

Remark: Born's representation of Schrödinger's Ψ-function had far-reaching consequences for physics and philosophy. However, Born has never formulated his probability postulate in a precise way; even his later discussions of the role of probability in quantum mechanics are rather confusing (compare e.g. Born, 1961). In particular, he did not make it clear that the probabilities of quantum mechanics are *conditional* probabilities, conditioned by a measuring process performed on the system. Likewise, it is disconcerting to find that Born even in his Nobel lecture (Born, 1955) speaks of a *statistical* explanation of quantum mechanics. What he means is *probabilistic*, and not statistical (compare also Fock, 1957). According to the orthodox interpretation, quantal probabilities are *primary*, and belong to a fundamentally different category than, say, the probabilities of population statistics. A truly statistical interpretation of quantum mechanics has been proposed by Einstein. This interpretation - which is *not* accepted by the majority of scientists - claims that a pure state refers to the statistical properties of an ensemble of individual systems but that it does not provide a complete description of an individual system. Born did not accept this view but confirmed that he did accept Bohr's view that quantum mechanics is a complete theory. Curiously, in a letter to Einstein (Born, 1969, letter no. 96, dated September 4, 1959), Born nevertheless claims that his interpretation is a statistical one, where he uses the notion "statistical" in the same sense as insurance agencies. So, *Born is a further example of a propounder of a revolutionary idea who does not recognize it as such.*

The discovery of the logical symmetries of quantum mechanics

A logical symmetry of a theory is a transformation that preserves the *essential* logical structure of the theory. The logical symmetries of *classical* mechanics are given by the *canonical transformations*, i.e. a one-to-one transformation in phase space that leaves the Poisson bracket of coordinates and momenta invariant. An attempt to transfer the classical theory of canonical transformations to quantum mechanics is known under the name *transformation theory*, introduced independently at about the same time by Dirac (1926a,d) and Jordan (1926a,b). The transformation theory showed that one could pass by a canonical transformation from any mode of description to any other by an appropriate choice of the basic variables. For example, one could transform the state vector $\Psi(q)$ in the configuration space representation into an equivalent description having a state vector $\hat{\Psi}(p)$ in the momentum representation so that the absolute square $|\hat{\Psi}(p)|^2$ was the probability density corresponding to the momentum distribution. However, at that time, the proper mathematical tools to formulate precisely the logical symmetries of the new mechanics were not yet available. Nevertheless, a clear and prophetic statement can be found in a paper by F. London (1926): "*Die Uebertragung der Transformationstheorie der Matrizenmechanik auf die Schrödingersche Eigenwerttheorie führt zu einer sehr generalisierten Auffassung derselben unter den allgemeinen Gesichtspunkten einer Invariantentheorie linearer Funktionaloperationen. Diese Operationen lassen sich darstellen als Drehungen im Hilbertschen Funktionenraume, welcher von dem Orthogonalsystem der* Ψ_k *aufgespannt wird*". In the modern language, this means that *the logical symmetries of pioneer quantum mechanics are the unitary transformations of the state space*. With this result the way was free for a general definition of the state space of the new mechanics.

The introduction of Hilbert space by John von Neumann

The final mathematical formulation of pioneer quantum mechanics is due to Johann von Neumann (1903-1957). During the winter term of 1926/27 David Hilbert (1862-1943) - one of the greatest mathematicians of the first half of the 20th century - gave a series of lectures on the mathematical foundations of quantum mechanics. In the published version (Hilbert et al., 1927) it is stated: "*Wichtige Stücke der mathematischen Durchführung rühren von J. v.Neumann her*". A short time later, von Neumann (1927a) introduced a new mathematical framework for quantum mechanics based on the concept of an abstract Hilbert space and its operator calculus. This new tool enabled von Neumann to develop a statistical ensemble formalism for quantum mechanics with an associated measuring theory (von Neumann, 1927b), and quantum statistical mechanics (von Neumann, 1927c). His spectral theorem for observables (v.Neumann, 1929a) allowed him to give a precise characterization of the mathematical status of observables and to prove the uniqueness of the irreducible representations of the Heisenberg-Schrödinger canonical operators P,Q fulfilling $QP-PQ = i\hbar$ (v.Neumann, 1931). His classic "*Mathematische Grundlagen der Quantenmechanik*" (v.Neumann, 1932) summarizes in a brilliant way pioneer quantum mechanics as a grand and well-thought-out mathematical theory.

Von Neumann's most spectacular contribution to pioneer quantum mechanics was the insight that the state space of the new mechanics had the structure of an infinite-dimensional vector space with an inner product characterizing its geometry. Two special realizations of Hilbert spaces were known long before the advent of quantum mechanics. In his famous theory of integral equations, David Hilbert (1904) introduced the sequence space (nowadays denoted by ℓ_2) as an infinite-dimensional extension of the notion of a Euclidean space. This space ℓ_2 consists of all infinite systems $\{x_n\}$ of numbers x_n for which $\Sigma_{n=1}^{\infty}|x_n|^2$ is finite, and for each pair of elements $x=\{x_n\}$ and $y=\{y_n\}$ the inner product is given by $\langle x|y\rangle = \Sigma_{n=1}^{\infty} x_n^* y_n$. In his further work from 1904 to 1910, Hilbert (1912) developed a rather complete spectral theory of bounded operators on the space ℓ_2. Hilbert's pupil Erhard Schmidt (1876-1959) introduced essential conceptual simplifications by introducing the concepts of Euclidean geometry (such as function space, subspace, orthogonal projection, norm, orthonormalization) into Hilbert's theory of functions with infinitely many variables (Schmidt, 1907a,b, 1908), so that a large part of the spectral theory necessary for quantum mechanics was ready by the end of the first decade of the twentieth century.

Remark: Hilbert called his studies "spectral analysis" and introduced the notions spectrum, point spectrum and continuous spectrum. Hilbert had no intention at all to connect these concepts with a theory of atomic spectra and, according to Constance Reid, he was taken by surprise when he learned that his theory had applications in the theory of atomic spectra: "*I developed my theory of infinitely many variables from purely mathematical interests, and even called it 'spectral analysis' without any presentiment that it would later find an application to the actual spectrum of physics*" (quoted from Reid, 1970, p.183).

At about the same time as David Hilbert was developing his theory of integral equations and ℓ_2-spaces, the French analyst Henri Lebesgue (1875-1941) was creating a new integral which now bears his name (Lebesgue 1901, 1902). Lebesgue's work is characterized by deep insight based on intuitive geometric concepts and he was, according to Norbert Wiener (1948, chapt.II) an "*exponent of the extremely exacting modern standards of mathematical rigor, and a writer whose works, as far as I know, do not contain one single example of a problem or a method originating directly from physics*". But Lebesgue's work has influenced almost every branch of modern mathematical physics. It may not be easy for the mathematical layman to grasp the fact that it is easy to measure the length of an interval of smooth curve but that one needs very sophisticated concepts to measure sets consisting of infinitely many points scattered irregularly on a curve. Lebesgue extended the notion of length, area or volume to the most extreme situations. *Measure* is a generalization of these simple concepts and applies to point sets which may be much more general than an interval, the inside of a square or a cube. As stressed by Norbert Wiener (1933, pp.4-5), it is impossible to establish a reasonable theory of natural phenomena on a less inclusive concept such as that of the popular Riemann integral. The Lebesgue integral is the cornerstone of modern measure theory, mathematical probability theory and functional analysis. The space of all summable and square-integrable functions in the sense of Lebesgue over a set Ω is denoted by $L_2(\Omega)$ with L standing for Lebesgue. Independently from each other but practically simultaneously, Ernst Fischer (1907, presented on March 5, 1907) and Friedrich Riesz (1907, presented on March 9, 1907) linked Lebesgue's integration theory and Hilbert's spectral theory by showing that Hilbert's sequence space ℓ_2 is isomorphic to the function space L_2 of the Lebesgue square-integrable functions.

Recognizing the fact that matrix mechanics used an ℓ_2-space and wave mechanics an L_2-space, and that the equivalence of these two theories was nothing but the Fisher-Riesz isomorphism of the ℓ_2- and L_2-spaces, von Neumann abstracted from the inessential special structures of these spaces and defined a *Hilbert space* abstractly as a space isomorphic to Hilbert's sequence space ℓ_2. Here isomorphic means (as always in mathematics) that the essential properties of these spaces are the same. More pedantically, we may describe an isomorphism as a one-to-one correspondence between the elements of these spaces which preserves the essential structure. The essential properties of these spaces were widely recognized to be those of a vector space with an inner product which was complete and separable (i.e. having a countable dense subset). Accordingly, von Neumann defined a Hilbert space axiomatically as an abstract, infinite-dimensional, separable complete inner product space. In this original definition the requirement of countable infinite dimensionality characterized a Hilbert space as a (within isomorphism) unique mathematical object. Nowadays, mathematicians have found it convenient to drop the axioms of dimensionality and of separability. According to the modern definition a Hilbert space is an inner-product space which is complete in the metric induced by the norm, so that von Neumann's Hilbert space of pioneer quantum mechanics is referred to as an infinite-dimensional, separable Hilbert space over the complex numbers.

The Hilbert-space model of pioneer quantum mechanics

The Hilbert-space formalism of pioneer quantum mechanics as introduced by von Neumann chooses as the basic mathematical object an abstract Hilbert space, called the *state space* H. Schrödinger, in his pioneering work, used a special realization of this state space, namely the Hilbert space $L_2(\mathbb{R}^{3n})$ of square-integrable functions over the configuration space $\mathbb{R}^{3n}$ of n particles, where $\mathbb{R}^{3n}$ denotes the 3n-dimensional Euclidean space. That is, Schrödinger's wave functions Ψ are normalized elements of $L_2(\mathbb{R}^{3n})$. Abstracting from this special realization of the state space, one calls a normalized vector $\Psi \in H$, $\|\Psi\|=1$, describing a quantum mechanical state a *state vector*. Since only $|\Psi|^2$ has a physical interpretation, a phase factor does not matter so that for all real α the vectors Ψ and $e^{i\alpha}\Psi$ represent the same physical state. From this viewpoint one should say that the state space is a *projective* Hilbert space. But it is more convenient to stick to the Hilbert-space structure and to *define a state as a one-dimensional subspace* of the state space H. Every normalized vector lying in this one-dimensional subspace is referred to as a state vector *representing the state*. Note that in the superposition $\Psi_1+\Psi_2$ of two state vectors Ψ_1, Ψ_2 an overall phase factor is immaterial while a relative phase between Ψ_1 and Ψ_2 is physically relevant.

Von Neumann (1929) recognized the need to extend Hilbert's spectral analysis from bounded to unbounded operators. Operators on a Hilbert space are transformations mapping one vector on another. A *bounded* linear operator T on a Hilbert space H is defined as a mapping of H into itself so that T assigns to *each* vector Ψ in H a vector $T\Psi$ in H such that $T(\Psi+\Phi) = T\Psi+T\Phi$ and $T(c\Psi) = cT\Psi$ for every vector Ψ, Φ and every number c. The product TS of two operators T and S is defined by $(TS)\Psi = T(S\Psi)$ for every vector Ψ in H. Von Neumann generalized this concept and defined an *operator* on a Hilbert space H as a linear transformation defined only on some subset of H, called the *domain* of the operator. Every linear operator with a dense domain has a unique *adjoint operator* T^*, defined by the relation $\langle\Psi|T\Phi\rangle = \langle T^*\Psi|\Phi\rangle$ for all vectors Φ in the domain of T. An operator is called self-adjoint, if $T=T^*$ (we use the modern terminology, von Neumann called self-adjoint operators *hypermaximal*).

The new concept of self-adjointness replacing the weaker concept of symmetric or Hermitian operators was crucial for a rigorous development of quantum mechanics. Von Neumann (1929a) successfully extended Hilbert's spectral theory to arbitrary (bounded or unbounded) self-adjoint operators. If T is a self-adjoint operator, then there exists a unique *spectral resolution* E on the spectrum Ω of T such that

$$T = \int_\Omega \omega E(d\omega) \;.$$

For every Borel subset B of Ω, the operator E(B) is a *projection operator*, i.e. $E(B) = E(B)^* = E(B)^2$, and the set of these projection operators form a *Boolean algebra* satisfying the relations

$$\begin{aligned} E(\Omega-B) &= 1-E(B) && \overset{\text{def}}{=} E^{\perp}(B) \\ E(A\cup B) &= E(A)+E(B)-E(A)E(B) && \overset{\text{def}}{=} E(A)\vee E(B) \\ E(A\cap B) &= E(A)E(B) && \overset{\text{def}}{=} E(A)\wedge E(B) \end{aligned}$$

(In addition, E is σ-additive in the strong operator topology, a technically important regularity condition which promotes E to the status of a *projection-valued measure.*

In classical mechanics an *observable* is represented mathematically by a real valued function of the position q and the momentum p. In pioneer quantum mechanics p and q are replaced by noncommutative quantities P and Q fulfilling Heisenberg's commutation relations. Von Neumann (1927a) recognized that the theory of operators acting on an abstract Hilbert space provides a most effective and elegant setting for formulating the new mechanics. He showed that the proper mathematical entities corresponding to the notion of observables are *self-adjoint* operators acting on the state space $\mathcal{H}$ (v.Neumann, 1932). If T is a self-adjoint operator corresponding to some observable, then its spectral values are interpreted as the possible values which one may obtain in an ideal measurement of this observable (ideal measurements are Pauli's (1933) *"measurements of the first kind"*). A real Borel function F of an observable T represents a new observable F(T) which can be measured by the very same apparatus used for T by replacing the scale of its meter by a new one in which every number ω is replaced by $F(\omega)$. In terms of von Neumann's spectral theorem, this means that the spectral resolution of T

$$T = \int_{\Omega} \omega E(d\omega) \quad ,$$

implies the spectral resolution of F(T)

$$F(T) = \int_{\Omega} F(\omega) E(d\omega) \quad .$$

If the system is in the state Ψ and if we perform an experiment to determine the value of the observable T, then the probability that the ideal measurement has as outcome a value within the Borel set B, is given by

$$\mu(B) = \langle \Psi | E(B) \Psi \rangle \quad .$$

Thus

$$\langle T \rangle \overset{\text{def}}{=} \int_{\Omega} \omega \mu(d\omega) = \langle \Psi | T \Psi \rangle$$

is the expected value of the observable T, and

$$\langle F(T)\rangle \overset{\text{def}}{=} \int_\Omega F(\omega)\mu(d\omega) = \langle\Psi|F(T)\Psi\rangle$$

is the expected value of the observable F(T), conditioned by a prior ideal measurement with a T-apparatus. An observable T is said to be *dispersion-free* with respect to a state vector Ψ if $\langle\Psi|T^2\Psi\rangle = \langle\Psi|T\Psi\rangle^2$ from which it easily follows that

$$T\Psi = \omega\Psi \quad \text{with} \quad \omega = \langle\Psi|T\Psi\rangle \ .$$

That is, an observable T has the value $\omega\in\Omega$ with certainty if and only if Ψ is an *eigenvector* of T with the eigenvalue ω.

Remark: In order to grasp the notion of an observable it is useful to ask why the self-adjointness of the operators representing observables is required. The standard answer is that measured values correspond to *real* numbers. This remark is grossly misleading because such a requirement is neither necessary nor sufficient. There is absolutely no problem to construct an apparatus which gives a complex number as the result of a measurement. On the other hand, the restriction to real valued expectation values leads us only to operators whose numerical range consists of real numbers. Any Hermitian operator fulfills this requirement but not every Hermitian operator is self-adjoint. The crucial point is *not* that a self-adjoint operator has a *real-valued* spectrum but that it generates a *Boolean* algebra. It is the Boolean character that leads to a possible classification in the sense of a *classical* logic, and these classical classifications can be related to physical operations in the sense of laboratory measurements. From a modern point of view, the essential representative of an observable is a Boolean algebra. The representation of an observable by a particular generating self-adjoint operator may be convenient but contains inessential elements. Because any *normal* operator generates a Boolean algebra, normal operators are (in spite of their complex-valued spectra) perfectly legitimate observables.

The decisive mathematical structure associated with an ideal measurement is a projection-valued measure and not a particular observable constructed from it. Relative to a fixed state vector Ψ, every projection-valued measure E defines via $\mu(B) = \langle\Psi|E(B)\Psi\rangle$ a *probability measure* μ in the sense of Kolmogorov's (1933) mathematical probability theory. An observable T with the spectrum Ω and having E as its spectral resolution can be regarded as a *random variable* on the sample space Ω with the probability measure μ. The expectation value of the observable F(T) relative to the state Ψ is uniquely determined by the probability measure μ

$$F(T) = \int_\Omega F(\omega)\mu(d\omega) \ .$$

Accordingly, pioneer quantum mechanics in the formulation of von Neumann contains as a *special case* the whole probability theory as axiomatized by Andrei Nikolaevič Kolmogorov (born 1903) in his epoch-making "*Grundbegriffe der Wahrscheinlichkeitsrechnung*" of 1933. But in contrast to classical probability theory the set of *all* observables of a quantal system in a given state cannot be regarded as a set of random variables on a Kolmogorov probability space.

Von Neumann's spectral resolution of observables allows the replacement of observables by the conceptually much simpler projection operators. In his book, von Neumann (1932, sect.III.5) proposed to consider projection operators as representing *propositions*. If a proposition E is *true* (e.g. by a verifying measurement) then one assigns to it the numerical value 1, if the proposition is *false*, one assigns to it the numerical value 0. This idea, first formulated in the framework of pioneer quantum mechanics, proved to be extremely fruitful and became later the starting point of a fundamentally new development (the so-called quantum logic) where propositions and not observables and states are basic for the description of natural phenomena.

Remark: Von Neumann invented the abstract Hilbert space and introduced it into quantum mechanics, but he did not regard Hilbert space as indispensable for quantum theory. In a letter to Garrett Birkhoff (dated Wednesday, November 13, presumably 1935), von Neumann wrote: "*I would like to make a confession which may seem immoral: I do not believe absolutely in Hilbert space anymore. ... Now we begin to believe, that it is not the vectors which matter, but the lattice of all linear (closed) subspaces. Because : (1) The vectors ought to represent the physical states, but they do it redundantly, up to a complex factor, only, (2) and besides, the states are merely a derived notion, the primitive (phenomenologically given) notion being the qualities, which correspond to linear closed subspaces. ...*" (quoted from Birkhoff, 1961, p.158).

3.3[#] OUTLINE OF THE HILBERT-SPACE FORMALISM OF PIONEER QUANTUM MECHANICS

The outline in this section is provided for readers who want to grasp the mathematical structure that underlies pioneer quantum mechanics. It is not necessary to study this section in order to follow the subsequent discussions.

We summarize the formalism of pioneer quantum mechanics from a modern point of view, correcting trivial slips in the original formulation and adding some newer results without going beyond the conceptual limits of pioneer quantum mechanics.

The mathematical structure of pioneer quantum mechanics is a category having Hilbert spaces as objects and unitary or antiunitary transformations as morphisms. Formerly, the preferred observables and the dynamics were determined by the correspondence principle, that is by a functor "quantization" from the category of classical mechanics (having symplectic manifolds as objects and symplectic transformations as morphisms) to the category of pioneer quantum mechanics. Nowadays, such quantization procedures are considered as obsolete and are replaced by the representation theory of the kinematical group of mechanics. In classical mechanics as well as in pioneer quantum mechanics the Galilei group acts as kinematical group. In what follows we explain this highbrow sketch in a less terse language.

The logical symmetries of pioneer quantum mechanics

The *state space* of a quantum system is an abstract Hilbert space of finite or countably-infinite dimension. A *state* (synonym, used in the statistical interpretation: *pure state*) is a one-dimensional subspace of the state space $\mathcal{H}$ (synonym: a *ray*). Any normalized vector in the one-dimensional subspace of a state can be used to represent this state, and is called a *state vector*. Two state vectors Ψ and Φ represent the same state if and only if $\Phi = c\Psi$, where c is a complex number of absolute value one.

The original formulation of pioneer quantum mechanics assumed a one-to-one correspondence of one-dimensional subspaces of the state space with physical states, implying the *unrestricted validity of the superposition principle* for state vectors. This requirement is equivalent to von Neumann's (1932) *irreducibility postulate*. In modern terminology we say that pioneer quantum mechanics has *no superselection rules*.

Remark: The physical nature of von Neumann's irreducibility postulate was first recognized by Paulette Destouches-Février (1951, sect.IV 12). Strangely, the unrestricted validity of the superposition principle was taken for granted until Wick, Wightman and Wigner (1952) in a famous paper pointed out the existence of superselection rules. A statement that selects some vectors, adding that they are physically unrealizable as state vectors is called a *superselection rule*. If there are superselection rules, then there exist subspaces of the state space that cannot be connected to each other by any observable. Not all self-adjoint operators on the state space are therefore observables. A subspace of the state space in which the superposition principle holds unrestrictedly, is called a *superselection sector*.

Wick et al. (1952) proved the existence of superselection rules for the intrinsic parity of elementary particles, conjectured a superselection rule that operates between states of different total charge and postulated that every observable has to commute with the charge operator. Similar superselection rules hold for the baryon number and, probably, for the lepton number. These superselection rules of elementary particle physics will not be of immediate concern to us. In pioneer quantum mechanics, the following three superselection rules have always been supposed to hold true: the superposition principle does not hold for two state vectors referring either to different instances of *time*, nor to particles having different *mass*, nor to particles having different *electric charges*. As a consequence, time, mass and charge are represented by *classical observables*, that is by observables which commute with every observable. A fourth superselection rule of pioneer quantum mechanics relates to the *symmetrization postulate* requiring the state vectors of a system of identical particles to be either symmetric or antisymmetric under permutations. This postulate is a consequence of the fact that all observables are invariant with respect to the interchange of identical particles (compare Galindo et al., 1962; Casher et al., 1965; Lüders, 1966; Kaplan, 1975).

These four superselection rules of pioneer quantum mechanics can easily be taken into account in an ad hoc manner. Most probably, von Neumann (1932) did not recognize the very profound *physical* meaning of his irreducibility postulate: it says that the quantum system considered has no classical properties at all. Jauch (1960) suggested to replace von Neumann's irreducibility postulate by a proper formulation of Dirac's completeness postulate. This recasting of pioneer quantum mechanics is necessary and easy but goes beyond the concepts of pioneer quantum mechanics. It will be discussed in chapter 4.

The structure-preserving mappings (i.e. the morphisms) of the formalism of a theory are called the *logical symmetries* of the theory. The set of all logical symmetries of a theory form a group, called the *logical group* of the theory. Ignoring superselection rules of pioneer quantum mechanics, the states of a quantum system span a projective Hilbert space. Every vector Ψ in the Hilbert space H determines a 1-dimensional subspace, called the ray $\overline{\Psi}$. The inner product of two rays $\overline{\Psi}$ and $\overline{\Phi}$ is defined by

$$\langle\overline{\Psi}|\overline{\Phi}\rangle \overset{\text{def}}{=} \frac{|\langle\Psi|\Phi\rangle|}{\|\Psi\|\cdot\|\Phi\|} \quad \text{where} \quad \Psi,\Phi\in H \ .$$

The set of all rays in H is called the projective Hilbert space $\overline{H}$ associated with the Hilbert space H (for technical details, compare Bargmann, 1954). A logical symmetry of pioneer quantum mechanics is an automorphism of the projective Hilbert space $\overline{H}$ associated with the state space H so that the logical group of pioneer quantum mechanics is given by the automorphism group $\mathrm{Aut}(\overline{H})$. A famous theorem by Wigner (1931) asserts that the automorphism group $\mathrm{Aut}(\overline{H})$ can be represented by the group of unitary operators acting on the state space H. Let H_1 and H_2 be Hilbert spaces and T be a mapping from H_1 into H_2. Then

(i) T is called *linear* if $T(a\psi+b\phi) = aT\psi+bT\phi$

(ii) T is called *antilinear* if $T(a\psi+b\phi) = a^*T\psi+b^*T\phi$

(iii) T is called *isometric* if $\|T\psi\| = \|\psi\|$

for all $\psi,\phi\in H_1$ and all complex numbers a,b. If the range of a linear isometric operator $T:H_1\rightarrow H_2$ is the whole space H_2, then T is called *unitary*. An *antiunitary* operator $T:H_1\rightarrow H_2$ is an antilinear isometric operator having the range H_2. Wigner's theorem implies that two realizations of pioneer quantum mechanics whose state spaces are connected by a unitary or antiunitary transformation are from a logical point of view equivalent.

Historical remarks: The fact that symmetries in quantum mechanics are described by projective unitary representations has been known since the beginning of quantum mechanics (compare Weyl, 1927). In his famous book on "Gruppentheorie und Quantenmechanik", Hermann Weyl states clearly: "*In der Quantentheorie finden die Darstellungen im Systemraum statt, dieser ist aber nicht als Vektor-, sondern als Strahlenkörper zu verstehen, weil der einzelne reine Fall nicht durch den Vektor, sondern durch den Strahl repräsentiert wird*" (Weyl, 1928, p.147). Wigner (1931) published his theorem in his classic "Gruppentheorie und ihre Anwendung auf die Quantenmechanik der Atomspektren" but without full proof. A complete proof was given by Bargmann (1954) (compare also: Hagedorn, 1959; Bargmann, 1964). Wigner's theorem characterizes symmetries as bijective transformations that preserve absolute values of inner products. Its physical meaning hinges on the probabilistic interpretation which represents the transition probability between two states by the absolute square of their inner product. Accordingly, Wigner considers a symmetry as a mapping that preserves the *probabilistic* structure of quantum mechanics. Since the probabilistic structure of quantum mechanics follows from a discussion of the measurement process, Wigner's definition of a symmetry is not quite satisfying. A conceptually much more satisfying definition of a *logical* symmetry is due to Uhlhorn (1962): *a logical symmetry is defined to be a one-to-one correspondence between the state vectors of a state space H_1 and the state vectors of a state space H_2 such that pairs of orthogonal vectors correspond to pairs of orthogonal vectors.* This definition also applies in the presence of superselection rules. Uhlhorn (1962) showed that if the dimensions of H_1 and H_2 are larger than two the logical symmetries are implemented by unitary or antiunitary operators.

The kinematical symmetry of pioneer quantum mechanics

Like Newtonian mechanics, pioneer quantum theory is a *mechanics* presupposing the existence of a space-time continuum whose symmetry is governed by a group, called the *kinematical group*. In contradistinction to Aristotelian mechanics, every mechanics considered in classical or modern physics is a *relativistic* theory. It is conceptually very misleading to call pre-Einsteinian classical mechanics a "nonrelativistic theory". This is wrong since Newtonian mechanics complies with Galilei's principle of relativity. It is illuminating insofar as it shows that the role of Galilean relativity has not been understood until the advent of Lorentz-invariant quantum mechanics.

In mechanics, space and time are primary concepts. In contrast to time, the concept of space is by no means a necessary precondition for language. Space is an empirical concept, an abbreviation for the way *solid bodies* can be assembled (compare Einstein, 1922, chapt.1). "*Si donc il n'y avait pas de corps solides dans la nature, il n'y aurait pas de géométrie*", says Henri Poincaré in his famous "La science et l'hypothèse" (Poincaré, 1902, chapt.IV). We speak of a *relativistic* theory if space and time cannot be considered independent of one another but must be combined to give a four-dimensional *space-time*. A space-time is characterized by the following three symmetry properties:

(i) all instants of time are equivalent,
(ii) all points in space are equivalent,
(iii) all directions in space are equivalent.

Note that the isotropy (iii) implies the homogeneity (ii), but not the other way round. It can be shown that the only kinematical symmetries compatible with this structure are the Aristotelian, the Galilean and the Einsteinian symmetry (compare Bacry and Lévy-Leblond, 1968; Gorini, 1971).

Space is the 3-dimensional Euclidean space E^3 carrying the usual Euclidean metric (defining the *spatial separation*) and possessing the 6-parameter *Euclidean group* (consisting of the 3-parameter *translation group* and the 3-dimensional *rotation group*) as its symmetry group. *Time* is the one-dimensional Euclidean space E^1 possessing the 1-parameter translation group as symmetry group, while the metric of E^1 defines *time difference*. A point $(\vec{q},t)$ $(\vec{q}\in E^3, t\in E^1)$ in space is called an *event*.

Aristotelian kinematics presupposes the *minimal symmetry* of space-time by assuming *absolute space* and *absolute time*, so that Aristotelian space-time is simply the Cartesian product $E^3 \times E^1$. The corresponding kinematical group is the direct product of the 6-parameter Euclidean group of E^3 and the 1-parameter dynamical group of E^1, i.e. a 7-parameter group. With this, in Aristotelian kinematics it is meaningful to speak of a spatial separation of two events $(\vec{q}_1, t_1)$ and $(\vec{q}_2, t_2)$, even if the time difference $t_1 - t_2$ is not zero. Just so, the time difference is always defined.

Galilean kinematics introduced the relativity of space but retained absolute time. In Galilean space-time a spatial separation of two events $(\vec{q}_1, t_1)$ and $(\vec{q}_2, t_2)$ is well-defined only if the events refer to the same instant of time, $t_1 = t_2$. On the other hand, the time difference is meaningful for every pair of events. The most general group of space-time transformations which respects the homogeneity of space-time and the isotropy of space, and which leaves the time interval between every pair of events invariant is a 10-parameter group, called the *Galilei group* (compare: Bacry and Lévy-Leblond, 1968; Berzi and Gorini, 1971; Lévy-Leblond, 1976). The minimal symmetry of Aristotelian space-time is enlarged by the addition of the so-called pure Galilei transformations

$$\vec{q} \longrightarrow \vec{q}\,' \overset{\text{def}}{=} \vec{q} + \vec{v}T \quad , \quad t \longrightarrow t' \overset{\text{def}}{=} t$$

where the 3-dimensional parameter $\vec{v}$ is a constant vector, having the meaning of a translation velocity. The Galilei group is generated by the time translations (1 parameter), the spatial translations (3 parameters), the rotations (3 parameters), and the pure Galilei transformations (3 parameters); it can be shown to equal the semidirect product of the direct product of the time-translation group and the rotation group with the direct product of the spatial translation group and the pure Galilei group.

Einsteinian kinematics relativized not only space but also time. In Einsteinian space-time the time difference between two events $(\vec{q}_1, t_1)$ and $(\vec{q}_2, t_2)$ is a priori only well-defined if the events occur at the same space point, $\vec{q}_1 = \vec{q}_2$. The most general group preserving this structure is a 10-parameter group, called the *Poincaré group* (synonym: inhomogeneous Lorentz group). Since any group of transformations leaving invariant the expression for the squared distance between infinitely near points in a space of n dimensions can have at most $n(n+1)/2$ parameters, the Einsteinian space-time has *maximal symmetry*. The Poincaré group contains a new universal constant c, having the meaning of a maximal velocity. In a very good approximation, c equals the velocity of light; if the rest mass of the photon vanishes, this equality is exact. In the limiting case $c \to \infty$, the Poincaré group contracts to the (mathematically more complicated) Galilei group. The so-called "nonrelativistic" theories refer to situations where the finiteness of the speed of light is negligible - the only case we shall discuss in this essay.

At the very beginning of quantum mechanics, Hermann Weyl (1927) proposed to describe the kinematical structure of a physical system by a subgroup of the logical symmetries of the theory. This group is called the *kinematical group, it reflects our space-time concepts.* Pioneer quantum mechanics decided for the same choice as Newtonian mechanics: the Galilei group. More precisely: *the kinematical group of pioneer quantum mechanics is given by a projective unitary representation of the Galilei group on the Hilbert space of state vectors.*

Historical remarks: The principle of relativity in classical mechanics is due to Galileo Galilei (1564-1642). In his masterly written *Dialogo sopra i due massimi sistemi del mondo* of 1632, Galilei considers a boat which is at rest and says: "*And particularly if you put on the roof of a vessel, a bucket with a hole, full of water, and then you mark to where the water is dripping on the floor, and then you imagine that the boat moves extremely fast, abstraction made from the waves which would disturb the movement, you will find that the water is dropping down exactly to the same place*". An earlier intuitive formulation of the law of inertia was found in 1614 by Isaac Beeckmann (1570-1637), a precise formulation was given in 1644 independently by René Descartes (1596-1650) in his *Les Principes de la Philosophie* and by Evangelista Torricelli (1608-1647) in his treatise *De Motu gravium naturaliter descendentium et projectorum*. Christiaan Huygens (1629-1695) generalized Galilei's principle of inertia to a principle of relativity, saying that it is impossible that an observer in uniform rectilinear motion can discover his own translation. Since Huygens rejected the concept of an absolute motion it is unclear what he means by a "uniform" motion. In contrast, Isaac Newton (1643 {1642 according to the Julian calendar} - 1727) assumed the existence of an absolute time and an absolute space which exist "*in its own nature, without regard to anything external*", and in his *Philosophiae naturalis principia mathematica* of 1687 he stated the principle of inertia as Lex I: "*Corpus omne perseverare in statu suo quiescendi vel movendi uniformiter in directum, nisi quatenus a viribus impressis cogitur statum illum mutare*", ("Every body continues in its state of rest, or of uniform motion in a right line, unless it is compelled to change that state by forces impressed upon it"). Newton was acutely aware that this law makes no sense without the concepts of absolute time and absolute space but also that "*the parts of that immovable space, in which those (absolute) motions are performed, do by no means come under the observation of our senses*". In contemporary astronomy, the concepts of cosmic time and a spatial frame of reference fixed to the average position of galaxies provide a substitute for Newton's absolute time and absolute space.

The replacement of the concept of absolute space by the modern notion of inertial systems is due to Ludwig Lange (1885). Such inertial systems are defined as reference systems in which the motion of an isolated particle is rectilinear and uniform, and they have been called *Galilean reference systems* by Albert Einstein (1916). The name *Galilei transformation* for the group of transformations in space and time connecting Galilean Cartesian reference systems was introduced by Philipp Frank (1909).

In 1882 and 1895 Hendrik Antoon Lorentz (1853-1928) published two fundamental papers showing that Galilei transformations do not play the same role in empty-space-electromagnetism as in mechanics, and in 1904 he introduced a new transformation associated with electromagnetism in a similar way as Galilei transformations are associated with Newtonian mechanics. This new transformation was named by Poincaré (1905) the *Lorentz transformation*. Already in 1877, the creator of the electromagnetic theory of light, James Clerk Maxwell (1831-1879) spoke of "*the doctrine of relativity of all physical phenomena*" and both Henri Poincaré (1854-1912) in 1904 and Albert Einstein (1879-1955) in 1905 postulated the principle of relativity as a general law of nature. Poincaré (1904) wrote: "*The principle of relativity, according to which the laws of physical phenomena should be the same whether for an observer fixed or for an observer carried along in a uniform movement or translation*", while Einstein (1905b) independently postulated: "*The laws by which the states of physical systems undergo change are not affected, whether these changes of state be referred to one or the other of two systems of coordinates in uniform translatory motion*". Since the acceptance of Einstein's special theory of relativity it has been generally acknowledged that the Galilean transformation is only an approximation to the Lorentz transformation, valid for velocities much smaller than that of light.

Newton's point mechanics has served as a paradigm for a physical theory for more than two centuries but its group theoretical structure was found only when the concepts of Newtonian mechanics had been superseded by those of the special theory of relativity and of quantum mechanics. In 1845/46 Carl Gustav Jacob Jacobi (1804-1851) showed in his lectures (published 1866) that within the canonical formalism of mechanics the invariance under displacements in time, position and angle gives rise to the conservation of energy, linear momentum, and angular momentum, respectively. Under the direct influence of Felix Klein (1849-1925) - whose enthusiasm for the group theoretical point of view was almost missionary - Emmy Noether (1882-1935) showed in 1918 the way in which invariance under a continuous group gives rise to constants of motion and proved for the first time that the well-known ten first integrals of Newtonian mechanics follow from the invariance properties of the Lagrangian under the infinitesimal transformations of the ten-parameter Galilei group. Little further work on the Galilei group was done in the next three decades.

In 1939 Eugene Paul Wigner (born 1902) showed that invariance arguments suffice to obtain a classification of the relativistic equations for elementary systems. In particular, he showed that the Poincaré group (i.e. the inhomogeneous Lorentz group) is a type I group, so that all its unitary representations can be decomposed into irreducible ones. With this result, Bargmann and Wigner (1948) could determine all relativistically invariant wave equations for free elementary particles.

A general intrinsic definition of an *elementary system* as a state belonging to an irreducible unitary projective representation of the appropriate symmetry group of space-time was introduced by T.D. Newton and E.P. Wigner (1949) in a paper attempting a Lorentz-invariant localization of elementary systems. The localization problem consists in finding localized states of elementary systems and the corresponding operator, representative of position such that the descriptions of the localization by observers in different inertial frames are equivalent. In a rigorous way, A.S. Wightman (1962) showed that every elementary Galilei system with positive mass is localizable in space-time in a Galilei-invariant manner. Independently, Wightman (1962) and Mackey (1963a; representing his Harvard lectures of 1960/61), related the existence of three commuting position operators to Mackey's imprimitivity theorem (Mackey, 1949). For a Galilei system, a system of imprimitivity for the action of the three-dimensional Euclidean group can be defined, and the three position operators can be obtained from the corresponding projection-valued measure.

While for the Poincaré group all projective representations are equivalent to two-valued vector representations, the Galilei group is more complicated. The simplicity of the Poincaré group has misguided even the experts. For example, Inönü and Wigner (1952) in their study of the vectorial representations of the Galilei group, wondered why position operators do not exist for vector representations of the Galilei group. They did not seem to be aware that they did in fact investigate Galilei systems of mass zero. Although the unitary projective representations of the Galilei group are built into Schrödinger's "wave mechanics", they were discussed explicitly for the first time by Valentine Bargmann (born 1908) in 1954. Bargmann's study led to an unexpected discovery: the invariance under Galilei transformations implies a superselection rule which guarantees the strict conservation of mass.

The fact that about thirty years elapsed between the publication of Maxwell's equations in 1861/64 and the proof by Lorentz, that these equations are invariant under the Lorentz group, may be explained by the slow emergence of a new point of view. Considering the great popularity of group theoretical methods to solve special problems in elementary quantum mechanics, it is much more surprising that again almost thirty years elapsed between the publication of Schrödinger's differential equation for a free nonrelativistic particle and the proof by Bargmann (1954) that this equation is uniquely given by a projective irreducible representation of the Galilei group. That Heisenberg's canonical commutation relations have a group theoretical origin, was clearly realized by Hermann Weyl (1885-1955) as early as 1927 when he stressed that *"the kinematical structure of a physical system is expressed by an irreducible Abelian group of unitary ray rotations in system space"* (quoted from Weyl, 1928, English translation of 1931, p.275). However, he apparently did not notice the relation of his Abelian group to the Galilei group.

Bargmann's superselection rule leads for the first time to a deeper understanding of the role of mass in mechanics. Bargmann's superselection rule says that the mass of an elementary particle in a Galilei-invariant theory is a classical observable. That is, the Galilei group gives the final explanation of the concept of the conservation of matter introduced into chemistry by Antoine Laurent de Lavoisier (1743-1794) and of the law of definite proportions due to Joseph Louis Proust (1754-1826) and John Dalton (1766-1844).

The corresponding discussion for classical mechanics was initiated by Loinger (1962) who studied representations of the Galilei group in Koopman's Hilbert space of classical mechanics. Only in retrospect has it become clear that the historically so important correspondence principle and the associated methods of quantization have their roots in the identity of the space-time symmetry groups in classical and quantum theories.

Additional material and references about the historical development of the concepts of mechanics and space-time can be found in Dugas (1950), Whittaker (1951/53), Jammer (1954, 1961) and Truesdell (1976).

Elementary objects in pioneer quantum mechanics

The problem of the existence and the nature of elementary objects is a very old one and the early Greek atomists anticipated many of the current ideas. Titus Lucretius Carus (ca. 97-55 B.C.) characterized elementary objects in his surprisingly modern *De rerum natura* as follows: *"I will set out to discourse to you on the ultimate realities of heaven and the gods. I will reveal those atoms from which nature creates all things and increases and feeds them and into which, when they perish, nature again resolves them. To these in my discourse I commonly give such names as the "raw material", or "generative bodies" or "seeds" of things. Or I may call them "primary particles", because they come first and everything else is composed of them"* (quoted from the English translation by Latham, 1951). We will call the "raw material" of modern physics *elementary objects*.

The prototype of an elementary object is the *electron*. Clearly, an electron is not an entity having colour, texture or smell. Fifty years ago, it was not yet plain that an electron is neither a particle nor a wave. That is not to say that in particular *states* an electron cannot behave as a wave or as a particle. Particle-likeness and wave-likeness are potential but not intrinsic properties of an electron. They need not be realized in every state, and we know that they cannot be realized in the same state at the same time. Early quantum theory introduced the particle-wave duality as two mutually exclusive and contradicting models of the electron ("particles" vs. "waves"), the final form of pioneer quantum mechanics eliminated it. The electron is neither a particle nor a wave, it has a nature of its own. Then, what is an electron really? In pioneer quantum mechanics *an electron is an irreducible projective representation of the Galilei group whose characterizing parameters (m,s) are given by* $m = 9{,}109...10^{-31}$ kg *and* $s = 1/2$.

This characterization calls for some elaboration. The kinematics of a theory is characterized by an abstract *kinematical group* which has to be represented as a subgroup of the logical symmetries of the theory. Pioneer quantum mechanics is (like Newtonian mechanics) a Galilei-relativistic theory. Its kinematical group is the *Galilei group*, so that we have to represent the Galilei group as a subgroup of the group of all logical symmetries. Each unitary operator acting on the state space of pioneer quantum mechanics defines a particular logical symmetry. Two unitary operators define the same logical symmetry if and only if one is a constant multiple of the other. The unitary operator associated with a logical symmetry is therefore uniquely determined up to multiplications by complex numbers of modulus unity. The set of all unitary operators of the form cU with a fixed operator U and arbitrary complex numbers c with modulus unity is called the *operator ray* $\bar{U}$ generated by U

$$\bar{U} = \{cU \mid c \in \mathbb{C} \ , \ |c|=1\} \quad .$$

The representation of a group by unitary operator rays is called a ray representation (compare e.g. Weyl, 1928). If we rephrase a ray representation in terms of its generating unitary operators, we speak of a projective representation.

A *projective representation* of a group G into the group $U(H)$ of all unitary operators acting on a Hilbert space H is a mapping $G \to U(H)$ such that

(i) the neutral element of G is mapped on the identity operator on H,

(ii) the mapping $g \to U_g \in U(H)$ fulfills

$$U_{gg'} = \sigma(g,g')U_g U_{g'}$$

for every g,g' in G, where $\sigma(g,g')$ is a complex number depending on g and g' (called a *multiplier* of G).

If $\sigma(g,g')=1$ for all $g,g' \in G$, one speaks of a *vector representation* (synonym: ordinary representation). For more details, compare e.g. Mackey, 1968, 1976.

The representations of physical interest in pioneer quantum mechanics are the projective representations of the Galilei group by unitary operators acting on the Hilbert space of the state vectors. Instead of projective representations one can equivalently but more conveniently consider the vector representations of the *extended Galilei group* which is the central extension of the covering group of the Galilei group by an Abelian group (compare Bargmann, 1954). The generator by which this extension can be made is called the *mass operator* M, it commutes with all other generators of the Galilei group.

The generators of a projective representation of the kinematical group are particularly interesting and are referred to as *kinematical observables*. The *algebra of the kinematical observables* of pioneer quantum mechanics is identified with the enveloping algebra of the infinitesimal generators of a unitary vector representation of the central extension of the Galilei group. The generators of the rotation subgroup are the angular momenta J_x, J_y, J_z, the generators of the pure Galilei transformations are the position operators Q_x, Q_y, Q_z, the generators of the space translations are the momenta P_x, P_y, P_z, and the generator of the time translation is the Hamiltonian H. Apart from the mass operator M, which generates the central extension of the Galilei group, there is only one additional central element in the enveloping algebra, namely

$$\vec{S}^2 = (\vec{J} - \vec{Q}\times\vec{P})^2 .$$

In an irreducible representation the central elements of the algebra of kinematical observables have a definite value. In an irreducible representation, the value of the mass operator M is denoted by m, while the value of $\vec{S}^2$ is denoted by $s(s+1)$; m can be any real number, while the possible values of s are $0,\frac{1}{2},1,\frac{3}{2},2,\ldots$. (For the representation theory of the Galilei group we refer to Bargmann, 1954; Lévy-Leblond, 1963; Voisin, 1965; Brennich, 1970. For a review, compare Lévy-Leblond, 1971).

Intuitively, the concept of a *Galilean object* requires that it preserves its individuality under Galilei transformations. If such an object cannot be decomposed into several Galilean objects, it is considered as an *elementary* Galilean object. Accordingly, an *elementary Galilean object of pioneer quantum mechanics is defined to be an irreducible projective unitary representation of the Galilei group on some Hilbert space.* It is characterized by a pair (m,s), $m\in\mathbb{R}$, $2s+1\in\mathbb{N}$, where the parameter m is called the *mass* and the parameter s the *spin* of the elementary object. There exists an antiunitary operator transforming the Galilean object (m,s) into the object (-m,s) so that these two systems are logically equivalent (i.e. they cannot be distinguished in isolation). By convention, we call an elementary Galilean object with positive mass a *Galileon*. To every Galileon (m,s) there is a twin system (-m,s), called *anti-Galileon*. In contradistinction to Einsteinian relativity, elementary Galilean objects with negative mass seem to play no role in science. The case of vanishing mass is somewhat pathological (Inönü and Wigner, 1952) but Sen and Zahavi (1972) have shown that the zero-mass reducible reprentations of the Galilei group have just the same properties as Landau excitations in superfluid helium.

The generator $\vec{J}$ has the meaning of total angular momentum, while $\vec{Q}\times\vec{P}$ is the external angular momentum (also called the orbital angular momentum). The difference $\vec{S} \overset{\text{def}}{=} \vec{J}-\vec{Q}\times\vec{P}$ is the internal part of the angular momentum, called the *spin angular momentum*. Historically, the proper incorporation of the spin angular momentum into quantum mechanics was obtained in the framework of Einsteinian-relativistic quantum theory. However, the often repeated statement that the spin is a purely Einsteiian-relativistic effect is wrong and very misleading (compare also the anthology collected by Lévy-Leblond, 1974, appendix A). A spin of a Galileon appears in a natural way from first principles, it is a compelling consequence of the isotropy of the 3-dimensional physical space. Lévy-Leblond (1967) has shown that for a Galileon with spin $s=\frac{1}{2}$, Pauli's equation for the electron follows in a natural way so that even the Landé factor g=2 of the electron is *not* an effect of Einsteinian relativity.

The group consisting of the spatial translations and the pure Galilei transformations forms a 6-parameter Abelian subgroup of the Galilei group. The projective irreducible representation of this Abelian group is given by a non-Abelian unitary group $\{U(\vec{p}),\ \vec{V}(\vec{q})\,|\,\vec{p},\vec{q}\in\mathbb{R}^3\}$ fulfilling the commutation relation

$$U(\vec{p})V(\vec{q}) = e^{i\vec{p}\cdot\vec{q}/\hbar}V(\vec{q})U(\vec{p}) \quad .$$

For the first time, this relation was derived by Hermann Weyl (1927) as the basic equation of quantum kinematics governed by an Abelian group of rotations of the rays associated with the state space. Accordingly, the operators U and V are referred to as *Weyl operators* and the above commutation relation as *Weyl's canonical commutation relation*. With a very weak continuity assumption (in the case of a separable state space, the Abelian groups U and V have to be weakly measurable) Stone's (1930, 1932) theorem warrants the existence of two (unbounded) self-adjoint operators $\vec{Q}$ and $\vec{P}$ such that

$$U(\vec{p}) = e^{i\vec{p}\cdot\vec{Q}/\hbar} \quad \text{and} \quad V(\vec{q}) = e^{-i\vec{q}\cdot\vec{P}/\hbar} \quad \text{for all} \quad \vec{p},\vec{q}\in\mathbb{R}^3 \quad .$$

This representation and Weyl's commutation relation imply the famous canonical commutation relations for the operators $\vec{Q}$ and $\vec{P}$,

$$Q_\nu P_\mu - P_\mu Q_\nu = i\hbar\delta_{\nu\mu} \quad , \quad \nu,\mu = 1,2,3 \quad .$$

Accordingly, *the origin of the canonical commutation relations of pioneer quantum mechanics is the assumption of a Galilean space-time*. The triad (Q_1,Q_2,Q_3) is the only triad of mutually commuting operators which are invariant under transformations to moving coordinate systems, to which a spatial displacement $\vec{a}$ adds the components of $\vec{a}$, and which transforms as a vector under rotations. Because of this, the operator $\vec{Q}=(Q_1,Q_2,Q_3)$ is called the *position operator*.

Technical remark: The generators of a projective unitary representation of a Lie group are in general *unbounded* operators that leads to various complications. In order to avoid these difficulties, it may be convenient to use the general eigenfunction expansion and unitary representation theory of locally compact groups in the associated *rigged Hilbert space* (compare Gel'fand and Vilenkin, 1961, and Maurin, 1966, 1967). In the case of the Galilei group, this amounts to replacing the topology induced by the Hilbert space of state vectors by a new topology such that the canonical operators $\vec{P}$ and $\vec{Q}$ become *continuous* operators (compare Kristensen et al., 1965). The corresponding rigged Hilbert space is the famous Gel'fand triple $S(\mathbb{R}^3) \subset L_2(\mathbb{R}^3) \subset S^*(\mathbb{R}^3)$, where $\mathbb{R}^3$ is the configuration space, $L_2(\mathbb{R}^3)$ the state space (i.e. the Hilbert space of square integrable functions on $\mathbb{R}^3$), $S(\mathbb{R}^3)$ the Schwartz space of rapidly decreasing smooth functions, and $S^*(\mathbb{R}^3)$ the space of tempered distributions. All kinematical observables are continuous on S (with respect to the topology of S) and essentially self-adjoint (with respect to the Hilbert space L_2) (compare Nelson and Stinespring, 1959; Nelson, 1959). Since S is a nuclear space, the nuclear spectral theorem assures that the dual space S^* contains all generalized eigenvectors of the kinematical observables. Every Hermite orthogonal function (i.e. every eigenvector of P^2+Q^2) is in S so that the Hermite orthonormal functions form a *kinematically preferred basis* for the state space $H=L_2(\mathbb{R}^3)$.

The fact that the representations of the Galilei group in pioneer quantum mechanics are of nontrivial projective nature has three physically most interesting consequences: (i) the possibility of a spin angular momentum, (ii) the noncommutativity of the operators $\vec{P}$ and $\vec{Q}$, (iii) Bargmann's superselection rule. Bargmann (1954) has shown that the way a state of a Galileon transforms under the Galilei group depends on its mass. *Bargmann's superselection rule* says that *in pioneer quantum mechanics it is impossible to have pure states which are superpositions of states describing Galileons of different mass*. Bargmann's superselection rule ensures the temporal stability of Galileons and guarantees the strict *conservation of mass* in pioneer quantum mechanics.

The concept of a *particle* comes from *classical* mechanics. In classical mechanics, a particle is the same as an elementary Galilean object. In classical mechanics an elementary Galilean object is a transitive representation of the Galilei group as a subgroup of the symplectic transformations on the phase space. Both in classical and in quantum mechanics elementary Galilean objects are purely theoretical constructs, useful as "raw material" in the sense of Lucretius. Both the raw material for building classical mechanics and the raw material of pioneer quantum mechanics stems from Galilean space-time symmetry but there are differences: *The Galileons of pioneer quantum mechanics are not particles in the sense of Newtonian mechanics.*

Historical remarks: The role of projective representations in quantum mechanics was recognized by Hermann Weyl (1885-1955) in 1927. Without referring to the Galilei group, he showed that both the canonical commutation relations for the position and the momentum operators and the commutation relations for the spin kinematics could be extracted from the projective representations of a 2-parameter Abelian group (Weyl, 1927, 1928). John von Neumann (1935) stressed that the factors of absolute value 1 in unitary projective representations are sometimes trivial, can be transformed away on other occasions, but that there exist groups "*for which neither is the case and the extra terms play an essential role*". Strangely, this crucial remark was not elaborated until Bargmann's (1954) classical paper (at least implicitly) recognized the importance of the cohomology theory of groups. The meaning of continuity of a unitary projective representation of a topological kinematical group was clearly stated in a fundamental paper by Wigner (1939). In the same paper, Wigner introduced into Einsteinian relativity the concept of an elementary "particle" as an irreducible representation of the Poincaré group. This definition is nowadays widely accepted in high-energy "particle" physics. The work of Bargmann and Wigner (1948), Newton and Wigner (1948) and Bargmann (1954) made it clear that in pioneer quantum mechanics the proper concept of an elementary object is given by the abstract notion of an equivalence class of irreducible unitary projective representations of the appropriate space-time symmetry group.

Interacting Galileons

The elementary object associated with the n-fold direct product of the Galilei group is called a *system of n noninteracting Galileons*. With this, it follows that the state space of n noninteracting Galileons is a subspace of the *tensor product* of the state spaces of the individual Galileons (Dixmier, 1964, prop.13.1.8). Elementary Galileons do not interact directly with each other but interact directly with a *field* that mediates the interaction between particles. In the macroscopic and molecular domain the only two mediating fields of importance are the electromagnetic field and gravitation. Moreover, physical space and interactions are not independent but in some way expressions for the same physical concept. The interactions between elementary objects depend on some parameter in such a way that the interaction is small if the parameter is large. We usually say that the electromagnetic interaction is weak if the spatial distance between two elementary objects is large. The theorist should have the courage to *define* spatial distance as an interaction parameter that is large if the interaction energy is small.

Some speculative remarks: It is not accidental that the only known two interactions relevant for molecular and macroscopic phenomena are governed by inverse-square law forces. In the language of quantum field theory this implies that these interactions are mediated by zero-mass boson fields. In fact, the electromagnetic field can be described by bosons having zero rest mass and spin 1 ("photons"), while gravitation can be described by bosons having zero rest mass and spin 2 ("gravitons"). The long-range character of these interactions (i.e. inverse-square force law) is due to the vanishing rest mass (the "nonlocalizability") of the photon. It is tempting to speculate that these bosons *generate* the space structure. The dynamical equations for the bosons of the electromagnetic fields are nothing else but the Maxwell equations. The field vectors of the electromagnetic fields are 3-dimensional vectors because photons have spin 1. So perhaps we should not say that the electromagnetic field is described by 3-dimensional vectors because physical space is 3-dimensional, but rather that *physical space is three-dimensional since photons have spin* 1. Note that electromagnetic interactions dominate in our everyday life and that the tensor character of the gravitation field (spin 2, corresponding to a 5-dimensional space) is unimportant. It is therefore to be expected that we can describe our experiences reasonably well in a 3-dimensional framework.

Galileons in external fields

Neglecting effects of radiation damping, the behavior of a particle of mass m, charge e and velocity $\vec{v}$ in an external electromagnetic field $(\vec{E},\vec{B})$ is in classical physics described by the Lorentz force $\vec{F}$, (in SI-units)

$$\vec{F} = e(\vec{E} + \vec{v}\times\vec{B}) \quad ,$$

and Newton's second law. The Hamiltonian H which yields the corresponding equation of motion cannot be written simply in terms of the field strength vectors $\vec{E}$ and $\vec{B}$ but one is forced to introduce the vector potential $\vec{A}$ and the scalar potential Φ,

$$H = \frac{1}{2m}(\vec{P} - e\vec{A})^2 + e\Phi \quad .$$

From these potentials the electromagnetic field can be calculated,

$$\vec{B} = \operatorname{curl}\vec{A} \quad , \quad \vec{E} = -\operatorname{grad}\Phi - \partial\vec{A}/\partial t \quad ,$$

but not conversely. The Lorentz force and the electromagnetic field are invariant under a gauge transformation of the potentials

$$\vec{A} \longrightarrow \vec{A} + \operatorname{grad}\Lambda \quad ,$$

$$\Phi \longrightarrow \Phi - \partial\Lambda/\partial t \quad ,$$

where $\Lambda:\mathbb{R}^4\to\mathbb{R}$ is an arbitrary single-valued smooth real function of space-time. The Lorentz force and the electromagnetic field are observable quantities. The potentials are not measurable, the gauge transformations represent a symmetry of electrodynamics taking into account the redundance of the potential. It is tempting but wrong to believe that the gauge invariance of electrodynamics has no genuine physical meaning but only ensures that the arbitrariness in the choice of the potentials does not affect the field strength vectors. The fact that the interaction Hamiltonian between matter and the electromagnetic field depends in an essential way on the gauge-dependent potentials is of decisive importance: it implies that not all aspects of our description of matter are gauge invariant. Those aspects of our description which are not invariant under gauge transformations are *not observable*. As stressed by Wigner (1964a), gauge groups refer to *internal symmetries*, they are very different from the external geometrical symmetries, they do not correlate events but they characterize *interactions*.

Traditionally, this situation is carried over into pioneer quantum mechanics by defining the time-dependent Schrödinger equation for a Galileon of mass m, charge e and spin zero as

$$i\hbar\, \partial\Psi/\partial t = \{\frac{1}{2m}(\vec{P} - e\vec{A})^2 + e\Phi\}\Psi \quad .$$

The result can be summarized by the following simple rule: *The influence of an external electromagnetic field on a Galileon of charge e can be expressed by the substitution*

$$\vec{P} \rightarrow \vec{P} - e\vec{A} \quad ,$$
$$\partial/\partial t \rightarrow \partial/\partial t + i(e/\hbar)\Phi \quad .$$

In the Schrödinger representation $\vec{P}=(\hbar/i)\partial/\partial\vec{q}$, the substitution can be written more elegantly as

$$\partial_\mu \rightarrow \partial_\mu - i(e/\hbar)A_\mu \quad .$$

where we have introduced $\partial_\mu = \partial/\partial x_\mu$, and the four-vectors x_μ and A_μ by

$$(x_1,x_2,x_3,x_4) = (\vec{q},ict) \quad ,$$
$$(A_1,A_2,A_3,A_4) = (\vec{A},i\Phi/c) \quad .$$

The Schrödinger equation of a Galileon in an external field is invariant under the simultaneous substitutions

$$\Psi \rightarrow e^{i(e/\hbar)\Lambda}\Psi ,$$

$$A_\mu \rightarrow A_\mu + \partial_\mu\Lambda ,$$

where Λ is an arbitrary smooth real function of space-time, $\Lambda:\mathbb{R}^4\rightarrow\mathbb{R}$. The group of all such substitutions is referred to as the *gauge group*. The invariance of the fundamental laws of pioneer quantum mechanics is called *gauge invariance*. The principle of gauge invariance indicates a dependence between the laws of matter and electromagnetism. The conservation of electricity is a double consequence of gauge invariance, "*it follows from the laws of matter as well as electricity*" (Weyl, 1929a). This fact suggests that there should be a deeper reason for the most successful recipe $\partial_\mu\rightarrow\partial_\mu - i(e/\hbar)A_\mu$ of pioneer quantum mechanics.

The electromagnetic interaction is a consequence of the principle of local invariance

The existence of electromagnetic interactions is related to invariance principles. Every time-dependent Schrödinger equation

$$i\hbar\,\partial\Psi/\partial t = H\Psi$$

is invariant under phase transformations

$$\Psi \rightarrow e^{i\varphi}\Psi$$

where φ is a real constant. Such a transformation is called a *global phase transformation* and defines a one-parameter Abelian Lie group, called the group U(1). The invariance under the global phase group U(1) implies by Noether's theorem the existence of a conserved quantity, called the *charge* (for molecular matter it turns out that it is the electric charge but for nuclear matter it could also be baryonic charge). Hence, *the invariance of pioneer quantum mechanics under global phase transformations implies the global conservation of the total charge*. This result is highly interesting but insufficient on physical grounds, since it would allow a situation in which the charge disappears in one place and appears simultaneously in some other place such that the total charge does not change. A tenet due to Albert Einstein says that such a thing is impossible but that the following *principle of local invariance* holds: "*if anything is conserved, it must be conserved locally*" (Feynman, 1965, p.63).

The requirement that charge is conserved locally requires that we can fix the phase of the state vector locally, without any reference to far-away distances. In the Schrödinger representation of a single Galileon, the charge density ρ is given by

$$\rho(\vec{q},t) = e|\Psi_t(\vec{q})|^2 \quad ,$$

where Ψ_t is the state vector at time t. This charge density is invariant under the *local phase transformation*

$$\Psi_t(\vec{q}) \rightarrow e^{i\varphi(\vec{q},t)}\Psi_t(\vec{q}) \quad ,$$

where $\varphi:\mathbb{R}^4\rightarrow\mathbb{R}$ is now a function of space-time. The principle of the locality of all conservation laws suggests to adopt local phase transformations as symmetries of the theory. The requirement of *local phase invariance* is much stronger than the invariance under the global phase group U(1).

Every symmetry principle says that some quantity is not observable. The invariance under the global phase group U(1) implies that the overall phase factor of the state vector ist not measurable, it can be chosen arbitrarily but it must be chosen the same over the entire universe and for all time. Local phase invariance would mean that phase differences are not observable for state functions at different points in space-time and allow us to fix the phase of the state vector locally. Since every experiment is confined to finite space-time regions, local phase invariance is not only desirable but almost inevitable. But is it possible? At first sight we seem to be in trouble since the Hamiltonian of a Galileon is *not* invariant under local phase transformations. It turns out that local phase invariance is possible provided we introduce a new *compensating field* whose only duty is to restore the invariance under local phase transformations. For Galileons, this compensating field is just the well-known electromagnetic field.

There is a standard way of restoring invariance under local transformations. As discussed in differential geometry, one has to replace the ordinary derivative ∂_μ by a *covariant derivative* D_μ which is a first-order differential operator transforming like the state function Ψ itself,

$$D_\mu \Psi(x) = e^{i\varphi(x)} D_\mu \Psi(x) \quad , \quad x \in \mathbb{R}^4 \ .$$

The covariant derivative D_μ is given by

$$D_\mu = \partial_\mu - i\lambda A_\mu \ ,$$

where the affine connection A_μ transforms as

$$A_\mu \rightarrow A_\mu + \frac{1}{\lambda}\partial_\mu \varphi \quad ,$$

and λ is some real nonvanishing constant. In our physical context, the function $A_\mu : \mathbb{R}^4 \rightarrow \mathbb{R}$ is called the *gauge field*, and the real number λ the *coupling constant*. The fields A_μ are not gauge invariant but the fields $F_{\mu\nu}$ defined by

$$F_{\mu\nu} = \partial_\nu A_\mu - \partial_\mu A_\nu$$

are. The simplest gauge invariant free Lagrangian for the gauge is the usual Lagrangian $-\frac{1}{4}F_{\mu\nu}F_{\mu\nu}$ for the free electromagnetic field $F_{\mu\nu}$.

The raison d'être for the electromagnetic field is the local phase invariance of quantum matter. The realm of electromagnetism is space-time since the interactions between Galileons are determined by local phase invariance. More precisely, electromagnetism is the consequence of pioneer quantum mechanics and the following two additional postulates:

(i) *The locality postulate*, saying that the local phase of the state vector in the Schrödinger representation is not observable, so that the transformation

$$\Psi(x) \rightarrow e^{i(e/\hbar)\Lambda(x)} \Psi(x) \ ,$$

with a differentiable real function $\Lambda : \mathbb{R}^4 \rightarrow \mathbb{R}$ is a symmetry causing only a permutation of actually observable quantities.

(ii) *The minimal coupling postulate.* The requirement of invariance under local phase transformations necessitates the existence of a compensating field. The principle of minimal electromagnetic interaction asserts that we have to include in the interaction Hamiltonian *only* those terms required by local phase invariance.

These two postulates imply that the derivative ∂_μ has to be replaced by the covariant derivation D_μ, determining completely the form of the interaction Hamiltonian. The gauge symmetry principle determines all properties of the electromagnetic field (including Maxwell's equation and the fact that the mass of the photon is zero) and its interaction with matter.

Historical remarks: The gauge group was introduced by Hermann Weyl (1918) in an ingenious but abortive attempt to unify the theories of gravitation and electricity, based on the idea that lengths of measuring rods and times of clocks depend on their history (compare also Weyl, 1923, §§ 40,41). After the advent of quantum mechanics, both London (1927) and Weyl (1928, 1929 a-c) recognized that the gauge group does not tie together electricity and gravitation but rather electricity and matter. Weyl also pointed out that global gauge invariance is analogous to Einsteinian special relativity while local gauge invariance is analogous to Einsteinian general relativity.

Local non-Abelian gauge transformations where first introduced by Yang and Mills (1954) in order to express the nonobservability of isospin phase differences at different space-time points, aiming at the theory of strong interactions. Nowadays, the term "Yang-Mills field" is used to denote any generalized gauge field accompanying a local conservation law. A general approach to a theory of compensating fields was given by Utiyama (1956), who also showed that the Yang-Mills field associated with the extension of the Lorentz group from global to local symmetry is the gravitational field. General relativity is associated with the Poincaré group (Kibble, 1961), whereby the vierbein components act as compensating gauge fields. This result is in accordance with the view of Fock (1959) who stressed that the invariance group of Einsteinian general relativity should not be interpreted as a geometrical symmetry.

The principle of local invariance leads to the appearance of compensating fields. The electromagnetic field is the compensating field of local phase transformations, the gravitational field is the compensating field of local Lorentz transformations. Such compensating fields cannot be scalars but are vector or tensor fields whose associated rest mass vanishes. Nowadays, it is widely believed (Bernstein, 1974; Weinberg, 1974) that the strong, weak and electromagnetic interactions can be described by gauge theories which combine gauge invariance with spontaneous symmetry breaking, so that the gauge vector bosons acquire a mass, as first suggested by Schwinger (1962). The phenomenon that the symmetry breaking of a gauge theory does allow the existence of non-zero-mass bosons is well illustrated in the case of superconductivity (Anderson, 1963). The theory of superconductivity is an Abelian gauge theory whose gauge symmetry is broken in the transition to the superconducting phase. The theorem of Goldstone (which would predict the emergence of a massless boson) does not apply to gauge fields, instead the photon acquires mass (this new boson is called *plasmon*) as manifested by the experimental fact that a magnetic field can penetrate only a short distance into a type I-superconductor (Meissner effect).

The minimal coupling rule $\partial_\mu \to D_\mu$ also allows us to derive the correct Hamiltonian for a Galileon with spin (cf. Lévy-Leblond, 1967; Celeghini et al., 1976). In the case of a Galileon of mass m, charge e and spin $s=\frac{1}{2}$, one gets the well-known Pauli Hamiltonian (again in SI-units) by the substitution $\vec{P}\to\vec{P}-e\vec{A}$, $\partial/\partial t\to\partial/\partial t+i(e/\hbar)\Phi$ in the free Hamiltonian $(\vec{\sigma}\vec{P})(\vec{\sigma}\vec{P})/2m$:

$$H = \frac{1}{2m}\{\vec{\sigma}(\vec{P}-e\vec{A})\}\{\vec{\sigma}(\vec{P}-e\vec{A})\} + e\Phi$$

$$= \frac{1}{2m}(\vec{P}-e\vec{A})^2 + e\Phi - \frac{e\hbar}{m}\vec{\sigma}\vec{B} \quad ,$$

where $\vec{B}=\text{curl}\vec{A}$, and $\vec{\sigma}=(\hbar/2)\vec{S}$ is the Pauli spin operator.

For a free Galileon, one interprets $\vec{Q}$ as the observable of position, $\vec{P}$ as the observable of momentum, and $\vec{J}$ as the observable of angular momentum. In the presence of an external field the Galilei symmetry is broken, neither $\vec{P}$ nor $\vec{J}$ is conserved, $\vec{P}$ does not equal $m\dot{\vec{Q}}$ and $\vec{J}$ does not equal $\vec{Q}\times\vec{P}+\vec{S}$, so that the physical interpretation of $\vec{P}$ and $\vec{J}$ is not plain if there are external fields. However, whether external fields are present or not, the observable $\vec{Q}$ has the interpretation of a position. Even if the Galilei symmetry is broken, the observable $\vec{Q}$ of position is characterized by the usual transformation behavior under a pure Galilei transformation with the velocity parameter $\vec{v}$:

$$\vec{Q} \to \vec{Q} \quad ,$$

$$\dot{\vec{Q}} \overset{\text{def}}{=} i[H,\vec{Q}]_- \to \dot{\vec{Q}} + \vec{v} \quad .$$

This principle of *restricted Galilean invariance* (Jauch, 1964a) is the trace left by breaking the full Galilean invariance with external fields. This remarkable behavior is due to the *locality* of the coupling.

Interaction Hamiltonians

Since matter is the source of electromagnetism, the interaction between Galileons can be evaluated by Maxwell's equations. Quite generally, the four-current operator J_μ asscociated with a system possessing the Hamiltonian H in an external field A_μ is given by

$$J_\mu(x) = -\delta H/\delta A_\mu(x) \quad , \quad x \in \mathbb{R}^4 \ .$$

According to Maxwell's equations, the four-current J_μ generates the four-tensor $F_{\mu\nu}=\delta_\mu A_\nu - \delta_\nu A_\mu$ of the electromagnetic field acting on the other Galileons,

$$\delta_\nu F_{\mu\nu} = \kappa J_\mu \quad ,$$

where in SI-units $\kappa=1/\varepsilon_0 c^2$. In this way the full interaction Hamiltonian (with all spin-orbit and spin-spin terms) can be obtained.

References: A good discussion of the form of the molecular Hamiltonian of interacting Galileons can be found in Moss (1973). Compare also Hirschfelder et al. (1954), Bethe and Salpeter (1957), and Slater (1960). Sometimes it is necessary to include Einsteinian-relativistic corrections; the most reliable method is to derive these terms from quantum electrodynamics as an approximation where the velocities of the Galileons are sufficiently small compared to the velocity of light. A good review of the results of these methods has been given by Itoh (1965); compare also Brodsky and Primack (1968), and Primack and Brodsky (1969).

The dynamics of pioneer quantum mechanics

The time evolution of a closed system is given by a strongly continuous one-parameter group $\{U_t | t \in \mathbb{R}\}$ of unitary operators U_t acting on the Hilbert space $\mathcal{H}$ of state vectors. In the Schrödinger picture, the time development of a state vector is given by

$$\Psi_t = U_t \Psi \quad ,$$

where Ψ is the state vector of the initial state at t=0. According to Stone's theorem, every dynamical group $\{U_t | t \in \mathbb{R}\}$ is of the form

$$U_t = e^{-iHt} \quad ,$$

where H is a unique (in general unbounded) self-adjoint operator on $\mathcal{H}$. In quantum mechanics, the generator H is called the *Hamiltonian* of the system. Conversely, every self-adjoint operator acting on the state space $\mathcal{H}$ generates some dynamical group $t \to U_t$.

In order that a Hamiltonian can serve as a generator of the dynamics, it has to be self-adjoint. The question of the self-adjointness of the n-dimensional Schrödinger operator has a long history. First of all, it is not obvious that the formal differential operator

$$H = \sum_{j=1}^{n} \sum_{k=1}^{n} a_{jk}\{-i\partial/\partial q_j - A_j(q)\}\{-i\partial/\partial q_k - A_k(q)\} + \Phi(q) \quad ,$$

where $q \in \mathbb{R}^n$, Φ and A_j are real-valued functions on $\mathbb{R}^n$, and (a_{jk}) is a positive definite real symmetric matrix, defines an operator on the Hilbert space $L_2(\mathbb{R}^n, dq)$. However, H is a well-defined operator on the set $C_0^\infty(\mathbb{R}^n)$ of all infinitely differentiable functions vanishing outside some bounded region, and is called the *minimal operator* H_{min}. If the A_j are continuously differentiable and if Φ is locally square integrable, the minimal operator H_{min} is a symmetric operator, hence has a smallest closed extension, called the closure of H_{min}. If the closure of H_{min} is self-adjoint, then H_{min} is called *essentially self-adjoint*. In this case, the formal operator H defines a *unique* self-adjoint extension on the Hilbert space $L_2(\mathbb{R}^n, dq)$, which we denote again by H. Note that a molecular Hamiltonian never is a differential operator (in the strict classical sense) but rather the self-adjoint *extension* of a differential operator.

Accordingly, the proper mathematical question is whether the formal differential operator H is essentially self-adjoint. The essential self-adjointness of the many-body Schrödinger Hamiltonian for the usual forces (e.g. the Coulomb force) remained an open problem for over twenty years. The first important result was obtained by Kato (1951) who showed that in the case of vanishing magnetic fields ($A_j=0$) it suffices to assume that each potential is the sum of a function in L_2 and a bounded function. Later Kato (1957) showed that the eigenfunctions of a molecular Hamiltonian are infinitely differentiable and satisfy the elliptic partial differential equation $H\Psi=E\Psi$ in the classical sense at each nonsingular point of the potential Φ. These results were generalized by Ikebe and Kato (1962) to the case of nonvanishing magnetic fields. The question of the essential self-adjointness of Schrödinger operators with more singular potentials aroused considerable interest among mathematicians and there is an extensive literature. These investigations were brought to a certain close by a definitive treatment again by Kato (1972, 1974).

References: A detailed treatment of the self-adjointness that covers most cases of interest to quantum mechanics can be found in the texts by Kato (1966) by Reed and Simon (1975), and rather comprehensively in Weidmann (1976).

Modes of reactions: the stochastic time evolution

In pioneer quantum mechanics, observables refer to *dispositional properties*, that is to "*induced stochastic transitions from one state to another one*" (Strauss, 1970). In this framework the stochastic time evolution is neither given by the unitary time evolution group nor explained in any other sense (attempts to do so go beyond the limits of pioneer quantum mechanics and are referred to as "theories of measurements"). Pioneer quantum mechanics does not explain why a particular transition occurs but gives the *probabilities* of transitions that are induced by the interaction with macroscopic classical systems.

It has proven to be convenient (but conceptually rather questionable) to discuss a class of stochastic state changes that behave in an experimentally unrealistic but mathematically simple way: the so-called *ideal measurements* (or, in Pauli's (1933) terminology: measurements of the first kind). Ideal measurements fulfill von Neumann's (1932) *repeatability assumption:*

> *After an ideal measurement of a physical quantity, an immediately subsequent measurement is constrained to give a result which agrees with the first.*

An ideal measurement is a procedure for obtaining a value of an observable that

(i) *changes* the state of the system such that
(ii) the final state is in some eigenspace of this observable
(iii) hence brings into existence the quantity measured.

There are no logical difficulties with this concept but it is important to acknowledge that all actual laboratory experiments (including the Stern-Gerlach experiment referred to by von Neumann) are *not* ideal measurements but tremendously more complicated operations. Moreover, the terminology is certainly not a fortunate one since a *classical measurement* is a procedure that

(i) can be performed without a significant change of the state,
(ii) only determines what is already the case.

Hence, a measurement in the classical sense is very different from the state changes called "ideal measurements".

From the definition of an ideal measurement, it follows that for observables with a nondegenerate discrete spectrum the measured values are eigenvalues of the observable, and that the state immediately after the measurement is represented by the eigenspace corresponding to the eigenvalue obtained. If this eigenspace is more than one-dimensional, the repeatability assumption does not fully specify the measurement. With an additional *assumption of minimal interference*, von Neumann's (1932) projection postulate can be phrased in the following sharpened form (Lüders, 1951):

> *Let* $E(\{\lambda\})$ *be the projector onto the eigenspace of an observable* A *corresponding to the eigenvalue* λ. *If the state before the ideal measurement of* A *is given by the vector* Ψ, *then after measuring* A *with outcome* λ *the state is given by* $E(\{\lambda\})\Psi$.

Probabilistic interpretation of pioneer quantum mechanics

An ideal measurement of an observable A with nondegenerate discrete spectrum leads to a particular eigenstate of A but even at the most fundamental level it is not determined *which* eigenstate will be realized. Born's probabilistic interpretation and the projection postulate give the *transition probability* for the state change $\Psi \to E(\{\lambda\})\Psi$ as

$$\|E(\{\lambda\})\Psi\|^2 = \langle\Psi|E(\{\lambda\})\Psi\rangle \ .$$

This result can be generalized to arbitrary self-adjoint observables A. Let the spectral resolution of A be given by

$$A = \int_\Lambda \lambda E(d\lambda)$$

where Λ is the spectrum of A whose Boolean σ-algebra of Borel sets will be denoted by Σ. The spectral measure E is a projection-valued set function on Σ having the properties:

(i) E is projection-valued, i.e.
$E(B) = E(B)^2 = E(B)^*$ for each $B \in \Sigma$,

(ii) E is nonnegative definite, i.e.
$E(B) \geq 0$ for each $B \in \Sigma$,

(iii) E is additive, i.e.
if $B_1 \cap B_2 = \emptyset$, then $E(B_1 \cup B_2) = E(B_1) + E(B_2)$,

(iv) E is multiplicative, i.e.
$E(B_1 \cap B_2) = E(B_1)E(B_2)$,

(v) E is continuous, i.e.
$\sup E(B_n) = E(B)$ whenever $\{B_n\}$ is an increasing sequence of sets in Σ whose union B is also in Σ.

The probabilistic interpretation of pioneer quantum mechanics can be phrased as follows:

Let A *be an observable with spectrum* Λ, *the* σ*-algebra* Σ *of Borel sets of* Λ, *and spectral measure* E. *The probability that an ideal measurement of* A *gives a value lying in the Borel set* B *is given by the probability measure* $\mu:\Sigma\to\mathbb{R}^+$,

$$\mu(B) \overset{\text{def}}{=} \langle\Psi|E(B)\Psi\rangle$$

where Ψ *is the normalized state vector before the measurement.*

This result implies that Kolmogorov's (1933) theory of probability is an inbuilt feature of pioneer quantum mechanics. Kolmogorov's mathematical probability theory rests on a triple, consisting of a sample space Λ, a Boolean σ-algebra Σ of subsets of Λ, and a probability measure $\mu:\Sigma\to\mathbb{R}^+$. Conditioned by an ideal measurement, the observable A (and every Borel function of A) can be considered as a *random variable* on the probability space (Λ,Σ,μ). The *expectation value* $\mathcal{E}(A)$ of this random variable is given by

$$\mathcal{E}(A) = \int_\Lambda \lambda\mu(d\lambda) = \langle\Psi|A\Psi\rangle ,$$

where, of course, Ψ refers to the state vector immediately before the measurement. Note that a family of observables can be regarded as a family of random variables on one and the same probability space if and only if they commute (for a rigorous proof, compare e.g. Nelson, 1967, p.117).

The concept of a statistical state in pioneer quantum mechanics

The outcome of ideal measurements with the same initial conditions exhibits variations from experiment to experiment which no amount of control can remove. Statistical probability theory discusses such stochastic experiments that can be repeated many times under well-controlled conditions and whose outcomes stabilize as the number of experiments is increased. These nonrandom regularities are called *statis-*

tical properties. The limit theorems of mathematical probability theory show that the relative frequency of the occurrence of events under large-scale repetitions is given by the mathematical concept of probability. Statistical probability theory is an *ensemble theory*, all statistical properties are to be obtained by averaging over an ensemble of a large number of identically prepared systems.

Let us consider an ensemble of quantum systems in Gibbs' sense, i.e. "*a large number of hypothetical systems which are introduced to describe one actual system of which our knowledge is only of a statistical nature*" (Dirac, 1929b). The state that characterizes an ensemble is called a *statistical state*. Note that a statistical state is a property of the virtual ensemble as a whole, it does *not* refer to an individual actual system. A statistical state may be thought to reflect the past history of the system. The notion of a statistical state is used to make predictions about the future behavior of the system. From experience, we know that we do not need the full past history in order to make optimal predictions, so that a statistical state should be defined in terms of the future behavior of the system (Husimi, 1937). Different experimental procedures which lead to the same future behavior are equivalent and define the same statistical state. Imitating the state definition Nerode (1958) has introduced into automata theory (and which is fundamental in mathematical system theory) we adopt the following conceptual definition of a statistical state:

> *A statistical state is the equivalence class of all histories of a system that give rise to the same predictions for all conceivable future experiments on the system.*

From an operational point of view "histories" may be replaced by "preparation methods".

In order to define a statistical state mathematically, we consider families of individual systems prepared in identical fashions so that a statistical state can be regarded as the result of a reproducible experimental preparation procedure. The most commonly discussed state preparation schemes are so-called "ideal measurements". Consider a homogeneous ensemble with all its members in the same state described by the state vector Ψ. Let A be an observable with a discrete spectrum, and let $\{\alpha_n\}$ be a complete orthonormal set of eigenvectors of A,

$$A\alpha_n = a_n\alpha_n \quad , \quad \langle\alpha_n|\alpha_m\rangle = \delta_{nm} \; ,$$

so that the state vector Ψ can be expanded as

$$\Psi = \sum_n \langle\alpha_n|\Psi\rangle\; \alpha_n \; .$$

An ideal measurement of the observable A leads with probability $p_n=|\langle\alpha_n|\Psi\rangle|^2$ to the eigenvalue a_n as the value measured and to the eigenvector α_n as final state. While every individual member of the ensemble is in some particular eigenstate $|\alpha_n\rangle\langle\alpha_n|$, the virtual ensemble contains this state with the relative frequency p_n, so that the statistical state of the ensemble after the measurement is given by the mixture

$$D \overset{\text{def}}{=} \sum_n p_n|\alpha_n\rangle\langle\alpha_n| \; .$$

The operator D is a nonnegative-definite self-adjoint operator fulfilling the normalization condition $\mathrm{tr}(D)=\Sigma_n p_n=1$; it is called the *statistical operator* of the ensemble after the measurement. A statistical operator characterizes the corresponding ensemble but it can also be used for the description of an individual member of which *our knowledge* is only partial and of a statistical nature.

More generally, a *statistical operator* (synonym: *density operator*) is defined as a normalized, self-adjoint, nonnegative-definite nuclear operator D acting on the separable Hilbert space $\mathcal{H}$ of state vectors,

$$D = D^* \geq 0, \; \mathrm{tr}(D) = 1 \; .$$

Remark: $D\geq 0$ means $\langle\Psi|D\Psi\rangle\geq 0$ for all $\Psi\in\mathcal{H}$. An operator T is called *nuclear* (synonym: an operator of the trace class) if it is compact and has a finite trace norm $\|T\|_1$. The trace norm of a compact operator is defined by $\|T\|_1=\Sigma_n\lambda_n$, where λ_n are the eigenvalues of $\sqrt{T^*T}$. An operator T is nuclear if and only if there is at least one orthonormal basis $\{\varphi_n\}$ of $\mathcal{H}$ such that $\Sigma_n\|T\varphi_n\|<\infty$.

Every statistical operator has a purely discrete spectrum consisting of nonnegative eigenvalues p_n, so that its spectral resolution can be written as

$$D = \sum_n p_n P_n$$

with projections $P_n = P_n^2 = P_n^*$ and

$$\mathrm{tr}(P_n) = 1 \quad , \quad \sum_n p_n = 1 \ , \ 0 \le p_n \le 1 .$$

A statistical operator D fulfills the inequality $D^2 \le D$, where the equality sign holds if and only if D is of rank one, i.e. $\lambda_1=1$ and $\lambda_j=0$ for $j \ne 1$. A statistical operator D contains the same information as an individual state if and only if D is idempotent, $D=D^2$, so that D is a projection operator, projecting on the one-dimensional subspace containing the state vector. In this case the ensemble is homogeneous (i.e. all its members are in the same individual state), and we speak of a *pure statistical state*. A statistical state which is not pure is said to be *mixed*. According to the spectral theorem, *every mixed state is a convex linear combination of pure states*. In contrast to the set of all individual states, the set of all statistical states is a *convex set*. That is, if D_1 and D_2 are two different statistical operators, then every convex linear combination $D=cD_1+(1-c)D_2$, $0<c<1$, is again a statistical operator. A statistical operator D which cannot be decomposed in this way is called a pure statistical operator. A pure state is an extremal point of the convex set of all statistical states.

Mixed states do not provide complete information about the individual states of the constituting elements of a Gibbsian ensemble. Moreover, ensembles that are created by mixing *different* collections of pure states may have the same statistical operator. That is, a statistical operator cannot distinguish between ensembles that are objectively different but nevertheless give rise to the same expectation values. *The knowledge of a mixed statistical operator does not imply the knowledge of the kind of ensemble to which it refers.* There exist infinitely many different ensembles having the same mixed statistical operator, so that a consistent statistical interpretation of statistical operators is possible only if we go back to the interpretation by individual states. The often used statistical interpretation of the statistical operator via its spectral decomposition is just one of many possible versions. There is no reason for requiring the mutual orthogonality of the individual states of the constituting subsytems of an ensemble. Clearly, it is possible to prepare a Gibbsian ensemble by mixing kinematically identical systems that are in arbitrary (i.e. nonorthogonal) individual states. This situation is characteristic for quantal systems and due to the fact that the cone of all unnormalized statistical states is *not* a simplex. In fact, *a system is classical if and only if every mixed state has a unique decomposition into pure states* (Neumann, 1974).

The duality of statistical states and observables

Conceptually, statistical states are directly related to statistical ensemble expectation values. For every bounded observable A the statistical expectation value $\rho(A)$ with respect to an ensemble characterized by the statistical operator D is defined by

$$\rho(A) = \mathrm{tr}(DA) \; .$$

A linear functional ρ on the algebra $\mathcal{B}(\mathcal{H})$ of all bounded operators acting on the Hilbert space $\mathcal{H}$ induces a σ-additive measure μ_A for every spectral resolution E_A,

$$\mu_A(B) = \rho\{E(B)\} \text{ for every Borel set } B,$$

if and only if ρ is generated by some density operator D,

$$\rho(A) = \mathrm{tr}(DA) \quad \text{for every} \quad A\in\mathcal{B}(\mathcal{H}) \; .$$

In pioneer quantum mechanics the set of all unnormalized statistical states is taken as the positive cone of the Banach space $\mathcal{B}_1(\mathcal{H})$ of all nuclear operators on some complex separable Hilbert space. The norm of the Banach space $\mathcal{B}_1$ is taken to be the trace norm $\|\cdot\|_1$. The totality of continuous linear functionals on $\mathcal{B}_1$ is a Banach space under the induced supremum norm, and is called the dual $\mathcal{B}_1^*$ of $\mathcal{B}_1$. It is well-known that the dual of $\mathcal{B}_1(\mathcal{H})$ equals the algebra $\mathcal{B}(\mathcal{H})$ of all bounded operators acting on $\mathcal{H}$,

$$\mathcal{B}_1(\mathcal{H})^* = \mathcal{B}(\mathcal{H})$$

where the norm induced on $\mathcal{B}(\mathcal{H})$ equals the operator norm $\|\cdot\|$ (compare e.g. Schatten, 1960, p.47). In pioneer quantum mechanics an *observable* is represented by a (bounded or unbounded) self-adjoint operator acting on the separable Hilbert space $\mathcal{H}$ of the state vectors. Conversely, it is assumed that every self-adjoint operator in $\mathcal{B}(\mathcal{H})$ represents an observable. With this, pioneer quantum mechanics contains the following *duality between statistical states and observables:*

The set of all unnormalized statistical states is given by the positive cone of the Banach space $\mathcal{B}_1(\mathcal{H})$ *of nuclear operators on a separable complex Hilbert space. The algebra of observables is the algebra* $\mathcal{B}(\mathcal{H})$ *of* all *bounded operators, which equals the dual* $\mathcal{B}_1(\mathcal{H})$ *of the statistical state space* $\mathcal{B}_1(\mathcal{H})$.

Warning: Mathematicians call an element ρ of the *dual* $\mathcal{B}^*$ of the algebra $\mathcal{B}$ a *state* if it is positive (i.e. if $A\geq 0$ implies $\rho(A)\geq 0$) and has norm 1 (so that $\rho(1)=1$.) Clearly, the statistical states of pioneer quantum mechanics are normalized positive elements of $\mathcal{B}^*$ but the converse is not true. The statistical states are precisely the so-called *normal* normalized positive elements of $\mathcal{B}^*$. A positive linear functional $\rho\in\mathcal{B}^*$ is said to be normal if it is completely additive, i.e. if $\rho(\Sigma_\alpha P_\alpha)=\Sigma_\alpha\rho(P_\alpha)$ for every family $\{P_\alpha\}$ of mutually orthogonal projectors in $\mathcal{B}$. This property is important for the probabilistic interpretation of linear functionals on the algebra $\mathcal{B}$ of observables since $\rho\in\mathcal{B}^*$ induces a σ-additive probability measure for every spectral resolution with values in $\mathcal{B}$ if and only if the state ρ is normal. The set of *normal* linear functionals on $\mathcal{B}$ is called the *predual* $\mathcal{B}_*$ of $\mathcal{B}$ since the dual of $\mathcal{B}_*$ equals $\mathcal{B}$. So we have

$$\mathcal{B}_1 = \mathcal{B}_* \subseteq \mathcal{B}^* \quad , \quad (\mathcal{B}_*)^* = \mathcal{B} \quad .$$

Moreover, $\mathcal{B}^*=\mathcal{B}_*$ if and only if the Hilbert space $\mathcal{H}$ is finite-dimensional.

The algebra $\mathcal{B}(\mathcal{H})$ is *irreducible*, that is, the only operators that commute with every operator in $\mathcal{B}(\mathcal{H})$ are the multiples of the identity operator. This property characterizes the algebra of observables of pioneer quantum mechanics, and is referred to as von Neumann's (1932) *irreducibility postulate of pioneer quantum mechanics.*

Remark: In the case of compound systems, the *principle of indistinguishability of identical particles* requires a modification of the formulation of the irreducibility postulate of pioneer quantum mechanics. Consider a system of n identical Galileons of spin s. If $\mathcal{H}_1$ is the Hilbert space associated with a single Galileon, then not every normalized vector in $\mathcal{H}_n=\mathcal{H}_1\otimes\mathcal{H}_1\otimes\ldots\otimes\mathcal{H}_1$ (n times) is a state vector. In agreement with the known experimental material, pioneer quantum mechanics restricts the state space associated with n identical Galileons with half-integer spin to the *antisymmetric tensor product* $\mathcal{H}_n^A\subset\mathcal{H}_n$ (containing all antisymmetric vectors of $\mathcal{H}_n$), and in the case of n Galileons with integer spin to the symmetric tensor product $\mathcal{H}_n^S\subset\mathcal{H}_n$ (containing all symmetric vectors of $\mathcal{H}_n$). Von Neumann's irreducibility postulate then means that the algebra of observables equals $\mathcal{B}_n^A=\mathcal{B}(\mathcal{H}_n^A)$ and $\mathcal{B}_n^S=\mathcal{B}(\mathcal{H}_n^S)$, respectively. With respect to $\mathcal{H}_n$ and $\mathcal{B}(\mathcal{H}_n)$, the algebras $\mathcal{B}_n^A$ and $\mathcal{B}_n^S$ are reducible, and there are nontrivial operators in $\mathcal{B}(\mathcal{H}_n)$ which commute with all operators of $\mathcal{B}_n^A$ and $\mathcal{B}_n^S$, so that we may speak of *superselection rules* due to the indistinguishability of Galileons.

The duality between statistical states and observables implies that states and observables of pioneer quantum mechanics transform contragrediently under time evolution so that the Schrödinger picture is equivalent to the Heisenberg picture. In the *Schrödinger picture* observables are fixed while the states vary with time. For a closed system, the time evolution of a statistical operator $D \in \mathcal{B}_1(\mathcal{H})$ is given by

$$D \to D_t = U_t D U_t^* \in \mathcal{B}_1(\mathcal{H})$$

or as differential equation by

$$i\hbar \dot{D}_t = [H, D_t]_- \quad , \quad D_0 = D \quad ,$$

where $\{U_t | t \in \mathbb{R}\}$ is the unitary time-evolution group, with $U_t = \exp(-itH/\hbar)$. The statistical expectation value $\langle A \rangle_t$ of an observable $A \in \mathcal{B}(\mathcal{H})$ at time t is given by

$$\langle A \rangle_t = \mathrm{tr}\{D_t A\} \quad .$$

The unitarity of U_t and the cyclicity of the trace imply that this relation can be written as

$$\langle A \rangle_t = \mathrm{tr}\{D A_t\}$$

with

$$A_t \overset{\mathrm{def}}{=} U_t^* A U_t \quad \in \mathcal{B}(\mathcal{H})$$

so that the statistical operator is fixed while the observables vary with time.

3.4 THE COPENHAGEN INTERPRETATION OF PIONEER QUANTUM MECHANICS

The basic ideas of the Copenhagen interpretation

The first physical interpretation of quantum mechanics is due to Heisenberg (1927) and Bohr (1928). These two papers, both written in Copenhagen, were rather vague and programmatic. The interpretation of quantum mechanics subsequently worked out on this basis by Bohr and his coworkers is often called *the Copenhagen interpretation*. This terminology is customary but unfortunate. *There is no uniform opinion of what the Copenhagen interpretation should be*. For example, Heisenberg's (1958) book "Physics and Philosophy" is often regarded as an authoritative presentation of the Copenhagen point of view. However, Leon Rosenfeld (1960) - one of the closest coworkers of Niels Bohr - has claimed that "*...certainly not one of the physicists now working in Copenhagen would subscribe to the general philosophical attitude underlying this account*". Not only did Bohr and Heisenberg defend rather different views in their first papers but these interpretations themselves went through essential alterations and developments in the course of time. The famous article by Einstein, Podolsky and Rosen (1935) and the critique by Fock (1951, 1957, 1958) led to a significant change of the viewpoint of Bohr (1935, 1958) (compare also Feyerabend, 1962; and Witmer, 1967), and Heisenberg's philosophical opinions changed (compare Heelan, 1965) as well.

Additional references: fundamental are the papers by Bohr (1928, 1935, 1936, 1948, 1949, 1958), Heisenberg (1927, 1930, 1955, 1956b, 1958), Pauli (1933, 1948, 1950, 1954a, 1955, 1957, 1958), Rosenfeld (1953, 1957, 1961), Fock (1951, 1957, 1958, 1965, 1969). For the history of the Copenhagen view compare the two books by Jammer (1966, 1974). An elaboration of the Copenhagen interpretation from a pragmatic S-matrix viewpoint is given by Stapp (1971). Stapp (1972) attempted to capture the essence of the Copenhagen view and added some approving letters by Heisenberg and Rosenfeld.

In the following we shall not consider the early versions of the Copenhagen view but rather discuss the result of the long-lasting critical revision and development of Bohr's ideas. It should be noted that expressions like "uncontrollable interaction", "duality of waves and corpuscles" which apparently seemed to be so important in the older literature, no longer occur in Bohr's papers, and that unnecessarily subjective phrases have been eliminated. Similarly, Heisenberg's uncertainty relations are no longer at the heart of the interpretation but are simple consequences of the basic mathematical formalism.

Bohr's own writings are remarkably careful but highly implicit, elusive and subtle; the clearest presentation of his view is his last paper (Bohr, 1958). We consider this paper together with the profound eludications by Pauli (1954a, 1957, 1958), Fock (1957, 1965) and Heisenberg (1958) as representative of the final Copenhagen view and compress some of the essential ideas into the following theses:

1. Quantum mechanics refers to individual objects.
2. The probabilities of quantum mechanics are primary.
3. The placement of the cut between observed object and the means of observation is left to the choice of the experimenter.
4. The observational means are to be described in classical terms.
5. The act of observation is irreversible and creates a document.
6. The quantum jump is a transition from the potentially possible to the actual.
7. Complementary properties cannot be revealed simultaneously.
8. Pure quantum states are objective but not real.

In the following, a short exposition of these main points is given.

First thesis: Quantum mechanics refers to individual objects

All varieties of the Copenhagen view assume that the state vector refers to an individual system and *not* to an ensemble of similarly prepared systems. According to the Copenhagen view, the probabilities of quantum mechanics are theoretical object probabilities, that is, *potential probabilities* (Fock, 1957).

Second thesis: The probabilities of quantum mechanics are primary

It is impossible to predict an individual quantal event. If the state of a system is known, we can in general make only probabilistic predictions about the outcome of future observations. Individual quantal events are not causally related. "*Probability is to be regarded as an essential element of the description and not as an indication of the incompleteness of our knowledge*" (Fock, 1969). Because the quantum probabilities are a basic trait of the world, they are called *primary* (Pauli, 1954a). Primary probabilities are *not reducible* to deterministic laws of nature and they have no bearings on the knowledge or ignorance of the observer.

At the fifth Solvay meeting in 1927, Dirac proposed to speak of a choice on the part of *nature*: "*Le choix, une fois fait, est irrévocable et affectera tout l'état futur du monde. La valeur de n choisie par la nature peut être déterminée par l'expérience et les résultats de toute expérience sont des nombres décrivant de pareils choix de la nature.*" (Solvay, 1928, p.262). However, this terminology was rejected by Heisenberg (ib., p.264) and by Bohr (1949, p.223) by the argument that there was no sense in talking about nature making a choice.

Primary probabilities refer to potential possibilities (Fock, 1965); thus, a series of experiments leads to a statistics with a well defined probability distribution. According to the Copenhagen view, *quantum mechanics allows for the maximal complete description of individual systems*; all information about an individual system can be deduced from its state vector. There does not exist a still more detailed (unknown to us until now) description of single systems. Pioneer quantum mechanics is *not* only a statistically correct but incomplete theory.

Third thesis: The placement of the cut between observed object and the means of observation is left to the choice of the observer

According to the Copenhagen view, the physical world has to be separated into two parts, the observed object and the observing system. Heisenberg claims that there exists a sharp *cut* between the observational means on the one hand and the observed system on the other. This separation can be made in an infinite variety of ways. The *placement* of the cut is arbitrary (within certain limits) and left to the *free choice* of the experimenter.

It should be noted that Heisenberg and Bohr have strikingly different notions about the measurement process (compare also Feyerabend, 1962, p.250 and p.271). Bohr's view was much more subtle than that of Heisenberg but also very radical and amazing. According to Bohr, quantum mechanics exclusively refers to propositions about macroscopic classical systems; for Bohr it is impossible to abstract from the experimental equipment and accordingly there is no atomic object to be observed. Bohr never refers to a "cut" but proposes to restrict "*the use of the word phenomenon to refer exclusively to observations obtained under specified circumstances, including an account of the whole experiment*" (Bohr, 1948). This leads to the necessity to consider an experiment as an *undivided whole* that includes the observational means. According to Bohr, quantum systems should not be thought of as having properties detached from the experimental arrangement so that in Bohr's view the state vector Ψ does not refer to an isolated microphysical system but to such a system plus the entire experimental apparatus used in making an observation. Bohr never speaks of quantum systems as constituted of interacting particles which exist separately from each other, he never mentions instruments of observing quantum systems. "*There is no quantum world*" is a famous saying, attributed to Bohr by Petersen (1963).

Fourth thesis: The observational means are to be described in classical terms

On the empirical level the concepts of naive realism and of plain language are to be used, " *since the word 'experiment' can in essence only be used in referring to a situation where we can tell others what we have done and what we have learned*" (Bohr, 1948). Bohr considered this fact as a *logical necessity*, while Heisenberg was more inclined to see it as a sociological fact. Our empirical knowledge has to be formulated classically and all the evidence of quantum phenomena has to be expressed in classical terms. From this Bohr concludes that all observational means (measurement apparatus, means of registration, human observers) have to be described in *purely classical terms*. While Bohr (1958) demands that the measurement apparatus has to be described with the aid of classical physics, explicitly excluding the quantum of action, Fock (1957) only requires that the measurement apparatus allows a sufficiently exact classical description for the purpose under consideration.

Fifth thesis: The act of observation is irreversible and creates a document

As only something that has actually happened can be communicated unequivocally, an observation can be regarded as complete only if there exists a *macrophysical document*. The observed result must therefore be registrated permantly in objective reality: *The act of observation shows an intrinsic irreversibility* (Bohr, 1958). If the final state of the measurement could be connected in an entirely causal manner with the initial state, then one could regain the initial state from the final state by interference or inversion of motion. The increase of entropy that occurs during the observation deprives us of precisely the information we would need to build up the proper interference. In contrast to classical measurements, the total entropy of object *and* measurement apparatus increases in quantum mechanics (Teller, 1961). On that account, the problem of the quantum-mechanical measurement process is closely related to the explanation of the occurence of irreversible phenomena in the framework of a reversible mechanics.

Sixth thesis: A quantum jump is a transition from the potentially possible to the actual

Every observation is an intervention of indeterminable extent both into the observational means as well as into the observed system, and interrupts the causal correlations (Pauli, 1950). Thus, the state of the object is in general changed by a measurement. An ideal measurement of an observable A with the (discrete and not degenerate) eigenvalues a_i and the eigenvectors α_i in a system in the state Ψ produces the result a_i with the probability $|\langle\alpha_i|\Psi\rangle|^2$. If the result a_i is realized, then after the measurement the system is in the pure state α_i. The quantum jump

$$\Psi = \sum_k \langle\alpha_k|\Psi\rangle\alpha_k \to \alpha_i$$

is also called the *reduction of the wave packet*. According to the Copenhagen view, this reduction does not follow from the quantal equation of motion but is to be interpreted *as a new type of natural law.*

In classical physics the time evolution is uniquely determined; all that is potentially possible is also realized. There is therefore no need to distinguish between the potentially possible and the actual. In quantum mechanics on the other hand one must carefully distinguish between potentiality and events. Deterministic laws are valid in the realm of the potentially possible but what is realized factually can be predicted only probabilistically (Heisenberg, 1958; Fock, 1957). From this results a new formulation of the causality principle: *In quantum mechanics, the causality principle is related to the potentially possible and not to the actually realized events.* Mathematically this causality principle is reflected in the well-posedness of the time dependent Schrödinger equation, which has a unique solution for every initial state. In contrast, the realization of one of the potentially possible results (the "quantum jump", the "reduction of the wave packet") is a stochastic process which according to the Copenhagen view, is not governed by the Schrödinger equation.

Seventh thesis: Complementary properties cannot be revealed simultaneously

It is known empirically that microphysical systems have potential properties that cannot manifest themselves simultaneously. Niels Bohr introduced *complementarity* as a way of accommodating concepts which were mutually exclusive but nevertheless necessary for a comprehensive description. Two potential properties are called complementary if their actualization is possible only under mutually exclusive conditions. The existence of complementary properties implies that every physical concept is only defined relative to a certain class of phenomena (Rosenfeld, 1961). Fock (1965) speaks of a *relativity with respect to the tools of observation*. The properties of a quantum object are revealed only in their interaction with classical observational means. Every property has the potential to appear, but its actual appearance depends on the external conditions. Invariably, "*every gain of knowledge of atomic objects by observations has to be paid for by a loss of other knowledge. ... Which knowledge is obtained and which other knowledge is irrevocably lost is left to the free choice of the experimentor, who may choose between mutually exclusive experimental arrangements*" (Pauli, 1948).

Eighth thesis: Pure quantum states are objective but not real

The classical ideal of an objective description of nature as suggested by Newtonian mechanics, electrodynamics and relativity theory, has been characterized by Einstein (1953a) as follows: "*Es gibt so etwas wie den realen Zustand eines physikalischen Systems, was unabhängig von jeder Beobachtung objektiv existiert und mit den Ausdrucksmitteln der Physik im Prinzip beschrieben werden kann*". This narrow view of the notion of reality is *not* fulfilled in quantum mechanics but according to Pauli (1955) there still remains "*an objective reality inasmuch as these theories deny any possibility for the observer to influence the results of a measurement once the experimental arrangement is chosen. Therefore particular qualities of an individual observer do not enter the conceptual framework of the theory. ... In this wider sense the quantum-mechanical description of atomic phenomena is still an objective description, although the state of an object is not assumed any longer to remain independent of the way in which the possible sources of information about the object are irrevocably altered by observations. The existence of such alterations reveals a new kind of wholeness in nature, unknown in classical physics, inasmuch as an attempt to subdivide a phenomenon defined by the whole experimental arrangement used for its observation creates an entirely new phenomenon."* The irreversibility inherent in the act of observation causes the description of quantum events to have an objective character (Bohr, 1958).

Heisenberg tends to a more subjectively colored interpretation: "... *we are finally led to believe that the laws of nature which we formulate mathematically in quantum theory deal no longer with the particles themselves but with our knowledge of the elementary particles*" (Heisenberg, 1956b). Now we must in any theory carefully distinguish between objective and subjective elements. Objective elements relate to the intrinsic properties while subjective elements refer to our knowledge about the systems. While according to Heisenberg the probabilities of a mixture of pure states are secondary and represent an incomplete knowledge, Heisenberg (1955) also accepts unequivocally the objective character of pure quantum states. But the identification of objective and real - permitted in classical theories - must be given up. According to Heisenberg, a pure state represents the potentially possible in an objective way, that is, in a way independent of the observer. That being so, "*the state of the closed system represented by a Hilbert vector is indeed objective but not real*" (Heisenberg, 1955).

Weizsäcker (1941, 1961) takes a logically possible but extremely subjective position: "*Man vermeidet also alle scheinbaren Paradoxien, wenn man den Zustandsvektor konsequent ... als eine Angabe eines Wissens deutet*". Furthermore: "*Menschen, die Verschiedenes wissen, benützen daher ohne Gefahr eines Widerspruchs verschiedene Zustandsvektoren*" (Weiszäcker, 1961). The same point of view is taken by Süssmann (1957, 1958). Accepting this view, one indeed eliminates most of the logical difficulties but only by introducing a new problem: Why should the subjective state vectors satisfy the time-dependent Schrödinger equation?

The subjective interpretation of the secondary probabilities is consistent with Heisenberg's (1958) view that irreversibility is not something entirely objective but ensues from the incomplete knowledge of the observer. The sudden change of the mathematical description on the occasion of a quantum jump then does not reflect anything but the sudden change of our knowledge.

The subjectitivistic way of speaking Heisenberg prefers is not necessarily in conflict with the objectivistic language of Pauli. The notions "objective" and "subjective" are problematic and do not essentially stand in an exclusive relation. Specifying all assumptions and qualifications, there exists in quantum mechanics an intersubjective agreement, so that one can also use an objectifying language.

A logical analysis of Scheibe (1964) shows that in the framework of classical logic the contingent propositions of quantum mechanics cannot be ontic but only epistemic. That is, if we assume the validity of Boolean logic for the properties, then the contingent propositions do not refer to properties of the physical object but to our *knowledge* of them.

Critique of the Copenhagen interpretation

Since its advent, the Copenhagen interpretation has been attacked from very different positions. The vast literature against the Copenhagen view is mostly polemic, often even superficial and boring (for a bibliography, compare Scheibe, 1967, and Nilson, 1976). Most attempts to deviate basically from the Copenhagen view have been motivated philosophically or ideologically and do not aim at an understanding of new scientific problems (compare the review given by Jammer, 1974). Again and again, the expositions of Bohr's view are distorted beyond recognition. An outstanding counterexample is the most instructive critique by Einstein (1936, 1948, 1949, 1953a, 1953b). The controversy between Einstein and Bohr ranks as one of the great philosophical debates in the history of science (compare also Bohr, 1949).

In spite of the fact that most criticism of the Copenhagen view is shallow in comparison with Bohr's insight, we should accept Benjamin Disraeli's dictum that *"it is the duty of Her Majesty's loyal Opposition to oppose"*. It can hardly be denied that Bohr tended to be one-sided and to emphasize excessively some novel aspects of quantum theory (compare also the critique by Feyerabend, 1958a, and Hooker, 1972b). In the following, we restrict ourselves to a discussion of those points that are of direct relevance for theoretical chemistry and theoretical biology. The most penetrating objection against quantum mechanics was formulated in a famous paper by Einstein, Podolsky and Rosen (1935). Because this so-called Einstein-Podolsky-Rosen paradox goes to the heart of any interpretation of quantum mechanics, we discuss it separately in section 3.7.

The role of the observer in the Copenhagen interpretation

Bohr insisted upon the need to include the observational means into the description of the experiment. It is undisputed that the observational means in quantum mechanics play a role that is utterly different from the role they play in classical physics. Nevertheless, many scientists have the impression that this logically incontestable standpoint is often maintained much too dogmatically. At any rate, Bohr's view hardly agrees with daily laboratory routine. Consider for example the decay of a radioactive nucleus as discussed by Blokhintsev (1968). Suppose we produce at time $t=0$ a radioactive nucleus with the mean lifetime τ and leave the system to its fate. From experience we know that we barely need to specify the environmental conditions. After a time T that is vastly greater than the mean lifetime τ, $T \gg \tau$, we can surmise with overhelming probability that the nucleus has disintegrated. Of course, we can carry out a control measurement at $t=T$. If we find the nucleus disintegrated, then it is intuitively absurd (though logically possible) to assume that the control measurement might have something to do with the radioactive decay. What, then, is the role of the observer? This is hardly a question of logic but of the admissibility of an easy way of speaking. It would be desirable if the interpretation of quantum mechanics would allow us to say "the nucleus is decomposed", without having to refer to an observer. This demand for a less operationalistic language becomes imperative in a theory of molecular biological evolution. This evolution happened before there were any observers and it happens today in spite of observers.

There exist objective and real properties of quantal systems

Bohr (1948) emphasized that "*an inherent element of ambiguity is involved in assigning conventional physical attributes to atomic objects*". This careful statement is often exaggerated, as in the following quotation: "*in the usual interpretation of quantum theory, an atom has no properties at all when it is not observed*" (Bohm, 1957, p.92). This overstatement is wrong. Bohr certainly does not deny - but neglects to stress - the objectivity and reality of such properties as the mass, the charge, the spin of an atomic system. These properties exist objectively and factually and are independent of any observation (compare also Fock 1951, 1958). Such properties that are invariant with respect to all particular observational contexts are called *invariants* by Born (1953). According to Born, the idea of invariants provides the key to a rational concept of reality.

The existence of objective and real properties is of outstanding importance for chemistry and biology. The larger a molecular system becomes, the more such properties can exist. The proper way of dealing with Born's invariants is to introduce classical observables into quantum mechanics (compare chapter 4).

What is a classical description?

The main flaw of the Copenhagen interpretation is the inadequate elucidation of the meaning of the word "classical". Rosenfeld (1957, p.52) considers it as altogether obvious that the fundamental concepts of quantum mechanics have to be formulated within the framework of a classical theory. But it is in no way clear a priori what a classical theory should be. Rosenfeld and Bohr regard it as obvious that the language of classical physics (in the sense of Newton and Maxwell) is the proper basis for the formulation of quantum mechanics. However, *this assumption is neither obvious nor undisputed.* It is imperative that the metalanguage used be compatible with classical logic. The only feature required of a classical mode of description is *"the assumption of complete independence of all physical processes of the conditions of observation"* (Fock, 1969).

Referring to the principle of correspondence, it is often maintained that classical mechanics is a limiting case of quantum mechanics. This assertion is wrong. Quantum mechanics and classical mechanics are fundamentally different theories. Bohr categorically requires that the observational means must *not* be described by quantum mechanics. By this prerequisite, the objective and real character of the measurement documents is guaranteed. Bohr considers the classical and the quantal description as mutually exclusive in the sense of a generalized principle of complementarity (compare also Shimony, 1963). According to Bohr, the classical nature of the means of observation is an indispensable condition for the formulation of quantum mechanics. From Bohr's standpoint it is basically impossible to deduce classical mechanics from quantum mechanics (compare also Feyerabend, 1968).

To many theoreticians, this view is not convincing. As the mathematical formalism does not compel us to distinguish conceptually between object and observational means, it has repeatedly been tried to merge object and observational means into a single quantal system. The best known attempts of this kind are those by von Neumann and by Everett to which we shall come back in detail. However, a quantum-mechanical description of classical systems hits deep-rooted difficulties so that for many theoreticians the consistency of Bohr's view is not obvious. Fock argues that one should not formalize the measurement process and states: "*Ich glaube, man muss mehr dialektisch denken, und wenn man das tut kann man auch in der gleichzeitigen Benützung der klassischen Beschreibung des Messapparats und der quantenmechanischen Beschreibung des beobachteten Systems keinen Widerspruch erblicken*" (Fock, 1971).

Does the state function describe a single system or an ensemble?

Since the interpretation of quantum mechanics in terms of individual systems leads to profound (but by no means insoluble) problems, there is a tendency among physicists to switch to an ensemble interpretation. An ensemble interpretation refers to the application of quantum mechanics to a statistical ensemble of infinitely many systems, all of them constructed, prepared and selected in exactly the same way. Of course, it may often be convenient *to imagine* that we have an ensemble of equivalent, uncorrelated systems, and are able to carry out a measurement on this ensemble. From the individual point of view, such a use of virtual statistical ensembles is always legitimate. However, the question at issue is whether quantum mechanics is *exclusively* an ensemble theory, that is, whether a state function never refers to an individual system at all. Surprisingly many physicists do not accept the view that quantum mechanics refers to single systems, and think that only an interpretation in terms of statistical ensembles is logically consistent.

References: The following authors fight for the exclusiveness of the ensemble interpretation: Einstein (1936), Kemble (1937, 1951), Blokhintsev (1953, 1966, 1968), Ludwig (1955, 1961a), Margenau (1963a, 1963b), Groenewold (1964, 1971), Park (1968a, 1968b), Ballentine (1970), Elsasser (1968, 1969a, 1971, 1973).

The usual argument is that quantum mechanics allegedly makes no predictions that bear on individual systems but that it predicts only statistical frequencies. But it is not true that quantum mechanics cannot predict anything for a single system. The inability to accept an individual interpretation often has its roots in a time-honored ideology, the determinism of the nineteenth century. Or in the words of Albert Einstein: "*Gott würfelt nicht*". From a historical point of view it may be understandable that it is less arduous to accept a statistical description than to accept a truly stochastic law. A further reason might be that the original amalgamation of the frequency interpretation of probability theory with quantum mechanics was not well thought out. Again, this is understandable historically: at the birth of quantum mechanics both the mathematical theory and the interpretation of probability were in a rather bad shape. The often heard remark that empirically only an ensemble interpretation can be corroborated is not correct. Because of the probabilistic nature of quantal events, a single observation is indeed insufficient. Yet, nobody is performing an infinite sequence of experiments as required by the ensemble interpretation. The usual laboratory procedure is to do just a *few* experiments. The empirical result is always a *finite random sequence*, a concept which makes no sense in the statistical interpretation but is a well defined and operational concept in the individual interpretation of algorithmic probability theory (compare e.g. Schnorr, 1971b).

Since the individual interpretation is of great importance for a rational theory of large molecular systems and for theoretical biology, we want to stress that *there are positively no logical difficulties connected with the individual interpretation of quantum mechanics* and that this point of view can be carried through consistently. "*The deeper reason for the circumstance that the wave function cannot correspond to any statistical collective lies in the fact that the concept of the wave function belongs to the potentially possible (to experiments not yet performed), while the concept of the statistical collective belongs to the accomplished (to the results of experiments already carried out)*" (Fock, 1952, 1957).

The hunt for hidden variables

Pioneer quantum mechanics seems unsatisfactory to some philosophers and to some physicists. They accept its elegant formalism, admit its outstanding predictive power but claim that it does not provide an intuitive picture of the world. Ever since the inauguration of quantum mechanics it has been suggested that the stochastic nature of quantal events is solely to be ascribed to an alleged incompleteness of quantum mechanics. In saying this it is assumed that there exist dynamical variables which are not observable and which are not considered in quantum mechanics but which would uniquely determine individual quantal events.

The ultimate aim of such theories with hidden variables seems to be the recovery of the determinism of classical physical theories of the nineteenth century. A large number of hidden variable theories have been proposed, but not one has been successful.

References: For a review and the deeper motivation of hidden variable theories compare : Freistadt (1957), Bohm (1962), Bub (1968b), Capasso et al. (1970), Belinfante (1973), Jammer (1974), Peña-Auerbach and Cetto (1977), Claverie and Diner (1977), Ghirardi et al. (1978).

However, there is one aspect of the quest for hidden variables that is of fundamental interest. What precisely are the conditions under which it is possible to show that quantum mechanics does not allow the introduction of hidden variables? The earliest result is due to von Neumann (1932) who has shown that under certain (and, as we know today, unnecessarily restrictive) assumptions, no hidden variable theory can reproduce the results of quantum mechanics. Many objections have been raised against this proof but there are several modern proofs that are free from these objections and that give exact mathematical conditions for the impossibility of embedding quantum mechanics into a classical theory.

References: Pauli (1953), Gleason (1957), Weidlich (1960), Jauch and Piron (1963, 1968), Kamber (1964), Dombrowski and Horneffer (1964), Bell (1964, 1966, 1971a), Zierler and Schlessinger (1965), Bohm and Bub (1966a, 1966b), Kochen and Specker (1967), Misra (1967), Gudder (1968a, 1968b, 1970a, 1972a), Plymen (1968c), Turner (1968), Bub (1969, 1973, 1974, 1976), Wigner (1970a), Ochs (1971, 1972b), Deliyannis (1971b), Bugajska and Bugajski (1972b), Demopoulos (1976).

These discussions have led to an essential clarification of the foundations of quantum mechanics; furthermore, a different point of view has evolved. Because ad hoc hidden variable theories *can* be constructed, it is pointless just to disprove the existence of such theories. The crucial question is: what is the price we have to pay for such an undertaking? A very interesting result by Bell (1964) shows that the price is very high indeed: *no theory can simultaneously comply with the demands of causality, individuality and locality, and be compatible with the statistical predictions of quantum mechanics* (to within a few percent). Bell's theorem is the deathblow to all reasonable hidden variable theories, provided the results predicted by quantum mechanics are approximately correct. So, an interesting feature of hidden variable theories is their potential to suggest new and stringent tests of quantum mechanics. All experimental tests carried out so far have confirmed quantum mechanics and given strong evidence against any kind of local hidden variable theories.

References: Papaliolios (1967), Kocher and Commins (1967), Clauser et al. (1969), Pearle (1970), Kasday (1970, 1971), Fox and Rosner (1971), Fox (1971), Freedman and Clauser (1972), Clauser (1972, 1974, 1976), Shimony (1971), McGuire and Fry (1973), Fry and Thompson (1976), Lamehi-Rachti and Mittig (1976).

3.5 THE VON NEUMANN-LONDON-BAUER INTERPRETATION OF PIONEER QUANTUM MECHANICS

Why another approach?

The Copenhagen interpretation is fundamentally pragmatic. According to Niels Bohr "*quantum theory merely offers rules of calculation for the deduction of expectations about observations obtained under well-defined experimental conditions specified by classical physical concepts*" (Bohr, 1963, p.60). The Copenhagen interpretation focuses the attention on systems that are first prepared in a specified manner and later examined in a specified manner. What is "really" happening is considered as metaphysical or as unimportant. "*In our description of nature the purpose is not to disclose the real essence of the phenomena but only to track down, so far as it is possible, relations between the manifold aspects of our experience*" (Bohr, 1931, in the English translation on p.18).

Bohr's view is logically consistent and cannot be rejected lightly. Moreover, the Copenhagen interpretation has proved to be adequate for describing numerous experimental investigations in atomic and nuclear physics. Nevertheless, it has unsatisfactory features. The Copenhagen interpretation *presupposes* the classical level in a form which makes it impossible to *derive* classical aspects of quantum systems, and even to discuss the logical consistency of the separation of the world into an observed system and a classical observing system. There are genuine scientific problems that cannot even be formulated in the framework of the Copenhagen view.

Perhaps the worst flaw of the Copenhagen interpretation is the view that a theory is nothing but an elaborate system of algorithms for computing experimentally measurable numbers. Such an approach is insufficient for *understanding* the behavior of matter; it is not false but it provides an inappropriate setting for a creative theoretician. A black-box theory which reduces science to the mere prediction of future events from past ones may appeal to man the machine maker - homo faber - but what about man the thinker - homo sapiens? Algorithms for calculating numbers are important, but they do not provide *aha!* reactions which are the basis of any creative activity. *A full-blooded interpretation of quantum mechanics should provide an ontological model of nature.*

The orthodox view

Bohr was very careful in his discussion of the epistemology of quantum mechanics but he always avoided formal mathematical discussions. The aversion of the Copenhagen school against any formalization is acridly defended by Rosenfeld (1965, 1968). We cannot share this attitude, for only a formalization protects us against contradictions and unwarranted preconceptions. That the Copenhagen view is free of contradictions is certainly not self-evident.

The Copenhagen view cannot be considered as an interpretation of the formalism of quantum mechanics because it does not enter into it. Furthermore, the Copenhagen view regards quantum mechanics as a generalization of the classical mode of description. According to the Copenhagen view, the measuring instruments *must* be described in classical terms. This attitude was challenged for the first time by John von Neumann in his famous book of 1932 on the mathematical foundations of quantum mechanics. The von Neumann interpretation aims at a *complete* quantal description of *every* system. That is, it is assumed that quantum mechanics can describe microphysical objects as well as all the observational means. However, this view creates a new riddle: the notorious *measurement problem of quantum mechanics*.

We must stress the fundamental difference between Bohr's view and von Neumann's interpretation of quantum mechanics. Von Neumann's interpretation is not a refinement of Bohr's arguments but he proposes an entirely different program. According to Bohr, there are no "microsystems" that can be separated from the measuring instruments; the measured "objects" have a theoretical status entirely different from that of the measuring instruments. Von Neumann's interpretation, by contrast, ascribes the same theoretical status to object systems and measuring instruments, and it makes now sense to speak of an interaction between these two systems. In Bohr's language it is impossible even to talk about a "measurement problem". But in von Neumann's interpretation the discussion of the measurement process is the central issue. Von Neumann's (1932, chapt.6) discourse on the measurement process is not yet a full interpretation of quantum mechanics. But accepting his ideas, we are almost bound to arrive at the fuller picture London and Bauer (1939) have given. Wigner (1963) proposed to call this interpretation the *orthodox view*. Note that some authors who speak of the "orthodox interpretation" have failed to realize the striking difference between Bohr's view and von Neumann's interpretation.

Apart from the fundamental difference concerning the theoretical status of measuring instruments, the orthodox interpretation is in accordance with the Copenhagen view. We shall give all formulations of the von Neumann-London-Bauer interpretation in terms of the individual interpretation while von Neumann sometimes prefers a statistical formulation. Despite the many assertions to the contrary, the use of statistical operators is not more general and brings no conceptual advantages. As stressed by Zeh (1970) - the use of statistical operators assumes a theory of the measurement process, and its use in the discussion of the measurement problem thus amounts to using a circular argument. Furthermore, it is well known that the structure of an ensemble cannot be rederived from its statistical operator alone. This fact leads to severe difficulties in the interpretation of the mixed state at the end of a measurement process. For example, many authors failed to distinguish between mixtures of the first and of the second kind (as introduced by d'Espagnat, 1966a) and so they have been led to erraneous interpretations. If we accept the individual interpretation (which is also accepted by von Neumann), then the only proper way to discuss the measurement problem is in terms of state vectors.

Deterministic and stochastic time evolutions

Pioneer quantum mechanics in the Copenhagen interpretation is a strangely dualistic theory. It has two types of fundamentally different time-evolutions. If the system is closed and does not interact with the observer, then the time-evolution is deterministic and given by a unitary time-evolution operator. As opposed to this deterministic time-evolution, the *stochastic time-evolution* refers to an irreversible measurement process, due to an intervention of the observer, and describes the *"reduction of the wave packet"*. Born's probabilistic interpretation of pioneer quantum mechanics refers to such *induced stochastic transitions* from one state to another state (the so-called "quantum jumps"). According to the Copenhagen view, the stochastic time-evolution does not follow from the formalism of pioneer quantum mechanics but has to be added as a new and independent postulate, called the *reduction postulate* (or, in the terminology introduced by von Neumann, 1932, *projection postulate*). In this interpretation, pioneer quantum mechanics is a *dualistic theory* with respect to the time-evolution. The idea that the stochastic time-evolution might be derived from the deterministic dynamics of the combined object and measuring systems is due to von Neumann (1932). The main result of von Neumann's theory of the measurement process was the proof that both types of time evolution are compatible wherever the cut between object and measuring instrument is placed. In addition, von Neumann (1932) clearly stated, but did not solve, the famous and technically difficult *"measurement problem of quantum mechanics"*.

Explanatory notes: (i) The *causal time evolution* is continuous, applies only to a *strictly closed system*, and is governed by the time-dependent Schrödinger equation. In other words, if Ψ is the state vector of a strictly closed system, then its time evolution is given by a continuous unitary one-parameter group $\{U(t) : t \in \mathbb{R}\}$, $\Psi(0) \rightarrow \Psi(t) = U(t)\Psi(0)$. According to Stone's theorem, this time evolution operator can be represented as $U(t) = \exp(-iHt)$, where H is a self-adjoint operator, called the Hamiltonian of the closed system. Since the inverse of the time evolution operator exists, $U^{-1}(t) = U(-t)$, the causal time evolution is *reversible.*

(ii) *The stochastic time evolution* of a system is *discontinuous* and applies only to *measurements* on this system. If a measurement takes place, the system is no longer separated from the rest of the world because there is an interaction between the object and the measuring apparatus. If the object system, in a state represented by the state vector Ψ, is subjected to a measurement (of the first kind) of an observable A with a simple discrete spectrum and the spectral representation

$$A = \sum_n a_n P_n ,$$

$$P_n = P_n^* \quad , \quad P_n P_m = \delta_{nm} P_n \, , \quad \sum_n P_n = 1 \quad ,$$

$$a_n = a_n^* \quad , \quad a_n \neq a_m \quad \text{for} \quad n \neq m \quad ,$$

then a discontinuous transition $\Psi \rightarrow \Psi_j$ occurs where $\Psi_j = P_j\Psi/\|P_j\Psi\|$. The probability that a measurement gives for the measured value the eigenvalue a_j so that the state after the measurement is Ψ_j, is given by $|\langle\Psi|\Psi_j\rangle|^2$. The stochastic time evolution is not invariant under time reversal and leads to the intrinsic irreversibility of quantal events. The stochastic time evolution represents the act of measurement, it reduces the state of the system from whatever it was prior to the measurement act to an eigenstate of the observable measured. This process describes a "quantal event" (or a "quantum jump"), and is called *"state reduction"* (or "reduction of the wave packet"). Lüders (1951) has given the appropriate modification of von Neumann's formalization for the case that the observable A has a degenerate discrete spectrum. The method proposed by Lüders corresponds to a measurement causing the least possible interference (compare also Süssmann, 1958; Goldberger and Watson, 1964; Bell and Nauenberg, 1966; Herbut, 1969, 1974).

Von Neumann's theory of quantal measurements

In von Neumann's theory a measurement is regarded as an *interaction* between the object system and the measuring apparatus. Both systems are described by quantum mechanics in the individual interpretation. Von Neumann's reasoning refers to a highly idealized abstraction, the so-called measurements of the first kind. A measurement is called of the first kind if it gives the same value when immediately repeated. A measurement of the first kind is therefore a state-preparing operation; *after* such a measurement the system is in eigenstate of the observable measured, and the observable *has* a certain numerical value. In contrast to such idealized measurements of the first kind, actual laboratory measurements as a rule change the state of the object even for an immediately repeated measurement. If this change of state by repeating the measurement is completely controllable, then such a measurement is said to be of the second kind (Pauli, 1933). There is no need to enter the theory of measurements of the second kind because it amounts to an inessential complication only (compare Landau and Peierls, 1931).

The substantial features of von Neumann's theory of quantal measurements can be discussed by using the special case of a measurement of the first kind of an observable represented by a self-adjoint operator A with a purely discrete and nondegenerate spectrum. At the outset ($t = o$), the object system and the measuring apparatus are assumed to be uncorrelated and in pure states.

Remark: This is a very strong assumption. As in von Neumann's theory there are distinct difficulties to get a disentangled final state; the presupposition of a disentangled initial state is not a trivial matter.

Let us assume that initially the object system is in the unknown pure state φ and the apparatus in a certain pure state Φ, so that the initial state of the combined system is given by $\Psi(t=o) = \varphi \otimes \Phi$. It will be expedient to expand the initial state φ of the object system in terms of the orthonormal eigenvectors α_n of the observable A,

$$A\alpha_n = a_n\alpha_n \ , \ \langle\alpha_n|\alpha_m\rangle = \delta_{nm} \quad .$$

so that the initial state is given by

$$\Psi(t = o) = \sum_n c_n \alpha_n \otimes \Phi \ , \text{ with } \ c_n = \langle\alpha_n|\varphi\rangle \ .$$

Assume that the measuring instrument contains a digital indicator with the values $\{b_1,b_2,b_3,...\}$. In quantum mechanics, we have to describe such an indicator by an observable B with the eigenvalues b_n

$$B\beta_n = b_n\beta_n \quad , \text{ where } \quad \langle \beta_n | \beta_m \rangle = \delta_{nm} \quad .$$

Von Neumann (1932) has shown that there always exists a measuring apparatus (i.e. a Hamiltonian H_M) that can interact (Hamiltonian V) with the object system (Hamiltonian H_O) in such a way that the time evolution of the combined system is given by

$$\Psi(t) = \exp\left\{- it\left(H_O + H_M + V\right)\right\}\Psi(o)$$

$$= \sum_n c_n(t)\alpha_n \otimes \beta_n \quad .$$

In this case, the n-th eigenstate of the observable A is uniquely correlated with the n-th eigenstate of the indicator B. That is, if we know the state of the measuring apparatus, we also know the state of the object system after the measurement. But the final state $\Psi(t)$ cannot be interpreted to mean that the digital indicator *is* in one of its possible states. The combined system is still in a quantum mechanical superposition of the pure states $\alpha_n \otimes \beta_n$. Hence the apparatus, *considered as a subsystem*, is *not* in a pure state and the observable B does not have a definite value. If we are interested in the value of the observable B, we have to measure it. That is we have to introduce another measuring instrument with an indicator described by an observable C (with the eigenvalues c_n and the eigenvectors γ_n, so that we get a new combined state

$$\Psi'(t) = \sum_n c_n(t)\ \alpha_n \otimes \beta_n \otimes \gamma_n \ .$$

This iteration shows the consistency of von Neumann's procedure: if we knew the state of the second measuring instrument, we would know the state of the first measuring instrument, hence the state of the object system. But we have not yet performed a measurement because the state $\Psi'(t)$ of the combined system still is an entangled state. A further addition of a new measuring instrument does not help, so it seems to be impossible to terminate this chain of measuring instruments. *According to von Neumann, London and Bauer, the measuring chain is terminated by an act of the consciousness.*

There is no cheap way out of the dilemma of the measuring instruments that observe the measuring instruments but again call for an observer for their results. This dilemma is the result of the superposition principle of quantum mechanics. Von Neumann, London and Bauer propose an ingenious and logically perfect but hair-raising solution. These authors claim that there is an ultimate end of this measuring chain: the consciousness of the observer. According to von Neumann (1932) any part of the world except the consciousness of the observer can be included into the measuring chain. London and Bauer (1939) explicitly have claimed that it is the conscious activity of the human mind which reduces the quantal state. This point of view has been endorsed by Wigner (1962a, 1963, 1969). According to the von Neumann-London-Bauer interpretation, the reduction of the state vector by consciousness is not attributable to the deterministic time evolution of the formalism of quantum mechanics. Quantum mechanics is not supposed to hold for consciousness; the "faculty of introspection" of the human mind is claimed to be responsible for the stochastic time evolution.

Accepting the von Neumann-London-Bauer interpretation of quantum mechanics means that we adopt the world view of epistemological solipsism. To assume that the wave packet is reduced by an act of the consciousness is logically irreproachable but hard to accept. The postulated capacity of the consciousness to reduce the wave packet is not supported by any empirical evidence and may be empirically wrong (compare also Shimony, 1963). On the other hand, at the present state of our knowledge of brain functions, such an assertion cannot be disproved empirically. *The von Neumann-London-Bauer interpretation is both logically consistent and not in contradiction with any known empirical fact.* Therefore, any genuine objection against the von Neumann-London-Bauer interpretation has to be extraphyiscal. Most scientists do not like to adopt the regulative principles of von Neumann, Bauer and London. We would prefer to posit in some sense the existence of an external real world. Certainly, we have to be prepared to find that a naive concept of reality contradicts quantum mechanics but we can still hope that a more sophisticated realism is feasible. In that case we attribute the reduction of the wave packet to some physical process in the measuring instrument rather than to the consciousness of the observer. However, a cheap modification of the von Neumann interpretation is not possible. In the framework of the traditional formalism of quantum mechanics, the following theorem can be proved: *there exists no measurement interaction that can leave an object-apparatus sys-*

tem in a mixture of states, each state being connected with a definite value of the apparatus observable.

References: Komar (1962), Wigner (1963), d'Espagnat (1966b), Earman and Shimony (1968), Fine (1970), Stein and Shimony (1971), Fehrs and Shimony (1974), Shimony (1974).

This theorem holds for pure and mixed states, for measurements of the first and the second kind as well as for the generalized measurements considered by Fine (1969); it is a quite trivial consequence of the superposition principle. The crucial ingredients of the proof are the linearity of the equation of motion and the irreducibility of the algebra of observables. This result rules out any reasonable explanation of the reduction postulate of pioneer quantum mechanics, the only loophole is a weakening of the superposition principle, say by superselection rules. Or, as expressed in plain words by Jordan (1949): "*We are unable to make a clock with a hand which does not always point to a definite figure on the dial. This is a well-known fact, but a fact of which present theory gives no sufficient account.*"

Objections against the reduction postulate

Some authors claim that the reduction postulate is untenable and should be dropped. Despite their assertions to the contrary, these authors do not present a complete and consistent alternative to Bohr's view or von Neumann's interpretation. However, in the framework of von Neumann's interpretation the reduction postulate is indispensable for a complete theory. Any theory with an external observer cannot operate without the reduction postulate, for without the stochastic time evolution there are no events! The only consistent interpretation without external observer and without a reduction postulate is the Everett interpretation, discussed in section 3.6.

A further critique refers to the specialized and idealized character of measurements of the first kind. The Margenau school stresses that measurements of the first kind do not correctly describe the state after measurement if the process of recording the result of the measurement destroys the object system. Although such situations often arise in laboratory experiments, this argument is not a valid objection against von Neumann's theory of measurements. The confusion is due to the very unfortunate terminology common in the quantum theory of measurement. At the risk of stressing the obvious, we may recall that what is meant by a measurement in quantum mechanics is what we have defined as a quantal measurement, and nothing else. Nobody is claiming that a "measurement of the first kind" is a typical example of a measurement as performed by an experimenter in the laboratory. The idea is, that once we have understood this simple situation, we might be able to go ahead and discuss the much more complex actual measurements, using measurements of the first kind as elementary building blocks. Landau and Peierls (1931) have given a theory of measurements of the second kind, a still more general theory of quantal measurements has been worked out by Fine (1969). For a discussion of some simple quantal experiments, compare Lamb (1969). A general and realistic theory of laboratory experiments, however, remains a great desideratum (compare also Primas and Müller-Herold, 1978).

References: The reduction postulate has been criticized by Popper (1935, 1967), Margenau (1936, 1937, 1949, 1958, 1963a, 1963b), McKnight (1957, 1958), Durand (1960), Margenau and Hill (1961), Mould (1962), Albertson (1963), Margenau and Park (1967), Bunge (1967a, 1967b), Park (1968a, 1968b, 1970), Park and Margenau (1968), Gsonka (1971), Fujiwara (1972), Moldauer (1972a, 1972b). Von Neumann's claim that the X-ray-electron collision experiment performed by Compton and Simon (1925) gives direct empirical evidence in favor of the reduction postulate, has been critically examined by Sneed (1966).

For a discussion of the measurement theory in a relativistic framework, compare Schlieder (1968), Hellwig and Kraus (1970b), Antoine and Gleit (1971).

More recently generalizations of von Neumann's measuring scheme have been discussed in a mathematically rigorous way by Davies (1969b, 1970a, 1970b, 1971. 1972, 1976a), Davies and Lewis (1970), Benioff (1972a, 1972b, 1972c, 1973a, 1973b, 1974), Friedman (1972). These studies are related to repeated measurement procedures with a finite or infinite number of steps and with observables having an arbitrary spectrum. Ali and Emch (1974) have introduced generalized observables that cannot account for the fuzziness inherent in actual measurement processes. These "fuzzy observables" reduce to the usual observables when the measurements to which they correspond are of infinite precision.

What is the measurement problem of quantum mechanics?

The measurement process has to explain the existence of quantum jumps, that is, the stochastic change of the state vector during a measurement. At first sight, the stochastic time evolution of the object system seems to be incompatible with the causal time evolution of the combined system consisting of object and measuring instrument. The measurement problem can be considered as resolved if - without adding new principles or ad hoc assumptions to quantum mechanics - we can deduce the stochastic time evolution of the object system from the unitary time evolution of the combined system. Of course, von Neumann, London and Bauer did not solve this problem. They introduced the ad hoc and extra-physical assertion that the reduction of the wave packet is effected by our consciousness. An adequate theory of the measurement process should stop von Neumann's infinite regress at the first true measuring instrument.

In a measurement one considers an object system with a Hilbert space H_1, a measuring system with a Hilbert space H_2, and the joint system with the Hilbert space $H_1 \otimes H_2$ as state space. A quantal measurement of the first or second kind consists of three consecutive stages:

(i) Preparation of an initial state in which the object state φ and the apparatus state Φ are disentangled. That is, after the preparation the initial state $\Psi(t=o)$ is given by the product state vector $\varphi \otimes \Phi$.

(ii) Next the object and the measuring system interact for some time, the effect of this interaction being represented by a *unitary* transformation of the state vector.

(iii) After this interaction the system should again be in an uncorrelated state, so that the final state of the apparatus can be determined by reading the scale of the measuring instrument. The correlation established during the interaction period then allows an inference about the object observables.

Can these three stages be explained by pioneer quantum mechanics? This puzzle is *"the measurement problem of quantum mechanics"*. It is a deep problem, both from the mathematical and the epistemological point of view. Many scientists have expressed the pessimistic view that the fundamental principles of quantum mechanics have only limited validity and cannot properly describe the measurement process. On the other hand, no interpretation of the formalism is complete if there is no satisfactory account of the measurement problem. Moreover, the mathematical questions involved in a proper solution of the measurement problem are crucial in many important topics of the theory of larger molecular systems.

Are the interference terms destroyed by an amplification process?

The relevance of quantum amplification processes in actual measuring instruments and biological systems has been discussed in detail by Jordan (1932b, 1936, 1938, 1941, 1944, 1949) and by Elsasser (1951).

The fact, often and properly commented upon, that all actual measurement instruments contain an amplification device has been confused with the essentials of the quantal measurement process. An amplification process is called for because a single quantal event is too weak to be directly and conveniently perceived by humans. Usually, the energetic amplification factor is so great that the measuring instrument behaves as a *metastable* system. Typical practical examples are cloud chambers, bubble chambers, spark chambers, scintillation counters, Čerenkov counters, photomultipliers, photographic emulsions, stimulation of the optic nerves by a few quanta acting upon the retina, X-ray-induced mutations. These amplification processes have to be clearly distinguished from the first appearance of a quantal event in the objective reality. For example, the exposure of a photographic film induces a primary photochemical process; its result is objectively fixed in reality, and is called the latent image. As we cannot directly see the latent image, we use an amplification process - the development of the film by processing in a bath containing a reducing agent like hydroquinone. Evidently, the development is crucial for practical photography but of no importance for the discussion of the measurement problem of quantum mechanics. The occurence of a quantal event and its amplification to a macroscopic scale are two conceptually quite distinct phenomena which should not be confused.

Again and again, a solution of the measurement problem has been attempted by appealing in some way to the metastable nature of most of the practical measuring instruments. Many authors have suggested to regard the measurement problem as a problem of statistical mechanics and ergodic theory with metastable equilibrium states of the detecting devices. All these investigations - as interesting as they may be in other contexts - do not solve the measurement problem of quantum mechanics. The claim (attributed to Rosenfeld by Daneri et al., 1966) that the amplifier is the key to an understanding of the reduction of the wave packet, is plainly wrong.

References: The following authors have related the measurement problem to the operation of metastable detectors or ergodic behavior: Jordan (1949), Ludwig (1953, 1954, 1955, 1957, 1958a, 1958b, 1961a, 1961b, 1963a, 1963b), Green (1958), Daneri et al. (1962, 1966), Blokhintsev (1966, 1968), van der Waerden (1966), Loinger (1968), Prosperi (1971a, 1971b), Lanz et al. (1971), George et al. (1972), Prigogine et al. (1973).

Compare also the critique by Tausk (1966), Jauch et al. (1967), Bub (1968a, 1971), Krips (1968, 1969a), d'Espagnat (1971, 1974).

The destruction of the interference terms is necessary but not sufficient for an explanation of the reduction of the state vector.

In the individual interpretation of quantum mechanics, the state reduction is given by

$$\sum_n \kappa_n \varphi_n \otimes \Phi_n \rightarrow \varphi_j \otimes \Phi_j \quad .$$

In the statistical interpretation, the state reduction is represented in the weaker form

$$\sum_{nm}\sum \kappa_n \kappa_m^* |\varphi_n \otimes \Phi_n\rangle\langle\varphi_m \otimes \Phi_m| \rightarrow \sum_k |\kappa_k|^2 |\varphi_k \otimes \Phi_k\rangle\langle\varphi_k \otimes \Phi_k| \quad .$$

The state reduction of the individual interpretation implies the state reduction used in the statistical interpretation, but not the other way round. In the statistical interpretation a state reduction is achieved if and only if all interference terms vanish,

$$|\varphi_n \otimes \Phi_n\rangle\langle\varphi_m \otimes \Phi_m| = 0 \quad \text{for } n \neq m \quad .$$

The reduced statistical state

$$\sum_k |\kappa_k|^2 |\varphi_k \otimes \Phi_k\rangle\langle\varphi_k \otimes \Phi_k|$$

is usually interpreted as a classical mixture of pure state. Such an interpretation is not justified because the decomposition of a mixed state into pure states is not unique. The uniqueness of the decomposition of a mixed state into pure states can be enforced if we *assume* that the pure states are pairwise orthogonal (provided all eigenvalues of the statistical operator are nondegenerate). But that is not a reasonable assumption since ensembles consisting of nonorthogonal pure states can easily be prepared. Furthermore, *all* the pure states of the decomposition of the final mixed state of a statistical state reduction exist with the same degree of reality. Hence one cannot say that the measuring apparatus *is* in some *definite* but unknown pure state. There exists a vast literature on models of the measurement process showing that interference terms in certain systems can be neglected for all practical purposes. These models for the measurement process however do not achieve what they pretend to achieve: the state reduction connected with an individual system is not explained at all. *We cannot disregard Aristotle's thesis that the orders of being and of knowing are different*. This deep difficulty is the reason why so many authors withdraw from an individual interpretation .

Although an explanation of the destruction of interference terms does not solve the measurement problem, it is still a crucial question of any theory of quantal measurements. It seems to be clear that the destruction of interference terms has something to do with the "macroscopic nature" and the "essential complexity" of actual measuring instruments. Indeed, it may be exceedingly difficult to distinguish between mixed and pure states of macroscopic systems (compare the examples given by Bohm, 1951, and by Peres and Rosen, 1964a). If two pure states differ from each other in a great many degrees of freedom, the interference terms between these two states are not well defined from an experimental point of view (compare also Wakita, 1960). Accepting a pragmatic point of view, it seems to be reasonable not to distinguish between states whose interference terms cannot be observed with the tools available to us. On that account, Jauch (1964) has proposed to collect all states that are indistinguishable for a classical observer into an equivalence class, called a macrostate. There are many detailed investigations and models that favor such a point of view. The general tenor of these studies is the assumption that interferences that cannot be measured in practice are not observable even in principle. Far from disavowing the substance of the basic idea, we should like to stress that one has to distinguish between *practice* and *principle*. As d'Espagnat (1966b) insists, such a distinction *"is of course necessary since otherwise we would be led to accept the idea that, for instance, neutrinos did not exist during the XIX century, when it was not possible to observe them"*. The effective disappearance of the interference terms is thought to be due to our limited ability to investigate the exceedingly complex correlations of the microstructure of a macroscopic instrument. However, as long as we have no sound theoretical guiding principle, we should be very cautious with claims of this type. We should never underrate the ingenuity of the experimenters of the future. That we do not see any possibility to measure extremely complex correlations does not imply that such correlations do not exist. How complex must a system be to prevent us from distinguishing between a coherent interference and a mixture?

References: The possibility of destructions of interference terms and its relevance for the state reduction has been discussed by: Elsasser (1937, 1953), Groenewold (1946, 1952, 1957, 1964, 1971), Bohm (1951, sect.22), Ludwig (1954, chapt.5), Feyerabend (1957, 1962), Süssmann (1958), Wakita (1960, 1962a, 1962b), Amai (1962, 1963, 1964), Wigner (1963), Shimony (1963) Peres and Rosen (1964a, b), Jauch (1964b, 1968a, 1971), Hack (1964), d'Espagnat (1965, 1971), Furry (1966), Blokhintsev (1966), Tausk (1966), Bohm and Bub (1966a), Weidlich (1967), Bub (1968a, 1971), Grib (1968), Haake and Weidlich (1968), Weidlich and Haake (1969), Krips (1969a, b), Jasselette and Voisin (1970), Freundlich (1972), Hepp (1972), Emch (1972b), Reece (1973), Prosperi (1974), Frigerio (1974), Whitten-Wolfe and Emch (1976).

In order that a system can qualify for a measuring instrument, it has to fulfill a set of stringent conditions. The presence of additively conserved quantities requires that a measuring instrument must be, in a well-defined sense, sufficiently large. Compare: Wigner (1952, 1962b), Araki and Yanase (1960), Yanase (1961, 1964, 1971), Stein and Shimony (1971).

Is the superposition principle universally valid?

The reduction of the state vector is a nonlinear operation. Accepting the reduction postulate implies at the least a reinterpretation of the superposition principle of quantum mechanics.

Von Neumann (1932) has remarked that the intellectual inner life of the individual is extra-observational by its very nature. London and Bauer (1939) have been much more explicit. According to London and Bauer, the observer has a property not shared by any other system in nature: the faculty of introspection. They have supposed that consciousness can indeed perform the reduction of superpositions of quantal states. Therefrom, Wigner (1962a, 1964b, 1969, 1970b) concludes that the superposition principle fails when the consciousness becomes involved in a quantal process. He even speculates that consciousness modifies the usual laws of physics and that the action of the mind on matter is associated with a grossly nonlinear time evolution equation for the state vector. Since the introduction of *ad hoc* modifications can ruin any theory, the proposals of nonlinear time evolutions for quantal states have no power to convince.

Schrödinger's cat

Accepting the unrestricted validity of the superposition principle of quantum mechanics, one is led to grotesque situations. In Schrödinger's (1935b) distinct prose: "*Eine Katze wird in eine Stahlkammer gesperrt, zusammen mit folgender Höllenmaschine (die man gegen den direkten Zugriff der Katze sichern muss): in einem Geigerschen Zählrohr befindet sich eine winzige Menge radioaktiver Substanz, so wenig, dass im Laufe einer Stunde vielleicht eins von den Atomen zerfällt, ebenso wahrscheinlich aber auch keines; geschieht es, so spricht das Zählrohr an und betätigt über ein Relais ein Hämmerchen, das ein Kölbchen mit Blausäure zertrümmert. Hat man dieses ganze System eine Stunde lang sich selbst überlassen, so wird man sich sagen, dass die Katze noch lebt, wenn inzwischen kein Atom zerfallen ist. Der erste Atomzerfall würde sie vergiftet haben. Die Ψ-Funktion des ganzen Systems würde das so zum Ausdruck bringen, dass in ihr die lebende und die tote Katze zu gleichen Teilen gemischt oder verschmiert sind*".

English translation: "A cat is locked up in a steel chamber, together with the following infernal machine (which must be protected against the direct grip of the cat): In a Geiger counter there is a minute amount of a radioactive substance, so little that within an hour may-be one of the atoms decays but equally probably none. If an atom decays, the counter triggers and activates a relay so that a little hammer breaks a flask containing prussic acid. If one has left this whole system for one hour, one would say that the cat is still alive if no atom has decayed. The first decay would have poisoned the cat. The Ψ-function of the whole system would express this in such a way that in it the living and the dead cat are mixed or smeared in equal parts."

Certainly, everybody would say that either the cat *is* alive or the cat *is* dead, but if the chamber is not opened yet, we just do not know. In the von Neumann-London-Bauer interpretation of quantum mechanics we cannot make such a statement because the state reduction is made by the consciousness of an observer. In this sense, we speak of *the paradox of Schrödinger's cat*. Note that the dilemma is not yet resolved by showing that the interference terms between the living and the dead cat are negligible small, because in the closed steel chamber no state reduction has been achieved. Yet, provided the interference terms are no longer observable, no contradiction to our experience can be constructed. *Schrödinger's paradox refers to the regulative principles adopted by von Neumann, London and Bauer.* In our everyday life we decide upon realism and reject solipsism as a regulative principle.

The crux of Schrödinger's paradox is the interpretation of the state $\Psi = \Psi_L + \Psi_D$, where Ψ_L is the state of the living cat and Ψ_D the state of the dead cat. Furthermore, we should be able to explain why the self-adjoint operator $A = |\Psi_L\rangle\langle\Psi_D| + |\Psi_D\rangle\langle\Psi_L|$ cannot be an observable. If A were an observable, we could transform a dead cat into a living cat, $|\Psi_L\rangle = A|\Psi_D\rangle$. Schrödinger's paradox should not be ridiculed, it focusses on the essentials of the measurement problem of quantum mechanics. To be sure, the cat was introduced by Schrödinger in order to dramatize the story. Nevertheless, there is an instructive feature gained by this dramatization. Is it really true that all events of our everyday life are strictly objectifiable? Consider the dying of a poisoned cat in terms of classical biology and medicine. If the concepts "living cat" and "dead cat" refer to an objective reality, then any cat is either living or non-living. Hence we are allowed to ask: at what time exactly does death occur? Admittedly, we can adopt a conventional definition of the moment of death. Any reasonable conventional definition of the moment of death will take into account the present state of medical art. In spite of the fact that "for all practical purposes" we can give a reasonable definition of the moment of death, there is no absolute definition that is independent of the tools available to the physician. Dying is not a fully objectifiable event. It may be worthwhile to ponder this problem, and to ask whether we do not require too much if we want fully objectifiable events in our theories.

3.6 THE EVERETT INTERPRETATION OF PIONEER QUANTUM MECHANICS

A monistic interpretation

Simplicity is the most difficult thing to achieve; thus the most straightforward interpretation of quantum mechanics had to wait until 1957 when Hugh Everett III in his Ph.D. thesis initiated a most fascinating and simple view, called by him the *relative state interpretation*. According to Bohr's view and according to the von Neumann-London-Bauer interpretation, the object of quantum mechanics cannot be the whole universe because the observer is distinct from it. The Everett interpretation asserts that it makes sense to talk about the state of the entire universe. In the following discussions, "universe" always means "universe of discourse", including all measuring instrument and observers. In the Everett interpretation, pioneer quantum mechanics is a self-contained theory, describing the universe with the observer *in* it.

References: The basic papers on the Everett interpretation are Everett (1957) and the assessment of Wheeler (1957); compare also the longer exposition of Everett (1973). Not knowing the papers by Everett and Wheeler, Cooper and Vechten (1969), and Zeh (1970, 1971) have put forth a similar interpretation. For a discussion of Everett's interpretation, compare also DeWitt (1968, 1970, 1971b).

Everett takes the mathematical formalism of quantum mechanics without adding the reduction postulate; no external observer is considered. In many respects, Everett's interpretation is nothing but a consistent formulation of von Neumann's theory of quantal measurements, in which the object plus the apparatus plus the observers are regarded as one single closed system. Measurement-like interactions within the entire system lead to a splitting of the universe into a number of branches, each of them being considered as real as any other one. In each branch a measurement gives a definite result, all possible outcomes of a quantal measurement are realized in separate but real worlds. The universe is constantly splitting into a stupendous number of mutually unobservable but equally real worlds. Of course, if we include ourselves into the universe under discussion, we are ourselves being split. The fact that we do not feel ourselves being split is explained by Everett in the simplest and most natural manner: the laws of quantum mechanics do not allow us to feel the split. The different branches cannot interact, each branch is unaware of the others. Somewhat overstated, we may say that the Everett interpretation aims at a quantum theory without quantum jumps while still explaining why observers encounter such jumps.

The philosophical implications of the Everett interpretation are astounding, one may regard them as ridiculous or as grandiose, but the great elegance with which the Everett interpretation describes scientifically interesting problems cannot be overlooked; it exhibits an internal consistency not present in any other interpretation of quantum mechanics.

The deterministic time evolution is universally valid

The Everett interpretation renounces any postulate concerning a stochastic time evolution. Without exception, all temporal changes are assumed to be governed by the deterministic time evolution, given by the Schrödinger equation

$$i \hbar \frac{\partial}{\partial t} \Psi(t) = H\Psi(t)$$

where the Hamiltonian H is a linear self-adjoint operator. There is no probability interpretation for $|\Psi(t)|^2$. According to Wheeler (1957) "*the word 'probability' implies the notion of observation from outside with equipment that will be described in classical terms. Neither these classical terms, nor observations from outside, nor a priori probability considerations come into the foundations of the relative state form of quantum theory*".

It is assumed that every system that is subjected to external observations can be regarded as a part of a larger isolated system. It is then claimed that it is possible to *derive* Born's probability interpretation from the framework of Everett's interpretation of quantum mechanics (compare Everett 1957, 1973; Hartle, 1968; Graham, 1970, 1973).

Relative states

A perfectly isolated system is in a pure state $\Psi \in \mathcal{H}$. If one divides this system into two subsystems,

$$\mathcal{H} = \mathcal{H}_1 \otimes \mathcal{H}_2 \quad ,$$

then in general the subsystems cannot be assigned individual states because Ψ cannot in general be decomposed such that $\Psi = \varphi \otimes \Phi$ with $\varphi \in \mathcal{H}_1$ $\Phi \in \mathcal{H}_2$. Hence a subsystem cannot be in a unique individual state. However, Everett defines for any arbitrarily chosen state of, say, the second subsystem a unique *relative state* for the first subsystem. Let $\Phi \in H_2$ be any possible state of the second subsystem. Then we define a state $\varphi_\Phi \in H_1$ of the first subsystem relative to the state Φ by

$$\varphi_\Phi \overset{\text{def}}{=} N_\Phi \langle \Phi | \Psi \rangle_{H_2} \quad , \quad \Phi \in H_2 \quad , \quad \Psi \in H_1 \otimes H_2 \quad ,$$

where the normalization constant N_Φ is determined by $\langle \varphi_\Phi | \varphi_\Phi \rangle = 1$. According to Everett, it makes sense to talk about the state vector for the whole universe of discourse but it is meaningless to ask for the state of a subsystem. However, one can say that a subsystem is in a state φ_Φ *relative* to a given state Φ of the rest of the world.

Memory states

In the Everett interpretation there are no external observers, but we can under certain circumstances give an implicit *a posteriori* definition of internal "observers". In this case we can investigate what quantum mechanics says about the *appearence* of phenomena to such observers (Everett, 1973). According to Everett quantum mechanics predicts that the subjective experience of the observers is in harmony with the predictions of the traditional external interpretations of quantum mechanics.

Everett characterizes observers by their *memories*. Memories are defined as "*parts of a relatively permanent nature whose states are in correspondence with past experience of the observers*" (Everett, 1957). The subjective experience of the observers described is given by the contents of their memory.

What is potentially possible will be realized

The Everett interpretation assumes that everything that is potentially possible will be realized. Consider a measurement-like interaction where time evolution of the combined system is given by

$$\Psi(t) = \sum_n c_n(t) \; \alpha_n \otimes \beta_n \quad ,$$

and the state β_j refers to the measuring instrument with the corresponding pointer value b_j ,

$$B\beta_j = b_j\beta_j \quad .$$

According to Everett, the state vector $\Psi(t)$ never collapses, say into $\alpha_j \otimes \beta_j$, but *all* branches $\alpha_n \otimes \beta_n$ are realized. The superposition $\Psi(t)$ describes a multitude of simultaneously existing real worlds and $\alpha_n \otimes \beta_n$ describes the state of the n-th parallel world. Accordingly, the measured subsystem is never in a pure state but we can speak of the *relative state* α_j. Since $\{\beta_n\}$ is an orthonormal set of vectors, the vector α_j indeed represents the relative state of the measured system, provided the measuring instrument is in the state β_j,

$$\alpha_j = c_j^{-1} \langle \beta_j | \Psi \rangle_{H_2} \quad .$$

Now if the states $\{\beta_n\}$ qualify as memory states, then an observer with a definite memory is only aware of his own branch, say $\alpha_j \otimes \beta_j$. According to Everett, such an observer cannot be aware of the other branches so that the splits of the universe are unobservable. In each branch, the memory contents of observers give a different but consistent picture.

My world is not the whole world

The Everett interpretation leads to a many-worlds view, the whole world consists of mutually unobservable worlds. The world as a whole is strictly deterministic but the inhabitants of a particular branch can only see a tiny part of it. What is realized factually in a particular branch can be predicted only probabilistically.

The psycho-physical parallelism is supposed to be such that an observer in a given branch of the Everett universe is *aware* only of what is going on in that branch. To such an observer, the world appears to consist of only one of the many existing branches. In the Everett interpretation, "my world" is not the "whole world". "My world" is defined by the set of all imaginable experiences I may have. What is not accessible to observation and measurement does not belong to "my world". However, there is a full intersubjective harmony between the various observers in "my world" - all documents, all persons

I can possibly encounter confirm my personal experience. In this sense, we obtain a fully objective description. But the subjective element cannot be eliminated: the subject itself is on a particular branch of the Everett world, and thereby determines *which* branch is to be discussed as being relevant. As an illstration, consider again the paradox of Schrödinger's cat. In the "whole world" Schrödinger's cat is described by the superposition $\Psi_L+\Psi_D$. Because there are no observables A with $\langle\Psi_L|A\Psi_D\rangle \neq 0$, the vectors Ψ_L and Ψ_D represent different branches of the Everett world. If I am on the branch with Ψ_L, then the cat is alive in "my world". To be sure, according to Everett there exists another branch, not accessible to me, in which the poor cat is dead.

Elaboration of the Everett interpretation

In the original expositions of Everett (1957, 1973) and Wheeler (1957), many important features of the proposed interpretation are discussed rather vaguely. So it is inevitable that later explications of the Everett interpretation led to divergent views. This goes so far that Bell (1971b, 1972) sees a close connection between Everett's ideas and de Broglie's pilot waves.

Unfortunately, some of the explanatory literature on the Everett interpretation is impaired by a rather unsophisticated philosophical point of view. For example, DeWitt (1968, 1970, 1971a) claims that "*the mathematical formalism of the quantum theory is capable of yielding its own interpretation*", and that therefore "*no metaphysics needs to be added to it*". Such statements are unfounded and grossly misleading - interpretative metaphysical assumptions are necessary for *any* interpretation of a theory and never follow from the formalism alone.

What is a splitting?

Everett gives neither a precise definition of the "splitting of the universe into parallel worlds, nor is he able to assign an instant when the universe splits. As an example consider an Everett decomposition of the world state Ψ of the type

$$\text{(i)} \qquad \Psi = \varphi_1 \otimes \Phi_1 + \varphi_2 \otimes \Phi_2 \,.$$

Instead of this decomposition, we might just as well consider the linear superposition

$$\text{(ii)} \qquad \Psi = \varphi_+ \otimes \Phi_+ + \varphi_- \otimes \Phi_- \quad ,$$

where $\varphi_\pm = (\varphi_1 \pm \varphi_2)/\sqrt{2}$ and $\Phi_\pm = (\Phi_1 \pm \Phi_2)/\sqrt{2}$. Do both decompositions (i) and (ii) represent splittings? The answer is intuitively clear, if we again consider Schrödinger's cat. If, say φ_1 represents the state of the living cat, and φ_2 the state of the dead cat, then the decomposition (i) describes a split world. That the decomposition (ii) is not appropriate does not follow from Everett's arguments.

While Wheeler (1957) claims that the Everett interpretation "*does not require for its formulation any reference to classical concepts*", Wakita (1962a), Cooper and Vechten (1969), and Graham (1970) have emphasized the necessity of introducing classically describable states. It is only when a quantal subsystem interacts with another subsystem having classical properties that the total state can be resolved into states that admit a reasonable Everett interpretation. Each classically inequivalent branch describes one parallel world. If the whole universe considered does not possess any classical properties, then there is only one branch. *The proper set for the Everett expansion are classically inequivalent states.*

The branches of the Everett world have to be classically inequivalent

An essential point in Everett's thesis is his introduction of *memories* into quantum mechanics. Memories are supposed to record the *results* of measurements. As examples of memories, Everett (1957) mentions "*automatically functioning machines*", "*recording devices*", etc. All this only makes sense in the terminology of classical physics. If Everett's memory states were not classical states then we would have the possibility of a coalescence of formerly different branches - a possibility never discussed by Everett, Wheeler, or DeWitt. In the original papers the important concepts "memories" and "instrument readings" are never theoretically explained. The whole matter is thrown into confusion by some additional remarks, like Everett's (1973) claims that it is not necessary to impose restriction on the complexity or number of degrees of freedom of measuring instrument or observers. Everett expands the world state vector $\Psi(t)$ in terms of a complete set of memory states $\{\Phi_n\}$,

$$\Psi = \sum_n \varphi_n \otimes \Phi_n \quad ,$$

but does not give necessary and sufficient criteria how the memory state Φ_n has to be chosen. This vagueness has led to many misunderstandings. Zeh (1973) for example, made the fallacious suggestion to make the Everett decomposition unique by requiring that it has to be the Schmidt decomposition. However, from the *context* of Everett's paper it is clear how the proper decomposition of the world state has to be defined. Memory states must represent definite values, alternative memory states have to be noncommutative and hence noninterfering. Furthermore, it must be possible to get access to the memory without changing the memory states. Thus, a straightforward refinement of the Everett interpretation is the requirement that Everett's memory states must be classical and mutually inequivalent. The trouble is that - according to the formalism of pioneer quantum mechanics - such classical states do not exist because they violate the superposition principle.

Ockham's razor

The most serious critique of the Everett interpretation relates to an extreme violation of William of Ockham's rule "*essentia non sunt multiplicanda praeter necessitatem*" , i.e. that entities should not be multiplied without necessity (compare e.g. Shimony, 1963). To be sure, it is a philosophically controversial question whether we should accept such a principle. From the viewpoint of a particular observer in the Everett world, all other branches are observationally inaccessible so that the crucial point of the Everett interpretation never can receive operational support in the laboratory. The Everett interpretation implies one observable world and myriads of unobservable worlds.

There is nothing new under the sun. The fiction "Many dimensions" by Charles Williams contains in a striking fashion the main conclusions of Everett's interpretation: "*He remembered having read somewhere once a fantastic theory that whenever a man made a choice, a real choice - whenever he definitely did one of two things he also did at the same moment the other and brought an entire new universe into being that he might do so. For otherwise an infinite number of potentialities would exist for ever unfulfilled - which, the writer had said, though Lord Arglay had forgotten his reasons, was absurd*" (Williams, 1931; in the new edition of 1963 on p.53).

The principal advantage of the Everett interpretation is its simplicity

It is difficult to imagine a laboratory experiment in the usual sense that could decide between Bohr's view, the von Neumann-London-Bauer interpretation, and the Everett interpretation. First of all, regulative principles are metaphysical and not operational. Nevertheless, we are not free in our choice of regulative principles and, moreover, the character of the whole theory crucially depends on the regulative principles. For example, DeWitt (1967, 1970) mentions that the assumption of the cosmological state vector and its time development during the "Big Bang" may have testable implications for cosmology.

The conclusions of the Everett interpretation may be considered as bizarre. The idea of split observers may be thoroughly repugnant to many scientists. However, the Copernican world view was condemned with very similar arguments. Novelty and repugnance are not valid arguments. Everett's view is to be judged by the theory it generates. The decisive argument in favor of the Everett interpretation is its simplicity, it is a *simpler theory* than quantum mechanics in the von Neumann-London-Bauer or Copenhagen interpretation. The dubious reduction postulate is eliminated, and there is no a priori distinction between object and subject. The Everett interpretation is fruitful for theoretical molecular biology because it eliminates the dualism microphysics/macrophysics, and allows us to attack the conceptual problems that large molecular systems pose.

The discomfort that the Everett interpretation like any new idea poses is overwhelmingly compensated by the insight and creative power this view provides. The Everett interpretation is superior in logical economy and gives us a more intelligible pattern of the world; it is the simplest interpretation of quantum mechanics, and a bold attempt to replace the traditional epistemic view by an ontic one. It is marred by the fact that the *formalism* of pioneer quantum mechanics does not allow a proper definition of memory states.

3.7 EINSTEIN-PODOLSKY-ROSEN CORRELATIONS

Albert Einstein discovers the nonseparability of quantum mechanics

In a famous paper Einstein, Podolsky and Rosen (1935) discussed correlations in spatially isolated quantal systems, and pointed out that according to quantum mechanics such systems still may interfere with each other. This interference gives rise to strongly nonlocal correlations. Einstein, Podolsky and Rosen (hereafter referred to as EPR) considered such a behavior of physical systems as unreasonable and advanced the argument that these nonlocal correlations are spurious and due to an incompleteness of quantum theory. EPR based the argument on a postulate asserting the existence of an *objective real state of a physical system* which is independent of any experimental operation that does not disturb the system at the time in question. Bohr (1935) responded immediately to this challenge and showed that Einstein's notion of "physical reality", as cautious as it may appear, is incompatible with the principles of quantum mechanics.

The argument of EPR was a great contribution to quantum mechanics. It did not refute quantum mechanics but it freed it from latent and unsuspected assumptions. The so-called EPR-paradox draws the attention directly to a central phenomenon: *quantal systems possess holistic properties*. In quantum mechanics, a system consisting of non-interacting spatially separated subsystems can be in pure state without the subsystems being in pure states. It is a mathematical property of classical mechanics that the states of all the subsystems of a given system determine the state of the whole system; this property is called *separability*. We call a system *holistic* if it does not possess the property of separability. Einstein discovered that pioneer quantum mechanics is a holistic theory. The existence of EPR-correlations is a compelling consequence of quantum mechanics and is experimentally well confirmed. So neither experimental facts nor quantum theory but the prejudices of common sense must be changed. In nature, *there are* peculiar phenomena; quantum mechanics is a peculiar theory because it describes nature *properly*.

References: A lucid logical analysis of the EPR-arguments is given by Reichenbach (1944). In the German translation of this book (Birkhäuser, Basel, 1949; pp. 192-193) an important clarification of Bohr's arguments due to Pauli is added. Compare also: Einstein (1936, 1948, 1949, 1953a), Schrödinger (1935a,b, 1936), Furry (1936a,b), Weyl (1949b), Bohr (1949), Margenau (1949), Bohm (1951), Bell (1964), d'Espagnat (1965, 1971), Jauch (1968a), Krips (1969c, 1971), Hooker (1970, 1971a,b, 1972a,b), Erlichson (1972), Scheibe (1973), Mirman (1973), Moldauer (1974), Mittelstaedt (1974).

The arguments of Einstein, Podolsky and Rosen

Einstein, Podolsky and Rosen (1935) consider a system consisting of two particles, each with one degree of freedom. Let Q_1, P_1 and Q_2, P_2 stand for the position and momentum of particles 1 and 2, respectively. The canonical commutation relations (we choose units such that $\hbar=1$)

$$Q_j P_k - P_k Q_j = \delta_{jk}$$

$$j,k = 1,2$$

$$Q_j Q_k = Q_k Q_j \quad , \quad P_j P_k = P_k P_j$$

imply that the total momentum $P \overset{\text{def}}{=} P_1+P_2$ commutes with the relative coordinate $Q \overset{\text{def}}{=} Q_1-Q_2$, $QP=PQ$. As a consequence, there exists a state of the two-particle system with a sharp value of the total momentum P (which we choose to vanish) and with a sharp value of the observable Q which we choose to be q. Accordingly, this state function Ψ equals

$$\Psi(x_1,x_2) = \delta(x_1 - x_2 - q),$$

where δ is Dirac's delta function. This state can be produced by an appropriate experimental procedure. The distance q between the two particles may be arbitrarily large, so that, once this state has been produced, we may assume that the forces between the two particles are vanishingly small. Consequently, the apparatus used to observe particle 1 can be chosen such that it does not interact at all with particle 2. It is left to the free choice of the experimenter what he wants to measure. For example, he can carry out a measurement of the first kind of the position of the first particle. Let the measurement of the observable Q_1 yield the value q_1, so that this particle is in the corresponding eigenstate φ_1 of the observable Q_1

$$\varphi_1(x_1) = \delta(x_1 - q_1) \quad .$$

The decomposition

$$\Psi(x_1,x_2) = \int_{-\infty}^{\infty} \delta(x_1 - x)\delta(x_2 + q - x)dx$$

taken together with the reduction postulate yields

$$\Psi(x_1,x_2) \to \delta(x_1 - q_1)\delta(x_2 + q - q_1) \quad .$$

That is, immediately after the measurement of the position of the first particle has been completed the second particle is also in the pure state φ_2

$$\varphi_2(x_2) = \delta(x_2 + q - q_1) \quad ,$$

and therefore sharply localized at the position $q_2=q-q_1$. On the other hand, the experimenter can perform a measurement of the momentum of the first particle. Let this measurement of the observable P_1 yield the value p_1, so that this particle is in the eigenstate $\tilde{\varphi}_1$ of the observable P_1,

$$\tilde{\varphi}_1(x_1) = (1/\sqrt{2\pi}) \int_{-\infty}^{\infty} \exp\{ix_1k\}\delta(k - p_1)dk$$

$$= (1/\sqrt{2\pi}) \exp\{ix_1p_1\} \quad .$$

From the Fourier representation

$$\Psi(x_1,x_2) = (2\pi)^{-1} \int_{-\infty}^{\infty} \exp\{i(x_1 - x_2 - q)k\}dk$$

taken together with the reduction postulate one obtains

$$\Psi(x_1,x_2) = \tilde{\varphi}_1(x_1)\tilde{\varphi}_2(x_2) \quad ,$$

where

$$\tilde{\varphi}_2(x_2) = (1/\sqrt{2\pi}) \exp\{-i(x_2 + q)p_1\} \quad .$$

That is, immediately after the measurement of the momentum of the first particle has been completed, the second particle is also in an eigenstate of the momentum with the eigenvalue $p_2 = -p_1$.

As a result, we can predict the position or the momentum of particle 2 without in any way interfering with this particle. Einstein, Podolsky and Rosen adopt the following metaphysical postulate: "*If without in any way disturbing a system, we can predict with certainty the value of a physical quantity then there is an element of physical reality corresponding to this physical quantity*". Accordingly, EPR argued that they have shown the physical reality of both the position and the momentum of particle 2. This is the EPR-paradox, for quantum mechanics prohibits simultaneous sharp values for noncommuting observables. As first stressed by Bohr (1935), there is no logical difficulty at all. EPR's criterion of "physical reality" reflects a *classical conception* of reality which cannot be incorporated into quantum mechanics. The laboratory arrangement is an essential element of the physical reality of a quantum event. No assertion concerning the values of an observable can be made *before* a measurement; it is only *after* a measurement on particle 1 that a statement about particle 2 can be made. The observational means for a measurement of position and of momentum are mutually exclusive. One is not allowed to say that a measurement of Q_1 *changes* the value of Q_2 because *before* the measurement the second particle is not in an eigenstate of Q_2. To be sure, the EPR-paradox is very much contrary to common sense but it is not self-contradictory in the logical sense. We do get a logical contradiction if we try - as Einstein, Podolsky and Rosen do - to *extend* quantum mechanics by a reality condition incompatible with quantum mechanics. (For more details, compare the analysis by Scheibe, 1973, and by Mittelstaedt, 1974).

Einstein accepted the logical selfconsistency of quantum mechanics but not its philosophical implications. In his autobiography, Einstein (1949) wrote: "*But on one supposition we should, in my opinion, absolutely hold fast: the real factual situation of the system S_2 is independent of what is done with system S_1, which is spatially separated from the former. According to the type of measurement which I make of S_1, I get, however, a very different Ψ_2 for the second partial system. Now, however, the real situation of S_2 must be independent of what happens to S_1. For the same real situation of S_2 it is possible therefore to find, according to one's choice, different types of Ψ-functions. One can escape from this conclusion only by either assuming that the measurement of S_1 (telepathically) changes the real situation of S_2 or by denying independent real situations as such to things which are spatially separated from each other. Both alternatives appear to me entirely unacceptable.*"

Einstein considered quantum mechanics to be an incomplete theory. Nowadays we know that locally hidden variables are impossible so that we have to accept the nonlocal EPR-correlations. The EPR-paradox shows that in quantum mechanics *a system without any interactions with other systems and moreover spatially separated from them, can yet be correlated and synchronized with these systems in a highly nontrivial way.*

Critique of the formulation of the EPR-paradox

If the discussion by Einstein, Podolsky and Rosen is correct, then it is doubtful whether one can speak legitimately of subsystems of a quantum system. Since this is a question of great practical importance, it is imperative to track down any eventual flaw in the argumentation of EPR. The following features of the argument have been criticized in the literature:

a) In formulating the EPR-paradox the reduction postulate is used. Margenau (1936, 1937) claims that the reduction postulate can be given up and so the paradoxical situation fades away.

b) The assumption of a vanishing interaction between two spatially separated systems is never exactly satisfied (Sharp, 1961).

c) For large distances nonrelativistic quantum mechanics becomes invalid as the retardation of the interaction must be taken into account (Schrödinger, 1927, 1935b, 1936). For large distances, the Hamiltonian has to be modified; the validity of the conservation laws could be questionable (Furry, 1936a, 1936b).

d) There are no self-adjoint representations for the momenta of separated systems (Cooper, 1950).

e) The quantum mechanical description is incomplete, quantum mechanics is to be completed by hidden variables (Einstein et al., 1935).

f) The EPR-thought-experiment cannot be realized and is therefore without interest (Breitenberger, 1965).

None of these objections is relevant. Objection a) which, by the way, has not been accepted by Einstein (1949, page 682), can only be advocated in an ensemble theory of quantum mechanics and is refuted by the successful EPR-experiments. Objections b) and c) are not relevant; it can be shown that the residual interactions and the retardation effects are not essential for the EPR-phenomena. A discussion of the EPR-phenomena in the framework of modern relativistic quantum field theory has been given by Schlieder (1968), Hellwig and Kraus (1970b), and Antoine and Gleit (1971). Objection d) is directed against the usual sloppy mathematical treatment of the hypothetical EPR-experiment and is well taken. Nevertheless the objection is not relevant, as a modification of the EPR-experiment proposed by Bohm (1951) avoids the use of unbounded operators and can be rigorously discussed in an elementary way. As Bell (1964) has shown, the proposal e) provides no acceptable alternative either. If one tries to complete quantum mechanics with the help of hidden variables, then these variables necessarily show a marked nonlocal behavior, so that Einstein's ideal is again violated. Objection f) has been refuted by modern experiments.

The existence of EPR-correlations is an experimental fact

The original formulation of the EPR-paradox involved the use of unbounded observables with a continuous spectrum, which leads to mathematical difficulties (nonnormalizable state vectors, problems of self-adjointness). A both conceptually and mathematically much simpler but legitimate example of the EPR-paradox is the case of the disintegration of a spin-zero particle into two spin-1/2 particles (Bohm, 1951).

Consider a system of zero total spin consisting of two particles each with spin 1/2 (for example a hydrogen molecule in a singlet s-state). Let this system be split by a method not influencing the spin of either particle. Then, the total spin remains zero while the particles are flying apart. When the particles have ceased to interact appreciably, the state function of the two-spin system is given by

$$\Psi(1,2) = (1/\sqrt{2})\{\varphi_+(1)\varphi_-(2) - \varphi_-(1)\varphi_+(2)\} \quad ,$$

where φ_+ and φ_- are the one-particle wave functions representing, respectively, a spin direction +1/2 and −1/2. Provided the particles 1 and 2 are sufficiently far apart, separate measurements can be carried out on particle 1 and 2. Let $\vec{S}_1$ be the spin operator of the first particle and $\vec{a}$ an arbitrary unit vector, and let us perform a measurement of the first kind of the observable $(\vec{a}\vec{S}_1)$. The eigenvalues of $(\vec{a}\vec{S}_1)$ are −1/2 and +1/2, so that the measurement yields either the value −1/2 or +1/2. As the total system is in a singlet state, a measured eigenvalue m ($m = 1/2$ or $m = -1/2$) of $(\vec{a}\vec{S}_1)$ implies that an immediately following measurement of $(\vec{a}\vec{S}_2)$ on the second particle must yield the eigenvalue $-m$. However, as the direction of the vector $\vec{a}$ is left to the choice of the experimenter, we are *not* allowed to say that the two particles have a well defined spin direction already before the measurement. Nevertheless, the two spins are correlated; by measuring any component of the spin of the first particle, we know this component of the second particle, without in a way interacting with it. Adopting a classical concept of reality, the situation is again paradoxical.

Experimental tests: It is not easy to carry out this experiment in the case of the hydrogen molecule. But several equivalent proposals for an experimental investigation of the EPR-phenomenon have been made:

a) Scattering of an electron at a proton (Jánossy and Nagy, 1956).

b) Correlation of the polarisation directions of the photon pair which is emitted in the annihilation of an electron-positron pair (Wheeler, 1946; Wightman, 1948; Yang, 1950; Bohm and Aharonov, 1957, 1960; Day, 1961).

c) Proton-proton scattering (Peres and Singer, 1960);

d) Autoionization of a helium atom in the $2s^2$ state;

$$He(2s^2) \rightarrow He^+(1s^1) + e^-$$

and the measurement of the spin correlations (Peres and Singer, 1960).

e) Correlations in a K-mesons pair $(K^0, \bar{K}^0)$ simultaneously produced by a proton-proton annihilation (Day, 1961; Inglis, 1961).

f) Correlation measurements with polarized photons in a split beam (Aharonov and Peterson, 1967).

g) Correlations of the linear polarization of two successively emitted photons in the cascade $6^0S_0 \rightarrow 4^1P_1 \rightarrow 4^1S_0$ of calcium (Kocher and Commins, 1967; Clauser et al., 1969; Fox and Rosner, 1971).

The experiments carried out until today belong to the group b) (Wu and Shakanov, 1950; Bertolini et al., 1955; Langhoff, 1960; Kasday, 1970, 1971; Kasday et al., 1970) and the group g) (Kocher and Commins, 1967; Freedman and Clauser, 1972; Clauser, 1976; Lamehi-Rachti and Mittig, 1976; Fry and Thompson, 1976). For a recent review and a careful evaluation of the experimental results, compare Clauser and Shimony (1978).

Beyond any reasonable doubt, the experimental tests of the EPR-phenomena have confirmed the predictions of quantum mechanics. Even if some scientists find this conclusion disturbing, *the existence of EPR-correlations has to be considered as an empirical fact.*

Entangled systems and holistic phenomena

The EPR-paradox is a special case of the remarkable behavior of entangled systems. *Entangled systems* are systems which have interacted in the past and have then been separated. It is characteristic for quantal systems that spatially separated parts of an isolated system are in general correlated, even if they no longer interact. Such holistic effects are unknown in classical physics. Quantum mechanics is the first mathematically formulated scientific theory where in a non-trivial way the *whole is more than the combination of its parts*.

If a system spontaneously decays into subsystems, then the dynamical interactions cause these subsystems to be in a correlated state. When the subsystems are separated, the dynamical interactions between them become more and more irrelevant. *Non-interacting systems with a common ancestor are therefore in general correlated*. Let us consider two systems, S_A and S_B, with the Hilbert space $\mathcal{H}_A$ and $\mathcal{H}_B$ as state spaces. If both these systems have a common ancestor, then the state function $\Psi \in \mathcal{H}_A \otimes \mathcal{H}_B$ of the total system is *not* an uncorrelated function

$$\Psi = \alpha \otimes \beta \quad , \quad \alpha \in \mathcal{H}_A \quad , \quad \beta \in \mathcal{H}_B \quad ,$$

but a superposition of the type

$$\Psi = \sum_n \sum_m \lambda_{nm} \alpha_n \otimes \beta_m \quad , \quad \alpha_n \in \mathcal{H}_A \; , \; \beta_m \in \mathcal{H}_B \; , \; \lambda_{nm} \in \mathbb{C} \; .$$

Consequently, *the subsystems* S_A *and* S_B *do not in general have a state vector*.

A convenient representation of correlated state vectors is the natural expansion, introduced in this context by Schrödinger (1935a):

$$\Psi = \sum_n \lambda_n \alpha_n \otimes \beta_n \quad , \quad \alpha_n \in \mathcal{H}_A \quad , \quad \beta_n \in \mathcal{H}_A$$

with

$$\langle \alpha_n | \alpha_m \rangle_{\mathcal{H}_A} = \langle \beta_n | \beta_m \rangle_{\mathcal{H}_B} = \delta_{nm}$$

The orthonormal functions α_n and β_n form a pair of eigenfunctions with the complex eigenvalue λ_n of the eigenvalue problem (in the sense of Erhard Schmidt, 1907a,b, 1908) with the kernel Ψ

$$\langle\alpha_n|\Psi\rangle_{H_A} = \lambda_n\beta_n \quad ,$$

$$\langle\beta_n|\Psi\rangle_{H_B} = \lambda_n\alpha_n \quad .$$

If this Schmidt decomposition yields more than one nonvanishing eigenvalue λ_n, then we speak of an *entangled system*. A system with only one nonvanishing eigenvalue λ_n is called a *disentangled system*. If we admit only interactions with S_A, then all statistical predictions for the outcome of experiments restricted to S_A can be expressed with the aid of the density operator D_A, defined by

$$D_A \overset{\text{def}}{=} \mathrm{tr}_{H_B}\{|\Psi\rangle\langle\Psi|\} = \sum_n |\lambda_n|^2 |\alpha_n\rangle\langle\alpha_n| \quad .$$

Analogously, the density operator D_B of the system S_B is given by

$$D_B \overset{\text{def}}{=} \mathrm{tr}_{H_A}\{|\Psi\rangle\langle\Psi|\} = \sum_n |\lambda_n|^2 |\beta_n\rangle\langle\beta_n| \quad .$$

Measurements on an ensemble of S_A-systems can at best determine the modulus $|\lambda_n|$ of the eigenvalues λ_n and the eigenfunctions α_n of the density operator D_A, and measurements on an ensemble of S_B-systems the modulus of the eigenvalues λ_n and the eigenfunctions β_n of the density operator D_B. As a result, the maximal information accessible by measurements on both subsystems can be represented with the aid of a density operator D'

$$D' \overset{\text{def}}{=} D_A \otimes D_B$$

with spectral resolution

$$D' = \sum_n |\lambda_n|^2 |\alpha_n\otimes\beta_n\rangle\langle\alpha_n\otimes\beta_n| \quad .$$

The complete description of the total system, however, is given by the state vector Ψ or, equivalently, by the density operator $D = |\Psi\rangle\langle\Psi|$ which can also be written as

$$D = \sum_n \sum_m \lambda_n \lambda_m^* |\alpha_n \otimes \beta_n\rangle\langle\alpha_m \otimes \beta_m| \quad .$$

The statistical operator D' represents the knowledge that can be obtained from measurements restricted to the two subsystems S_A and S_B. The fact that this knowledge is incomplete is reflected by the fact that D' is not idempotent, $(D')^2 \neq D'$, and therefore does not represent a pure state. Thus, *the information that can be obtained from measurements on the two subsystems does not allow the reconstruction of the full statistical operator* D. The missing information is contained in the phases φ_n

$$\lambda_n \overset{\text{def}}{=} |\lambda_n| \exp(i\varphi_n)$$

of the eigenvalues λ_n. These phases can be determined only by measurements of the *correlations* between the two systems. A measurement of the first kind of the observable $P_j = |\alpha_j\rangle\langle\alpha_j|$ leads to the state reduction

$$\Psi \rightarrow \alpha_j \otimes \beta_j \ ,$$

and hence to the creation of a pure state β_j in the subsystem S_B. Accordingly, we have the following *generalized EPR-phenomenon* (Schrödinger, 1936): *Let* S_A+S_B *be an entangled system with two spatially separated subsystems* S_A *and* S_B *in a pure state* $\Psi = \sum \lambda_n \alpha_n \otimes \beta_n$. *An appropriate measurement apparatus interacting only with the subsystem* S_A, *can put the subsystem* S_B *into any pure state that is an element of the subspace spanned by the natural vectors* $\{\beta_1, \beta_2, \ldots\}$.

The formal reason for this peculiar situation is the superposition principle of quantum mechanics, or, in other words, the fact that in quantum mechanics the composition of two systems S_A, S_B into a joint system $S = S_A + S_B$ involves the *tensor product* of the corresponding Hilbert spaces while in classical mechanics we use the *direct sum* (or equivalently, the Cartesian product) of the corresponding phase spaces. The density operators D and D' differ by the interference terms $|\alpha_n \otimes \beta_n\rangle\langle\alpha_m \otimes \beta_m|$, $n \neq m$: *all EPR-correlation effects are interference effects.* If for any reason the information about the phase relations between the complex eigenvalues λ_n is lost, the interference terms are no longer observable and the experimenter cannot find EPR-correlations. However, it would be fallacious to believe that the phase φ_n would be smeared out in all macroscopic phenomena. The best known counterexamples are the coherence effects in superconductors (compare the review by Mercereau, 1969). Modern precision measurements of e/h with the aid of the Josephson effect (compare Parker et al.,1969) provide an impressive proof of the extremely exact phase relationship that can be maintained by more than 10^{20} electrons. These experiments refute the frequently uttered opinion that in macroscopic systems the individual phase φ_n would "of course" be averaged, and that only the phase average of D, that is D', would be of interest. *Quantum systems can show holistic effects even if their subsystems are separated by macroscopic distances.*

Summing up

We have to accept the inevitable: *the EPR-correlations are a genuine characteristic of nature.* The EPR-phenomena are puzzling only in a classical context, i.e. if we have not yet developed appropriate forms of thought and intuition. Far from bringing any difficulties to quantum theory, the EPR-correlations provide a new and important tool for understanding some of the holistic phenomena occuring in nature. A quantum subsystem adapts its behavior to its surroundings. Entangled systems have properties which are not present in anyone of its subsystems. Furthermore, the EPR-correlations bring a hereditary element into quantum mechanics. An entagled system shows "recognition phenomena"; the mutual reactivity of two systems may depend on whether they have a common ancestor or not. All these aspects show a deep formal similarity with the behavior of biological systems. Quantum mechanics is the first mathematically formulated theory that shows phenomena we know from theories working on a higher hierarchical level, like biology and psychology.

Nowadays, the main problem is not to understand the EPR-correlations in some classical sense. Once we have accepted the facts, there is a new problem, however: why are the EPR-correlations so rarely observed? *It will be an important task to clarify by rigorous mathematical theorems the conditions under which interference terms are no more relevant in large molecular systems.* This problem cannot be solved merely by plausibility arguments and heuristic considerations. The future development of theoretical chemistry and biology will crucially depend on a proper solution of this problem.

3.8 CONCLUSION: THE STATUS OF PIONEER QUANTUM MECHANICS

What is quantum mechanics about?

The *referent of a theory* stands for that what a theory is about (cf. Bunge, 1973, chapt.4). A theory may be about

the sensations of observers,
the ideas of observers,
the results of experiments and measurements,
individual physical systems,
Gibbsian ensembles of physical systems,
the qualities of physical systems,
objectively existing ideal entities,

and so on. Anyone of these possibilities may be valid, interesting and useful. There is no such thing as the "correct choice" of the referent but there are bad choices. There are choices that are logically inconsistent and incompatible with the remaining theoretical structure, and there are choices that do not fit into the cultural basis of our science. Since "*the filling of minds with technical concepts without establishing their relationship is a form of pollution*" (Hammer, 1971), we have to know what the referent of a theory is. In spite of its overwhelming pragmatic success, *there is no general agreement about the referent of pioneer quantum mechanics.*

Most text-book authors try to avoid this problem by adopting an *operational point of view.* They say that science is a tool for making predictions about what will be observed in different situations, and consider a physical quantity as being defined when the procedures for measuring that quantity are specified. Sometimes they even claim that they do not use any philosophy at all, and that philosophy is irrelevant to science. These authors do not seem to be aware of the fact that they defend a kind of operationalism or logical positivism - highly controversial and probably untenable philosophical positions.

Historical remarks: The view that a physical quantity acquires its meaning by a procedure for measuring its value probably goes back to Ernst Mach (1838-1916). The definition of physical quantities by operations of measurement was stated explicitly by Hugo Dingler (1881-1954) (compare his early work, Dingler 1907) and was used by Constantin Carathéodory (1873-1950) in his important study on the foundation of thermodynamics (Carathéodory, 1909). In his famous "The Logic of Modern Physics" of 1927, the American physicist Percy William Bridgman (1882-1962) popularized the view that all scientific concepts have to be linked to experimental procedures and stated an influential program of operationalism. Although

Bridgman himself modified his early extreme doctrine in an essential way (cf. Bridgman, 1959), many scientists still regard "The Logic of Modern Physics" as the bible of *the* scientific method, indispensable to a correct understanding of modern science.

According to operationalism science should be cleansed from operationally undefinable concepts. That the operationalist's maxim is untenable and that operationalism is inadequate as a philosophy of science is recognized even by those modern empiricists who do not deviate too much from the basic viewpoints of logical positivism (cf. Hempel, 1965, 1966). In fact, not one existing physical theory complies with the basic tenets of operationalism. The primary error of operationalism is the confusion between the referent of a theory and the way of testing it (cf. Bunge, 1973). If we insist that science is more than a set of useful rules for classifying experimental data, then physics and chemistry are not basically experimental sciences but sciences that are concerned with the study of matter and its interactions.

The mathematical concepts used in the formalism of pioneer quantum mechanics have euphonious names like state space (for the underlying Hilbert space), state vector (for a unit vector in the state space), Hamiltonian (for the generator of the dynamical group), observables (for the self-adjoint operators acting on the state space). Of course, a self-adjoint operator becomes neither observable nor physically meaningful just by receiving a name. If we acknowledge that pioneer quantum mechanics is about molecular matter, then we have to give up operationalism and cannot define fundamental concepts like states or observables operationally. "What *is* an observable?" is not a romantic question that can be ignored in favor of an operationalistic definition.

Category mistakes

Most chemists probably will agree that a chemical theory should be about the objective material world. The more specific claim that chemical theories should be about molecules and their interactions would, however, be careless since molecules and chemical substances are categories of a different logical type.

Individual objects, classes of individual objects, classes of classes etc. are categories of different logical type Different theories may have referents standing for categories of different logical type. When categories of different logical types are treated as being on par, we speak of *category mistakes* (Ryle, 1949). Many jokes, as Ryle observes, are type-pranks but many category mistakes are not obvious and have misled philosophers and scientists. A widespread category mistake in chemistry is taking the concept of substance as being on equal footing with molecules, or the confusion of phenomenological thermodynamics with statistical mechanics. For example, a steam engine may belong to the reference class of thermodynamics but not of statistical mechanics. In the traditional formulation of statistical mechanics the corresponding reference class is a Gibbsian ensemble of infinitely many identical noninteracting and uncorrelated steam engines.

The quest for the referent of modern scientific theories is nontrivial and the danger of making category mistakes is serious. For example, it is not at all clear that collision theory (as used for the description of molecular beam experiments) and phenomenological kinetics (as used to describe enzyme kinetics in biochemistry) have reference categories of the same logical type.

A serious category mistake in many contemporary expositions of quantum mechanics and quantum chemistry concerns the confusion between theoretical and experimental concepts. The theoretical quantities called "observables" in pioneer quantum mechanics do by no means refer to direct measurements in the sense of the experimenter. Things are not that simple! Every realistic description of actual measuring devices uses highly developed classical physical theories, such as classical continuum mechanics, classical optics or the Maxwell-Lorentz theory of electrodynamics in an essential way. Together with mathematical system theory and the theories of filtering, interpolation and extrapolation of stochastic processes they allow a well-founded and concise representation of modern experiments by *phenomenological theories*. Such phenomenological theories - which are not covered by pioneer quantum mechanics - enable us to design the experimental apparatus and to describe the results of the experiments in theoretical terms.

Epistemic vs. ontic interpretations

There are two basically different views on the referents of scientific theories. First we have the philosophical viewpoint of *homo sapiens*: science deals with a world-to-be-contemplated. In contradistinction, the modern technological view of *homo faber* considers science as dealing with a world-to-be-acted-on. The first viewpoint has the grandiose aim to tell us how the world *is* and interpretations of this kind will be called *ontic interpretations*. The more modest second view does not claim to tell us how the world is but only what results we can expect if an observer interferes with the system; interpretations of this kind are called *epistemic interpretations*.

Examples: Newtonian celestial mechanics is the paradigm for an ontic interpretation, Modern mathematical system theory (cf. Kalman et al., 1969) aims at the elucidation of the mathematical structure of the experimentally accessible input-output map of arbitrary black-box systems; it may be taken as a paradigm for an epistemic interpretation. The referents of Bohr's interpretation of quantum mechanics are laboratory measurements (and not microphysical objects!); it is an epistemic interpretation referring to individual macroscopic objects. In its statistical version, the von Neumann-London-Bauer interpretation of quantum mechanics is an epistemic interpretation referring to a Gibbs ensemble of microphysical objects. The Everett interpretation is an ontic interpretation having the whole universe as its referent. Note that not one of the traditional interpretations of pioneer quantum mechanics has microphysical objects (like molecules) as referents.

The objective of theories in an ontic interpretation is to understand the fundamental nature of things. The proper place of epistemic theories is in engineering science. Fundamentalistic and engineering sciences have different goals but supplement each other. Modern engineering science has a strong theoretical foundation, its aim is the creation of new systems, its essence is a theory of model building. According to Kalman (1968a) "*a model is a summary of experimental data; repeating an experiment on the model should yield exactly the same data as was assumed in constructing the model*". A *minimal model* is one which is simplest in some sense, e.g. has the minimal number of parameters. Mathematical system theory is the study of dynamical relationships, it also deals with the art of model building. Its basic concept is the input-output map representing an experimentally accessible stimulus-response relationship. A *realization* of an input-output map is a dynamical system that has the same stimulus-response relationship. The task of mathematical system theory is to explain the experimental data in dynamical terms, in the sense that *all* possible realizations of the empirical input-output map are discussed and a mathematically well-defined minimal realization is selected (compare Kalman, 1968b).

The appropriate tool for the design and the description of any kind of physical or chemical experiments is mathematical system theory. Note that even for experiments on microphysical systems quantum mechanics is *not* necessary. A famous example are the phenomenological dynamical equations proposed by Felix Bloch (1946) for the description of nuclear magnetic resonance. Without using quantum mechanics, Bloch's equation can be modified in such a way that they cover every conceivable detail of modern nuclear magnetic resonance experiments in terms of a system theoretical description.

Remark: It is sometimes said that Heisenberg's uncertainty relations are characteristic for quantum mechanics. This view is not only historically wrong but also conceptually very misleading. Heisenberg's inequality and the canonical commutation relations are associated with the harmonic analysis of the two-parameter Abelian group $\mathbb{R}^2$. This group arises naturally in many contexts, e.g. in communication theory. Communication theory deals with the transmission of messages, where the messages are represented by continuous functions of time $t \in \mathbb{R}$. The signal $t \to f(t)$ may be the amplitude of an electric current in a communication circuit, or a function describing a sound wave pattern, etc. Let $t \to f(t)$ be a continuous, quadratically integrable signal, and $\hat{f}$ its Fourier transform,

$$\hat{f}(\nu) = \int_{\mathbb{R}} f(t) \exp(-2\pi i \nu t)\, dt \quad .$$

The simple mathematical fact that a signal f and its Fourier transform cannot be both very small at infinity is of crucial engineering importance. The fact that not both of the functions $|f|$ and $|\hat{f}|$ can be highly concentrated can be expressed in terms of the variances $(\Delta t)^2$ and $(\Delta\nu)^2$

$$(\Delta t)^2 \overset{\text{def}}{=} \int_{\mathbb{R}} (t-\bar{t})^2 |f(t)|^2 dt \Big/ \int_{\mathbb{R}} |f(t)|^2 dt \quad ,$$

$$(\Delta\nu)^2 \overset{\text{def}}{=} \int_{\mathbb{R}} (\nu-\bar{\nu})^2 |\hat{f}(\nu)|^2 d\nu \Big/ \int_{\mathbb{R}} |\hat{f}(\nu)|^2 d\nu \quad ,$$

where we introduced the mean values $\bar{t}$ and $\bar{\nu}$ by

$$\bar{t} \overset{\text{def}}{=} \int_{\mathbb{R}} t |f(t)|^2 dt \Big/ \int_{\mathbb{R}} |f(t)|^2 dt \quad ,$$

$$\bar{\nu} \overset{\text{def}}{=} \int_{\mathbb{R}} \nu |\hat{f}(\nu)|^2 d\nu \Big/ \int_{\mathbb{R}} |\hat{f}(\nu)|^2 d\nu \quad .$$

The quantity Δt is a measure of the *effective time duration* and $\Delta\nu$ a measure of the *effective bandwidth* for the signal f. It is a simple mathematical exercise to prove that

$$(\Delta t)(\Delta\nu) \geq 1/4\pi$$

This inequality corresponds exactly to Heisenberg's inequality $(\Delta p)(\Delta q) \geq h/4\pi$, and says that we cannot arbitrarily limit bandwidth and duration of a signal. This inequality is much older than the celebrated Heisenberg inequality. The fact that a signal which is confined to a small time interval cannot be confined to a small frequency interval, was used by Sommerfeld (1914) to discuss the coherence length of wave-trains, and by Küpfmüller (1924) for an inequality between the settling time and the band width of a band-pass filter. According to a private communication of Norbert Wiener to Barnes (1964, 1970), Wiener had given a mathematically precise formulation of the variance-type time/frequency inequality at a Göttingen mathematics seminar in 1924, with Heisenberg present.

Apart from inequalities of the Heisenberg type, modern mathematical system theory also makes use of the Heisenberg-Weyl canonical commutation relations. For example, to every regular (i.e. purely nondeterministic), wide-sense stationary stochastic process there is associated an *energy operator* H (generating the time evolution of the process) and a *time operator* T, fulfilling rigorously the commutation relation HT-TH=i (compare Hanner, 1950; Tjøstheim, 1975, 1976; Gustafson and Misra, 1976).

The mathematical analysis of signals and stochastic processes therefore leads in a natural way to the basic mathematical tools used in pioneer quantum mechanics.

Mathematical system theory combined with the related theory of stochastic processes correlates a set of experimental observations with the results of other observations; it searches for phenomenological correlations rather than for the essential properties of matter. In view of the progress these theories have made in recent times, the epistemic interpretation of quantum mechanics as advanced by Bohr has to be considered as inadequate and obsolete. It is true that denying the existence of underlying realities, we can get rid of all philosophical difficulties of quantum mechanics. However, nowadays the task of *representing* the world is dealt with by mathematical system theory and related phenomenological theories - here we do not need quantum mechanics. As opposed to system theory, a *fundamental* theory of matter not only must represent it must also *explain*. In order to fulfill this task, quantum mechanics must be more than a theory of relations between observable quantities. The ultimate referent of a fundamental theory has to be matter and radiation, and not experiments or experience. That is, *from a fundamental point of view the only adequate interpretation of quantum mechanics is an ontic interpretation.*

Regulative principles

Pioneer quantum mechanics has brought about two new aspects of profound philosophical importance:

(i) Quantum events are *indeterminate*. Pioneer quantum mechanics was the first theory in which primary probabilities played a crucial role.

(ii) In classical theories, the possibility of assigning sharp values to all observables is the basis for the *objectifiability* of physical properties. Due to the existence of complementary quantities, the *traditional* association of a value to every observable (measured or not) is no longer possible.

No interpretation of quantum mechanics can abstain from dealing with the philosophical problems of determinism and realism. Of course, an interpretation which is logically inconsistent or which contradicts accepted empirical facts must be rejected. Yet, the mathematical formalism together with the empirical facts are not sufficient to determine the interpretation. The normative rules needed in addition are called *regulative principles* (cf. Körner, 1957). By definition, normative rules are not susceptible of experimental proof or disproof; they are concerned with the beautiful, the true and the useful (cf. Naville, 1920).

The fact that regulative principles are of metaphysical nature does not mean that we can eliminate them. Everybody does adopt some regulative principles. Not stating the regulative principle adopted does not imply that one has a scientific theory free from metaphysics. The positivistic doctrine that metaphysical propositions are meaningless is itself a metaphysical regulative principle. Regulative principles express the philosophical prejudices we adopt. They are important since they function as rules of conduct (say for the creative activity of theory construction) and since they play a crucial role in our reasoning and explications.

No particular regulative principle is logically or empirically necessary, we can reject it if we wish to do so. The adopted regulative principles can answer the question whether a particular scientific theory gives a "better explanation" of empirical facts than another. The vital spark of a theory depends very much on the regulative principles adopted. Convincing regulative principles inspire scientists to speculative theoretical developments that can be corroborated experimentally. Since the natural sciences cannot be isolated from human thought and culture, we are always guided by our past experience and by the way of thinking induced by our cultural basis so that we are not entirely free in adopting or rejecting normative rules. On the other hand, it is not too astonishing that the interpretation of a scientific theory should change in the course of history. We may realize that some of the regulative principles adopted in earlier times are no longer compelling for us. What was a necessity of thought for Kant may be an inconvenient choice of a regulative principle for us.

In the early days of quantum theory the regulative principle of Kantian causality seemed to be the main obstacle to a proper understanding of quantum mechanics. Nowadays, the reconciliation of primary probabilities with our cultural basis no longer is an essential problem for the younger generation. On the other hand, the proper formulation of regulating principles relating to the idea of realism has become more and more an urgent problem.

Realism as a philosophy of science

Pioneer quantum mechanics was the first scientific theory to seriously question the philosophical idea that the world exists and has definite properties independently of any observer. The orthodox interpretation of quantum mechanics seems to deny that quantum systems are objectifiable. According to many text-books, a quantum system has no state until a state has been determined experimentally. This is a disquieting situation since most scientists explore nature in order to find out *what really is*, and not in order to find out what the effect of their exploration is.

A realist holds that a world of things exists independently of people, and that science is an intellectual and experimental activity aiming at an understanding of this world. Scientific realism is the view that an external world exists and that theories describe entities having reality beyond the observing mind and independently of it. The idea of an objective reality is not a hypothesis that can be verified or disproved experimentally but a regulative principle. We cannot ask whether scientific realism is true or false, but we have to investigate whether and in which form it is possible to adopt realism as a regulative principle. Realism is as much a problem of philosophy as of science. The crux of any interpretation of pioneer quantum emchanics is that it is impossible to take over all the regulative principles generally accepted for classical theories.

In the older literature on the interpretation of quantum mechanics it is claimed that adopting the regulative principle of realism leads to epistemological contradictions. This view is not correct; epistemological contradictions occur only if one uses hidden assumptions which are taken to be trivially or evidently true. For example, one gets a contradiction if one assumes the validity of Boolean logic for the temporal propositions about properties.

We speak of an *ontic interpretation of quantum mechanics* if the framework of this theory permits us logically to assume that matter and radiation have a reality beyond the observing mind and independently of it. Even if this realism is possible, its rejection is logically incontestable. It is sterile but possible to assume that all our experience "is but a dream within a dream". The crucial point is which compatible regulative principles are better suited to our understanding of phenomena.

Digression: The problem of reality is splendidly discussed in Bertolt Brecht's Turandot:
DER LEHRER *Si Fu, nenne uns die Hauptfragen der Philosophie.*
SI FU *Sind die Dinge ausser uns, für sich, auch ohne uns, oder sind die Dinge in uns, für uns, nicht ohne uns.*
DER LEHRER *Welche Meinung ist die richtige?*
SI FU *Es ist keine Entscheidung gefallen.*
DER LEHRER *Zu welcher Meinung neigte zuletzt die Mehrheit unserer Philosophen?*
SI FU *Die Dinge sind ausser uns, für sich, auch ohne uns.*
DER LEHRER *Warum blieb die Frage ungelöst?*
SI FU *Der Kongress, der die Entscheidung bringen sollte, fand wie seit zweihundert Jahren im Kloster Mi Sang statt, welches am Ufer des Gelben Flusses liegt. Die Frage hiess: Ist der Gelbe Fluss wirklich, oder existiert er nur in den Köpfen? Während des Kongresses aber gab es eine Schneeschmelze im Gebirg, und der Gelbe Fluss stieg über seine Ufer und schwemmte das Kloster Mi Sang mit allen Kongressteilnehmern weg. So ist der Beweis, dass die Dinge ausser uns, für sich, auch ohne uns sind, nicht erbracht worden.*
DER LEHRER *Gut. Die Stunde ist zu Ende.*
(quoted from the Werkausgabe Edition Suhrkamp, Brecht, 1967).

Everyone believes in some kind of realism. The world of the experimenter is realistic and materialistic. It would be very disturbing and inconsistent to expect that a scientist believes in the existence of an external reality in the world of everday experience and in the laboratory but not when he is working on his theories. *It is the duty of the theoretician to make such a choice of regulative principles that the reasonable requirements of realism are fulfilled.* Neither the Copenhagen interpretation nor the von Neumann-London-Bauer interpretation are acceptable as a fundamental theory of matter.

The missing link is a quantum theory of classical systems

Pioneer quantum mechanics in any of the proposed interpretations has an agonizing shortcoming: *it cannot describe classical systems.* The basic trouble with every enlargement of the domain of pioneer quantum mechanics is that it preserves the linearity of the equations of motion and the superposition principle. This is not the fault of the interpretations; it is a built-in feature of the *formalism* of pioneer quantum mechanics.

In the pioneer time, quantum mechanics was considered as containing classical mechanics in the limit $\hbar \to 0$. Besides conceptual and mathematical difficulties with the existence of this limit, it should not be forgotten that in Bohr's view the classical concepts are *prerequisites* for establishing quantum mechanics.

In the framework of the von Neumann-London-Bauer interpretation the central problem is the description of measurements within the formalism of pioneer quantum mechanics. This *measurement problem of quantum mechanics* can be solved if and only if we are able to *derive* the existence of classical properties from quantum mechanics. The problem is insoluble for pioneer quantum mechanics of systems having only finitely many degrees of freedom.

The basic concepts of the Everett interpretation are the *branches* and the *splitting* into branches. These notions refer to *classically inequivalent states*, a concept that has no solid foundation in the formalism of pioneer quantum mechanics.

Classical properties are defined as properties for which incompatible properties do not exist. A quantum system can have classical properties if and only if its algebra of observables has a nontrivial center, that is if there exist observables that commute with *all* observables of the system. These observables span an Abelian subalgebra of the algebra of all observables, they are called *classical observables*. The innocently looking irreducibility axiom introduced by von Neumann (1932) implies that pioneer quantum mechanics has no (nontrivial) classical observables. This assumption is empirically wrong. Every molecule possesses at least two classical properties: the mass and the electric charge. More complicated and more interesting classical observables emerge in complex molecules. Chirality is an example of a classical observable that can emerge already in small molecules, the tertiary structure of biomolecules is a classical property requiring a rather complex molecule. But note that the largeness of a system has a priori nothing to do with the existence of classical properties. There exist very small systems having classical properties (e.g. the charge of the electron) and very large systems exhibiting typical quantum phenomena (e.g. superfluids and superconductors).

Fortunately, these difficulties are restricted to the traditional formulation of pioneer quantum mechanics and fade away if we give up von Neumann's irreducibility postulate. *The development of a theory of quantal systems that possess both quantal and classical properties is a problem of the formalism and not a problem of the interpretation of the formalism.* In order to inaugurate such a theory, we have to explain what we mean when we claim that a large system can be isolated from its surroundings. Because large molecular systems are as a rule strongly coupled with their surroundings (Zeh, 1970, 1971, 1973; Baumann, 1970, 1972), we have to investigate the stability of any isolation procedure. We can no longer assume that a large system is isolated and is in a pure state; the state concept has to be discussed in connection with the dynamical interactions of the system with the external world. This is an exceedingly difficult mathematical problem, since it has to be *proved* by a stability analysis that the weak coupling to the environment (which is a system with infinitely many degrees of freedom!) leaves the *qualitative* behavior of the molecular system unchanged. If a stability analysis validates the behavior expected by ab-initio quantum chemistry, we call the molecular system a *robust* one. A thorough analysis shows that there are both robust as well as nonrobust molecular systems. The *qualitatively* new behavior of nonrobust systems can be summarized by introducing *superselection rules* and *classical observables*. In particular, *large molecular systems can acquire classical properties by their residual interactions with the rest of the world.* Such a derivation goes beyond the scope of pioneer quantum mechanics of systems with finitely many degrees of freedom; we must use the methods of modern algebraic quantum mechanics.

Comment: The reason is that the *von Neumann-Stone* uniqueness theorem for the representations of the canonical commutation relations by operators on a Hilbert space is valid only for systems with finitely many degrees of freedom. There are uncountably many unitarily inequivalent representations of the canonical commutation relations for systems with infinitely many degrees of freedom (as the electromagnetic radiation field of a molecule). In each of these inequivalent representations the molecular system shows qualitatively different features.

To conclude, we repeat: *the nonexistence of classical properties within pioneer quantum mechanics is a problem of the formalism only, no epistemological questions are involved.* The measurement problem can be solved, classically inequivalent branches can be introduced, and a consistent ontic interpretation of quantum mechanics is possible if and only if von Neumann's irreducibility postulate is given up. We shall have to say more about this in the following chapters.

Summing up and outlook

- The correspondence principle is outdated. The fundamental laws of nature are invariance principles. The success of quantization is due to a common kinematical symmetry of classical mechanics and pioneer quantum mechanics. The proper way to handle this common group structure is the representation theory of groups. In pioneer quantum mechanics the space-time structure is introduced by the Galilei group, the internal symmetries are given by an Abelian gauge group leading to the electromagnetic interaction between elementary Galilean objects.
- The great success of pioneer quantum mechanics is due to the fact that practitioners hardly worry about logical inconsistencies and unsettled philosophical problems.
- The limitations of pioneer quantum mechanics are related to many open conceptual problems which go deep into the logic and philosophy of science.
- The operationalism adopted in many presentations of pioneer quantum mechanics is a pseudo-philosophy and fails as a basis for a theory of matter. Age-old philosophical problems cannot be settled by operationalistic definitions.
- Classical mechanics and pioneer quantum mechanics are incompatible. Yet, classical physics is indispensable for establishing a relation between theoretical quantities and empirical data. Pioneer quantum mechanics is not a fundamental, unified theory of molecular matter explaining why nature with all its complexity is the way it is.
- The Copenhagen interpretation in the strict version of Niels Bohr has no measurement problems and therefore avoids the profound consistency problems of the orthodox or the Everett interpretation. But Bohr's epistemic interpretation expresses merely states of knowledge and misses the point of a genuine scientific inquiry. Chemists never have adopted Bohr's view that microphysical objects do not exist.
- If we suppose that science investigates the nature of things,

and that matter and radiation exist and have properties, we have to use an ontic interpretation of quantum mechanics.

- If we assume that pioneer quantum mechanics is a universal theory of molecular matter, then an ontic interpretation of this theory is impossible. In the framework of pioneer quantum mechanics of systems having finitely many degrees of freedom, there exist no measurement instruments for the so-called observables. There exist no classical records since in pioneer quantum mechanics every observable is accompanied by incompatible observables.

- The regulative principles for a consistent interpretation of the formalism of pioneer quantum mechanics cannot be chosen to reproduce scientific realism. This fact leads to a grave dilemma, for a chemist and a biologist should be permitted to assume that they are investigating something that has a real existence independently of their investigations.

- Fortunately, these difficulties are restricted to the traditional formalism of pioneer quantum mechanics. The modern investigations into the foundations and the axiomatics of quantum mechanics have resulted in a *generalized quantum mechanics* which contains in a natural way both pioneer quantum mechanics as well as classical mechanics. It is now possible to choose the regulative principles in such a way that in the case of a classical theory the interpretation of generalized quantum mechanics is identical with the generally accepted interpretation of classical theories. There is universal agreement that classical physical theories refer to objectively existing, individual systems. It is not a trivial matter that such an ontic interpretation is possible for quantum mechanics as well; indeed it was rejected as inconsistent in the early years of the development of pioneer quantum mechanics. We know today, that the obstacle was the unfounded belief in the unrestricted validity of the superposition principle, formalized in von Neumann's irreducibility postulate.

4. BEYOND PIONEER QUANTUM MECHANICS

"It must be emphasized that it is not a question of accepting the correct theory and rejecting the false one.
It is a matter of accepting that theory which shows greater formal adaptibility for a correct extension. This is a formalistic, esthetic criterion, with a highly opportunistic flavor".

John A. von Neumann (1955b)

4.1 INTRODUCTION

The new era of quantum theory began around 1932, and witnessed a maturation, many extensive developments, and a striking turn to an abstract structural approach. Important results of modern quantum theory, although easily accessible, are as yet not very well known. In spite of its urgency, the conceptual recasting of pioneer quantum mechanics has been slow. Even slower is the transference of successful reformulations into our textbooks which are still full of archaisms and inconsistencies; they hardly reflect the important progress made in the last two decades. In this chapter we shall attempt to give a brief outline of the main trends in the development of quantum theory since 1932. In this period, the new results have been so voluminous that it is necessary to limit our discussion and to select a few topics which are of special importance for the theory of molecular matter. Naturally the selection is in part guided by my prejudices, so my apologies are due to all those whose work has been ignored or insufficiently discussed.

The logical and philosophical foundation of quantum mechanics has interested many writers and has been the subject of hundreds of articles. These discussions were mainly critical and had but a small influence on the evolution of quantum theory. What scientists prefer is a creative approach, not just a critical one. The works that changed radically our basic point of view came not from philosophy but from mathematical physics.

One of the most notable features of the new era is the trend to higher levels of abstractions. There is a distinct tendency to move away from specific problems and to prefer comparative studies of the structure of physical theories. Nowadays, physical theories are characterized by their abstract structural components. Quantum theory has become more abstract, hence simpler (for the initiate). The abstract of yesterday is the concrete of today. The classical pictorial models have been replaced by algebraic mathematical concepts, so that abstract algebra and functional analysis have become highly influential in modern quantum theory. Mathematics is no longer a mere tool for solving specific problems but the basic weapon to generate deep structurally based insight. Clearly, the increasingly mathematical character of science draws our attention again to the philosophical and practical aspects of the connection between science and mathematics. Moreover, there are important educational and sociological problems. According to Wightman (1969), "*the older generation almost always regards the younger as 'too mathematical'*".

Certainly, there are no *a priori* reasons to assume a fundamental parallelism between mathematical and scientific concepts. Nevertheless, the concubinage between physics and mathematics has never been an unhappy one. Probably the most important result of this joint undertaking is the gradual emergence of the idea that *science is a symbolic construction of man: "... anstatt eines realen räumlich-zeitlich-materiellen Seins behalten wir nur eine Konstruktion in reinen Symbolen übrig"* (Weyl, 1949a). If there is any unifying concept, it is very likely to be connected with symmetries. According to the modern point of view, we always should look for the underlying symmetry of a theory and amplify on the role played by the relevant group. More and more, theoretical scientists connect the ultimate reality to the underlying group itself.

The modern versions of quantum theory have been developed not out of a mere desire for greater generality but they have become imperative for a better understanding of thermodynamics and of the classical aspects of molecular matter. Recall that

(i) pioneer quantum mechanics cannot describe classical systems, hence it cannot deal in a proper way with experiments,

(ii) pioneer quantum mechanics is not appropriate to deal with systems having infinitely many degrees of freedom.

Several attempts have been made to analyze quantum mechanics by axiomatization. The basic aim of any axiomatization is to elucidate the logical foundation of the theory and to find the basic structures of physical theories like classical mechanics, pioneer quantum mechanics and thermodynamics. Not all of the currently popular approaches to an *axiom-*

atic quantum theory are conceptually well-founded but most of them have produced deep insight into the mathematical structure of a future fundamental theory of matter.

The choice of an appropriate set of axioms for a generalized quantum mechanics is not a well-posed problem, there are many different feasible possibilites. One would like the axioms to be independent and their number to be small. More importantly, one would like every axiom to have a clear intuitive meaning which desideratum may contradict the mathematician's requirement of independence. There are three main approaches to an axiomatic quantum theory:

(i) the algebraic approach,
(ii) the quantum-logic approach,
(iii) the convex-state approach.

These three axiomatizations are pairwise distinct but each one leads to a very general theoretical framework that contains both Newtonian mechanics and pioneer quantum mechanics as special cases. In spite of the fact that classical mechanics and pioneer quantum mechanics are extreme views, they can be treated in every version of modern generalized quantum theory on an equal footing.

It must be emphasized that there is no question of accepting one of these axiomatizations as the correct one and rejecting the others as false ones. Notwithstanding, it is gratifying that we have not to make an arbitrary decision. The intersection of the three main approaches is nonempty and gives rise to a conceptually well-defined and mathematically rich theoretical framework: the so-called theory of W*-systems which will be discussed in greater detail in chapter 5. There are good reasons to believe that the theory of W*-systems contains the best ideas of the algebraic approach, of the quantum-logic approach and of the convex-state approach.

Remark: *On the impossibility of full axiomatizations*
It may be helpful to recall that it is impossible to characterize the system of natural numbers axiomatically. That is, given any consistent countable system of axioms satisfied by the system of natural numbers, there always exists another linearly ordered system satisfying all these axioms but which is not isomorphic to the system of natural numbers as an ordered system (Skolem, 1934). This example shows the limitations of the formalist's method and should warn us to expect too much from axiomatics. Nevertheless, to formulate *some* (never: *all*) physical principles as axioms in clean mathematical form and to explore their consequences may be an extremely helpful strategy to get some control over our intuitive ideas. However, complete formalization never can be the aim of modern axiomatic scientific theories.

4.2 ALGEBRAIC QUANTUM MECHANICS

Observables form a Jordan algebra

Algebraic quantum mechanics studies the algebraic structures associated with observables. It starts with a remark by Pascual Jordan (1932a, 1933a-b, 1934) to the effect that the statistics of measurements can be understood in terms of a nonassociative algebraic structure which nowadays is called a *Jordan algebra* (compare e.g. Braun and Koecher, 1966). Jordan noted that in pioneer quantum mechanics the sum of the two observables is again an observable but that in general the product AB of two observables A and B has no conceptual meaning. The product AB is an observable if and only if the two observabels A and B commute (otherwise AB is not self-adjoint). However, the combination $A\circ B \overset{\text{def}}{=} (AB+BA)/2$ always is meaningful and expressible in terms of sums and squares,

$$2A\circ B = (A+B)^2 - A^2 - B^2 \quad .$$

Jordan posed the problem of finding an algebraic characterization of the bilinear mapping $A\circ B$ independently of an underlying associative algebra. In pioneer quantum mechanics, this *Jordan product* $A\circ B$ is commutative, $A\circ B=B\circ A$, and distributive, $A\circ(B+C)=A\circ B+A\circ C$, but in general non-associative, $A\circ(B\circ C)\neq(A\circ B)\circ C$. However, the observables of pioneer quantum mechanics satisfy the weaker relation $A\circ A\circ(B\circ A)=(A\circ A\circ B)\circ A$, which implies the power associativity $A^{(m)}\circ A^{(n)}=A^{(m+n)}$, where $A^{(m)}=A\circ A\circ\ldots\circ A$ (m times). Abstracting this algebraic structure of the observables of pioneer quantum mechanics, one defines: a real algebra J is called a *Jordan algebra* if the following three conditions are satisfied for every A,B,C in J:

$$\text{(i)} \quad A\circ B = B\circ A \quad ,$$

$$\text{(ii)} \quad A\circ(B+C) = A\circ B+A\circ C \quad ,$$

$$\text{(iii)} \quad A\circ A\circ(B\circ A) = (A\circ A\circ B)\circ A \quad .$$

A Jordan algebra J is called formally real if $A^2+B^2=0$ implies $A=B=0$ for every pair $A,B\in J$. A Jordan algebra which arises from an associative algebra $\mathcal{A}$ by defining $A\circ B=(AB+BA)/2$ is called a *special Jordan algebra*. A Jordan algebra that is not special is called *exceptional*. Jordan, von Neumann and Wigner (1934) showed that with one exception, every formally real simple finite-dimensional Jordan algebra is special, arising from the ordinary associative matrix algebras. The exception is at least mathematically interesting, it is the Jordan algebra M_3^8, the algebra of all hermitean 3×3-matrices over the Cayley numbers. Albert

(1934) proved that this Jordan algebra cannot be represented be any associative matrix algebra, so that M_3^8 is the only finite-dimensional exceptional Jordan algebra, and in principle it could be used for a new kind of quantum mechanics. The idea that some internal degrees of freedom could be associated with the exceptional Jordan algebra M_3^8 has never been settled but is still intriguing contemporary theoreticians trying to unify weak, electromagnetic and strong interactions (compare e.g. Horwitz and Biedenharn, 1979).

While all finite dimensional Jordan algebras with the single exception of M_3^8 lead to the same results as pioneer quantum mechanics, the introduction of Jordan algebras in quantum mechanics was an important step towards algebraic quantum mechanics. Jordan's idea that a non-associative but commutative algebra should supersede the associative but noncommutative algebra of observables of pioneer quantum mechanics is rich in consequences even in the case of special Jordan algebras. Differing assumptions on the structure of the algebra of observables affect the size of its automorphism group which describes the *symmetries* of the system. Since the Jordan-algebra structure is *weaker* than the associative algebraic structure of pioneer quantum mechanics, the automorphism group of a special Jordan algebra is *larger* than the automorphism group of its associative algebra. The additional symmetries allowed in a Jordan system correspond to the anti-unitary transformations of pioneer quantum mechanics.

Jordan morphisms #

A linear mapping $\alpha: J \to \tilde{J}$ between two Jordan algebras J and $\tilde{J}$ is called a *Jordan homomorphism* if (i) $\alpha(A \circ A) = \alpha(A) \circ \alpha(A)$ for every A in J, and (ii) $\alpha(A \circ B \circ A) = \alpha(A) \circ \alpha(B) \circ \alpha(A)$ for every A,B in J. A bijective Jordan homomorphism (i.e. if α is one-to-one and onto) is called a *Jordan isomorphism*. A Jordan isomorphism of a Jordan algebra onto itself is called a *Jordan automorphism*.

Observables form a Banach space

Jordan's original approach was restricted to finite-dimensional algebras. It is plain that for infinite-dimensional algebras (like the ones used in pioneer quantum mechanics) some topological conditions have to be added. Von Neumann (1936) started the study of infinite-dimensional Jordan algebras but never completed it.

Remark #

Von Neumann's paper of 1936 is truly remarkable and it is a pity that it remained almost unnoticed. Von Neumann introduced an *abstract* Jordan algebra (without any reference to a relization as an algebra of operators acting on a Hilbert space) and an abstract $\mathcal{T}$-topology acting like the weak topology in the Hilbert-space case. Von Neumann's Jordan algebras resemble the JBW-algebras introduced recently by Shultz (1979). A JBW-algebra is a JB-algebra which is the dual space of a Banach space (the definition of a JB-algebra is given below). A JBW-algebra has the same relation to a JB-algebra as a W*-algebra to a C*-algebra. This implies that von Neumann's Jordan algebras and JBW-algebras allow a quantum-logical interpretation while JB-algebras and C*-algebras in general do not.

Irving Ezra Segal (1947) was the first to introduce the norm topology into Jordan algebras and to propose axioms for a Jordan-algebraic quantum mechanics, using the then developed theory of normed algebras (compare also Sherman, 1951). Segal assumed that the observables formed a real Banach space (i.e. a complete normed vector space) $\mathcal{A}$ such that if $A \in \mathcal{A}$, then the powers of A are well-defined and in $\mathcal{A}$, $A^n \in \mathcal{A}$ for $n=0,1,2,..$, and such that the usual rules for operating with polynomials in a single variable are valid. With this structure, one can define a Jordan product $A \circ B$ between two observables by

$$A \circ B = \tfrac{1}{2}\{(A+B)^2 - A^2 - B^2\} \quad .$$

In addition, Segal required three additional postulates describing the connection between algebraic and topological properties, namely:

(i) $\|A^2 - B^2\| \leq \max\{\|A^2\|, \|B^2\|\}$,

(ii) $\|A^2\| = \|A\|^2$,

(iii) A^2 is a continuous function of A.

An algebraic system having these properties is called a *Segal system.* Segal required neither a distributive postulate nor a weak topology, but he was still able to develop a spectral theory of observables of a Segal system.

The most important examples of Segal systems are the so-called *C*-systems*. The set of all self-adjoint elements of a C*-algebra fulfill all axioms of a Segal system. In particular, pioneer quantum mechanics is a C*-system (hence a Segal system), whereby the C*-algebra is taken to be the algebra of all bounded operators acting on some Hilbert space.

C-algebras and W*-algebras: definition and historical remarks* #

The algebra of all bounded operators acting on a Hilbert space is a Banach algebra with respect to the operator norm induced by the inner product of the Hilbert space. In addition, this algebra possesses an involution $A \rightarrow A^*$ that satisfies (i) $(A+B)^* = A^* + B^*$, (ii) $(cA)^* = c^*A^*$, $c \in \mathbb{C}$, (iii) $(AB)^* = B^*A^*$, (iv) $(A^*)^* = A$. A Banach algebra with an involution is called a Banach *-algebra. A routine computation shows that every norm-closed *-subalgebra of the algebra of all bounded operators on a Hilbert space fulfills the relations $\|A\| = \|A^*\|$ and $\|A^*A\| = \|A^*\|\|A\|$, hence $\|A^*A\| = \|A\|^2$. Such an algebra is called a concrete C*-algebra.

Two famous papers by Gelfand (1939, 1941) initiated the investigations of the Russian school on abstract Banach algebras. Gelfand and Naǐmark (1943) succeeded in characterizing C*-algebras of operators on a Hilbert space by an elegant system of intrinsic postulates. Gelfand and Naimark defined a B*-algebra as a Banach algebra with an involution $A \to A^*$ which satisfies (i) $\|A\|=\|A^*\|$, (ii) $\|A^*A\|=\|A^*\|\ \|A\|$, (iii) $1+A^*A$ has an inverse, and they showed that every B*-algebra is isometrically *-isomorphic with a C*-algebra of operators on some Hilbert space. Moreover, they conjectured that the axioms (i) and (iii) are redundant. Many mathematicians have tried to weaken the axioms for a B*-algebra but the final proof for the conjecture by Gelfand and Naimark had to wait for 24 years! Glimm and Kadison (1960) proved the redundance of conditions (i) and (iii) for algebras having an identity, and Vowden (1967) the same for the general case. Since this question is now settled, the distinction between B*- and C*-algebras has become pointless, so that nowadays it is usual to define abstractly a C*-algebra as a Banach *-algebra satisfying the usual algebraic properties and in addition axiom (ii), or equivalently, the axiom $\|A^*A\|=\|A\|^2$.

A *W*-algebra* is a C*-algebra which is the dual space of a Banach space. This Banach space is called the *predual* of the W*-algebra $\mathcal{A}$ and is denoted by $\mathcal{A}_*$, so that $(\mathcal{A}_*)^*=\mathcal{A}$. W*-algebras are the abstract versions of *von Neumann algebras* (defined as weakly closed sub-*-algebras of operators acting on some complex Hilbert space and containing the identity operator). Every von Neumann algebra is a W*-algebra, and every W*-algebra is W*-isomorphic to a von Neumann algebra.

Standard reference works on C- and W*-algebras:* Dixmier (1957, 1964), Sakai (1971), Strătilă and Zsidó (1979), Pedersen (1979), Bratteli and Robinson (1979), Takesaki (1979).

A Segal system is called *special* if it is isomorphic to the set of self-adjoint elements of a C*-algebra, otherwise it is called *exceptional*. Special Segal systems are distributive, but there exist many exceptional Segal systems which are nondistributive (Sherman, 1956; Lowdenslager, 1957). Nondistributive Segal systems do not seem to have any physical interpretation so that it is reasonable to exclude them by requiring that a Segal system forms a Jordan algebra under the Jordan product. Jordan algebras fulfilling Segal's axioms are called *JB-algebras*. Every C*-algebra is a JB-algebra in the Jordan product $A \circ B=(AB+BC)/2$; such a JB-algebra is called a JC-algebra. Another example of a JB-algebra is M_8^3, the hermitean 3×3-matrices over the Cayley numbers; it is not a JC-algebra. An important result by Alfsen et al. (1978) shows that every JB-algebra can be constructed from these two examples in a canonical way, but it is not true that every JB-algebra is a direct sum of a JC-algebra and a M_8^3-type algebra.

JB-and JBW-algebras #

In analogy with B*-algebras, a real Jordan algebra $\mathcal{A}$ is called a JB-algebra if it is also a real Banach space, the norm of which satisfies the condition

$$\|A\|^2 = \|A^2\| \le \|A^2+B^2\| \quad \text{for each pair } A,B\in\mathcal{A} \ .$$

Every JB-algebra is formally real. In analogy with W*-algebras, a JB-algebra which is the dual space of a Banach space is called a JWB-algebra. Every JBW-algebra $\mathcal{A}$ admits a unique decomposition $\mathcal{A}=\mathcal{A}_{sp}\oplus\mathcal{A}_{ex}$ into a special part $\mathcal{A}_{sp}$, and a purely exceptional part $\mathcal{A}_{ex}$. Here $\mathcal{A}_{sp}$ is isomorphic to a weakly closed Jordan operator algebra (a so-called JW-algebra), and $\mathcal{A}_{ex}$ is isomorphic to the algebra $C(X,M_3^8)$ of the continuous functions from a hyperstonean space X into M_3^8.

References: Alfsen et al. (1978), Shultz (1979).

The C-algebraic approach to generalized quantum mechanics*

Segal's axiomatization has acted as a seminal force for the development of the modern C*-algebraic approach to generalized quantum theories, in particular for the theory of systems having an infinite number of degrees of freedom. The purely algebraic viewpoint where the nature of elements does not play any role has turned out to be most fruitful in mathematics and theoretical science. Since the norm of a C*-algebra is algebraically determined, Segal's approach has the appealing feature of being purely algebraic.

The basic mathematical object of the C*-algebraic approach is an *abstract algebra* $\mathcal{A}$ which, for technical reasons, is chosen to be a C*-algebra. The self-adjoint elements of $\mathcal{A}$ form a Jordan algebra, and are called observables. No reference is made to a realization as an algebra of linear transformations acting on some Hilbert space. Segal (1947) proposed to interpret the positive linear functionals on the algebra $\mathcal{A}$ of observables as statistical states and introduced the following *mathematical* notion of state: a *state* is defined as a normalized positive linear function on the C*-algebra $\mathcal{A}$. That is, every state ρ assigns to each $A \in \mathcal{A}$ a complex number $\rho(A)$ such that

(i) $\rho(1) = 1$ (normalization),

(ii) $\rho(A^*A) \geq 0$ (positivity),

(iii) $\rho(aA+bB) = a\rho(A)+b\rho(B)$ (linearity),

for every $A,B \in \mathcal{A}$ and $a,b \in \mathbb{C}$. A state ρ is called *mixed* if it can be written as

$$\rho = c\rho_1+(1-c)\rho_2$$

where ρ_1 and ρ_2 are distinct states and $0<c<1$. A state which is not mixed is called *pure*.

Critical remarks on the terminology

Unfortunately, Segal's proposal to call a normalized positive linear functional (on a Jordan algebra or a C*-algebra) a *state* has been adapted universally by mathematicians and mathematical physicists. This is bad since it prematurely identifies a mathematical notion with a physical concept. The notion of a *state* is a physical and not a mathematical one. Moreover, it is neither obvious nor established that a physical state has to be represented by a positive linear functional (for example: why should the functional representing a physical state be *linear?*). Indeed, the proponents of the C*-algebraic approach tacitly adopt a *statistical ensemble interpretation* and claim that giving a state is synonymous with giving the statistical ensemble expectation value for all observables. In mathematical system theory or in an ontic interpretation it is not true that every normalized positive linear form is a possible physical (or ontic) state.

To call non-pure states mixed states is even worse. *If* the state space is a simplex, each state is the resultant of a unique measure supported by the pure states. However, the state space is a simplex if and only if the algebra of observables is commutative, i.e. in classical theories. In a non-commutative C*-system (e.g. in pioneer quantum mechanic), the knowledge of a non-pure statistical state does not imply the knowledge of the kind of ensemble to which it refers. There exist infinitely many different ensembles having the same non-pure state, so that the use of the name "mixed state" is conceptually confusing because in nonclassical systems one does not know what the components of the mixture are.

However, this terminology has become so common that we will accept it in spite of its foolishness. Since its inception, quantum mechanics has been plagued by a misleading terminology. No end of this disease is in sight.

Even in the familiar case of pioneer quantum mechanics, Segal's C*-algebraic approach brings out some new features. In pioneer quantum mechanics the algebra of observables is taken as the algebra $\mathcal{B}(\mathcal{H})$ of all bounded linear operators acting on some complex Hilbert space, so that $\mathcal{A}$ is a concrete C*-algebra, $\mathcal{A}=\mathcal{B}(\mathcal{H})$. Every normalized vector $\Psi\in\mathcal{H}$ defines a state ρ_Ψ by $\rho_\Psi(A)=\langle\Psi|A\Psi\rangle$ for every $A\in\mathcal{A}$. Such a state ρ_Ψ is called a *vector state*. More generally, every density operator $D\in\mathcal{B}(\mathcal{H})$ defines a state ρ_D by $\rho_D(A)=\mathrm{tr}(DA)$ for every $A\in\mathcal{A}$. Such states are called *normal states*. A normal state is pure if and only if D is a projection, $D^2=D$, hence if and only if it is a vector state. Note that there exist non-normal states, even pure ones. An example of such non-normal states can be given as follows. Let $A\in\mathcal{A}$ be a self-adjoint operator with a purely continuous spectrum. Segal (1947) has shown that for every point λ of the spectrum of A there exists a pure state ρ_λ such that $\rho_\lambda(A)=\lambda$. Since A has no eigenvectors in $\mathcal{H}$, the pure state ρ_λ cannot be a vector state, hence ρ_λ is not normal. The traditional version of pioneer quantum mechanics does not admit non-normal states, so that the C*-algebraic version of pioneer quantum mechanics leads to a sensible generalization of the mathematical framework.

The existence of superselection rules finds a natural description in the algebraic approach. A pure state ρ is said to be *coherent* with another pure state ρ', if there exists an element B in the algebra of observables, such that

$$\rho'(A) = \rho(B^*AB) \quad \text{for all} \quad A\in\mathcal{A} \quad .$$

The set of all mutually coherent states is called a *superselection sector* (or, for short, a *sector*). If two states are not related in this way, then we say that they are separated by a *superselection rule*.

Coherent superposition of pure states #

A *coherent superposition* of two pure states is defined as follows (Roberts and Roepstorff, 1969). Let ρ be a pure state from some sector, and let J_ρ be the left ideal in A generated by ρ,

$$J_\rho \overset{\text{def}}{=} \{A : A \in A,\ \rho(A^*A) = 0\} \quad .$$

For any $B \in A$, $B \notin J_\rho$, the normalized linear functional ρ_B defined by

$$\rho_B(A) \overset{\text{def}}{=} \frac{\rho(B^*AB)}{\rho(B^*B)} \quad \text{for all} \quad A \in A \quad ,$$

is a pure state of the sector chosen. Let B, C be two modulo J_ρ linearly independent observables from A, then ρ_{B+C} is defined to be the *coherent superposition* of ρ_B and ρ_C. According to this definition, coherent superpositions are possible only between pure states of the *same* superselection sector. Any coherent superposition is again a *pure* state which lies in the *same* sector.

It may be worthwhile to exemplify this definition for the Hilbert-space model of pioneer quantum mechanics. In the traditional Hilbert-space formalism, a pure state $\rho(A)$ equals $\langle\Psi|A|\Psi\rangle$ with a normalized state vector Ψ. If we define:

$$\Psi_B = B\Psi/\|B\Psi\| \quad , \quad \Psi_C = C\Psi/\|C\Psi\| \quad , \quad \Psi_{B+C} = (\Psi_A + \Psi_B)/\|\Psi_A + \Psi_B\| \quad ,$$

we have $\rho_B(A) = \langle\Psi_B|A|\Psi_B\rangle$, $\rho_C(A) = \langle\Psi_C|A|\Psi_C\rangle$,

and $\rho_{B+C}(A) = \langle\Psi_{B+C}|A|\Psi_{B+C}\rangle$.

C-Algebras for infinite systems*

C*-algebras appear naturally in the representation theory of the canonical commutation relations. Already in the very first papers on pioneer quantum mechanics, Heisenberg's canonical commutation relation $QP-PQ=i\hbar$ between the position observable Q and the momentum observable P was used to characterize the kinematics of the system. The formally equivalent Weyl commutation relations

$$U(p)V(q) = V(q)U(p)\exp(ipq/\hbar) \quad , \quad p,q \in \mathbb{R}$$

in terms of the unitary operators $U(p)=\exp(ipQ/\hbar)$ and $V(q)=\exp(-iqP/\hbar)$ are nowadays considered as the mathematically proper formulation of the canonical commutation relation. For systems having finitely many degrees of freedom, Stone (1930) and von Neumann (1931) have shown that up to trivialities (that is, up to unitary equivalence and multiplicity) there is one and only one system of unitary operators U(p), V(q) $(p,q \in \mathbb{R}^f,\ f<\infty)$ fulfilling Weyl's canonical commutation relations. It took quite some time before theoretical physicists realized that for systems having an infinite number of degrees of freedom the canonical commutation relations do *not* specify the kinematics uniquely. In fact, there is a striking difference between systems of a finite number and infinite number of degrees of freedom: *there exist uncountably many unitarily in-*

equivalent irreducible representations of Weyl's commutation relations for f=∞. This fact makes a C*-algebraic approach indispensible.

Historical remarks

The existence of a representation of the canonical (anti)commutation relations for systems of an infinite number of degrees of freedom was first shown in a heuristic way by Fock (1932) and in a mathematically rigorous form by Cook (1953). The Fock representation (or, more precisely, the Fock-Cook representation) is the only representation having a total number operator and a no-particle state (vacuum). In the old formulations of quantum field theories (like quantum electrodynamics), this Fock representation was considered exclusively.

The occurrence of inequivalent representations is implicitly contained in the fundamental paper by von Neumann (1938) on infinite tensor products of Hilbert spaces. Later, Cameron and Martin (1947) proved the singularity of trivial scale transformations in Wiener space which implies the non-implementability of the corresponding canonical transformation by a unitary map. However, these works were not appreciated by physicists; non-Fock representations were regarded as pathological phenomena of no physical interest. To the surprise of the pyhsicists, Friedrichs (1951), Miyatake (1952a, 1952b), and van Hove (1952) discovered that there exist several tangible unitarily inequivalent representations of the canonical commutation relations for fields. These non-Fock representations were called "strange representations", while Friedrichs used the term "Myriotic fields". Both, Miyatake and van Hove pointed out that the Hilbert space spanned by the eigenvectors of the free Hamiltonian H_0 of a neutral scalar field is perpendicular to the Hilbert space spanned by the eigenvector of the total Hamiltonian H_0+V of the meson field interacting with a fixed point. Therefore the states of H_0+V cannot be expanded in terms of the eigenstates of H_0 so that perturbation theory with respect to V breaks down (i.e. gives divergent results). Accordingly, the algebra of the (anti)commutation relations is not sufficient to specify the kinematics. In addition one has to specify a Hilbert space, or equivalently, an appropriate irreducible representation. Gårding and Wightman (1954a, 1954b) and Wightman and Schweber (1955) proved the existence of uncountably many unitarily inequivalent irreducible representations of the canonical (anti)commutation relations. The work by Haag (1955) and by Hall and Wightman (1957) has made it plain that inequivalent representations must appear in any nontrivial relativistic field theory. At least some of the divergences appearing in the old formulation of quantum field theory are definitely connected with an incorrect choice of the representation. The first general criteria for the unitary implementability of an important class of canonical transformations were given by Segal (1958).

In the C*-algebraic approach, one considers the objects U(p)V(q) (p,q∈Λ, where Λ is a finite-dimensional or infinite dimensional real inner product space) as abstract elements of a C*-algebra, fulfilling the relations

$$U^*(p) = U(-p) \quad ,$$
$$V^*(q) = V(-q) \quad ,$$
$$U(p)V(q) = V(q)U(p)\exp(ipq/\hbar) \quad .$$

The abstract C*-algebra generated by $\{U(p)V(q)\,|\,p,q\in\Lambda\}$ is called the *Weyl algebra* over Λ, it is simple so that every of its representations is faithful.

Representation of C-algebras* #

If a C*-algebra $\mathcal{A}$ is abstractly given, then it can be represented as an algebra of operators acting on a Hilbert space. A *representation* π of an abstract C*-algebra

$\mathcal{A}$ in a Hilbert space $\mathcal{H}$ is a map $A \to \pi(A)$ assigning a bounded operator $\pi(A)$ acting on $\mathcal{H}$ to each element $A \in \mathcal{A}$ such that:

(i) $\pi(A^*) = \pi(A)^*$,

(ii) $\pi(c_1A_1+c_2A_2) = c_1\pi(A_1)+c_2\pi(A_2)$, $c_1,c_2 \in \mathbb{C}$,

(iii) $\pi(A_1A_2) = \pi(A_1)\pi(A_2)$.

A famous theorem by Gelfand, Naĭmark and Segal (the so-called GNS-theorem) says that every state ρ on a C*-algebra $\mathcal{A}$ generates a representation π_ρ on a Hilbert space $\mathcal{H}_\rho$ (which can be constructed), such that there is a cyclic vector $\Omega_\rho \in \mathcal{H}_\rho$ with

$$\rho(A) = \langle \Omega_\rho | \pi_\rho(A)\Omega_\rho \rangle \quad \text{for every} \quad A \in \mathcal{A} \quad .$$

This representation is called the GNS representation of the C*-algebra $\mathcal{A}$ with respect to the state ρ. The triple $(\mathcal{H}_\rho,\pi_\rho,\Omega_\rho)$ is unique up to unitary equivalence. The GNS representation is irreducible if and only if ρ is pure.

References: Dixmier (1964), Sakai (1971).

An abstract C*-algebra $\mathcal{A}$ of observables can be represented by a C*-algebra of bounded linear operators acting on some Hilbert space. In the simplest possible case, all faithful representations of $\mathcal{A}$ are unitarily equivalent, so that the traditional Hilbert-space formalism of quantum mechanics can be recovered. However, most C*-algebras have an uncountable number of unitarily inequivalent representations. In particular, for systems with infinitely many degrees of freedom, an entirely new situation occurs which is not covered by pioneer quantum mechanics.

Every representation of the Weyl algebra is faithful. If the configuration space Λ of the Weyl system $\{U(p)V(q) | p,q \in \Lambda\}$ is finite-dimensional, then the uniqueness theorem of Stone (1930) and von Neumann (1931) implies that all representations of the Weyl algebra are unitarily equivalent. However, if the configuration space Λ is infinite-dimensional, there are uncountably many unitarily inequivalent representations of the Weyl algebra. Since the canonical commutation relations are not of type I (in the sense of Mackey, 1963b), these representations cannot be fully classified with the presently known techniques.

C*-algebraic quantum mechanics has been contrived in order to eliminate the need for this myriad of inequivalent representations. Originally it was hoped that physical theories could be formulated independently of any representation. In particular, Haag and Kastler (1964) emphasized that only the C*-algebraically invariant properties should have a physical meaning. They adopted the view that no actual experimental arrangement can prepare a statistical state precisely, but only specify a weak neighborhood in the space of positive linear forms on the algebra of all observables. From this point of view, all faithful representations of the algebra of observables are physically equivalent, no matter whether these representations are unitarily equivalent or not. Unitarily inequivalent but faithful representations have states that behave diffently with respect to observations made infinitely far away, what is considered as physically irrelevant by Haag and Kastler. However, it has proven to be extremely useful to distinguish unitarily inequivalent faithful representations as different models for idealized physical systems. Choosing different inequivalent faithful representations amounts to making different idealizations.

The relevance of inequivalent representations

One may object that all actual systems are finite. Even if this were true, there are great advantages in considering infinite systems as idealizations of finite systems. The best known example is equilibrium statistical thermodynamics. In the thermodynamic limit,

number of particles	$n \to \infty$	,
volume	$V \to \infty$	,
density	n/V finite	,

the systems considered become infinite systems. The use of the thermodynamic limit is an idealization which permits a rigorous mathematical characterization of such properties as phase transitions which would otherwise be masked by finite-size effects. It is only in the thermodynamic limit that phase transitions exhibit a discontinuity or singularity in the thermodynamic functions. Moreover, the thermodynamic limit is an essential ingredient to describe coexisting pure phases properly. Rather than performing the thermodynamic limit at some intermediate stage, it is conceptually more appealing and mathematically more rigorous to study thermodynamic systems directly as systems of an infinite number of degrees of freedom.

Furthermore, there are simple molecular systems having only a few manifest degrees of freedom but which require for their proper description either the introduction of ad hoc phenomenological concepts, or, in a strictly fundamental approach, the use of infinite systems. Most molecular systems are strongly entangled with their surroundings. Even a so-called free molecule is interacting with the rest of the world by the unavoidable coupling to the electromagnetic radiation field. Due to the peculiar nature of this coupling, the effect of this coupling can be of qualitative importance. In order to discuss such effects we are forced to include the electromagnetic quantum field so that the model for the molecular system is an infinite system. The inclusion of the surroundings can have dramatic effects, even if the individual interaction constants are extremely small. The resulting molecular state may be well disentangled from the environment; nevertheless it is not the same state one would get by neglecting the effects of the environment. The actually realized molecular state (the "dressed state") is accompanied by an infinite number of photons whose degrees of freedom are hidden, manifesting themselves only by a few parameters which can be described either by phenomenological concepts or the corresponding theoretical notions (like "classical observables", "molecular superselection rules"). From a fundamental point of view, the crucial phenomenon is a change of the equivalence class of the representing Hilbert space of the total system and its environment.

From a mathematical viewpoint, the theory of infinite systems is quite different from the corresponding theory for finite systems. So we have to ponder the difference between a finite system having very many degrees of freedom and an infinite system. First of all, a very large but finite system is as a rule extremely complicated. From a physical point of view, the finite accuracy of any feasible experiment implies that we can never distinguish empirically between an infinite system and a system having sufficiently many degrees of freedom. For a given experimental situation one can always approximate an infinite system by a sufficiently large finite system. On the other hand, an expedient way to avoid the unobservable complexity of a large but finite system is to introduce an appropriate limit to a system having infinitely many degrees of freedom. The limiting procedure is appropriate if and only if the corresponding finite and infinite systems are *physically equivalent* in the sense that all theoretical predictions observable *under the given circumstances* coincide for the two systems. As stated already by Friedrichs (1951, in the book of 1953 on page 142), "*different limit processes may be appropriate for different experimental situations*". *The uniqueness of the characterization of the kinematics via the Weyl group we find with finite systems does no longer hold due to the different possibilities of passing from finite to infinite systems.*

Infinite systems are in many respects much simpler than finite but very large systems. The enormous complexity of very large systems is replaced by myriads of physically inequivalent infinite systems. The equivalence classes of infinite systems are a useful tool for describing complex systems in which entirely new properties can emerge. Infinite systems display a new level of complexity, a change of the equivalence class of the representation corresponds to a change in the qualitative behavior of the system.

The qualitatively different behavior of infinite systems is reflected in the algebra of observables and in the symmetry of the system. Infinite systems have new symmetries not possessed by the corresponding finite systems. For example, in an infinite thermodynamic system, the total energy E and the total number n of particles are infinite so that both the Hamiltonian and the total number operator are global operators not belonging to the algebra of actually measurable observables. Accordingly, such a system is invariant with respect to a finite change in energy or particle number. This is a new symmetry unkown in finite systems; it implies the existence of *new* physically relevant quantities, the temperature T and the chemical potential μ. In a thermodynamic sense, the pairs (E,T) and (n,μ) define conjugated quantities. The new quantities T and μ have full physical significance only when the conjugated quantities E and n have lost their usual meaning. In fact, the existence of global operators implies the decomposability of the algebra of observables; in thermodynamic systems the new quantities T and μ are necessary to label the continuously many unitarily inequivalent representations. Another example for the emergence of a fundamentally new physical quantity is provided by the Josephson effect in superconductors and superfluids. The algebra of observables generated by the quantal state of a superconductor or a superfluid does not contain the particle number operator but its canonically conjugated operator, the phase operator. As the modern measurements of the Josephson effects show (cf. Parker et al. 1969), the phase operator is one of the most precisely measurable physical quantities.

Infinite systems are an important tool to investigate complex systems in an asymptotic limit. Any asymptotic analysis tends to simplify the system considered. The many inequivalent representations of an infinite system correspond to a decomposition of the system of interest into many independent autonomous subsystems. To select a particular representation means to concentrate the attention to a particular phenomenon and to forget effects of secondary importance. With such a choice (which is possible if we adopt an appropriate topology in our mathematical formalism) we lose irrelevant information so that the system becomes simpler. To decide which representation one has to use is the same as to decide what is relevant for the problem at hand. This cannot be done *a priori* but only when the relevant patterns have been decided upon.

Symmetries and broken symmetries of C-systems*

A *symmetry* of a C*-system is a transformation leaving all conceptually significant features of the system invariant. In the C*-algebraic approach symmetries are represented by *Jordan automorphisms*. A Jordan *-automorphism of a C*-algebra $\mathcal{A}$ is a bijection $\alpha:\mathcal{A}\to\mathcal{A}$ such that for all $A,B\in\mathcal{A}$ and all $a,b\in\mathbb{C}$, we have

$$\text{(i)} \quad \alpha(aA+bB) = a\alpha(A)+b\alpha(B) \quad ,$$

$$\text{(ii)} \quad \alpha(A^*) = \alpha(A)^* \quad ,$$

$$\text{(iii)} \quad \alpha(A\circ B) = \alpha(A)\circ\alpha(B) \quad ,$$

where $A\circ B$ denotes the Jordan product $(AB+BA)/2$. Particularly important Jordan *-automorphisms are the *-automorphisms $\alpha:\mathcal{A}\to\mathcal{A}$ which preserve the *whole* structure of the C*-algebra $\mathcal{A}$, i.e.

$$\text{(i')} \quad \alpha(aA)+\alpha(bB) = a\alpha(A)+b\alpha(B) \quad ,$$

$$\text{(ii')} \quad \alpha(A^*) = \alpha(A)^* \quad ,$$

$$\text{(iii')} \quad \alpha(AB) = \alpha(A)\alpha(B) \quad .$$

However, not every Jordan *-automorphism of $\mathcal{A}$ is a *-automorphism of $\mathcal{A}$. If we reverse the order of the terms in the product (iii'), we get a Jordan automorphism which is not an automorphism, but is called a *-antiautomorphism of $\mathcal{A}$:

$$\text{(i'')} \quad \alpha(aA)+\alpha(bB) = a\alpha(A)+b\alpha(B) \quad ,$$

$$\text{(ii'')} \quad \alpha(A^*) = \alpha(A)^* \quad ,$$

$$\text{(iii'')} \quad \alpha(AB) = \alpha(B)\alpha(A) \quad .$$

As shown by Kadison (1951), every Jordan *-automorphism is the direct sum of a *-isomorphism and a *-antiisomorphism (for a simple proof, see Simon, 1976). This theorem is the appropriate generalization of the theorem by Wigner (1931) and Uhlhorn (1962), saying that in pioneer quantum mechanics every symmetry is implemented by a unitary or antiunitary transformation.

If α is a symmetry of an abstract C*-algebra A of observables, and if π is a representation of A, then one cannot guarantee that α induces a symmetry of the concrete algebra $\pi(A)$. Nowadays, this mathematical possibility is considered as physically uninteresting. However, even if α induces a symmetry of $\pi(A)$, it need not be implementable by a unitary or antiunitary operator acting on the representation space. Haag (1962) has suggested that a *breakdown of symmetry* is characterized by an automorphism group of the algebra of observables that is not implementable in the representation of interest. Accordingly, one says that a *symmetry is spontaneously broken* if a symmetry of the C*-algebra of observables does not define a symmetry in Wigner's sense, i.e. if in the representation π there exists *no* unitary or antiunitary operator U acting on the representation space H_π such that

$$\pi\{\alpha(A)\} = U\pi(A)U^{-1} \quad \text{for all} \quad A \in A \quad .$$

Examples of spontaneously broken symmetries are provided by crystals, ferromagnets, superfluids, superconductors, and phase transitions in thermal equilibrium systems.

Technical remark

If ρ is a state invariant under the automorphism α, i.e. if

$$\rho\{\alpha(A)\}=\rho\{A\} \qquad \text{for all} \quad A \in A \quad ,$$

then α is unitarily implemented in the Hilbert space H_ρ in the GNS-representation π_ρ.

Time evolution of C-systems*

The proper formulation of dynamics is the principal headache of C*-algebraic approach to generalized quantum mechanics. Originally, it was assumed that the time evolution is given by a one-parameter group $t \to \alpha_t$ of automorphisms α_t of the C*-algebra A of observables (Haag et al., 1967). This assumption implies that the time evolution should be specified in the Heisenberg picture where the dynamics is given by the map $A \to A_t = \alpha_t(A)$ for all $A \in A$. For some simple systems (like lattice systems with short range interactions, non-interacting fermions), it is well-established that the dynamics is given by a one-parameter group of automorphisms of the algebra of observables. But there exist examples demonstrating the opposite, for example the ideal Bose gas and the BCS-model of superconductivity (Dubin and Sewell, 1970). Even very simple Hamiltonians of pioneer quantum mechanics of finite systems do not necessarily induce a *-automorphism of the Weyl algebra (Fannes and Verbeure, 1974). In general, the time evolution is an automorphism of an algebra *larger* than the C*-algebra of observables (Ruskai, 1971). One may hope that the time evolution always is an automorphism of some natural enlargement of the C*-algebra of observables, however, the general situation is not yet clear (compare also Sewell, 1973; Rudin, 1973; Maksimov, 1975, 1976,a,b; Bratteli and Robinson, 1976; Narnhofer, 1978; Roos, 1978). The time evolution can always be described as an automorphism of the von Neumann algebra generated by the GNS-representation or as an automorphism of an enveloping W*-algebra so that in a W*-algebraic formulation the time-evolution is always an automorphism.

Instead of enlarging the C*-algebra of observables, one can equally well restrict the state space. This point of view has been introduced in a basic paper by Kadison (1965; compare also Haag et al., 1970). A Kadison-type dynamical C*-system is a triple $(A, S_o, t\rightarrow\eta_t)$ consisting of a C*-algebra A of observables, a full set $S_o \subset A^*$ of distinguished states on A and a weakly continuous dynamical group $t\rightarrow\eta_t$ acting on S_o. The self-adjoint elements in A are supposed to represent the bounded observables of the system, the elements of S_o represent the statistical states, and $t\rightarrow\eta_t(\rho)$ represents the motion of the state $\rho\in S_o$. This description corresponds to the Schrödinger picture in pioneer quantum mechanics. Kadison (1965) has given necessary and sufficient conditions that η_t is given by a Jordan *-automorphism α_t of A, i.e. conditions for the equivalence of the Schrödinger and the Heisenberg pictures. Important examples of Kadison systems are the W*-systems $(A, t\rightarrow\alpha_t)$ where A is a W*-algebra and $t\rightarrow\alpha_t$ is a σ-weakly continuous one-parameter automorphism of A. A W*-algebra is a C*-algebra which is the dual of some Banach space, called the predual A_* of A. If we put $S_o=A_*$ and $\rho_t(A)=\rho(A_t)$ with $\rho_t=\eta_t(\rho)$ for every $\rho\in A_*$, and $A_t=\alpha_t(A)$ for every $A\in A$, then $(A, S_o, t\rightarrow\eta_t)$ is a Kadison system.

Remark: Time evolution and measurement process

The measurement problem of quantum mechanics is not soluble in the framework of pioneer quantum mechanics. In a very interesting paper, Hepp (1972) has rephrased the measurement problem in the C*-algebraic framework. The main result is, that if the time evolution is indeed an automorphism of the algebra of observables, then it cannot destroy coherence in any finite interval of time. That is, if $\alpha_t\in\mathrm{Aut}(A)$, and if ρ and ρ' are two mutually coherent states, then for every $|t|<\infty$ the two states $\rho\circ\alpha_t$ and $\rho'\circ\alpha_t$ are mutually coherent. However, in the limit $t\rightarrow\infty$, coherent states may converge weakly towards incoherent states - a necessary condition for the solution of the measurement problem of quantum mechanics.

Critique of the C-algebraic approach*

C*-algebraic quantum mechanics presupposes a statistical interpretation. However, an *exclusively* statistical interpretation of a quantum theory is not well founded. In any nonclassical system (with a noncommutative algebra of observables), a mixed statistical state cannot determine the ensemble, since the convex set of the states is not a simplex.

Furthermore, C*-algebraic quantum mechanics uses unashamedly an entirely unjustified operationalistic language. Segal (1959) characterizes observables as quantities that in principle can be measured directly. Such an explanation is necessarily naive since in the framework of the postulated theory actual measurements are not discussed at all. The alleged connection between algebraic quantum mechanics and operationalism does not exist. In fact, it is not at all clear how the self-adjoint elements of a C*-algebra should be related to experimental measurements. Jauch (1971) reports that "*today Segal does no longer maintain that the elements of the algebra have anything to do with observables*".

For the development of practically all of the impressive results of algebraic quantum mechanics it is crucial that the statistical expectation functional on the algebra of observables is *linear* (i.e. a so-called "state" in the mathematical sense). This far-reaching assumption is not motivated at all, it is just required by an axiom which is not as innocent as it looks.

Of course, these severely weak points of C*-algebraic quantum mechanics do not impair it as a most powerful tool which can take full advantage of the highly developed theory of C*-algebras. Fortunately, algebraic quantum theory can be linked with the conceptually much higher developed quantum logic (compare section 4.4).

4.3 # ALGEBRAIC STATISTICAL MECHANICS

General set-up of statistical mechanics

Equilibrium thermodynamics (whose proper name is, of course, "thermostatics") is a self-contained logical structure. The aim of statistical mechanics is to explain the thermodynamic behavior of macroscopic systems, one of its main tasks is the characterization of equilibrium and near-equilibrium states of thermodynamic systems. Such states can be described by means of a *small* number of parameters, like the temperature and the chemical potential. These parameters describe new symmetries and are in one-to-one correspondence with one-parameter groups of automorphisms. In equilibrium thermodynamics, the number of such parameters is small since there are few one-parameter groups of automorphisms commuting with the time-evolution automorphism - a fact which is related to the alleged ergodic nature of macroscopic systems.

According to the traditional view, a *macrostate* is an equivalence class consisting of a vast number of microscopically different states. Those observables that characterize the equivalence class of a macrostate are called *macroobservables*. Temperature and chemical potential are examples for such macroobservables. In order to be in agreement with phenomenological thermodynamics (which is a *classical* theory), all macroobservables have to be represented by *classical observables*. A macrostate is considered as *macroscopically pure* if all macroobservables take dispersionfree values with respect to this state. The goal of any macroscopic kinetic theory is the specification of an appropriate set of macrostates and macroobservables such that every valuation of the macroobservables determines a macrostate, and that the change of the macroobservables in time is described by means of a dynamical equation that contains only macroquantities.

The traditional approach to statistical mechanics

Traditional statistical mechanics is neither conceptually nor mathematically adequate. It requires for its mathematical formulation that the system is enclosed in a finite box of volume V. In the framework of pioneer quantum mechanics, one chooses as Hilbert space $\mathcal{H}_V$ the Lebesgue space of square-integrable functions having support in a bounded region $V \subset \mathbb{R}^3$. The algebra $\mathcal{A}_V$ of observables is taken to be the algebra of all bounded operators acting on $\mathcal{H}_V$. The time evolution is given by

$$A \to A(t) = \exp(itH_V)\, A \exp(-itH_V) \quad , \; t \in \mathbb{R}, \quad A \in \mathcal{A}_V \quad ,$$

where H_V is the Hamiltonian. The thermodynamic equilibrium state of a finite-volume system is defined via a Gibbsian statistical ensemble and is taken to be given by the Gibbs state ρ_V, defined by

$$\rho_V(A) = \mathrm{tr}(D_V A) \quad , \quad A \in \mathcal{A}_V \quad ,$$

with

$$D_V = \exp(-\beta H_V)/\mathrm{tr}\{\exp(-\beta H_V)\} \quad ,$$

where $\beta=1/kT$ with the Boltzmann constant k and the absolute temperature T.

We cannot avoid the use of a box since in general $\exp(-\beta H)$ is not a nuclear operator, so that $\mathrm{tr}\{\exp(-\beta H)\}$ is not well-defined and a Gibbs state does not exist. Moreover, for an *infinite* system, in general there exists no operator H associated to the algebra of observables such that the time evolution $A \to A(t)$ is given by $A(t)=\exp(itH)A\exp(-itH)$. Again, one cannot use Gibbs states directly to describe thermodynamic equilibrium states of infinite systems. The intensive thermodynamic quantities have sharp values with vanishing fluctuations around the ensemble average only in the limit of infinite systems. Moreover, typical thermodynamic properties like phase transitions, ergodic behavior, approach and return to equilibrium exist only in the limit of truly infinite systems.

The traditional way out of this dilemma is to calculate expectation values for finite systems, and subsequently to take the *thermodynamic limit* where the volume V and the number n of particles go to infinity while the density n/V is kept finite and nonvanishing.

The Kubo-Martin-Schwinger boundary conditions

The Gibbs states of pioneer quantum mechanics enjoy a characteristic analyticity property first discovered by Kubo (1957) (fluctuation-dissipation theorem), and extensively used by Martin and Schwinger (1959) as a periodic boundary condition in imaginary time for thermodynamic Green functions. The invariance of the trace under cyclic permutations of its arguments implies that for every Gibbs state ρ_V

$$\rho_V\{A(t)B\} = \rho_V\{BA(t+i\beta)\}$$

for all $A,B \in A_V$. Moreover, it is easy to show that there exists a function $G_{AB}(z)$ which is holomorphic in the open strip $0<\mathrm{Im}(z)<\beta$, continuous and bounded on the closed strip $0\le\mathrm{Im}(z)\le\beta$ and which satisfies the boundary conditions

$$G_{AB}(t) = \rho_V\{AB(t)\}$$

and

$$G_{AB}(t+i\beta) = \rho_V\{B(t)A\}$$

for all $t\in\mathbb{R}$. These boundary conditions are referred to as Kubo-Martin-Schwinger conditions, or for short as the *KMS conditions*.

The rise of algebraic quantum statistical mechanics

Algebraic quantum statistical mechanics has been growing out of the highly developed formalism of imaginary-time Green functions (compare e.g. Bonch-Bruevich and Tyablikov, 1961; Kadanoff and Baym, 1962; Abrikosov et al., 1962). The characterization of thermodynamic equilibrium states by the KMS condition was inspired by the corresponding property of finite-volume Gibbs states of pioneer quantum mechanics. In a most important pilot study on the thermodynamics of the infinite free boson gas, Araki and Woods (1963) established the importance of the KMS conditions for infinite systems, and discovered the relevance of non-Fock representations of the canonical commutation relations for statistical mechanics. This work marked the turning point in the resistance against non-Fock representations in nonrelativistic quantum mechanics. It made clear that the occurrence of inequivalent representations is by no means a mathematical pathology but of great physical relevance. Non-Fock representations give us a new powerful tool for the investigation of complex molecular systems. Nowadays it is generally accepted that for infinite systems a fundamentally new situation occurs which is not covered by the formalism of pioneer quantum mechanics.

In the C*-algebraic approach, the role of Green's functions is taken over by the time correlation functions $\rho\{AB(t)\}$ where ρ is a state on the C*-algebra A of observables, $A,B\in A$, and $B(t)=\alpha_t(B)$ is generated by the dynamical one-parameter group $t\to\alpha_t\in\mathrm{Aut}(A)$. For finite volume-systems it is easy to prove that the Gibbs states can be characterized by their KMS property. In a fundamental paper, Haag, Hugenholtz and Winnink (1967) proved that in the thermodynamic limit the limiting state still satisfies the KMS condition. For the special case of lattice systems,

Araki and Ion (1974) have established the equivalence of the Gibbs and KMS conditions. It seems to be an open question whether every KMS state is the thermodynamic limit of Gibbs states, but nowadays it is widely accepted that equilibrium states should be characterized by states fulfilling the KMS condition. The C*-algebraic approach together with the KMS conditions avoids the involved limiting procedures of traditional statistical mechanics and brings with it a great descriptive simplification.

Additional references

The work of Haag, Hugenholtz and Winnink (1967) and the reinforcement by Kastler, Pool and Poulsen (1969) have provided a rigorous mathematical framework for the study of thermodynamic systems. For a more recent review of the state of the art, compare Kastler (1976), The texts by Ruelle (1969), Emch (1972a), Dubin (1974) and Bratteli and Robinson (1979) are outstanding but certainly not elementary. As an introduction we recommend the fine textbook by Thirring (1980).

Properties of KMS states on C-algebras*

A state $\rho\in\mathcal{A}^*$ of a C*-algebra $\mathcal{A}$ is said to be a β-KMS state for the inverse temperature β and with respect to a dynamical group $t\mapsto\alpha_t\in\mathrm{Aut}(\mathcal{A})$ if for every pair of elements $A,B\in\mathcal{A}$ there is a function $G_{AB}(z)$ which is holomorphic in the open strip $0<\mathrm{Im}(z)<\beta$, continuous and bounded on the closed strip $0\leq\mathrm{Im}(z)\leq\beta$, and which satisfies the KMS boundary conditions $G_{AB}(t)=\rho\{A\alpha_t(B)\}$ and $G_{AB}(t+i\beta)=\rho\{\alpha_t(B)A\}$ for all $t\in\mathbb{R}$.

Let $(\pi,\mathcal{H},\Omega)$ be the GNS-representation of the C*-algebra $\mathcal{A}$ on the Hilbert space $\mathcal{H}$ with respect to the β-KMS state ρ, so that $\rho(A)=(\Omega|\pi(A)|\Omega)$, where $\Omega\in\mathcal{H}$. The von Neumann algebra generated by $\pi(\mathcal{A})$ will be denoted by $\mathcal{N}$, $\mathcal{N}=\{\pi(A)|A\in\mathcal{A}\}''$. If $\mathcal{N}$ is a factor (i.e. if $\mathcal{N}$ has a trivial center), then ρ is called a *factor state*.

KMS states have a number of remarkable properties (for simple proofs, compare Hugenholtz, 1972). The following properties are of particular importance for statistical mechanics:

(i) Every KMS state is time-translation invariant, i.e. we have $\rho\{\alpha_t(A)\}=\rho(A)$ for all $A\in\mathcal{A}$ (Winnink, 1968).

(ii) The center of $\mathcal{N}$ is elementwise time-translation invariant, so that the central decomposition yields again invariant KMS-states (Araki, 1968).

(iii) Any decomposition of a KMS-state finer than the central decomposition yields states which are not KMS (Araki and Miyata, 1968).

(iv) For $\beta<\infty$, the von Neumann algebra $\mathcal{N}$ is *reducible*, and $\mathcal{N}$ is in a one-to-one correspondence with its commutant $\mathcal{N}'$ (Haag et al. 1967). More precisely: There exists a conjugation J (i.e. an antiunitary operator $J:\mathcal{H}\to\mathcal{H}$ with $J^2=1$) such that $J\mathcal{N}J=\mathcal{N}'$ and $J\Omega=\Omega$.

(v) There exists a family of unitary operators $U_t\in\mathcal{B}(\mathcal{H})$, strongly continuous in t, which implements the time automorphism, and such that $U_t\Omega=\Omega$ and $U_tJ=JU_t$. If we define a state $\tilde{\rho}$ on $\mathcal{N}$ by $\tilde{\rho}(N)=\langle\Omega|N\Omega\rangle$, $N\in\mathcal{N}$, and an automorphism group by $\tilde{\alpha}_t(N)=U_t^*NU_t$, $N\in\mathcal{N}$, then $\tilde{\rho}$ is a *normal* state on the W*-algebra $\mathcal{N}$ satisfying the KMS condition with respect to the dynamics $t\mapsto\tilde{\alpha}_t$.

(vi) A β-KMS state ρ with $0<\beta<\infty$ and corresponding to a pure phase of an infinite system generates a factor $\mathcal{N}$ of type III (Hugenholtz, 1967). The limiting case $\beta=0$ of infinite temperature represents a maximally chaotic state; in this case ρ is an invariant trace and $\mathcal{N}$ is a factor of type II_1. If the limit of KMS states when the temperature goes to zero exists, then the algebra $\mathcal{N}$ is of type I and the time automorphism is implemented by a positive Hamiltonian

(Sirugue and Testard, 1971). For *finite* systems, the algebra N is of type I for *all* temperatures.

(vii) Two type III-KMS states on the same C*-algebra A but referring to different temperatures generate *disjoint* GNS-representations (Takesaki, 1970b). Since disjoint representations have orthogonal central supports this result implies the existence of a self-adjoint operator affiliated to the center of N and taking the dispersionfree value β for a β-KMS state. That is, the temperature is represented by a classical observable.

(viii) There exists a continuous family of KMS states for temperatures $0<T<\infty$, giving rise to a continuous family of non-isomorphic type III factor representations (Herman and Takesaki, 1970).

(ix) To every faithful normal state ρ of a W*-algebra there exists a unique one-parameter automorphism group with respect to which ρ satisfies the KMS condition for $\beta=1$ (Takesaki, 1970a).

These properties show that the KMS condition says a good deal about the structure of the algebras involved. This is due to the fact that thermodynamic equilibrium states are modular states in the sense of the theory of modular automorphisms of von Neumann algebras. After many years of effort, Tomita (1967) succeeded in generalizing the theory of Hilbert algebras for general von Neumann algebras. Tomita's theory was further developed by Takesaki (1970a). The new tool was the introduction of the modular automorphism group - equivalent to the KMS conditions - which measures the non-trace like behavior of a state. The KMS condition for a state ω and a one-parameter automorphism $t\to\alpha_t$ implies the existence of a well-behaved map $\omega\{A\alpha_t(B)\}\to\omega\{\alpha_t(B)A\}$. This Tomita-Takesaki theory gave rise to an enormous progress in the purely mathematical structure theory of W*-algebras. The astonishing connection between Tomita's theory of modular Hilbert algebras and the KMS condition of statistical mechanics was first recognized by Winnink (1968) in his thesis.

Physical deduction of the KMS conditions

An important problem for algebraic statistical mechanics is the deduction of the KMS condition from first principles, without recourse to the Gibbs states of finite systems. It is instructive that there are several different ways to do so.

A notable result of modern algebraic quantum mechanics is the discovery of the role the KMS conditions play as a *stability criterion* for dynamical systems. Consider any C*-dynamical system, that is a pair $(A,t\to\alpha_t)$ consisting of a C*-algebra A and a one-parameter group $t\to\alpha_t$ of automorphisms α_t of A. Araki (1973) has shown that KMS states on a C*-dynamical system $(A,t\to\alpha_t)$ are not only stationary but also *stable* under small changes of the dynamical law. A dynamics $t\to\beta_t$ is called a *local perturbation* of the dynamics $t\to\alpha_t$ if the perturbation

$$\delta = \{\partial(\alpha_t^{-1}\beta_t)/\partial t\}_{t=0}$$

is inner, that is if $\delta(A)=i[V,A]_-$ for all $A\in A$, where the perturbing Hamiltonian V is a self-adjoint element of the algebra A itself. Araki's result implies that KMS states are stable in the sense that for each small local perturbation of the dynamics there is another state close to the original one which is a KMS state with respect to the perturbed dynamics. In a basic paper, Haag, Kastler and Trych-Pohlmeyer (1974) have shown that the converse is also true, so that one can use the stability property of KMS states as the defining properties of thermodynamic equilibrium states.

These results show that a thermodynamic equilibrium state of a physical system can be characterized by the following three requirements:

(i) a thermodynamic equilibrium state ρ shall be *stationary* with respect to the dynamical group $t \to \alpha_t$ of the system, i.e. $\rho\{\alpha_t(A)\}=\rho(A)$ for all $A \in \mathcal{A}$,

(ii) a thermodynamic equilibrium state ρ shall have sharp values on all macroscopic observables, that is ρ shall be a *factor state*,

(iii) a thermodynamic equilibrium state shall be stable under all local perturbations of the dynamics by a perturbing Hamiltonian $V=V^* \in \mathcal{A}$.

The results of Haag et al. (1974) imply that a state ρ of a C*-system $(\mathcal{A}, t \to \alpha_t)$ fulfilling the conditions (i), (ii) and (iii) is a KMS state.

Additional references
For further developments and refinements, compare Kastler, (1975, 1976), Bratteli and Kastler (1976), Haag and Trych-Pohlmeyer (1977). A more dynamical stability theory has been initiated by Robinson (1973) by investigating the return to equilibrium; compare also Narnhofer and Robinson (1975) and Bratteli et al. (1978).

There is another characterization of KMS states using the following formulation of the second law of thermodynamics: no work can be obtained from an adiabatically isolated system in thermodynamic equilibrium by varying external parameters. In system theory, an input-output system that only can absorb and not generate energy is called a *passive system*. For a linear system with input $t \to u(t)$ and output $t \to y(t)$, the total energy $E(t_0,t)$ supplied by the input during the time interval (t_0,t) is *positive*. If the system is initially empty, then $E(t_0,t)$ equals the sum of the stored energy and the dissipated energy and is given by

$$E(t_0,t) = \int_{t_0}^{t} u(\tau)\dot{y}(\tau)d\tau \quad ,$$

so that the condition of passivity can be written as

$$\int_{t_0}^{t} u(\tau)\dot{y}(\tau)d\tau \geq 0 \quad .$$

This characteristic property of a passive linear system was first introduced by Meixner (1954) (compare also König and Meixner, 1958; Dolph, 1963). In an important new approach to nonequilibrium thermodynamics, Meixner (1969) proposed to replace the Clausius inequality $dS \geq \delta Q/T$ and the Clausius-Duhem inequality of traditional thermodynamics by the weaker passivity condition. The reason for this proposal is that in the framework of classical thermodynamics no definition of an entropy in nonequilibrium ever has been given. Meixner (1969a,b) has developed an entropy-free thermodynamics based on the passivity condition as replacement of the traditional formulation of the second law of thermodynamics.

In the theory of dynamical C*-systems, a state is called passive if energy cannot be extracted from the system in any cyclic process effected by the application of local perturbations during a finite interval of time. Pusz and Woronowicz (1978) have shown that KMS-states are uniquely characterized by a passivity condition. Consider a dynamical C*-system $(A, t\to\alpha_t)$ whose free dynamics $t\to\alpha_t = \exp(t\delta)$ is perturbed by a time-dependent external force $u(t)$. The perturbed dynamics $t\to\beta_t$ is then given by

$$d\beta_t(A)/dt = \beta_t\{\delta(A)\} + i\beta_t\{[u(t)V,A]_-\} \quad ,$$

for every $A \in A$, where V is a self-adjoint operator in A. A state ρ of the dynamical C*-system $(A, t\to\alpha_t)$ is called passive if the work done by the external force $t\to u(t)$ is positive. The energy $E(t_0,t)$ transmitted to the system is given by

$$E(t_0,t) = \int_{t_0}^{t} \dot{u}(\tau)\rho\{\beta_\tau(V)\}d\tau \quad .$$

A dynamical C*-system $(A, t\to\alpha_t)$ is called passive with respect to the state ρ if $E(t_0,t) \geq 0$ for every cyclic change of the external force $t\to u(t)$, $V \in A$. It follows (Pusz and Woronowicz, 1978) that every passive state of a dynamical C*-system is stationary, and that all KMS-states are passive. Since every mixture of KMS-states of different temperature also is passive, the condition of passivity alone does not characterize KMS-states. However, for pure phases, the physically relevant states are the factor states so that we have the following characterization: *a factor state is a KMS-state if and only if it is passive.*

Yet another characterization of KMS-states has been given by Kossakowski et al. (1977) by showing that the property of *detailed balance* is characteristic of dynamical semigroups acting on a W*-algebra describing the relaxation to thermal equilibrium.

The local structure of chemical thermodynamics

The science of thermodynamics is primarily concerned with transformations of various forms of energy. The crucial concepts of phenomenological thermodynamics are the *inner energy* and the *available energy*. Available energy is that part of the system's inner energy which can be converted (under the given constraints) into useful work (e.g. mechanical work or electrical work). The first law of thermodynamics expresses the conservation of energy in a closed system, while the second law of thermodynamics can be viewed as the fact of inevitable degradation of available energy in real processes. The thermodynamic entropy is a measure of the mechanical unavailability of energy.

This general framework of thermodynamics is entirely independent of the molecular structure of matter, it does not even need an elaborate notion of physical space. A fortiori, *thermodynamics is in general not a theory of infinite systems* (since "infinite" refers either to the number of particles or to the spatial extension of the system, that is to concepts that are not even defined in general thermodynamics). For example, in the framework of mathematical system theory one can introduce in a natural way the concepts of free energy and entropy for dissipative input-output systems such that the rules of general thermodynamics are fulfilled (compare Willems, 1972a,b; 1974).

Chemical thermodynamics is a special case of general thermodynamics and has an additional rich structure. First of all, we have to distinguish between the various *chemical constituents* of a chemical system. A macroscopic description of a chemical system can be made in terms of the mass densities of the various species together with the energy and entropy densities. In order to know what these concepts mean we have to introduce a *local structure* into thermodynamics, say via the Euclidean group which describes the 3-dimensional physical space. In the simplest case, one considers macroscopically homogeneous regions, called *phases*, so that one needs infinitely extended systems.

In infinite systems, the distinction between *local* and *global* aspects becomes important. Chemical thermodynamics deals with systems in which one can distinguish *finite subsystems*. In algebraic statistical mechanics these finite subsystems are put into evidence by a family of local subalgebras of the algebra of observables.

Since measurements can be made simultaneously for disjoint regions, one assigns each bounded region $V \subset \mathbb{R}^3$ of the 3-dimensional physical space $\mathbb{R}^3$ a W*-algebra $A(V)$ of observables in such a way that

$$A(V_1)A(V_2) = A(V_2)A(V_1) \quad \text{if} \quad V_1 \cap V_2 = \emptyset \quad .$$

This condition is called *locality*. The algebra $A(V)$ is generated by the observables having support in V, so that the local algebras have the property of *isotony*, that is

$$A(V_1) \subseteq A(V_2) \quad \text{if} \quad V_1 \subseteq V_2 \quad .$$

The isotony implies that the union $\cup A(V)$ of all algebras $A(V)$ (where the union is performed over all finite regions of $\mathbb{R}^3$) is again a normed *-algebra, called the algebra A_L of strictly local observables. For mathematical convenience, one adds its limit points and obtains a C*-algebra A, called the *algebra of quasi-local observables* of the infinite system.

Remark: Quasi-local C-structures*

A C*-algebra A is said to have a quasi-local structure if A possesses a family $\{A(V)\}$ of W*-subalgebras $A(V)$, where V are bounded subsets of $\mathbb{R}^n (n=1,2,\ldots)$ such that

(i) each $A(V)$ contains the identity,

(ii) the elements of $A(V_1)$ commute with all elements of $A(V_2)$ whenever V_1 and V_2 are disjoint,

(iii) for every disjoint V_1, V_2, $A(V_1 \cup V_2)$ is the smallest W*-algebra containing both $A(V_1)$ and $A(V_2)$,

(iv) $A = \lim A(V)$ as C*-inductive limit.

An important consequence of the condition (ii) is the fact that every quasi-local C*-algebra A is asymptotically Abelian with respect to space translations, i.e.

$$\|\alpha_a(A)B - B\alpha_a(A)\| \to 0 \quad \text{for} \quad |a| \to \infty$$

for all $A, B \in A$, where $\alpha_a \in \mathrm{Aut}(A)$ means a spatial translation by the vector $a \in \mathbb{R}^n$. Moreover, physically relevant quasi-local algebras are simple (compare Bratteli and Robinson, 1979, p.134-135), so that all nontrivial representations of A are faithful.

The typical classical observables of a thermodynamic system do *not* belong to the quasi-local algebra, but they can be generated from it. Consider the algebra $A_L^{\perp}(V)$ of strictly local observables outside V,

$$A_L^{\perp}(V) = \{A \mid A \in (V^{\perp}),\ V^{\perp} \cap V = \emptyset\} \quad .$$

The norm closure of $A_L^{\perp}(V)$ is a C*-algebra, called the *algebra* $A^{\perp}(V)$ *of quasi-local observables outside* V. Let π be a *-representation of the

C*-algebra A of all quasi-local observables on a Hilbert space H_π. The weak closure of $\pi(A)$ in H_π is a von Neumann algebra, called the *algebra N_π of global observables*. Denote the weak closure of $\pi\{A^\perp(V)\}$ in H_π by $N_\pi^\perp(V)$. The intersection of all von Neumann algebras $N_\pi^\perp(V)$ is the algebra of global observables outside of every bounded open set, it is called *the algebra N_π^∞ of observables at infinity* in the representation π. The local commutativity implies $[A^\perp(V),A(V)]_-=0$, so that $[N_\pi^\infty,N_\pi]_-=0$. Since $N_\pi^\infty \subset N_\pi$, it follows that all observables at infinity belong to the center Z_π of the algebra N_π of global observables in the representation π,

$$N_\pi^\infty \subseteq Z_\pi \qquad .$$

The typical classical observables of a thermodynamic system appear as self-adjoint elements of the algebra N_π^∞, where the representation π is generated by the thermodynamic equilibrium state via a GNS-construction.

References:
Haag et al. (1967), Lanford and Robinson (1968), Lanford and Ruelle (1969), Nandts (1970), Ruelle (1970), Hepp (1972), Robinson (1975).

The quasi-local structure relevant for statistical mechanics is further restricted by the fact that thermodynamic systems have in each finite subsystem only a *finite* number of particles. That is, thermodynamic states must correspond to finite particle densities, or in the technical jargon, to locally normal states. A state ρ over a quasi-local C*-algebra A is called *locally normal* if its restriction to a W*-subalgebra $A(V)$ of strictly local observables can be represented by a normal state ρ_V

$$\rho(A) = \rho_V(A) \quad , \quad A \in A(V) \quad , \quad \rho_V \in A(V)_* \qquad .$$

The representing Hilbert space determined by locally normal states in the GNS-construction is always *separable* (Hugenholtz and Wieringa, 1969). Every thermodynamic limit-state is locally normal (Sakai, 1957; Dell'Antonio, 1967). On the other hand, Takesaki and Winnink (1973) have shown that on practically all C*-algebras of interest to statistical mechanics the KMS states are locally normal.

Additional references
The concept of locally normal states has been introduced in a study on the canonical commutation relations of infinite systems by Dell'Antonio, Doplicher and Ruelle (1966). Their importance for quantum statistical mechanics has been investi-

gated by Ruelle (1966), Lanford and Ruelle (1967, 1969), Hugenholtz and Wieringa (1969) and Robinson (1970, 1975). An important generalization of locally normal states to C*-inductive limits of nets of von Neumann algebras is due to Haag, Kadison and Kastler (1970).

Temperature and chemical potential are classical observables

Thermodynamic states are never pure states, but they are "macroscopically pure" in the sense that every macroscopic thermodynamic quantity has a sharp value. In technical terms, this means that thermodynamic states representing a pure phase are KMS *factor states* (also called primary KMS-states). The parameters which distinguish disjoint KMS factor states are the equilibrium parameters like temperature and chemical potential. In algebraic quantum statistics these parameters are given by the sharp values of the corresponding classical observables in the thermodynamic equilibrium state.

The C*-algebraic theory of the chemical potential has been developed by Kastler (1976), Araki and Kishimoto (1977) and by Araki et al. (1977). Before describing the algebraic background of the chemical potential, let us recall the description of a grand canonical ensemble in the statistical mechanics of finite systems. The traditional grand canonical Gibbs state ρ_V to the temperature $1/k\beta$ and the chemical potential μ_j of the j-th species is defined by

$$\rho_V(A) = \mathrm{tr}\{D_V A\} \quad , \quad A\in\mathcal{A}_V \quad ,$$

$$D_V \overset{\mathrm{def}}{=} \exp(-\beta G_V)/\mathrm{tr}\{\exp(-\beta G_V)\} \quad ,$$

$$G_V \overset{\mathrm{def}}{=} H_V - \Sigma_j \mu_j N_{jV} \quad ,$$

where H_V is the Hamiltonian, and N_{jV} the particle number operator of the j-th species of the system enclosed in a box of volume V. The state ρ_V does *not* satisfy the KMS condition relative to the time translation group generated by the Hamiltonian H_V. But it is easily shown that ρ_V is a β-KMS state with respect to the one-parameter automorphism group $t\to\tau_t$, where

$$\tau_t(A) = \exp(itG_V)A\exp(-itG_V) \quad , \quad A\in\mathcal{A}_V \quad .$$

We write

$$\tau_t = \alpha_t \gamma^{(1)}_{\mu_1 t}\gamma^{(2)}_{\mu_2 t}\cdots\gamma^{(k)}_{\mu_k t}$$

where $t \to \alpha_t$ represents the time-translation group,

$$\alpha_t(A) = \exp(itH_V)A\exp(-itH_V) \quad , \quad t \in \mathbb{R} \quad , \quad A \in \mathcal{A}_V \quad ,$$

and $\varphi \to \gamma_\varphi^{(j)}$ represents the circle group $\mathcal{T}$,

$$\gamma_\varphi^{(j)}(A) = \exp(-i\varphi N_{jV})A\exp(i\varphi N_{jV}) \quad , \quad \varphi \in \mathcal{T} \quad , \quad A \in \mathcal{A}_V \quad .$$

The conservation of the number of particles of each kind implies that H_V and N_{jV} commute, hence we have

$$\alpha_t \gamma_\varphi^{(j)} = \gamma_\varphi^{(j)} \alpha_t \quad , \quad t \in \mathbb{R} \quad , \quad \varphi \in \mathcal{T} \quad , \quad j=1,..,k \quad .$$

The transformation $\varphi \to \gamma_\varphi^{(j)}$, $\varphi \in \mathcal{T}$, is called a *gauge transformation* (of the first kind) with respect to the j-th species.

For a chemical system with k different chemical components, the group of generating all possible gauge transformations is the direct product of k circle groups $\mathcal{T}$, i.e. the k-dimensional torus group $\mathcal{T}^k$,

$$\mathcal{G} = \mathcal{T}^k \overset{\text{def}}{=} \mathcal{T} \times \mathcal{T} \times \cdots \times \mathcal{T} \qquad (k \text{ times}) \quad .$$

Accordingly, *a system with* k *different chemical components is characterized by a compact and Abelian* k*-parameter Lie group acting as a (generalized) kinematical group*. This group $\mathcal{G}$ is called the *gauge group* of the system.

In order to describe processes that do not conserve the particle number of the various species of a chemical system, the algebra $\mathcal{A}$ of observables has to be extended to the so-called *field algebra* $\mathcal{F}$. The field algebra $\mathcal{F}$ is a C*-algebra generated by the creators $a_j^*(\vec{r})$ and the annihilators $a_j(\vec{r})$ of the j-th species, $j=1,..,k$, $\vec{r} \in \mathbb{R}^3$. These operator-valued distributions are characterized by their behavior under the gauge group of the system,

$$\gamma_g\{a_j^*(\vec{r})\} = \exp(i\varphi_j)a_j^*(\vec{r}) \quad ,$$

$$\gamma_g\{a_j(\vec{r})\} = \exp(-i\varphi_j)a_j(\vec{r}) \quad ,$$

where $g = (\varphi_1, \varphi_2, .., \varphi_k) \in \mathcal{G}$.

The algebra $\mathcal{A}$ of observables is the gauge-invariant part of the field algebra $\mathcal{F}$

$$\mathcal{A} = \{A | A \in \mathcal{F} \ , \ \gamma_g(A) = A \quad \text{for all } g \in \mathcal{G}\} \quad .$$

The algebraic description of the grand canonical ensemble is made in terms of a C*-system $\{F, tg \to \alpha_t \gamma_g\}$ where $\alpha \times \gamma$ is a continuous representation of the direct product of the time translation group $\mathbb{R}$ and the gauge group G by *-automorphisms $\mathbb{R} \ni t \to \alpha_t \in \mathrm{Aut}(F)$ and $G \ni g \to \gamma_g \in \mathrm{Aut}(F)$. The requirement that a stationary factor state ρ should be stable under local perturbations $V \in F$ of the dynamics $t \to \alpha_t \in \mathrm{Aut}(F)$ implies (as in the case of the canonical ensemble) the condition

$$\int_{\mathbb{R}} dt \rho\{[\alpha_t(V), A]_-\} = 0 \qquad ,$$

which characterizes ρ as a KMS state with respect to $t \to \alpha_t$ (Haag et al., 1974). A grand canonical equilibrium state is characterized by the additional requirement that the local perturbation does not cause a change in the number of particles of each kind so that V has to be gauge invariant, i.e. $V \in A \subset F$. Since every gauge-invariant operator V can be written in the form

$$V = \int_G \gamma_g(F) dg \in A \qquad , \qquad F \in F \quad ,$$

the stability condition relevant for the grand canonical ensemble can be written as

$$\int_{\mathbb{R}} dt \int_G dg \rho\{[\alpha_t \gamma_g(F), A]_-\} = 0$$

which is a KMS condition for the one-parameter group $t \to \tau_t$, $t \in \mathbb{R}$, where again $\tau_t = \alpha_t \gamma^{(1)}_{\mu_1 t} \gamma^{(2)}_{\mu_2 t} \cdots \gamma^{(k)}_{\mu_k t}$ (Kastler, 1976). In the GNS-construction associated with ρ, this automorphism group is unitarily implemented by the infinitesimal generator

$$G = H - \sum_{j=1}^{k} \mu_j N_j \qquad ,$$

where H is the Hamiltonian and N_j the particle number operator of the j-th species. For infinite systems, the infinitesimal generators H and N_j are no longer associated to the von Neumann algebra of the GNS-representation of the algebra of observables, so that neither H nor N_j can be considered as observables.

Instead of the commutative gauge group, more general kinematical groups can be considered. Araki et al. (1977) has discussed the case of an arbitrary compact group. For example, an equilibrium superfluid in a rotating container, or a superconductor in a constant magnetic field (which is by Larmor's theorem equivalent to the same system in a rotating frame of reference) has the space rotation group SO(3) as an additional symmetry. In these examples, an angular momentum can be introduced by a parametrization of the KMS condition with respect to the compact group SO(3), similar to the chemical potential (Loupias and Mebkhout, 1979).

To summarize: Thermodynamic equilibrium states are KMS factor states which are labelled by the inverse temperature β and additional macroscopic variables which are directly related to the corresponding kinematical group. KMS states with different temperatures, and β-KMS states with different additional labels (such as the chemical potential) are disjoint, so that such states belong to different superselection sectors. Since disjoint representations have orthogonal central supports, there exist self-adjoint operators affiliated to the center of the von Neumann algebra generated by the GNS-construction of the thermodynamic equilibrium state which take just the values of the labels of the KMS states as eigenvalues. Accordingly, there exist classical observables for the temperatures, for the chemical potentials etc. such that in a pure thermodynamic phase these classical observables take dispersion-free values.

Phase transitions and symmetry breakdown

To get the historical perspective, we recall that the first statistically derived example of a phase transition is due to Albert Einstein. In his two papers "Quantentheorie des einatomigen idealen Gases", Einstein (1924, 1925) predicted that below a certain critical temperature a gas of noninteracting bosons would show a gradual condensation into the ground state. The phenomenon that a macroscopic number of particles of a quantum statistical equilibrium system occupies the lowest energy state is nowadays known as the *Einstein condensation*. Einstein (1925) did not call this condensation phenomenon a phase transition. Even as late as 1937 doubts where raised as to whether phase transitions can be explained at all by statistical mechanics since phase transitions manifest themselves as singularities of thermodynamic functions. The elementary fact that sharp phase transitions can only occur in the thermodynamic limit $n\to\infty$, $V\to\infty$, n/V fixed (where n is the number of particles, V the volume of the container, and n/V the number density) seems to have emerged in a debate at the van der Waals Centenary Conference in 1937 (compare Pais,

1979, sect.VI-D). London (1938a,b) has suggested the Einstein condensation as an explanation for the superfluid behavior in superconductors and liquid helium. Today, it is generally accepted that the Einstein condensation is a truly fundamental phenomenon which can serve as a caricatural prototype for many systems exhibiting broken symmetry. The λ-point of liquid He^4, the critical point of a superconductor, the threshold of a laser are thought to be the onset of an Einstein condensation in an *interacting* many-particle system.

Rigorous discussion of the Einstein condensation

Einstein's argument was simple, sound and convincing, but from a mathematical point of view, his demonstration was not rigorous. The seemingly simple problem of a proper mathematical treatment of a system of noninteracting particles proved to be rather tricky. Most of the older literature is marked by a sloppy mathematical analysis. The first convincing analysis is due to Lewis (1972) and Lewis and Pulé (1974), and is based on the pioneering work by Araki and Woods (1963). Compare also Robinson (1965), Cannon (1973), Dubin (1973, 1974), Davies (1973), Moya (1975).

Since finite systems have a unique KMS state at a given temperature, there are no phase transitions in finite systems. In contrast, infinite systems may have a plurality of β-KMS states. The set of all β-KMS states of a given dynamical C*-system form a Choquet simplex so that every β-KMS state has a unique barycentric decomposition in terms of extremal elements of this set. A β-KMS state that cannot be written as a nontrivial convex combination of two different β-KMS states is called an *extremal β-KMS state*. Extremal KMS states are factor states and correspond to pure thermodynamic phases.

References

For many models it has been rigorously established that the extremal KMS states correspond to pure thermodynamic phases, and it appears that this correspondence holds generally. Compare: Ruelle (1966, 1969), Kastler and Robinson (1966), Doplicher et al. (1967), Lanford and Ruelle (1967), Robinson and Ruelle (1967), Doplicher and Kastler (1968), Araki and Miyata (1968), Knops and Verboven (1969), Emch et al. (1970), Emch and Knops (1970), Emch (1972a,c).

The KMS states which are obtained by the traditional limiting procedure need not be extremal. The extremal decomposition of such a KMS state amounts to a reconstruction of pure thermodynamic phases. The extremal decomposition of a KMS state can break a symmetry the original KMS state possessed. If an extremal KMS state has a lower symmetry, we say that the equilibrium system undergoes a *symmetry breakdown* in this state. Breaking the symmetry is a typical feature of a phase transition.

The best known examples of such states with broken symmetry are given by crystals, ferromagnets, superfluids and superconductors. The symmetries which are broken may be the spatial translation symmetry (crystals), the spin-rotation symmetry (ferromagnets), the gauge symmetry (superfluids and superconductors). These broken-symmetry states are distinguished by the emergence of a *macroscopic order parameter*. Since states with different order parameters are disjoint, the order parameters obey the laws of classical physics. The fact that algebraic quantum statistics offers a conceptually transparent and mathematically rigorous way to describe phase transitions with an associated spontaneous symmetry breaking is one of the most important results of the modern development of quantum mechanics.

4.4 THE DEVELOPMENT OF QUANTUM LOGICS

The discovery of quantum logic

John von Neumann has suggested that the revolutionary character of quantum mechanics involves even the level of logic. He has called attention to the striking analogies between the algebraic relations of the projectors on the Hilbert space of pioneer quantum mechanics and the propositional connections of classical logic (von Neumann, 1932).

Classical logic: Once over lightly

Every historical language has its grammar and many exceptions. Synthetic languages are less colorful, very specialized and of restricted applicability, but they are precise. The language used to talk about scientific theories is called a *metalanguage*, it is a formal language consisting of a *syntax* (i.e. the grammar) and a *logic* (i.e. a theory of conclusive reasoning). Logic in this sense is a theory of inference, it regulates the ways in which one reasons from given propositions, irrespective of their truth or falsity. Note that the task of logic is *not* to determine what is true.

The history of formal deduction goes back to Aristotle (384 B.C.-322 B.C) but only around 1850 the logicians George Boole (1815-1864) and Gottlob Frege (1848-1925) gave a definitive form to what formal deduction actually is. Boole gave us an algebraic model of Aristotle's propositional calculus; this algebra of logic (in its definite form due to Ernst Schröder) is called a *Boolean algebra*. The extension of Boole's "Laws of Thought" leads to rules for making deductions in algebraic logic; his theory is called the predicate calculus, and is the logical basis for all of today's mathematics.

A basic doctrine of classical logic is that every proposition is either true or false. Such propositions are said to have one or the other (but not both) of two *truth-values*. Hence classical logic is a *two-valued logic*. Propositions can be combined by means of sentential connections, the most important being "not", "and", "or", "if-then", and "if and only if". Let $\perp$ be the sign for "not", $\wedge$ the sign for "and", $\vee$ the sign for "or", and = the sign for "if and only if". Let F and G be any two propositions, then $F^{\perp}$ is the proposition which is true whenever F is false, $F\wedge G$ the proposition which is true if F *and* G are true, $F\vee G$ the proposition which is true if F *or* G is true. $F=G$ means that F is true if and only if G is true. If the relation $F\wedge G=F$ holds, then we write $F\leq G$ and say that F logically entails G. A set of propositions which is closed under the operation $\perp$ and the two binary operations $\wedge$ and $\vee$, and fulfilling the standard rules of classical logic is called a *Boolean algebra* (synonym: *Boolean lattice*).

Boolean logic is the basis of mathematical logic, but it does not apply to everything. To say that classical logic is universally valid is stupid; it is clearly restricted to a right and proper application (whatever this precisely means).

Reference: The study of logic is rendered complicated by the various notational systems currently used by mathematicians, logicians, computer scientists and engineers. The dictionary of logical terms by Greenstein (1978) may be helpful in this respect, it includes also a glossary of logical terms and a useful bibliography. For a vivid introduction into mathematical logic on a sophisticated level, try Manin (1977).

Classical logic uses a Boolean algebra of propositions in which the

Boolean lattice operations $\wedge$ (called *meet*), $\vee$ (called *join*) and $^{\perp}$ (called *orthocomplementation*) correspond to the logical operations of conjunction, disjunction and negation, respectively. The partial ordering in this Boolean algebra is interpreted as the logical relation of implication between the propositions.

In pioneer quantum mechanics observables are represented by self-adjoint operators. Observables whose spectra consist of at most the two values 1 and 0 are called *projectors*, they represent yes-no questions, indicating whether or not some event has occurred. The spectral theorem says that every observable can be recovered from its spectral projections, that is from yes-no questions.

Quantum logic originates with the seminal work of Birkhoff and von Neumann (1936). They pointed out that in pioneer quantum mechanics the projectors describe the empirically verifiable propositions about the system, but that they do not form a Boolean algebra. Both classical theories (like Newtonian mechanics) and pioneer quantum mechanics have a natural lattice structure with a meet operation $\wedge$, a join operation $\vee$, an orthocomplementation $^{\perp}$, and a partial order relation $\leq$ defined by

$$F \leq G \quad \text{if and only if} \quad F \wedge G = F \quad .$$

or, equivalently, by

$$F \leq G \quad \text{if and only if} \quad F \vee G = G$$

In pioneer quantum mechanics, there is a one-to-one correspondence between projectors and closed subspaces of the Hilbert space $\mathcal{H}$ of state vectors. A linear operator F acting on the Hilbert space $\mathcal{H}$ is said to be a projector from $\mathcal{H}$ onto a closed subspace $\mathcal{F} \subset \mathcal{H}$ if $F\Psi \in \mathcal{F}$ and $\Psi - F\Psi \in \mathcal{F}^{\perp}$ for every $\Psi \in \mathcal{H}$, where $\mathcal{F}^{\perp}$ denotes the orthogonal complement of $\mathcal{F}$. Recall that $\mathcal{F}^{\perp}$ is defined as the set of all $\Phi \in \mathcal{H}$ such that $\langle \Psi | \Phi \rangle = 0$ for all $\Psi \in \mathcal{F}$, that $\mathcal{F}^{\perp}$ is a closed subspace of $\mathcal{H}$, and that $(\mathcal{F}^{\perp})^{\perp} = \mathcal{F}$. From this it follows that every projector is self-adjoint and idempotent, $F = F^* = F^2$. The collection of all subspaces of a Hilbert space is partially ordered by inclusion and forms a complete lattice $\mathcal{L}(\mathcal{H})$ under this ordering so that one can define the lattice operations $\wedge$ and $\vee$ in the usual way. The meet $\mathcal{F} \wedge \mathcal{G}$ is the largest subspace contained both in $\mathcal{F}$ and $\mathcal{G}$, and equals the set-theoretical intersection,

$$\mathcal{F} \wedge \mathcal{G} = \mathcal{F} \cap \mathcal{G} \quad .$$

The join $F \vee G$ is the smallest subspace of H containing both F and G. The lattice $L(H)$ admits an orthocomplementation $F \to F^{\perp}$ induced by the inner product of H, where $F^{\perp}$ is the orthocomplement of F, $F \oplus F^{\perp} = H$. With this one can express the join by the relation

$$F \vee G = \{F^{\perp} \cap G^{\perp}\}^{\perp} \quad .$$

The one-to-one correspondence between the projectors acting on a Hilbert space H and the closed subspaces of H implies that the lattice structure of H can be transferred to the projectors acting on H. Let F be the projector from H onto the closed subspace F of H, and G the projector from H onto the closed subspace G. Then we define

$$\mathrm{F} \leq \mathrm{G} \quad \text{iff} \quad F \subseteq G$$

$$\mathrm{F} \wedge \mathrm{G} \quad \text{iff} \quad F \wedge G$$

$$\mathrm{F} \vee \mathrm{G} \quad \text{iff} \quad F \vee G$$

$$\mathrm{F} = \mathrm{G} \quad \text{iff} \quad F = G$$

so that we have

$$\mathrm{F}^{\perp} = \mathrm{E} - \mathrm{F}$$

where E is the identity operator. The set of all projections acting on the Hilbert space H is denoted by $P(H)$; clearly $P(H)$ and $L(H)$ are ortho-isomorphic lattices. The largest projector in $P(H)$ equals the identity operator, the smallest projector equals the zero operator.

Some notion of lattice theory #

Recall that a *lattice* L is a partially ordered set $(L, \leq)$ which has a least element (called *zero* and denoted by O), a greatest element (called *unity* and denoted by E), and in which every pair of elements has a supremum and infimum. A lattice is said to be *complete* if every nonempty subset of it has a supremum and infimum. We write $F \wedge G = \inf(F,G)$, $F \vee G = \sup(F,G)$, $O = \wedge L$, $E = \vee L$, and we have $F \leq G$ iff $F \wedge G = F$ or equivalently $F \leq G$ iff $F \vee G = G$. Clearly we have $O \leq F \leq E$ for every $F \in L$. A lattice is called *finite* if the number of its elements is finite.

An element $A \neq O$ in a lattice L such that $F \leq A$ implies either $F = O$ or $F = A$ is called an *atom*. A lattice is called *atomic* if every nonzero element has an atom under it. A lattice containing no atoms is called *atom-free*. Note that there are lattices which are neither atomic nor atom-free.

A lattice L is called *complemented* if for every $F \in L$ there exists a *complement* $G \in L$ such that $F \wedge G = O$ and $F \vee G = E$. An *involution* is a map $F \to F'$ satisfying $(F')' = F$, $(F \wedge G)' = F' \vee G'$ and $(F \vee G)' = F' \wedge G'$. An *orthocomplementation* is an involution $F \to F^{\perp}$ such that $F^{\perp}$ is a complement of F.

An *orthocomplemented lattice* L is characterized by the following properties:

(i) The commutative law holds for meet and join, that is: $F \vee G = G \vee F$ and $F \wedge G = G \wedge F$ for all $F, G \in L$.

(ii) The associative law holds for meet and join, that is: $F \vee (G \vee H) = (F \vee G) \vee H$ and $F \wedge (G \wedge H) = (F \wedge G) \wedge H$ for all $F, G, H \in L$.

(iii) The absorption law holds between meet and join, that is: $F \vee (G \wedge F) = F$ and dually

$F\wedge(G\vee F)=F$ for all $F,G\in L$.

(iv) The element $O\in L$ fulfills $O\vee F=F$ for all $F\in L$, and the element $E\in L$ fulfills $E\wedge F=F$ for all $F\in L$.

(v) The orthocomplementation $F\to F^{\perp}$ satisfies:

(a) $F=(F^{\perp})^{\perp}$ (double negation)

(b) if $F\leq G$ then $G^{\perp}\leq F^{\perp}$ (contraposition)

(c) $F\wedge F^{\perp}=0$ (noncontradiction)

(d) $F\vee F^{\perp}=E$ (excluded middle)

Two elements F,G of an orthocomplemented lattice L are said to be *orthogonal*, written $F\perp G$, if $F\leq G^{\perp}$. This is equivalent to $G\leq F^{\perp}$ so that $F\perp G$ if and only if $G\perp F$. Two elements $F,G\in L$ are said to form a *modular pair* if

$$(H\vee F)\wedge G = H\vee(F\wedge G) \quad \text{for every} \quad H\leq G \quad .$$

A lattice is called *modular* if every pair of elements is a modular pair. A lattice is called *semimodular* if the fact that (F,G) is a modular pair implies that (G,F) is a modular pair. An orthocomplemented lattice is called *orthomodular* if every orthogonal pair is modular. Equivalently, an orthocomplemented lattice L is orthomodular if and only if $F\leq G$ implies $F\vee(F^{\perp}\wedge G)=G$ for all $F,G\in L$. A lattice L is *distributive* if the distributive law $F\wedge(G\vee H)=(F\wedge G)\vee(F\wedge H)$ holds for all $F,G,H\in L$. An orthocomplemented distributive lattice is called a *Boolean lattice* (synonym: a *Boolean algebra*). Every distributive lattice is modular but not the other way round. Every Boolean lattice is orthomodular but not the other way round.

In a general orthomodular lattice, the distributive law does not hold globally but there are many Boolean sublattices so that there is some local distributivity. This local distributivity can be studied through a compatibility relation. Two elements F,G of an orthomodular lattice L are called *compatible*, written $F\leftrightarrow G$, if the sublattice generated by $F,F^{\perp},G,G^{\perp}$ is distributive, i.e.

$$F\leftrightarrow G \quad \text{iff} \quad F = (F\wedge G)\vee(F\wedge G^{\perp}) \quad .$$

In a Boolean lattice, all elements are mutually compatible. The set of all elements in an orthomodular lattice L which are compatible with every element in L is called the *center* of L; it is a Boolean algebra. If L represents a nonclassical logic, the center of L represents its classical part.

An orthocomplemented lattice L is orthomodular if and only if the compatibility relation $\leftrightarrow$ is symmetric, that is, if $F\leftrightarrow G$ implies $G\leftrightarrow F$ for every $F,G\in L$. In an orthomodular lattice $F\leftrightarrow G$ implies $G\leftrightarrow F$, $F\leftrightarrow G^{\perp}$, $F^{\perp}\leftrightarrow F$ and $F^{\perp}\leftrightarrow G^{\perp}$. Moreover, in an orthomodular lattice *comparable elements are compatible*, that is $F\leq G$ implies $F\leftrightarrow G$.

References:

The comprehensive monograph by Birkhoff (1940) is the standard reference work, the modernized third edition of 1967 can be highly recommended. Of the more specialized monographs, the treatises on Boolean algebras by Sikorski (1960), on modular lattices and continuous geometries by Maeda (1958), and on modular and orthomodular lattices by Maeda and Maeda (1970) are of special importance. A fascinating review of the theory of orthomodular lattices - which are of crucial importance for quantum theory - has been given by Holland (1970).

The structure of the projection lattice $P(H)$

The set of all closed subspaces of any topological vector space forms a complete lattice under set-theoretical inclusion. A deep theorem due to Kakutani and Mackey (1946) states that the possibility of an orthocomplementation of the lattice of all closed subspaces of a Banach space is a necessary and sufficient condition that the topology of the Banach space coincides with a Hilbert space topology.

Both the lattice $L(H)$ of all closed subspaces of a complex Hilbert space H and the lattice $P(H)$ of all projectors acting on H are *complete, irreducible, atomic, orthomodular and semimodular* (Kaplansky, 1951; Topping, 1967; Holland, 1970). Both $L(H)$ and $P(H)$ are *modular* if and only if H is *finite-dimensional*; they are distributive if and only if H is one-dimensional, i.e. $H=\mathbb{C}$.

Birkhoff and von Neumann (1936) postulated that every measurement of every physical quantity can be reduced to experiments which decide the truth or falsehood of certain statements, called *experimental propositions*. They proposed to choose a closed subspace $\mathcal{F}$ of the Hilbert space $\mathcal{H}$ of state vectors, or, equivalently, the projector F from $\mathcal{H}$ onto $\mathcal{F}$ as the mathematical representative of an experimental proposition. of pioneer quantum mechanics, the projection F is an *observable*, and *has* the value one if and only if the state vector Ψ lies in the range of F, $\Psi \in \mathcal{F}$, or equivalently if $F\Psi = \Psi$. We say that in the state Ψ the proposition F is *true* if $F\Psi = \Psi$, and we say that F is *false* if $F\Psi = 0$. Since $F \leq G$ and $F\Psi = \Psi$ implies $G\Psi = \Psi$, the relation $F \leq G$ means that the proposition F implies the proposition G in the sense that whenever one can predict F with certainty, one can predict G with certainty. Generalizing this result, Birkhoff and von Neumann (1936) postulated that *"the physical qualities attributable to any physical system form a partially ordered system"*.

In pioneer quantum mechanics this partially ordered system equals the lattice $P(\mathcal{H})$ of projections on the Hilbert space $\mathcal{H}$. The orthocomplementation $F \rightarrow F^{\perp} \stackrel{def}{=} E - F$ of $P(\mathcal{H})$ corresponds to the passage from an experimental proposition to its negation: the proposition F is true if and only if the proposition $F^{\perp}$ is false, and F is false if and only if $F^{\perp}$ is true, so that the same measurement which is used to decide the validity of F is also used to decide the validity of $F^{\perp}$. Unlike the Boolean lattices associated with classical theories, the lattice of the experimental propositions of pioneer quantum mechanics is *not distributive* so that the relation

$$F \wedge (G \vee H) = (F \wedge G) \vee (F \wedge H)$$

is *not* generally valid. *The non-distributivity of the propositional calculus of pioneer quantum mechanics reflects the fact that quantum mechanics allows incompatible experiments*. Therefore one cannot proceed as in classical mechanics where *"one can easily define the meet or join of any two experimental propositions as an experimental proposition - simply by having independent observers read off the measurements which either proposition involve, and combining the results logically"* (Birkhoff and von Neumann, 1936). In quantum mechanics, the sentential connectives "or" and "and" can be defined only in special cases, i.e. only when the corresponding experiments are compatible. If F and G are two *compatible* propositions, then $F \vee G$ is the proposition which is true whenever F *or* G is true, and $F \wedge G$ is the proposition which is true whenever F *and* G are true.

To summarize, the main idea of Birkhoff and von Neumann was to associate to every physical system an orthocomplemented partially ordered set $(L, \leq, ^{\perp})$ where the members of the set L correspond to the experimental propositions concerning the system, the ordering $\leq$ in L corresponds to the relation of implication, and the orthocomplementation $^{\perp}$ corresponds to the relation of negation. For classical systems, L is a Boolean algebra, while for quantum systems L is a nondistributive lattice resembling a projective geometry. Such lattices are non-Boolean but they contain many different interlocking Boolean algebras. A specific Boolean sublattice of L can be singled out by specifying a specific experimental arrangement.

Modular logic

In the special case of pioneer quantum mechanics with a *finite-dimensional* Hilbert space, the lattice L of propositions is *modular*. Although pioneer quantum mechanics with an *infinite-dimensional* Hilbert space has a *non-modular* projection lattice, Birkhoff and von Neumann (1936) suggested to rephrase quantum mechanics in terms of a lattice that fulfills the axiom of modularity. In an important but not widely known paper, Husimi (1937) took up the problem of founding a modular logic directly on empirical facts. This program has been pursued by Watanabe and has led to a generalization of empirical logic and probability theory based on a modular lattice (Watanabe 1961, 1965, 1966, 1969a,b).

Modularity is a continuity condition #

The fact that the lattice $L(H)$ of closed subspaces of a Hilbert space is in general not modular is connected with the fact that in infinite-dimensional vector spaces there always exist closed subspaces F, G whose algebraic sum $F+G$ is *not* closed. In fact, modularity is a *continuity condition* (Kaplansky, 1955) and is essentially the same as the condition that the sum of closed subspaces is closed (Mackey, 1950). Of course, if H is finite-dimensional, then every subspace of H is closed, hence $L(H)$ is modular.

Birkhoff and von Neumann (1936) did not regard Hilbert spaces as an indispensable tool for quantum mechanics. The projection lattice of pioneer quantum mechanics generates a W*-algebra of type I, while the proposal of Birkhoff and von Neumann leads to a lattice which is isomorphic to the projection lattice of a type II_1 W*-algebra and which is closely related to a continuous geometry (von Neumann, 1960). Von Neumann speculated that continuous geometries might be the proper arena for quantum mechanics or related theories (compare von Neumann 1937a,b, 1942, 1961; Murray and von Neumann, 1936; Birkhoff, 1961). This suggestion is attractive since every orthocomplemented modular lattice is a continuous geometry, and the dimension function in a continuous geometry has all formal properties of a probability.

The further development of quantum logics did not follow this idea of John von Neumann. Though Birkhoff and von Neumann argued in favor of the modular law as a replacement of the distributive law, subsequent researchers have rejected modularity in favor of the yet weaker axiom of orthomodularity.

The logic of complementarity

Independently of Birkhoff and von Neumann, Martin Strauss (1936) also suggested that pioneer quantum mechanics requires a modification of classical logic. His proposal differs from that of Birkhoff and von Neumann insofar as he admits the distributive law but denies that the logical connectives "F and G" and "F or G" are meaningful whenever F and G are meaningful. Strauss suggested to associate *dispositional properties* with the projectors of pioneer quantum mechanics, and proposed a predicate logic with a *partial Boolean algebra* as a logical codification of Bohr's ideas on complementarity (compare also Strauss 1938, 1967, 1970, 1973). In classical theories, propositions refer to ordinary properties whose predicate calculus is a Boolean algebra. A dispositional property, however, is a *mode of reaction* that manifests itself only in an appropriate interaction process. According to Strauss, dispositional properties are as a rule not subject to classical Boolean logic but governed by a quantum logic with restricted sentential connectivity whose propositional calculus is given by a partial Boolean algebra.

What is a partial Boolean algebra? #

A partial Boolean algebra is given by a set L, a binary relation $\leftrightarrow \subseteq L\times L$ (called *compatibility*), a unary relation $^{\perp}: L\to L$ (called *orthocomplementation*), a partial binary function $\vee$ whose domain Dom($\vee$) is contained in $L\times L$, $\vee:\mathrm{Dom}(\vee)\to L$. In addition, an element $E\in L$ is distinguished (called the *unit* of the partial algebra). There are the following axioms ($F,G,H\in L$):

(i) The binary relation $\leftrightarrow$ is symmetric and reflexive,

(ii) $E\leftrightarrow F$ for every $F\in L$,

(iii) if $F\leftrightarrow G$, then $F\leftrightarrow G^{\perp}$

(iv) if F,G,H are mutually compatible then $F \vee G \leftrightarrow H$,
(v) if F,G,H are mutually compatible, then the Boolean polynomials from F,G,H form a Boolean algebra with the unity E.

Every partial Boolean algebra is made up of a collection of maximal compatible subsets each of which is a Boolean algebra. Of course, every Boolean algebra is also a partial Boolean algebra.

References: Kochen and Specker (1965a, 1967).

Strauss' important contribution remained hidden to most scientists for a long time. Nowadays Strauss' work is easily accessible in a volume collecting his most important papers, all in English translation, some with new postscripts (Strauss, 1972). The calculus of partial propositional functions based on partial Boolean algebras has been investigated in detail by Specker (1960), Kamber (1964), and by Kochen and Specker (1965a,b; 1967). The interrelations of the lattice approach and the partial-Boolean algebra approach to quantum logics have been investigated by Czelakowski (1974, 1975). For an important and large class of quantum logics both approaches coincide.

Is quantum logic really logic?

Birkhoff's and von Neumann's (1936) suggestion that the system of logic is not determined by a priori considerations, and that pioneer quantum mechanics is best formulated in terms of a logic in which the distributive law fails, has led to extensive philosophical discussions about the status of classical and quantum logics. Birkhoff and von Neumann have suggested a revision of the system of logic on which science is based, but they have given no indication how quantifiers might be introduced so as to get an adequate theory of inference.

Some logicians claim that "*in order to base a branch of physics on a non-classical logic, it would be necessary to present a completely formalized theory, which would set forth the logical and mathematical, as well as the physical, principles to be used. ... Thus it would by no means be sufficient merely to give axioms and rules of the formal system; it would also be necessary to specify which sentential connectives were to be used in defining identity, whether more than one kind of identity was to be treated, and so on. We should be confronted with the formulation of a vast number of definitions, and the proof of an enormous number of theorems, of classical mathematics; this part of the project would be somewhat analogous to the writing of Principia mathematica, though vastly more onerous*" (McKinsey and Suppes, 1954).

The critique by McKinsey and Suppes reflects Russell's Peanesque logic which incorporates the metatheory of the system. In such a system it is strictly impossible to talk *about* the logical system. For example, if we reject the law of excluded middle, we cannot deduce that it is false, for such a deduction would illegimately move outside the bounds of logical discourse. Or in Bertrand Russell's words, "*we cannot conclude, from the proposition that the law of excluded middle is not true, that it is false, without using it*" (in a conversation with Philip Jourdain on April 20, 1909; quoted in Grattan-Guiness, 1977, p.112). Russell's conception of logic precludes the use of metatheoretical investigations. Recall that a metalanguage is the language used to talk *about* an object language, and that a metatheory of a theory is concerned with the analysis of the theory itself. However, *it has never been the intention of quantum logic to embrace the metatheory of quantum theory*. Moreover, the mathematical deductions used in a theoretical framework of a natural science do not belong to the scientific theory itself but to its metatheory. Since there is no reason to assume that a theory and its metatheory have to use the same logic, the remark that quantum logic is (as yet) without a fully developed system of quantifiers is not a relevant criticism. Quantum logic refers to the *properties* of the systems discussed in quantum theory, it is non-Boolean. The metalogic of quantum theory is still the traditional Boolean logic (together with the usual but informal dialectical handling of different contexts).

Unfortunately, it has become common mathematical usage to apply the term "logic" to any orthomodular lattice or orthomodular partially ordered set, whether or not its elements are interpreted as sentences. Jauch and Piron (1970) have argued that a partially ordered set should be called a logic only if it admits the algebraic counterpart of a *modus ponens* (which is a mode of reasoning from a hypothetical proposition stating that if the antecedent is confirmed the consequent is affirmed).Doubts have been raised by Jauch and Piron (1970) and by Greechie and Gudder (1971) whether it is possible to define such a conditional as a logical proposition of a quantum logic. Their claim that a quantum logic does not admit an implication operation by means of which an inference scheme can be incorporated does not seem to be justified. The so-called Sasaki hook $F^{\perp}\vee(F\wedge G)$ on an orthomodular lattice satisfies the axioms of an implication algebra as given by Finch (1970), so that the implication $F\geq G$ turns out to be the Sasaki hook (Piziak, 1974). Accordingly, the binary operation $\rightarrow$

$$F \rightarrow G = F^{\perp} \vee (F \wedge G)$$

can be interpreted as material implication in such a way that $F \rightarrow G$ is a proposition of the logic. Independently, Mittelstaedt (1970, 1972, 1973) has shown that the Sasaki hook allows an interpretation of the lattice of quantal propositions as an operative logic.

Additional references

Kotas (1963) has shown that the Birkhoff-von Neumann quantum logic allows us to define six operations which in a Boolean algebra are identical with the classical implication relation. The various admissible candidates for the role of an implication connective are discussed by Kalmbach (1974). The material implication connective in orthomodular logics has been discussed by Kunsemüller (1964), Herman and Piziak (1974), Herman et al. (1975), Hardegree (1974, 1975a,b) and Bugajski (1978).

Is logic empirical?

According to the traditional view, logic is an a priori discipline. Traditional logicians assume they can know the laws of logic through pure reflection, and assert that logical principles, therefore, cannot be confirmed or rejected by experiments. They say that principles of logic are necessarily true when true, and necessarily false when false. Consequently, the realm of logic should be independent of the results of empirical research. This traditional view has been challenged by many mathematicians, philosophers and scientists on the basis of widely different reasons.

Aristotle (384-322 B.C.) defended the principles of identity, of contradiction, and of excluded middle as "laws of thought". However, according to Łukasiewicz (1951) the identification of two-valued logic with Aristotelian logic is not correct and the principle of two-valuedness of propositions was probably formulated for the first time by the Stoic philosopher Chrysippus (279-206 B.C.).

Classical Boolean logic is a two-valued logic, it accepts the doctrine that every proposition is either true or false. It is admitted that there are intermediate possibilities between being known to be true and being known to be false, but it is claimed that there are none between truth and falsehood themselves. This *law of excluded middle* (or: *tertium non datur*) has been vigorously discussed from the time of the Greeks; many eminent philosophers (including Aristotle) have entertained the view that there may be other possibilities.

Even in mathematics the law of excluded middle has to be called in question. The founder of mathematical intuitionism, Luitzen Egbertus Jan Brouwer (1881-1966) defended the thesis that no mathematical proposition is true unless we can know it to be true in a nonmiraculous way. According to Brouwer, a proposition is true if and only if it can be proved. But a proof of the negation $F^{\perp}$ of a proposition F is a construction which obtains an absurdity from the assumption that there is a proof of F: the mere fact that F cannot be proved does not yet imply the falsity of F. Since we have no reason to suppose quite generally that we shall ever be in possession either of a proof of an arbitrary proposition F, or of the deduction of an absurdity from F, the dogma of the universal validity of the principle of the excluded middle has to be rejected. Brouwer also stressed that the thoughtless use of the principle of excluded middle leads to the erroneous idea that every mathematical problem can be solved.

Brouwer's critique of the principle of excluded middle in classical logic led to a nonclassical logic, called the *intuitionistic logic*. While according to Russell mathematics is to be reduced to logic which is taken as fundamental, the intuitionists take mathematical thought to be primary and think of logic as a form of inference constructed by an *a posteriori* investigation. More radically, Quine (1953) has suggested that the laws of logic are not analytic, and according to Putnam (1969, 1975) logic is a natural science rather than a body of a priori necessary truth. Watanabe (1961) also takes the view that classical logic has to be modified in order to categorize empirical facts in a comprehensive form, and points out that many forms of human inference do not fit into Boolean logic (Watanabe, 1969a). He claims that the structure of ordinary language is essentially given by a non-Boolean modular lattice, and that a Boolean language is not sufficiently comprehensive to tell the "whole story". It is an empirical fact that the propositions of human experience cannot be fully comprehended by the rules of classical logic, and Blau (1973, 1977) has shown that the syntax of colloquial speech is not Booolean. Von Weizsäcker (1971a,b; 1973) has reminded us of an old lacuna of classical logic, namely its insufficient description of temporal statements, and he has suggested that temporal propositions might obey logical laws different from those of classical logic.

Since the two-valuedness of the truth functional of a propositional calculus implies the distributive law, a logic with a nondistributive lattice cannot have just the two truth values "true" and "false". For this reason, Reichenbach (1944, 1951) proposed a three-valued logic. (Compare also the critique by Hempel, 1945, and the commentary by Putnam, 1957). Reichenbach's proposal never became popular, just as little as Weizsäcker's idea to use a complex-valued logic for quantum mechanics (Weizsäcker, 1955, 1958; Weizsäcker et al. 1958). Finkelstein (1962, 1969, 1972) proposed not to change the classical meaning of $\wedge$, $\vee$, and $^{\perp}$, so that it is not the law of excluded middle that breaks down but the distributive law $F\wedge(G\vee H)=(F\wedge G)\vee(G\wedge H)$. Similarly, Putnam (1969) claims that if the operations $\vee$, $\wedge$ and $^{\perp}$ of quantum logic *"are not the operations of disjunction, conjunction and negation, then no operations are"*. However, this claim is unsubstantiated; for example, Finch (1969a, 1970) has proposed with good reasons the operation $(F\vee G^{\perp})\wedge G$ as the appropriate operation of logical conjunction between F and G (compare also Clark, 1973).

Predicates in quantum logics

For the definition of binary predicates in quantum logics, compare also Richter (1964), Kunsemüller (1964), Kamber (1964, 1965), Mittelstaedt (1970, 1973), Moroz (1971), Goldblatt (1974).

The idea that quantum mechanics uses a revised logic has been criticized harshly. Bunge (1967b, p.241) thinks that *"the mistake regarding the so-called quantum logic has been perpetuated only because it was introduced and endorsed by a few eminent people"*. In a rather arrogant paper, Sir Karl Raimund Popper (1968) severely criticized the Birkhoff-von Neumann (1936) paper claiming that it *"culminates in a proposal which clashes with each of a number of assumptions made by the authors"*. Popper's paper is trivially wrong; it is an amusing exercise to find all the technical errors of this paper (the reader may find some help in Jauch and Piron, 1970, footnote on page 171; and in Scheibe, 1974).

Additional references

For critical investigations of the logical approach to quantum mechanics, compare also: Bohm and Bub (1966b), Lenk (1969), Heelan (1970a,b), Gardner (1971), Fine (1972). For the view that quantum logic finds a rightful place among modal logics, compare Fraassen (1972, 1973, 1979), Herman and Piziak (1974), Dalla Chiara (1977), Dishkant (1977, 1978). However, as Bub (1973) stresses, the modal interpretation *"is no more than a philosopher's formulation of the statistical interpretation. It is in no sense a rival quantum logical interpretation"*. But even some proponents of quantum logic claim that the logical approach to quantum mechanics does not embody a deviant logic (Piron, 1964; Jauch 1968a on p.77). A systematic study of the structure and philosophical implications of deviant logics has been given by Haack (1974).

To summarize: As usual, there is no agreement about the most fundamental questions. Since every view depends on the interpretation of the concepts used, no simple answer can be expected. A simple-minded yes-no answer cannot be a good answer. One reason for the confusion is the prevailing dogma that logic has to be universal, applicable to all propositions, propositions about logic included. Once it is recognized that this dogma is a metaphysical prejudice, we can adopt the view of Paulette Destouches-Février (1951) that *there is no logic which works for every subject matter.* Adopting this point of view makes many differences of opinion fade away.

Empirical logic

In order to straighten out the muddle of the meaning of quantum logics we have to know the conceptual characterization of their propositions. It is often not clear whether the propositions of a particular quantum logic are to be understood as contingent or a priori, as temporal or timeless, as ontic or epistemic. Sometimes, theoretical propositions are confused with empirical ones. It is, of course, an undisputed fact that the closed subspaces of a Hilbert space form a non-distributive, complete orthomodular lattice. Clearly, the elements (also called *propositions*) of such a lattice obey a nonclassical logic. Sticking to the (baseless) operationalistic jargon of pioneer quantum mechanics, Birkhoff and von Neumann (1936) called such propositions *experimental* ones. In a large part of the modern literature, the propositions of quantum logics are directly identified with *yes-no experiments* (Jauch and Piron, 1963, 1969; Jauch, 1968a), with *experimental yes-no questions* (Mackey, 1963a), or with experimentally verifiable statements (Varadarajan, 1968). According to this view, the *theoretical propositions* of pioneer quantum mechanics (represented by projection operators, or equivalently by closed linear subspaces) are identified with *empirical propositions* (represented by yes-no experiments). However, these two notions are conceptually very different. According to Heelan (1970a), yes-no experiments are *not* isomorphic to theoretical propositions, and Foulis and Randall (1978) stressed that quantum logic is *not* a faithful image of laboratory procedures.

Birkhoff (1961) has put forward the view that the logic of quantum mechanics should draw its authority directly from experiments. This goal has *not* been achieved by contemporary quantum logics. A large part of the investigations on quantum logics is based on an extreme operationalistic view which is philosophically untenable. This fact has not influenced the *mathematical* development of quantum logic; nonetheless the claim that quantum logic is empirically founded comes from a confusion of theoretical and empirical concepts, and from the failure to distinguish between the meaning of a statement and the method of testing it empirically. Some clarification can be achieved by using the concepts of *operative logic* developed by Lorenzen (1955, 1962). This approach uses protological properties of propositions only; all we have to know is the availability or nonavailability of propositions in a dialogue between two partners. If the dialogue between the proponent and the opponent is restricted to the available propositions, then the calculus of these propositions is called an *operative logic*. A proposition is called *empirical* if it is related to an experiment. An empirical proposition is called *available* if we have performed an experiment which resulted in the empirical event claimed by the proposition. For classical systems all empirical propositions are supposed to be available, while for quantal systems the availability of empirical propositions is severely restricted. *The operative logic of quantum mechanics is a restricted logic.*

References: on empirical and operative logics

The empirical logic of general physical systems has been discussed by Grawert (1953), Suppes (1965), Finkelstein (1969) and Putnam (1969).

A detailed investigation of the operative logic of quantum mechanics has been given by Mittelstaedt (1959, 1960, 1962, 1963, 1970, 1972, 1973, 1977, 1978, 1979), Mittelstaedt and Stachow (1974), and Stachow (1976, 1978). From the context it is clear that Mittelstaedt uses a strictly epistemic interpretation of quantum mechanics in terms of an intuitionistic operative logic. Unfortunately, experimental and theoretical propositions are identified without any deeper discussions (assuming the validity of discontinuous state transitions associated with the measuring process of pioneer quantum mechanics).

A genuine *empirical logic* has been developed by Foulis, Randall and their coworkers. They did not attempt to advocate a particular physical theory, rather they conceived empirical logic as a language which allows to discuss any empirical science with great precision. By a *physical operation* they mean a collection of instructions which describe a well-defined, physically realizable and reproducible procedure, including the specification of what must be observed and recorded. These actual physical operations are taken as the unanalyzable primitives. The collection of admissible physical operations is represented mathematically by a formal structure called a *manual*, which defines an empirical universe of discourse. All other concepts are defined in terms of the physical operations of the manual. *Operational propositions* are by definition propositions that can be confirmed or refuted by the evidence available from a sequence of experiments. Foulis and Randall neither require nor discourage a realistic view of the world, they intend to give a universal mathematical foundation for all empirical science, independent (as far as possible) of any particular philosophical point of view.

The empirical logic of Foulis and Randall can be regarded as a generalization of traditional statistics whereby the notion of a sample space is generalized from the classical Kolmogorov setting to a set X together with a symmetric and irreflexive relation $\perp$. Two outcomes f and g are said to be *orthogonal*, written $f \perp g$, if f and g reject each other operationally in the sense that the execution of an operation which yields the outcome f cannot yield the outcome g, and vice versa. The pair $(X,\perp)$ is referred to as an *orthogonality space*. While Kolmogorov's probability theory produces only classical Boolean logic, the generalized approach of Foulis and Randall produces generalized logics. The operational logic of a manual forms a so-called *orthologic*: a bounded partially ordered set together with an orthogonality relation. Every orthogonal partially ordered set (hence every orthogonal lattice) is an orthologic.

Empirical logic is sufficiently general to allow a description of every quantum experiment, but *empirical logic cannot be identified with generalized quantum mechanics*. Under reasonable conditions, the operationally accessible empirical propositions form an orthomodular lattice L_{emp}. However, the σ-completion of L_{emp} is *not* the lattice of the proposition of quantum mechanics. The reason is that L_{emp} of Foulis and Randall is defined truly and strictly operationally while the propositions of generalized quantum mechanics refer (in spite of the operationalistic jargon of pioneer quantum mechanics) to a *theory*. Since no theory can be derived from empirical data only, we cannot expect that the completion of L_{emp} is isomorphic to the lattice of theoretical propositions of quantum mechanics. However, there are some similarities. For example, the lattice of theoretical propositions of pioneer quantum mechanics is non-Boolean, and the empirical logic (in the sense of Foulis and Randall) of quantum experiments is also non-Boolean. The empirical logic approach by Foulis and Randall includes any empirical-statistical procedure such as psychological testing or opinion polling. In comparison to general empirical logic, Wright (1978b) considers pioneer quantum mechanics still as "*very classical*".

References on the Foulis-Randall approach

The basic papers of the empirical logic approach are by Randall (1966), Randall and Foulis (1970, 1973, 1976) and Foulis and Randall (1972). The connection between empirical logic and quantum mechanics has been discussed by Randall (1966), Foulis and Randall (1974a,b, 1978) and Randall and Foulis (1979). A discussion of the Stern-Gerlach experiment in terms of spin manuals has been given by Wright (1978a). For applications to fields having a structure very different from quantum mechanics, compare Wright (1978b). Fischer and Rüttimann (1978) have shown that the empirical logic approach is closely connected with the convex-state approach.

Axiomatic quantum logic

Quantum logic takes *propositions* (synonyma: *events, questions*) as primitive elements. States and observables enter as derived concepts.

General treatises on axiomatic quantum logic

Useful sourcebooks on the development of the quantum logic approach are the anthologies edited by Hooker (1975, 1979a-b). Additional material concerning the historical development can be found in chapter 8 of Jammer (1974). In his excellent text, Jauch (1968a) bases the foundation of pioneer quantum mechanics on quantum logic. The two-volume text by Varadarajan (1968, 1970) gives a thorough development of quantum mechanics from a quantum logical and group theoretical point of view. Reviews of quantum logic from diverse positions have been given by Gudder (1970c, 1977c), Beltrametti and Cassinelli (1973, 1976), van Fraassen (1974), Bub (1974), Rüttimann (1977b), and Mittelstaedt (1978). A rather ambitious program to understand the unity of physics on the basis of the logic of temporal propositions has been initiated by Weizsäcker (1971a-b, 1973), carried on by Drieschner (1979).

In axiomatic quantum logic it is assumed that the propositions form at least an orthomodular partially ordered set L (often called a *logic*). An *observable quantity* A is represented by a σ-homomorphism $E_A:\Sigma_{\mathbb{R}}\to L$ from the σ-algebra $\Sigma_{\mathbb{R}}$ of Borel sets of the real axis into the logic L. For every Borel set $B\in\Sigma_{\mathbb{R}}$, the element $E_A(B)\in L$ is interpreted as the proposition that a measurement of A produces an outcome in B. A *statistical state* is represented by a σ-orthoadditive map $\rho:L\to[0,1]$ of the logic L into the closed unit interval. The value of ρ at the proposition $E_A(B)$ is interpreted as the probability $\mu(B)$ that a measurement of the quantity A on a system in the state ρ gives an outcome in the Borel set B,

$$\mu(B) = \rho\{E_A(B)\} \quad .$$

The expectation value $\langle A\rangle$ of the observable quantity A with respect to the statistical state ρ is then defined as

$$\langle A\rangle = \int_{\mathbb{R}} \lambda\mu(d\lambda) \quad .$$

If L is a Boolean algebra then one obtains Kolmogorov's (1933) formulation of classical probability theory which among other things is the basis of classical statistical mechanics. If one takes L as isomorphic to the lattice of all projection operators on a Hilbert space, then one recovers the logical structure of pioneer quantum mechanics.

This scheme has been proposed and axiomatized by Mackey (1957, 1960, 1963a). His first six axioms apply to classical mechanics as well as to pioneer quantum mechanics. Mackey's seventh axiom singles out pioneer quantum mechanics by requiring that L be isomorphic to the lattice of all closed subspaces of an infinite-dimensional separable Hilbert space over the complex numbers. Clearly this axiom (later referred to as Mackey's Hilbert space axiom) is entirely ad hoc, so that Mackey posed the problem of deriving this axiom from plausible and physically meaningful axioms.

References on Mackey's axiomatization

Improvements of Mackey's axiomatic system have been discussed by Gleason (1957), Varadarajan (1962), Ramsay (1966), Maçzyński (1967, 1971, 1972, 1973b, 1974), Deliyannis (1971a), Guz (1971, 1975, 1978a,b, 1979).

The first realization of Mackey's program is due to Zierler (1961, 1966) but it has been objected by Bugajska and Bugajski (1972a) that Zierler's axioms are not independent and have little physical justification.

Generalizations of the approach by Mackey, and axioms for a Hilbert-space model of quantum mechanics have been discussed by:
McLaren (1964), Kamber (1964, 1965), Nikodým (1966) [consult the review of D. Babbitt in Mathematical Reviews 36, #7382], Gunson (1967, 1972), Gudder (1965, 1966, 1967a,b,c; 1968c, 1969a,b; 1970b,c,d; 1972b, 1973a), Pool (1968a, 1968b), Varadarajan (1968, 1970), Finch (1969b,c), Gudder and Mullikin (1973), Beltrametti and Cassinelli (1973), Greechie and Gudder (1973), Bugajska and Bugajski (1972a, 1973a, 1973b, 1973c).

One of the best known axiomatizations of pioneer quantum mechanics is due to Piron (1964). Piron's main contribution to axiomatic quantum theory was his characterization of the logic of pioneer quantum mechanics within the category of all logics. He made the physically plausible assumption that the set of events is given by an orthomodular complete lattice L, and established that if L is also irreducible, atomic and fulfills a covering law, then L can be represented by the closed linear subspaces of a vector space over a division ring with involution. This approach leads to the interesting question, how far the logical structure of quantum mechanics singles out the familiar choice of complex numbers as a number field for the carrier space of physical states. The investigations by Eckmann and Zabey (1969), Beltrametti and

Cassinelli (1972) and by Ivert and Sjödin (1978) have shown that any algebraic extension of p-adic fields and finite fields have to be excluded. Accordingly, the admitted fields have to be algebraic extensions of the unique Archimedean, non-p-adic valuation of the rational field. One is, therefore, left to choose between the real numbers, the complex numbers, and the quaternions. In pioneer quantum mechanics, the complex numbers are chosen as number field. Quantum mechanics over the field of real numbers turns out to be equivalent to traditional quantum mechanics if one introduces an antisymmetrical classical observable J with $J^2=-1$ (Stueckelberg, 1960; Stueckelberg and Guenin, 1961, 1962; Stueckelberg et al., 1961). The feasibility of a generalized quantum mechanics over the quaternions has also been investigated (Jordan, 1934; Finkelstein et al., 1962, 1963; Emch, 1963, 1965; Jauch, 1968b). The fact that there is no natural tensor product of quaternion Hilbert spaces has the interesting consequence, that a generalized quantum mechanics with a quaternion Hilbert space as a state space exhibits a complementarity between the qualities of different noninteracting subsystems. Since such a strange phenomenon is empirically unknown, the quaternion field can be excluded.

References on embedding quantum logics in Hilbert space

Piron's (1964) proof of his representation theorem is incomplete, a correct proof has been given by Amemiya and Araki (1966). An excellent exposition of representation theorems of quantum logics in Hilbert space can be found in the text by Varadarajan (1968). Compare also the related work by Rüttimann (1970), Gudder and Piron (1971), Ingleby (1971), Toyoda (1973), Chen (1973), Maçzynski (1973a, 1976), Cirelli and Cotta-Ramusino (1973), Bugajska (1974), Jones (1976), Wilbur (1977), Gudder and Michel (1979), Aerts and Daubechies (1979).

Jauch and Piron (1969) have inaugurated a new motivation for Piron's (1964) axiomatic formulation of quantum logics. In this version, states are defined without recourse to probabilistic statements. Accordingly, there are no difficulties in discussing *individual* systems.

References on the Jauch-Piron approach

Jauch and Piron (1963, 1968, 1969), Piron (1964, 1969, 1971, 1972, 1976), Jauch (1968a, 1968b, 1970, 1971), Emch and Piron (1963), Emch and Jauch (1965), Guenin (1966), Amemiya and Araki (1966), Plymen (1968b), Pool (1968b), Ochs (1972a, 1972c), Jenč (1972), Bugajska and Bugajski (1973b), Hellwig and Krausser (1974a, 1974b, 1977), Pulmannova (1975, 1976), Zabey (1975), Rüttimann (1977a), Aerts and Daubechies (1978,a,b).

Note that there is no unique Jauch-Piron axiomatic system, and that the view of Jauch and the view of Piron are not identical.

Jauch and Piron (1969) have given a set of axioms which imply that the system of propositions of a physical system forms a logic L with the following properties:

(i) L is a lattice,

(ii) L is complete,

(iii) L is orthomodular,

(iv) L is semimodular,

(v) L is atomic.

They have dropped the earlier postulate that L has to be irreducible since this requirement would imply that there are no superselection rules.

Jauch and Piron have regarded their axioms as intuitively well motivated. However, they provide too strong a justification for their postulates since we know nowadays that there are reasonable physical theories not fulfilling all their axioms. For example, in the algebraic approach to thermodynamics L is given by the projection lattice of a type III W*-algebra which is not atomic. Moreover, there exist simple but reasonable logical systems that cannot be described in terms of a Hilbert space model (Mielnik 1968, 1969; Davies 1972b). To assume that the axioms of Jauch and Piron are necessary would therefore be a premature conclusion. For this reason it is worthwhile to discuss the evidence for each of their postulates.

Orthomodular logics

With a few exceptions (e.g. Mielnik, 1976; Guz, 1980), most theorists agree that the system of proposition of any physical system is at least an orthocomplemented partially ordered set $(L,\leq,\perp)$, where L is the set of all propositions of the system, the relation $\leq$ the partial order of logical implication, and $F\to F^{\perp}$ the orthocomplementation of L induced by the logical negation. If the supremum of two elements $F,G\in L$ exists, it is unique and we denote it by $F\vee G$; if the infimum exists, it is unique and we denote it by $F\wedge G$. The basic presupposition that the logic is an orthocomplemented partially ordered set means that for every proposition $F\in L$ there exists a unique proposition $F^{\perp}\in L$ that is false whenever F is true, and furthermore fulfills the law of double negation, $(F^{\perp})^{\perp}=F$, the law of excluded middle, $F\vee F^{\perp}=E$, and de Morgan's law $(F\wedge G)^{\perp}=F^{\perp}\vee G^{\perp}$. The last relation implies that the orthocomplementation is compatible with the partial order in L, that is, if $F\leq G$, then $G^{\perp}\leq F^{\perp}$.

A fundamental problem of axiomatic quantum logic is whether L is a *lattice*, that is: does $F \vee G$ and $F \wedge G$ exist for *all* $F, G \in L$? Sometimes it has been tried to justify the lattice structure of L phenomenologically (compare e.g. Jauch, 1968a) but a closer analysis shows that it is not possible to defend the lattice structure in terms of experimentally realizable operations. For this reason, some theoreticians do not assume that L is a lattice (compare the reviews by Gudder, 1970c; and by Greechie and Gudder, 1973).

In their new approach, Jauch and Piron (1969) have introduced propositions as classes of empirically equivalent yes-no experiments. It is remarkable that Jauch and Piron have been able to *derive* the existence of arbitrary suprema and infima, so that their logic L is a complete lattice. However, not everybody finds Jauch and Piron's definition of the infimum of a set of experimental statements convincing (compare e.g. Varadarajan, 1977); moreover their reasoning is not conclusive either (compare for example the critique by Ochs, 1972c). In particular, their claim that L is even a *complete* lattice is much too strong since it excludes classical statistical mechanics which uses only a *σ-complete* Boolean lattice (recall that a lattice L is called *complete* if *every* nonvoid subset has a supremum and an infimum in L; a lattice L is called *σ-complete* if every *countable* set has a supremum and an infimum in L). Hence the lattice postulate of quantum logic should be weakened by requiring that *L is a σ-complete lattice* (Plymen, 1968b; a less attractive alternative defines statistical states only on a σ-sublattice of a complete lattice, as proposed by Guenin, 1966).

We shall adopt the view that it is not reasonable to consider the lattice of a logic as conceptually inevitable but that it is mathematically *convenient* to construct a natural *extension* of the conceptually motivated logic into a lattice such that the extended logic possesses all the essential features of the original one (Bugajska and Bugajski, 1973c). In what follows we shall assume that L is a σ-complete orthocomplemented lattice. The logic of every *classical* system is a σ-complete *Boolean* lattice (recall that a Boolean lattice is by definition an orthocomplemented distribution lattice). In order to find the proper substitute for the distribution law characteristic for classical systems, we introduce the notion of *compatible* propositions. *Two elements* $F, G \in L$ *of an orthocomplemented lattice are said to be compatible if the ortho-*

complemented sublattice generated by $\{F,F^\perp,G,G^\perp\}$ *is Boolean*. We denote the relation of compatibility by the symbol $\leftrightarrow$. The following characterization is equivalent

$$F \leftrightarrow G \quad \text{iff} \quad (F\wedge G)\vee(F\wedge G^\perp) = F.$$

The compatibility relation is obviously reflexive and symmetric, i.e.

$$F \leftrightarrow F \quad ,$$

$$F \leftrightarrow G \quad \text{iff} \quad G \leftrightarrow F \quad .$$

Furthermore we have:

$$F \leftrightarrow F^\perp \quad ,$$

$$F \leftrightarrow E \quad ,$$

$$F \leftrightarrow 0 \quad ,$$

$$F \leftrightarrow G \quad \text{iff} \quad F \leftrightarrow G^\perp \quad ,$$

$$F \perp G \quad \text{implies} \quad F \leftrightarrow G \quad .$$

In his axiomatics, Piron (1964) made the intuitively most reasonable *postulate that comparable elements are compatible*, i.e. that

$$F \leq G \quad \text{implies} \quad F \leftrightarrow G \quad .$$

The class of orthocomplemented lattices fulfilling this relation are called *orthomodular lattices* (synonyma: weakly modular lattices, CROC= canoniquement relativement orthocomplémenté).

Additional references on the postulate of orthomodularity

There are many different but equivalent definitions of an orthomodular lattice; compare Piron (1964), Rose (1964), Fáy (1970), Maeda and Maeda (1970). For further motivations for the postulate of orthomodularity in quantum logics, compare Finch (1969a-c, 1970), Maçzyński (1970), Fáy (1970), Mittelstaedt (1970), Guz (1971, 1979).

Orthomodular lattices are non-distributive generalizations of Boolean lattices. Clearly, every Boolean algebra is an orthomodular lattice. Non-Boolean orthomodular lattices are of particular importance in the decomposition theory of operator algebras. The self-adjoint and idempotent elements of an arbitrary W*-algebra form an orthomodular lattice. According to Holland (1970) "*von Neumann algebra theory is the Mother Theory, continuous geometry is the first born son, and orthomodular lattice theory is the second son*".

Orthomodular logics provide a very broad framework, but for physical applications additional restrictions are certainly necessary. For example, there exist finite orthomodular lattices not admitting a full set of states (Bennett, 1970). Furthermore, there exist orthomodular lattices admitting no states at all (Greechie, 1971), and orthomodular partially ordered sets with a full set of states which cannot be embedded in the lattice of all closed subspaces of a complex Hilbert space (Greechie, 1969). There are orthomodular lattices that make sense as an orthomodular logic in the framework of the empirical logic of Foulis and Randall (and which could be of interest in the social sciences) but which support states that cannot be interpreted in a Hilbert-space projection lattice (Wright, 1978b). Hence the problem is to find physically plausible axioms so that the resulting axiomatic system is useful for physics and chemistry.

All presently known physical theories fall into a rather special class of orthomodular logics, having a considerably richer logical structure. For example, the well-examined W*-systems (which encompass among other theories also classical mechanics, pioneer quantum mechanics and quantum statistical mechanics) have an orthomodular logic which is in addition also semimodular. Orthomodular lattices and semimodular lattices are distinct generalizations of modular lattices. Recall that two elements F,G of a lattice L are said to form a *modular pair* (F,G) if $(H\vee F)\wedge G=H\vee(F\wedge G)$ for every $H\leq G$. A lattice L is called *modular* if every pair of elements in L is a modular pair. A lattice is called *semimodular* if the fact that (F,G) is a modular pair implies that (G,F) is a modular pair. An ortholattice $(L,\leq,^{\perp})$ is *orthomodular* if every orthogonal pair is a modular pair, i.e. if $F,G\in L$ and $F\perp G$ implies that (F,G) is a modular pair. Clearly, every Boolean lattice is both orthomodular and semimodular.

Since Newtonian mechanics, pioneer quantum mechanics and statistical mechanics are all governed by *semimodular* orthomodular logics, it would be interesting to have an intuitively well-motivated justification for an axiom of semimodularity. All we know at present is that semimodularity in pioneer quantum mechanics is closely related to the possibility of so-called "measurement of the first kind" (Pool, 1968b; Jauch and Piron, 1969; Ochs, 1972a; Bugajska and Bugajski, 1973b; Cassinelli and Beltrametti, 1975). A measurement of a proposition $F\in L$ is said to be *of the first kind* if the answer "yes" implies that F is true immediately after the measurement. A measurement of F is said to be *ideal* if every true proposition compatible with F is also true after the measurement of F. If L is a complete, atomic orthomodular lattice (as in pioneer quantum mechanics) and if for every proposition $F\in L$ there exists an ideal measurement of the first kind, then it follows that L is semimodular.

Remark: The "covering law" of Jauch and Piron

Piron and Jauch postulated the following covering law: "If $F\in L$ is an atom, then the relation $G\leq H\leq G\vee F$ implies either $H=G$ or $H=G\vee F$. The following formulation of the covering law is equivalent: "If $F\in L$ is an atom, and $G\in L$ a non-absurd proposition ($G\neq 0$) such that $G^{\perp}\wedge F=0$, then $(F\vee G^{\perp})\wedge G$ is an atom. In án *atomic* orthomodular lattice L, the covering law is equivalent to the semimodularity of L (Pool, 1968b). Originally the covering law was introduced by Piron (1964) as a highly technical axiom in order to prove the isomorphism between his quantum propositional system and the lattice of closed subspaces of a Hilbert space.

Since pioneer quantum mechanics has an *atomic logic*, it is often assumed that the logic of any physical system is atomic (Piron, 1968a; Jauch and Piron, 1969; Jenč, 1972, 1974). Recall that a logic is called atomic if every non-zero proposition has an atom under it. In an atomic logic every non-absurd proposition is implied by an atomic proposition (which by definition, is not implied by any other non-absurd proposition). There is no general agreement on the question whether or not classical mechanics has an atomic logic. For example, Jauch claims that "*only the atomic lattice represents really classical mechanics as it always has been understood*" (Jauch, 1968a, p.80), while Birkhoff and von Neumann (1936) say that it is unrealistic and absurd to identify each subset of the phase space of classical mechanics with a proposition. Von Neumann proposed to use the Lebesgue-measurable subsets of the phase space modulo sets of Lebesgue measure zero as representants of experimental propositions. This proposal is closely related to Kolmogorov's (1933) foundation of mathematical probability theory and leads to an *atom-free* logic for classical mechanics (for more details, compare Primas, 1980, sect.2). For the purpose of classical statistical mechanics atomic lattices are not appropriate (Varadarajan, 1977). Moreover, the postulate of atomicity would exclude type III W*-systems which are a main tool of modern algebraic quantum statistical mechanics. It seems to be fair to admit that historically atomicity has been required in order to obtain the formalism of pioneer quantum mechanics from quantum logic. For these reasons we consider an axiom requiring the atomicity for the logic of an *arbitrary* physical system not as a reasonable assumption.

Additional remark: Dynamics in atomic Boolean lattices

There is an additional grave difficulty with classical dynamical systems having an atomic logic. Every classical system with a Boolean logic L can be embedded into a commutative W*-algebra A. If L is atomic, then A is atomic. The time evolution of a dynamical W*-system is represented by a one-parameter automorphism group $t\to\alpha_t\in\mathrm{Aut}(A)$, $t\in\mathbb{R}$. The weakest reasonable continuity assumption for the dynamics is that $t\to\alpha_t$ is σ-weakly continuous. It is easy to show that every σ-weakly continuous dynamics leaves every atom of a commutative W*-algebra invariant. Hence the time evolution of a classical dynamical system with an atomic logic cannot be represented in the usual way by differential equations of motion.

The lattice of propositions of pioneer quantum mechanics is *irreducible*. This additional structure reflects the fact that pioneer quantum mechanics does not take superselection rules into account. In a general orthomodular logic L, the set of those propositions that are compatible with every proposition of L is called in the center $Z(L)$ of L,

$$Z(L) \overset{\text{def}}{=} \{Z \mid Z\in L \quad , \quad Z \leftrightarrow F \quad \text{for all} \quad F\in L\} \quad .$$

The center of an orthomodular lattice is a Boolean algebra. A system is called *classical* if its logic L is Boolean, i.e. if $L=Z(L)$. A system is entirely nonclassical if there are no nontrivial compatible propositions, i.e. if the center consists only of the two trivial elements O and E. Such a system is called *irreducible*. A system with an irreducible and atomic logic L satisfies the *superposition principle* which can be expressed as follows (Jauch, 1968a): "For every pair of distinct atomic propositions $F,G\in L$ there exists a third atomic proposition $H\in L$, $H\neq F$, $H\neq G$, such that $F\vee G=G\vee H=H\vee F$". The atomic proposition H plays the role of a superposition of F and G, it can never be compatible with both F and G. An immediate consequence is that the sublattice generated by F,G and H cannot be Boolean, so that *the superposition principle implies the irreducibility of the logic*. Furthermore, Gudder (1970b) has shown that if a superposition principle holds, then the logic is a complete atomic lattice. The logics of general physical systems are in general *reducible*. In this case the superposition principle has only a restricted validity, and with Wick, Wightman and Wigner (1952) we say that the system allows *superselection rules*, a concept introduced into quantum logic by Paulette Destouches-Février (1951).

To sum up, it seems to be a sound procedure to formalize the logics of physical systems by *σ-complete orthomodular and semimodular lattices*. However, the logic of a physical system is not necessarily Boolean, atomic or irreducible. Boolean logics, atomic logics and irreducible logics represent important *special cases* of general quantum logic.

Two orthomodular σ-complete logics L_1 and L_2 are considered to be logically equivalent if there exists a bijective map φ from L_1 into L_2 which preserves the partial ordering (i.e. the logical implication), the orthocomplementation (i.e. the logical negation), and countable suprema and infima. Such a map $\varphi: L_1 \to L_2$ from a σ-complete orthomodular lattice L_1 onto a σ-complete orthomodular lattice L_2, fulfilling

(i) $\varphi(O) = O$

(ii) $\varphi(F) \le \varphi(G)$, iff $F \le G$,

(iii) $\varphi(F^{\perp}) = \varphi(F)^{\perp}$ for every $F \in L_1$,

(iv) $\varphi(\vee F_i) = \vee\varphi(F_i)$ for every disjoint sequence $\{F_i\} \subset L_1$,

is called an ortholattice σ-isomorphism. Every ortholattice σ-isomorphism maps semimodular lattices into semimodular lattices, atomic lattices into atomic lattices, and irreducible lattices into irreducible lattices. An ortholattice σ-isomorphism of an orthomodular σ-complete lattice L onto L is called an *automorphism* of L. Speaking intuitively, a *logical symmetry* of a system is a transformation which leaves all significant logical features invariant; in the theory of orthomodular logics such a transformation is represented by an automorphism of the lattice L of propositions (Emch and Piron, 1963; Pulmannová, 1977).

The kinematical and dynamical symmetries of physical systems are given by locally compact groups. In quantum logic, a representation of a group G is defined as a homomorphism α from G into the group Aut(L) of automorphisms of L. The mathematical problem of representing topological groups continuously as automorphisms of an orthomodular lattice has been discussed by Gudder (1971), Gallone and Mania (1971), and Leveille and Roman (1975). However, this theory is not yet as well developed as the theory of unitary group representations in Hilbert spaces, or as the theory of automorphic group representations on W*-algebras. This fact is one important reason why in practical applications orthomodular logic is often restricted to the more limited context of projection logics of W*-algebras where more powerful analytical tools are available.

4.5 # NON-BOOLEAN PROBABILITY THEORY

Ensembles and collectives

The concept of probability is of central importance for quantum theory. Classical statistical decision theory is often used to deal with experiments that can be repeated many times under similar and well-controlled conditions. Although we are often interested in individual phenomena, the emphasis of every statistical theory is shifted to the *distribution* of individual events. *The referent of statistical probability theory is a family consisting of an extremely large number of noninteracting, uncorrelated and identically prepared individual systems.* We must clearly distinguish between theory and experiment, and the corresponding notions "ensemble" and "collective". An *ensemble* is a concept of statistical *theory*, it refers to an infinite family of noncorrelated individual systems, and is theoretically characterized by a statistical state. With von Mises (1931, in the English version of 1964 on p.11-12), "*we shall apply the term collective to a long sequence of identical observations or experiments, each experiment leading to a definite numerical result, provided that the sequence of results fulfills the two conditions: existence of limiting frequencies and randomness*". The notion of *probability* belongs to statistical ensemble theory, it is a measure on the lattice of *theoretical* propositions. The concept of *relative frequency* is operationally defined and refers directly to *empirical* propositions. Some scientists take it for granted that theories can be either refuted or verified by comparison with experimental data. However, the frequencies resulting from experiments on the individual systems of a collective have to be *inferred* from the probabilities of a statistical ensemble theory. Here one must employ again a theory. It is usually assumed that the classical statistical estimation theory and statistical decision theory (compare e.g. Lehmann, 1959) provide the necessary techniques.

Remark: Modern views on Randomness

Von Mises' condition of the "*existence of limiting frequencies and randomness*" has led to many and often discussed difficulties. However, the use of a precise concept of an algorithm has made it possible to overcome the inadequacies of the traditional foundations of statistical theories. The most exciting new result of these developments is the feasibility of a conceptually simple and mathematically rigorous definition of the notion *complexity* which provides an enthralling alternative to the well-established measure-theoretic approach to probability. According to algorithmic probability theory, *high complexity is the same as randomness.*

The roots of the new developments can be traced to the pioneering work of von Mises (1919) where he proposed his *principle of excluded gambling systems.*

The formalization of this principle as proposed by von Mises turned out to be mathematically untenable. The further investigations by Wald (1937) were brought to their culmination by Church (1940). Von Mises wanted to exclude "all" gambling systems but he did not properly specify what he meant by "all". Church pointed out that a gambling system which is not effectively calculable is of no practical use. Accordingly, a gambling system has to be represented not by an arbitrary function but by an effective algorithm for the calculation of the values of a function. In accordance with von Mises' intuitive ideas and Church's refinement, a sequence is called random if no player who calculates his pool by effective methods can raise his fortune indefinitely when playing on this sequence.

The *computational complexity* of a mathematical object reflects the difficulty of its computation. Kolmogorov (1965) proposed a measure of complexity based on the "size of a program" which, when processed by a suitable universal computing machine, yields the desired object. An adequate formalization of the notion of a universal computing machine is that given by Turing (1936). A Turing machine is essentially an automatic computer having an infinitely expandable memory; it can be taken as a precise definition of the concept of an algorithm (for texts discussing computability by means of Turing machines compare e.g. Boolos and Jeffrey, 1974, or Brainerd and Landweber, 1974). According to Kolmogorov (1963, 1965) and Chaitin (1966) the randomness of a finite binary sequence is measured by its complexity, which is defined as the length of the shortest binary program that produces this sequence in a given Turing machine. A finite sequence of maximal complexity has no patterns and the basic idea is to regard patternless finite sequences as *finite random sequences*.

A slightly different but more powerful variant of algorithmic probability theory is due to Schnorr (1969). According to Schnorr (1971a) *a sequence fails to be random if and only if there is an effective process in which this failure becomes evident*. The tests considered by Schnorr (1970a-c, 1973) are constructive to such an extent that it is possible to approximate infinite random sequences to an arbitrary degree of accuracy by computable sequences of high complexity (so-called pseudo-random sequences). The approximation will therefore be better the greater the effort required to reject the pseudo-random sequence as being truly random.

For an excellent review of modern algorithmic probability theory, compare Schnorr (1971b).

Boolean probability theory

Because of its significance for almost all aspects of modern science, as well as its important historical position, we give a short account of classical mathematical probability theory. Probability theory is based on three fundamental notions: *event, probability of an event,* and *random variable*.

The outcome of a repeatable experiment is called an *event*. The *sure event* E is the event that occurs whatever the outcome of the experiment. The impossible event O is the event that never occurs. By definition, both E and O are observable events. The non-occurence of an event F is itself considered as an event, and called the *complementary event* $F^{\perp}$ of the event F. The event of realizing at least one of the events F,G is the event "F *or* G", written as the join $F \vee G$. The event of realizing both F and G is the event "F *and* G", written as the meet $F \wedge G$. Two events F,G exclude each other if $F \wedge G=0$, in this case F and G are called *disjoint events*. An event F *implies* the event G, written $F \leq G$, if $F=F \wedge G$. If F implies G, and G implies F, then F and G are called *equivalent events*, and we write F=G. The nature of two equivalent events may be different but they are identified within the abstractions of probability theory.

In the framework of classical probability theory it is assumed that for every F and every G, $F^{\perp}$, $G^{\perp}$, $F \wedge G$. $F \vee G$ are also observable events, and that these combinations of events are governed by the rules of classical logic. Accordingly, the family L of all events associated with an experiment is closed with respect to the operations of disjunction $\vee$, conjunction $\wedge$, and negation $^{\perp}$, so that L is a complemented distributive lattice, i.e. a *Boolean algebra*.

If the number of possible events is infinite, an additional characterization of L is necessary. A lattice L is called complete if every subset of L has a supremum and an infimum. If L has more than countably many disjoint elements, the requirement of completeness is too strong. In probability theory it is replaced by the weaker requirement of σ-completeness. A lattice L in which every *countable* subset has a supremum and infimum is called *σ-complete*. In classical probability theory, the family L of all events is assumed to be a *σ-complete Boolean algebra* (synonyma: σ-complete Boolean lattice, Boolean σ-lattice, or σ-algebra).

In Boolean probability theory, *probability* is defined as a countably additive normed measure on a Boolean lattice L.

Warning:
It is not claimed that a normed measure can be introduced into every Boolean lattice. There exist Boolean σ-algebras without any nontrivial probability measure.

The probability is therefore a real-valued function $p: L \to \mathbb{R}$ on the elements of L satisfying the following three conditions

(i) p is *strictly positive*, i.e. $p(F) \geq 0$ for every $F \in L$, and $p(F)=0$ implies $F=0$,

(ii) p is *normalized*, i.e. $p(E)=1$,

(iii) p is *σ-additive*, i.e. $p(\vee F_j)=\Sigma p(F_j)$ for every finite or countably infinite sequence $\{F_j\}$ of pairwise disjoint events in L.

These postulates imply $0 \le p(F) \le 1$ for every $F \in L$.

To summarize: Classical probability theory is based on classical Boolean logic, and from a mathematical point of view it is fully characterized by the pair (L,p), where L is a σ-complete Boolean lattice, and $p:L \to [0,1]$ is a σ-additive probability measure. This abstract lattice-theoretical approach to classical probability theory is conceptually transparent and makes it clear that classical probability is based on the propositional calculus of classical logic where every proposition is either true or false. For the further mathematical development and for applications it is nevertheless convenient to consider isomorphic embeddings of this basic structure into richer structures such as sets or algebras. Of course, such embeddings are not necessary. A detailed study of the abstract lattice-theoretical approach to classical probability can be found in the monograph by Kappos (1969).

Following Kolmogorov (1933), traditional Boolean probability theory represents events by sets. This representation is possible without restricting the generality since every Boolean algebra can be realized by an algebra of subsets of a certain set Ω. The main advantage of such a set-theoretical realization of Boolean probability theory is that the study of mathematical probability is linked to the highly developed framework of measure theory with its powerful analytical techniques. The fundamental representation theorem is due to Loomis (1947) and Sikorsky (compare Sikorsky, 1960, §29), and says that every Boolean σ-algebra is σ-isomorphic to a σ-algebra Σ of point sets of some set Ω modulo a σ-ideal. For every Boolean probability system (L,p) there exists a Kolmogorov probability space (Ω,Σ,μ) such that L is abstractly identical with a σ-algebra Σ of subsets of a set Ω (called the *sample space*), reduced by identification according to sets of measure zero, and such that $p(F)=\mu(F)$ for every event $F \in L$ and the corresponding subset $F \in \Sigma$. The probability measure $\mu:\Sigma \to [0,1]$ is defined for every subset of Σ, and normalized such that $\mu(\Omega)=1$. Note that the points $\omega \in \Omega$ need not have a probabilistic significance. In the basic Boolean probability system (L,p), the only event of zero probability is the impossible event. In the set-theoretical realization (Ω,Σ,μ) every element of Σ of μ-measure zero is an event of zero probability. Therefore, equality or convergence in the set-theoretical formulation is always understood up to a set of events of measure zero.

Intuitively, a *random variable* is a quantity whose numerical values are determined by chance. The principal concern of probability theory are quantities which take on values depending on the outcome of an experiment. Such quantities describing a stochastic experiment which measures certain numerical quantities are called real-valued (or complex-valued) random variables. In Kolmogorov's formalization of classical probability theory random variables are defined as *measurable functions* on some probability space. More precisely, a random variable on the probability space (Ω,Σ,μ) is an equivalence class of measurable functions on Ω, the equivalence relation being equality almost everywhere with respect to μ. For example, a complex-valued random variable can be represented by a function $x:\Omega\to\mathbb{C}$ on Ω which is Σ-measurable on (Ω,Σ). The mean value or mathematical expectation $\mathcal{E}\{\varphi(x)\}$ of a Borel function φ of a random variable x is defined by

$$\mathcal{E}\{\varphi(x)\} = \int_\Omega \varphi\{x(\omega)\}\mu(d\omega) \quad ,$$

insofar the integral exists. The expectation $\mathcal{E}\{x^p\}$ is called the p-th moment of the random variable x. The set of all equivalence classes of complex-valued μ-measurable functions with finite absolute p-th moment is called the Lebesgue space $L_p(\Omega,\Sigma,\mu)$,

$$L_p(\Omega,\Sigma,\mu) \overset{\text{def}}{=} \{x|x(\omega)\in\mathbb{C}, x \text{ is } \mu\text{-measurable}, \|x\|_p<\infty\}$$

where

$$\|x\|_p \overset{\text{def}}{=} \{\int_\Omega |x(\omega)|^p\mu(d\omega)\}^{1/p} \quad , \quad 1\le p<\infty \quad ,$$

and $\|\cdot\|_\infty$ is the supremum norm,

$$\|x\|_\infty \overset{\text{def}}{=} \inf\{c| \quad |x(\omega)|\le c \text{ for } \mu\text{-almost all } \omega\in\Omega\} \quad .$$

For $1\le p\le\infty$ the Lebesgue space $L_p(\Omega,\Sigma,\mu)$ is a Banach space with the norm $\|\cdot\|_p$. For a probability space (Ω,Σ,μ), the measure μ is finite, hence we have

$$L_\infty\subseteq L_p\subseteq L_q\subseteq L_1 \quad \text{for} \quad 1\le q\le p\le\infty \quad .$$

In many respects it is advantageous and also more natural to base classical probability theory on the family of all random variables together with an expectation functional $\mathcal{E}$ rather than on Kolmogorov's probability space (compare Segal, 1954). Taking random variables as the primary notion, it is convenient to assume that all moments exist. The restriction to bounded random variables $x \in L_\infty(\Omega,\Sigma,\mu)$ is not severe since unbounded random variables can be treated in terms of bounded ones. If we take (L_∞, E) as a starting point, the probability measure $\mu:\Sigma \to [0,1]$ can be recovered by $\mu(A)=E(\chi_A)$ where χ_A is the indicator function of the subset $A \in \Sigma$ (defined by $\chi_A(\omega)=1$ if $\omega \in A$, and $\chi_A(\omega)=0$ if $\omega \notin A$). Moreover, the set of all characteristic functions in $L_\infty(\Omega,\Sigma,\mu)$ is isomorphic to the Boolean lattice L of the Boolean probability system (L,p).

The Banach space $L_\infty(\Omega,\Sigma,\mu)$ of all complex-valued, essentially bounded measurable functions on Ω is a commutative algebra with respect to ordinary linear operations and multiplications of functions. Under the supremum norm $\|\cdot\|_\infty$, $L_\infty(\Omega,\Sigma,\mu)$ is a Banach algebra, and with the involution $x \to x^*$ defined by $x^*(\omega)=\{x(\omega)\}^*$, the algebra L_∞ becomes a Banach *-algebra. Since the supremum norm fulfills the relation $\|x^*x\|_\infty=\|x\|_\infty^2$, L_∞ is a C*-algebra. Moreover, the dual space of the Banach space $L_1(\Omega,\Sigma,\mu)$ of all integrable functions equals $L_\infty(\Omega,\Sigma,\mu)$ so that L_∞ is the dual of a Banach space, $L_\infty=L_1{}^*$. Recall that a C*-algebra which is the dual of a Banach space is called a W*-algebra. Accordingly, the space $L_\infty(\Omega,\Sigma,\mu)$ of all essentially bounded complex-valued random variables on the probability space (Ω,Σ,μ) is a *commutative W*-algebra*. On the other hand, every commutative W*-algebra A with a separable predual A_* is *-isomorphic to a W*-algebra $L_\infty(\Omega,\Sigma,\mu)$ on some probability space (Ω,Σ,μ). Due to this isomorphism, Kolmogorov's probability theory can be rephrased in an abstract algebraic way in terms of a probability algebra (A,E) where A is a commutative W*-algebra and E an expectation functional on A. The expectation E can be characterized without recourse to the Kolmogorov formalism as a normal state on A (i.e. a σ-additive, normalized positive linear functional).

Remark: *Separability and Σ^*-algebras*

If the sample space Ω is not separable, the W*-algebraic formulation does not give a perfect correspondence with Kolmogorov's approach. The reason is that the lattice of projectors of every W*-algebra is complete. In general completeness is too strong a requirement and has to be weakened to σ-completeness. In this case, the appropriate algebraic tool is a Σ^*-algebra, introduced by Davies (1968, 1969a). Every W*-algebra is a Σ^*-algebra, but not the other way round. The projection lattice of a Σ^*-algebra is only σ-complete. Since a W*-algebra with separable predual contains at most countably many mutually orthogonal projectors, the notions of W*-algebras coincide with that of Σ^*-algebras in this important special case.

In all applications of physical interest, the measure space (Ω,Σ,μ) is separable (for more details and references, see Primas, 1980), so that the Boolean lattice L of classical physics contains at most countably many mutually orthogonal elements. This implies that both the Hilbert space $L_2(\Omega,\Sigma,\mu)$ and the predual $L_1(\Omega,\Sigma,\mu)$ of the W*-algebra $L_\infty(\Omega,\Sigma,\mu)$ are separable. Accordingly, every classical system of physical interest can be represented either by a separable Kolmogorov space (Ω,Σ,μ) or, equivalently, by a commutative W*-algebra A having a separable predual A_* .

Quantum mechanics is not a classical stochastic theory

The scope of classical probability theory is restricted to cases where the usual Boolean logical combinations of statements have a meaning. In quantum theories, the domain of validity of Kolmogorov's probability theory has to be restricted to sets of *compatible* events which may always be assumed to form a Boolean σ-algebra. As stressed by Koopman (1957) "*the necessity of such a restriction was implicit in classical science, well before quantum physics*". Since quantum mechanics deals also with incompatible events, classical probability theory is not general enough to encompass all the situations which arise in quantum mechanics. Nevertheless, *Kolmogorov's mathematical probability theory is an inbuilt feature of pioneer quantum mechanics* (compare sect.3.3). Let A_A be the commutative W*-algebra generated by some observable $A=A^*\in A$, where A is the noncommutative algebra of observables of pioneer quantum mechanics, and let ρ_A be the restriction of any normal state $\rho\in A_*$ to the algebra A_A, then (A_A,ρ_A) is a W*-algebraic realization of classical probability theory. Therefore, the probability distribution of every single observable with respect to any normal state is well-defined in pioneer quantum mechanics. The same is true for W*-algebraic generalized quantum mechanics.

Remark: *What is the meaning of expectation values in quantum mechanics?*
Clearly, the probabilities of quantum mechanics are conditioned by measurements of the first kind, and it is clear that these conditional probabilities fulfill Kolmogorov's axioms of classical probability theory. Unfortunately, many philosophers have difficulties understanding physics textbooks, and write hundreds of pages about the most elementary concepts. For example, Stegmüller (1974, p.462) claims that viewing quantum probabilities as conditional probabilities is a new proposal due to Sneed (1970)! It is strange how often even eminent philosophers are unable to understand the language of scientists. In his standard text, Dirac (1930, in the fourth edition on p.47) states: "*We ... speak of the probability of [an observable] having any specified value for the state, meaning the probability of this specified value being obtained when one makes a measurement of the observable*". I think this is a very clear characterization of a *conditional* probability.

Quantum mechanics has a new feature that distinguishes it from classical probability theory. While in Kolmogorov's probability theory any two random variables have always a unique *joint probability distribution*, joint probability distributions (with the usual properties) for two quantum mechanical observables exist if and only if the observables are compatible. Since incompatible observables depend on mutually exclusive measurements of the first kind, we have no reason to expect that probabilities associated with incompatible observables should be the marginal probabilities of a common joint probability distribution.

Remark: A theorem by Nelson

There exists a vast literature (not worth reviewing) on the alleged problem of the simultaneous measurement of incompatible observables, and on the supposed existence of joint probability distributions of noncommuting observables. Certainly, one cannot base such a discussion on the so-called Heisenberg uncertainty relations. Using neither concepts from measurement theory nor a particular interpretation, the following theorem by Nelson (1967) shows that in general classical joint probability distributions do not exist in pioneer quantum mechanics.

Theorem: Let $(A_1,..,A_n)$ be an n-tuple of operators on a Hilbert space H such that for all $x\in\mathbb{R}^n$ the operator $x\cdot A=x_1A_1+\cdots+x_nA_n$ is essentially self-adjoint. Then either $A_1,..,A_n$ commute pairwise, or there exists a state vector $\Psi\in H$ with $\|\Psi\|=1$ such that there do not exist random variables $a_1,..,a_n$ on some Kolmogorov probability space with the property that for every $x\in\mathbb{R}^n$ and $\lambda\in\mathbb{R}$

$$\text{probability } \{x\cdot a\geq\lambda\} = \langle\Psi|E_\lambda\Psi\rangle \quad ,$$

where E_λ are the spectral projectors of the closure of $x\cdot A$ (Nelson, 1967, p.117).

Nelson's theorem says that n observables of pioneer quantum mechanics may be regarded as classical random variables, in all states, if and only if they commute. Note that Nelson's theorem does not claim that there are no states having a joint probability for incompatible observables. Such states exist and are of special importance in quantum communication theory. Nelson's theorem asserts that such states are exceptional and not the rule.

Non-Boolean probability structure

Since the probability structure of quantum mechanics is not Boolean, Kolmogorov's probability theory is not sufficiently general to encompass the non-Boolean situations that are possible in quantum theory. It has long been recognized that Kolmogorov's formulation of classical probability can be expressed in a way which emphasizes the mathematical structures used in pioneer quantum mechanics (Segal, 1953, 1954; Mackey 1957, 1960, 1963a).

A mathematically straightforward generalization is a *probability theory on orthomodular lattices*. Here, the classical Boolean lattice of events is replaced by an appropriate nondistributive orthomodular lattice L. Since the definition of a probability measure ρ on a Boolean lattice makes no reference to the distributivity of the underlying lattice, the generalization for a general orthomodular lattice is trivial. A probability measure $p:L\to\mathbb{R}$ on an orthomodular lattice L is characterized by the following three postulates:

(i) p is *strictly positive*, i.e. $p(F)\geq 0$ for every $F\in L$ and $p(F)=0$ implies $F=0$,

(ii) p is *normalized*, i.e. $p(E)=1$,

(iii) p is *σ-additive* for compatible and disjoint events, i.e. $p(\vee F_j)=\Sigma p(F_j)$ for every countable sequence $\{F_j\}$ of pairwise orthogonal elements $F_j \in L$, $F_j \perp F_k = 0$ for $j \neq k$.

Remark:

In an orthomodular lattice, the relations $F \leftrightarrow G$ and $F \wedge G=0$ imply orthogonality of F and G, defined by $F \perp G$ iff $F \leq G^{\perp}$.

From these postulates it follows that $0 \leq p(F) \leq 1$ for all $F \in L$, and that $F \leq G$ implies $p(F) \leq p(G)$. For the further development of generalized probability theory on abstract orthomodular lattices, compare Kappos (1974, 1976), Hackenbroch (1976), and Kalmár (1978).

Another generalization of Kolmogorov's probability is the *noncommutative probability*, where the commutative W*-algebra $L_\infty(\Omega,\Sigma,\mu)$ is replaced by a noncommutative W*-algebra, and the expectation with respect to μ is replaced by a normal state on A. The self-adjoint elements of A correspond to the bounded random variables of Kolmogorov's theory, and are called *observables*. The expectation value of an observable $A \in A$ in a state $\rho \in A_*$ is given by $\rho(A)$. The set of all projectors in A forms an orthomodular lattice L,

$$L = \{F \mid F \in A \ , \ F=F^2=F^*\} \quad .$$

Every normal state ρ on A defines a semidefinite probability measure p on L by

$$p(F) = \rho(F) \quad \text{for every} \quad F \in L \subset A \quad .$$

Does every probability measure on L arise in this way? Using the spectral theorem, it is a matter of routine to extend a probability measure on a projection lattice L to the selfadjoint operators of the algebra A. However, if A is not commutative, it is highly nontrivial to show that this extension is additive. Hence the crucial question is whether all probability measures on a projection lattice of a W*-algebra A are genarated by *linear* functionals on A.

A classical result by Gleason (1957) says that every σ-additive measure p on the lattice of all projectors acting on a separable Hilbert space H of dimension >2 has the representation $p(F)=\mathrm{tr}(DF)$ for every projector $F=F^2=F^* \in B(H)$, where $D \in B(H)$ is a density operator (i.e. a linear, positive, self-adjoint, nuclear operator of trace 1). Gleason's theorem implies that if A is an atomic W*-algebra of type I having a separable predual A_*, and if A has no type I_2 factor as a direct summand, then *every* probability measure $p: L \to [0,1]$ on the projection lattice L of A has a *unique* extension to a normal state ρ on A, that is p has the form $p(F)=\rho(F)$, $F \in L \subset A$, with $\rho \in (A_*)^1_+$. It is remarkable that the assumptions of this theorem refer to *compatible* propositions only, a very desirable situation from an intuitive viewpoint. Gleason's theorem settles the question of the *additivity* of the expectation functional in pioneer quantum mechanics; its generalization to arbitrary W*-algebras is a cornerstone for the conceptual foundation of noncommutative probability theory.

Remark: Generalizations of Gleason's theorem

Gleason's theorem has been extended to type I W*-factors with nonseparable preduals by Eilers and Horst (1975); compare also Drisch (1979). For a hyperfinite factor A of arbitrary type, Gunson (1972) has shown that every probability measure on the projection lattice L of A is the restriction of a normal state on A to L. A generalization of Gleason's theorem for type I W*-algebras is due to Tischer (1980).

Lodkin (1974) claimed that Gleason's theorem is true for every W*-algebra having a separable predual (and no direct summand of type I_2) but there is a gap in his proof so that this generalization is not yet settled.

From a slightly different point of view, the linearity problem of the expectation functional on a noncommutative algebra of observables has been discussed by Aarnes (1969, 1970), and Akemann and Newberger (1973).

The development of a viable noncommutative probability depends crucially on the possibility of introducing concepts of conditional expectation entropy and information. The investigations concerning the possibility of a noncommutative probability theory on W*-algebras are closely connected with the study of general physical theories which include pioneer quantum mechanics (in its statistical ensemble interpretation) and classical statistical mechanics as special cases.

References on noncommutative information and probability theory

In the development of a noncommutative probability theory on W*-algebras the generalized Radon-Nikodym theorem for finite W*-algebras by Dye (1952) and the noncommutative integration theory (in which the measure is required to be unitarily invariant) by Segal (1953) played a decisive role. Extensions and simplifications of Segal's integration theory are due to Nelson (1974). Using Gleason's theorem, Gudder and Marchand (1972) extended Dye's Radon-Nikodym theorem to W*-algebras of type I_∞, and discussed sufficient statistics and coarse-graining for semifinite W*-algebras.

W*-algebras of *finite* (and sometimes even of *semifinite*) type provide a rather straightforward extension of classical probability and information theory. In this framework, the important concept of conditional expectation has been discussed by Umegaki (1954, 1956, 1959, 1962), Nakamura and Turumaru (1954), Nakamura et al. (1960), and Nakamura and Umegaki (1961b). A general entropy concept for normal states on semifinite W*-algebras has been introduced by Segal (1960), and has been further investigated by Nakamura and Umegaki (1961a), Umegaki (1961, 1962), Davis

(1961), Ruskai (1973), Chen (1977), Lieberman (1978), Ochs and Spohn (1978). Many basic theorems of measure theory and the laws of large numbers have generalizations in noncommutative W*-algebraic probability theory; compare for example Gootman and Kannan (1976), Batty (1979), Driessler and Wilde (1979), Padmanabhan (1979).

The generalization of probability theory to *arbitrary* W*-algebras (including type III algebras) is closely related to the famous Tomita-Takesaki theory of modular automorphisms (compare: Takesaki, 1970a). The Radon-Nikodym theorem - a fundamental building block of any probability theory - has various generalizations to arbitrary W*-algebras. A positive linear functional ρ on a W*-algebra $\mathcal{A}$ is said to be absolutely continuous with respect to φ, written $\rho \leq \varphi$, if for every $A \in \mathcal{A}$ the relation $\varphi(A^*A)=0$ implies $\rho(A^*A)=0$. The first general result is due to Sakai (1965) and says that for every positive normal linear functional ρ on a W*-algebra $\mathcal{A}$ such that $\rho \leq \varphi$, $\varphi \in \mathcal{A}_*$, there exists a positive element $T \in \mathcal{A}$, with $0 \leq T \leq 1$, such that $\rho(A) = \varphi(TAT)$ for every $A \in \mathcal{A}$. If ρ is faithful, then T is unique. If $\mathcal{A}$ is commutative, then the Sakai operator T coincides with the positive square root of the measure-theoretical Radon-Nikodym derivative $d\rho/d\varphi$. Another formulation says that there exists a positive element $D \in \mathcal{A}$, with $0 \leq D \leq 1$, such that $\rho(A) = \frac{1}{2}\varphi(DA+AD)$ (Sakai, 1971, p.77). This theorem has been generalized to weights by van Daele (1975). The operator D is called a Radon-Nikodym derivative, it is unique if ρ is faithful. If ρ is a faithful normal state and φ is a faithful semifinite and normal trace, then the Radon-Nikodym derivative D equals the modular operator so that the modular automorphism group $\{\sigma_t{}^\rho : t \in \mathbb{R}\}$ associated with ρ is given by $\sigma_t{}^\rho(A) = D^{it}AD^{-it}$, $A \in \mathcal{A}$, $t \in \mathbb{R}$, (Takesaki, 1973, theorem 8.1). Accordingly, the logarithm of the Radon-Nikodym derivative D is related to the canonically associated Hamiltonian of the state ρ (relative to the trace φ). For the special case where ρ is invariant under the modular automorphism group associated with φ, the Radon-Nikodym derivative D also is invariant, so that $\rho(A) = \rho(DA) = \rho(AD)$ for every $A \in \mathcal{A}$, a special case of a different Radon-Nikodym theorem by Pedersen and Takesaki (1973) (compare also Tischer, 1979). The Pedersen-Takesaki theorem does not require that ρ is majorized by φ but assumes that ρ is invariant under the modular automorphism of φ. For related work on Radon-Nikodym theorems compare also Takesaki (1972), Araki (1974), Gudder (1979a).

The concept of sufficient statistics has been generalized to W*-algebras by Hiai et al. (1979); they have shown that the concept of sufficiency characterizes the KMS-states with respect to a modular automorphism group.

A quite general noncommutative probability theory in which no boundedness, finiteness or tracial conditions are imposed, has been developed by Gudder and Hudson (1978).

Quantum measurements as statistical inference

The measurement process of pioneer quantum mechanics can be considered from the viewpoint of information theory. Nakamura and Umegaki (1962) proposed to consider a measurement of the first kind as a *conditional expectation*, conditioned by the commutative W*-algebra generated by the observables measured. According to this viewpoint, the density operator is nothing but the Radon-Nikodym derivative of the expectation functional with respect to the trace, where the trace is interpreted as an a priori state reflecting the information about the system prior to a measurement of the first kind. With this, the measurement process can be related to statistical inference.

Statistical inference is a mathematical technique which allows to estimate the most probable statistical state of a stochastic system about which some partial information is available. In generalized quantum theories this situation can be discussed as follows. The observables of the system are given by the self-adjoint elements of some W*-algebra A. Assume that we perform a partial measurement (for example, a measurement of a set of pairwise compatible observables, or a measurement of macroscopic observables in statistical mechanics), and let B be the smallest W*-subalgebra of A containing these selected observables. Let $\rho\in A_*$ be a normal state on A reflecting our a priori information about the system *prior* to the measurement of the observables in B, and let $\nu_B\in B_*$ be a normal state on B representing the information obtained from a measurement of the observables in B. What is the most likely state of the system which is compatible with new partial information, and does not contain more information about the system than the a priori knowledge and the knowledge of the outcome of the partial measurement? In the classical case (i.e. if A is commutative) the answer can be expressed in terms of a Radon-Nikodym derivative. In the general case, we have to assume that the state ν_B is dominated by the restriction ρ_B of ρ to B, $\nu\le\lambda\rho\,(\lambda\ge 0)$ (that is, $\nu_B(B^*B)\le\lambda\rho_B(B^*B)$ for every $B\in B$). Then a generalized Radon-Nikodym theorem for W*-algebras (compare Sakai, 1965) says that there is a unique positive operator $T\in B$, with $0\le T\le 1$, such that

$$\nu_B(B) = \rho_B(TBT) \quad \text{for every } B\in B \quad .$$

With this, we can define a normal state $\nu\in A_*$ by

$$\nu(A) \overset{\text{def}}{=} \rho(TAT) \quad \text{for every } A\in A \quad .$$

This state ν *on the original algebra* A *is called the statistical inference from the partial measurement* ν_B *with respect to the a priori state* ρ (Marchand, 1977; compare also Benoist et al., 1977; Marchand and Wyss, 1977; Benoist and Marchand, 1979; Gudder et al., 1979). Since statistical inference is an information-theoretical concept, the state ν is an *epistemic state* (i.e. it describes the state of our knowledge of the system rather than its real state).

If A is a commutative W*-algebra, we can write $\nu(A)=\rho(T^2A)$ where $T^2=d\nu/d\rho$ is the classical Radon-Nikodym derivative of ν with respect to the a priori state ρ.

If A equals the algebra of all bounded operators on a separable

Hilbert space H, then the normal states ρ and ν admit density operators D_ρ, D_ν such that $\rho(A)=\mathrm{tr}(D_\rho A)$ and $\nu(A)=\mathrm{tr}(D_\nu A)$ for all $A\in A$. In this case D_ν is given by $D_\nu = TD_\rho T$. Conversely, if D_ρ^{-1} exists (perhaps as an unbounded operator), then the Sakai operator T of ν relative to ρ is given by the unique continuous extension of $D_\rho^{-1/2}(D_\rho^{1/2}D_\nu D_\rho^{1/2})^{1/2}D_\rho^{-1/2}$ from the domain of $D_\rho^{-1/2}$ to H (Pedersen and Takesaki, 1972).

Quantum statistical communication theory

With the advent of lasers and optical communication systems it has become important to investigate quantum effects in communication systems. While thermal noise as a rule dominates other noise sources in the microwave range, the maximum amount of information transmissible by an electromagnetic field in the optical range is given by quantum noise (since the quantum energy $\hbar\omega$ is larger than the thermal energy kT). In this case classical estimation and detection theory cannot be applied to *design* an *optimal* communication system. "*Of course, if we are concerned with the data already obtained by a given measurement, then the classical theory of statistics is fully applicable; but the question of an optimal choice of such a measurement is out of the scope of classical statistics and needs special quantum-mechanical consideration*" (Holevo, 1973a).

References on quantum statistical signal detection theory

The development of a quantum mechanical detection and estimation theory was initiated by Helstrom (1967, 1968, 1969). Compare also the review by Helstrom et al. (1970) and the textbook by Helstrom (1976). The theory has been extended on a sound mathematical basis by Holevo (1972a-c, 1973a-c, 1974, 1975a-b, 1978). This treatment necessitates the use of generalized quantum measurements where the orthogonal projectors of orthodox measurement theory are replaced by positive operator-valued measures. This generalization of the concept of an observable was introduced by Davies and Lewis (1970).

There exists a voluminous literature on the problem of the simultaneous measurability of incompatible observables; since it is mostly neither really correct nor plainly wrong but just dull, we will not review it. Quantum statistical estimation theory solved this alleged philosophical problem in an imaginative and constructive manner. Neither Heisenberg's inequality nor any first principle of quantum mechanics excludes the possibility of simultaneous measurement operations of incompatible observables (warning: do not confuse so-called "measurements of the first kind" with laboratory measurements!). What is interesting in engineering are never measurements of the first kind but statistical estimations of the state of the system. Of particular importance is the estimation of a state through the optimal estimation of parameters within a class of parametrized states. The question of a simultaneous measurement of the coordinate Q and its canonically conjugate momentum P, fulfilling $QP-PQ=i\hbar$, was already discussed qualitatively by von Neumann (1932, sect. V.4), where he also noted the role the coherent states play in simultaneous measurements of P and Q. The simultaneous measurement of noncommuting observables has become an engineering problem of great practical importance in quantum electronics. It is known that the technique of heterodyning is equivalent to an optimal statistical measurement of the Schrödinger pair (P,Q) or, equivalently, to an optimal measurement of the photon annihilation operator Q+iP (compare Arthurs and Kelley, 1965; Gordon and Louisell, 1966; She and Heffner, 1966; Personick, 1971). Holevo (1973a-c) has shown that the optimal joint estimation of the canonical observables (P,Q) can be effected by a measurement of the first kind of the commuting observables $P\otimes 1+1\otimes P_o$ and $Q\otimes 1-1\otimes Q_o$, where (P_o,Q_o) is a canonical pair of an auxiliary system related to the measuring apparatus performing the heterodyning (compare also the related work by Helstrom 1973, 1974a-c, and by Helstrom and Kennedy, 1974). More generally, Holevo (1975, 1977, 1978) has shown that every optimal joint estimation of incompatible observables can be described by a positive operator-valued measure in the sense of the operational approach of Davies and Lewis (1970).

Quantum statistical system theory

Interest in communication and information theory has been stimulated by radio engineers. A system-theoretic description of a general communication system distinguishes an information source together with an encoder (called *input*), a system through which information is conveyed (called *channel*), and a receiver together with a decoder (called *output*). If the transmitting channel alters the signal by noise, the channel is called a *noisy channel*, otherwise it is called a *noiseless channel*. In a noisy channel it is possible that two different input states are mapped to the same output state.

In the *classical case*, a statistical communication system is described by a measurable space (Ω_1,Σ_1) of input messages, a measurable space (Ω_2,Σ_2) of output messages, and a transition probability $T:(\Omega_1,\Sigma_2)\to\mathbb{R}^+$. For every fixed $\omega\in\Omega_1$, $B\to T(\omega,B)$ is a probability measure on the output space (Ω_2,Σ_2), while for every fixed $B\in\Sigma_2$, $\omega\to T(\omega,B)$ is a measurable function of $\omega\in\Omega_1$. If $\mu_1:\Sigma_1\to\mathbb{R}^+$ is a probabiliy measure on the input space, then the transition probability T defines by

$$\mu_2(B) = \int_{\Omega_1} T(\omega,B)\,\mu_1(d\omega) \quad , \quad B\in\Sigma_2 \quad ,$$

a probability measure $\mu_2:\Sigma_2\to\mathbb{R}^+$ on the output space. Accordingly, the transition probability uniquely defines an operator τ from the convex set of all probability measures on (Ω_1,Σ_1) into the convex set of all probability measures on (Ω_2,Σ_2),

$$\mu_2 = \tau(\mu_1) \quad .$$

This affine mapping τ characterizes the channel of the communication system.

These concepts can be generalized to noncommutative probability theory. Since probability measures can be considered as the normal states on the commutative W*-algebra of all bounded measurable functions on the output and input spaces, a *quantum channel* can be defined as a linear transformation which maps every normal state on an input W*-algebra into a normal state on an output W*-algebra. Echigo and Nakamura (1962) have shown that a positive linear mapping $\tau:(A_1)_*\to(A_2)_*$ maps the normal states of a W*-algebra A_1 into the normal states of a W*-algebra A_2 if and only if the dual mapping $\tau^*:A_2\to A_1$ is a positive normal linear mapping of A_2 into A_1 preserving the identity. Recent work has shown that for general W*-systems one should require a stronger positivity condition, namely that a dual channel $\tau^*:A_2\to A_1$ is a *completely positive* normal linear mapping (Holevo, 1972c, 1975b, 1977b).

Remark: Completely positive mappings

Recall that a linear mapping φ of one W*-algebra A_2 into another W*-algebra A_1 is called *positive* if $\varphi(A)$ is positive for all positive $A\in A$. Let M_n denote the W*-algebra of all complex n×n-matrices. A linear mapping $\varphi:A_2\to A_1$ of one W*-algebra A_2 into another W*-algebra A_1 is said to be *n-positive* if the linear mapping $\varphi\otimes id:A_2\otimes M_n\to A_1\otimes M_n$ is positive, where id denotes the identity mapping of M_n. The mapping α is called *completely positive* if it is n-positive for every n=1,2, ... Note that there exist n-positive mappings that fail to be (n+1)-positive (Choi, 1972). Two important classes of completely positive mappings are the homomorphisms and the projections of norm one (for a review, compare Størmer, 1974).

Completely positive linear mappings are the natural generalization of positive mappings of commutative algebras. Stinespring (1955) has shown that on a commutative C*-algebra every positive linear mapping is completely positive. In the noncommutative case, complete positivity is an essential restriction. For example, every *-antiautomorphism of a noncommutative W*-algebra is positive, but not 2-positive. (Gorini et al., 1976). Another example of a positive mapping which is not completely positive in the transposition of a complex matrix.

Complete positivity of operations is a necessary consistency requirement for composite systems (Kraus, 1971). Consider two noninteracting systems whose algebra of observables is given by the W*-tensor product $A_1 \bar{\otimes} A_2$. Since the tensor product of two positive mappings is not necessarily positive, positivity is not a sufficient requirement. On the other hand, Stinespring (1955) has shown that the tensor product of completely positive mappings is again completely positive.

A mapping $\tau:(A_1)_* \to (A_2)_*$ is called completely positive if the dual mapping $\tau^*: A_2 \to A_1$ is completely positive.

A *W*-communication system* can be characterized by three W*-algebras A_1, A_2, A_3 and three channels τ_{13}, τ_{33}, τ_{32}, where $\tau_{jk}:(A_j)_* \to (A_k)_*$ is a completely positive normal linear mapping which preserves the norm of positive elements. The algebra A_1 refers to the observable of the input system, A_2 refers to the output system, and A_3 refers to the transmitting system. The input-output mapping $\tau:(A_1)_* \to (A_2)_*$ can be represented as $\tau = \tau_{13} \circ \tau_{33} \circ \tau_{32}$ where τ_{13} is the encoding mapping, τ_{31} is the decoding mapping while τ_{33} describes the dynamics of the transmitting system.

Inputs and outputs must have some characteristics demanded by human nature, cultural tradition and engineering common sense. Since experimental data are supposed to be unambiguous (at least in principle), they have to be expressible in terms of a two-valued Boolean logic. This requirement leads to the choice of a *commutative* W*-algebra A_2 for the output system. In order to investigate a system by external stimuli we must be able to create well-defined input signals. Experimentally useful stimuli must again fulfill the requirements of two-valued Boolean logic, so that the input algebra A_1 has also to be a *commutative* W*-algebra. All scientific experiments can be subsumed under this classical input-output scheme which is typical for contemporary mathematical system theory (compare Kalman et al., 1969). As a paradigmatic example for a modern scientific experiment we can take a nuclear magnetic resonance experiment where the input is given by a classical time-varying magnetic field while the output is given by the voltage induced in the receiver coil by the spin magnetization. If we observe a system without stimuli due to the experimentalist (as with telescopic observations in astronomy), we speak of a *free system*.

While the input algebra A_1 and the output algebra A_2 of every W*-communication system are commutative, the transfer algebra A_3 is noncommutative in a quantum communication system. In a nuclear magnetic resonance experiment, the algebra A_3 is generated by the spin operators (i.e. A_3 is a factor of type I_n with $n<\infty$), it describes the system under investigation. In a laser communication system A_3 is related to the algebra of quasi-local observables of the electromagentic field (a factor of type I_∞ or of type III_1), in this case A_3 describes the inevitable noise of the transmission channel. As Davies (1977) remarked, "*it is rather amusing to note that, in an interplanetary laser communication system, the classically described input and output devices have dimensions of a few meters, while the channel, which is the relevant part of the quantised electromagnetic field, has dimensions of order 10^{11} m*".

The *encoding mapping* $\tau_{13}:(A_1)_*\to(A_3)_*$ transforms the classical input signal into a quantum signal. The commutative input algebra A_1 can be taken as Lebesgue space $L_\infty(\Omega_1,\Sigma_1,\mu_1)$ where $(\Omega_1,\Sigma_1,\mu_1)$ is some Kolmogorov space, so that $(A_1)_*=L_1(\Omega_1,\Sigma_1,\mu_1)$. Let $\chi_B\in L_1(\Omega_1,\Sigma_1,\mu_1)$ be the indicator function of the set $B\in\Sigma_1$, and define a Σ_1-measurable family $\{\rho_\omega|\omega\in\Omega_1\}$ of normal states $\rho_\omega\in(A_1)_*$ by

$$\int_B \rho_\omega \mu_1(d\omega) = \tau_{13}(\chi_B) \quad \text{for every} \quad B\in\Sigma_1 \quad .$$

With this, the encoding mapping can be represented as

$$\tau_{13}(f) = \int_{\Omega_1} \rho_\omega f(\omega)\mu_1(d\omega) \quad , \quad f\in L_1(\Omega_1,\Sigma_1,\mu_1) \quad .$$

The *decoding mapping* $\tau_{32}:(A_3)_*\to(A_2)_*$ converts the transmitted quantum signal into a classical signal. If we take the classical output algebra A_2 as $L_\infty(\Omega_2,\Sigma_2,\mu_2)$, where $(\Omega_2,\Sigma_2,\mu_2)$ some Kolmogorov space, we have $(A_2)_*=L_1(\Omega_2,\Sigma_2,\mu_2)$. Consider the dual channel $\tau_{32}^*:A_2\to A_3$, and define a positive operator-valued measure $F:\Sigma_2\to A_3$ by $F(B)=\tau_{32}^*(\chi_B)$, where $\chi_B\in L_\infty(\Omega_2,\Sigma_2,\mu_2)$ is the indicator function of the set $B\in\Sigma_2$. We have $F(\Omega_2)=1$, and the dual decoding mapping can be represented as

$$\tau_{32}^*(x) = \int_{\Omega_2} x(\omega)F(d\omega) \quad , \quad x\in L_\infty(\Omega_2,\Sigma_2,\mu_2) \quad .$$

Hence, the decoding mapping τ_{32} is uniquely determined by a positive operator-valued measure on the spectrum Ω_2 of the commutative output algebra A_2. That is, the dual channel τ_{32}^* is a generalized quantum measurement in the sense of Davies and Lewis (1970) and Holevo (1973b). The class of decoding mappings is considerably larger than the class of "measurements of the first kind" (considered in pioneer quantum mechanics), but in the special case when $F:\Sigma_2\to A_3$ is a projection-valued measure, τ_{32} represents an ideal measurement in the sense of von Neumann (1932).

The input-output mapping $\tau:(A_1)_* \to (A_2)_*$ can be described by *classical* mathematical system theory, even if the transmission algebra A_3 sandwiched between the input algebra A_1 and the output algebra A_2 is nonclassical. An example for such a classical system-theoretical description of a quantum system is Bloch's equation for nuclear magnetic resonance experiments on single-spin systems (Bloch, 1946; Bloch et al. 1946). Since such classical system-theoretical descriptions are highly successful, one may ask why we do not use always the classical input-output description $(A_1)_* \to (A_2)_*$ instead of the apparently more complicated quantum-theoretical description $(A_1)_* \to (A_3)_* \to (A_2)_*$? There are many reasons, probably the most cogent one being that sometimes we would like to understand a system and not only to describe it. Classical mathematical system theory is a general theory of model building (compare Kalman, 1968a-b), it is a powerful tool for *homo faber*, the tool making animal. Since it does not use any specific physical laws it cannot give the more detailed insight *homo sapiens* (man the thinker) likes to have. Important problems of natural science cannot even be stated within the purely classical system-theoretical description $(A_1)_* \to (A_2)_*$ since they refer to the quantum observables (i.e. the elements of A_3) which are not amenable to a classical system-theoretical representation.

A *full description* of the transmitting system assumes that the transmission algebra A_3 is a W*-subalgebra of some W*-algebra A of a large *closed* W*-system $(A, t \to \alpha_t)$ whose dynamics is given by a σ-weakly continuous *group* $\{\alpha_t | t \in \mathbb{R}\}$ of automorphisms $\alpha_t \in \mathrm{Aut}(A)$. Such a full description includes possibly unobserved quantities (like thermal reservoirs, noises etc.), and may be unreasonably involved. If one prefers to focus on the evolution of some observables of special interest (by definition: the elements of A_3), the transmitting system has to be considered as an *open system* whose dynamics can be obtained by reducing the assumed dynamical group of the transmitting system together with its relevant environment to the transmitting channel proper. Such a *reduced description* can be obtained via a conditional expectation $\varepsilon: A \to A_3$ from A onto A_3.

Remark: Conditional expectations in operator algebras

By a *conditional expectation* in an operator algebra A, conditioned by a certain fixed subalgebra $B \subseteq A$ we mean a projection of norm one from A onto B. A *projection of norm one* from a W*-algebra A to a W*-subalgebra B is a mapping $\pi: A \to B$ fulfilling the conditions

(i) $\pi(A) = B$,

(ii) $\pi^2 = \pi$,

(iii) $\|\pi\| = 1$,

where $\|\pi\| \overset{\text{def}}{=} \sup\{\|\pi(A)\| \,|\, A \in A ,\ \|A\|=1\}$.

A conditional expectation $\pi: A \to B$ is called *normal* (synonym: σ-weakly continuous) if $\sup_\alpha\{\pi(A_\alpha)\} = \pi(\sup_\alpha\{A_\alpha\})$ for each uniformly bounded directed set $\{A_\alpha\}$ of positive elements A_α of A. A conditional expectation $\pi: A \to B$ is called *faithful* if $\pi(A^*A)=0$ implies $A=0$, $A \in A$.

A classical result by Tomiyama (1957) says that with exception of normality a projection of norm one enjoys all properties which characterize the classical probabilistic concept of a conditional expectation (as characterized by Moy, 1954), namely:

(i) $\pi(B_1 A B_2) = B_1 \pi(A) B_2$ for every $B_1, B_2 \in B$, $A \in A$,

(ii) $\pi(A^*) = \pi(A)^*$ for every $A \in A$

(iii) $\|\pi(A)\| \le \|A\|$,

(iv) $\pi(A^*A) \ge \pi(A)^*\pi(A) \ge 0$,

(v) π is order preserving .

Moreover, every conditional expectation is completely positive.

Let $(A, t \to \alpha_t)$, $t \in \mathbb{R}$, be a reversible dynamical W*-system and $\varepsilon: A \to A_3$ a normal conditional expectation such that $\varepsilon\{\alpha_t(A)\} \in A_3$ for every $t \ge 0$. Then the W*-system $(A_3, t \to \beta_t)$ with $\beta_t = \varepsilon \circ \alpha_t$, $t \ge 0$, describes an *open system*, or what is the same, a *quantum stochastic process*. A system-theoretical description requires that the dynamics $t \to \beta_t$ of the transmission system is given by a *semigroup* $\{\beta_t | t \ge 0\}$ with $\beta_s \beta_t = \beta_{s+t}$ for all $s,t \ge 0$. Since automorphisms and conditional expectations are completely positive, the mapping $\beta_t: A_3 \to A_3$ is completely positive. In the theory of quantum communication channels, the dual channel τ_{33}^* equals the dynamical mapping β_t.

Remark: Quantum stochastic processes

There is a large literature concerned with the extension of the theory of classical stochastic processes to a theory of stochastic processes over noncommutative C*- or W*-algebras. Pioneering investigations are due to Davies (1969b, 1970a-b, 1971, 1972a); for an introductory account compare Davies (1976a). Important further work on quantum stochastic processes are due to Accardi (1975, 1976, 1978, 1979), Lewis and Thomas (1975), Chen (1976), Emch (1975, 1976), Evans and Lewis (1977a), Cycon and Hellwig (1977), Cockroft and Hudson (1977), Emch et al. (1978), Hudson (1979), Ekhaguere (1979), Lindblad (1979a-b).

In the further development of the theory of quantum stochastic processes the concept of a dynamical semigroup has emerged as the central mathematical object. A dynamical semigroup on a W*-algebra A is a σ-weakly continuous one-parameter semigroup $\{\varphi_t | t \ge 0\}$ of completely positive, normal identity preserving mappings of A into itself. The infinitesimal generator δ of a dynamical semigroup $\{\varphi_t\}$ is given by $\varphi_t = \exp(t\delta)$; an important problem is to find some canonical representation for δ. The first result is due to Gorini et al. (1976) who proved that for a factor A of type I_n, $n < \infty$, the derivation δ of any dynamical semigroup admits the representation $\delta(A) = KA + AK^* + \Sigma_n K_n A K_n^*$, $A \in A$, where $K, K_n \in A$. This result has been generalized to norm-continuous semigroups on factors of type I_∞ by Lindblad (1976a). The general case is not yet settled but important progress has been made by Lindblad (1976b), Evans and Lewis (1976, 1977a-b) and Davies (1979a).

As a paradigmatic example for an open system we can take the Markovian relaxation of a system consisting of a single spin one-half.

The W*-algebra A_3 of this system is generated by the three Pauli matrices τ_1, τ_2, τ_3, so that A_3 equals a factor of type I_2. The semigroup $t \to \beta_t$ can be derived from a Hamiltonian dynamics of the spin and its environment by a normal conditional expectation onto A_3. It is given uniquely by Bloch's equations of motion which can be taken in their standard form

$$\{d/dt + 1/T_\perp\}\ \sigma_1(t) = -\ \omega\sigma_2(t) \quad ,$$
$$\{d/dt + 1/T_\perp\}\ \sigma_2(t) = +\ \omega\sigma_1(t) \quad ,$$
$$\{d/dt + 1/T_{||}\}\ \sigma_3(t) = \quad 0 \quad ,$$

where m is the equilibrium value of τ_3, $-1<m<1$. The constant ω is real and represents the Zeeman frequency. The relaxation times $T_\perp$ and $T_{||}$ satisfy the relation $0<T_\perp \le 2T_{||} \le \infty$ (compare Emch and Varilly, 1979).

The convex-state approach to generalized probability

Many mathematical models used in the behavioral, social and natural sciences have a common feature: there is a natural notion of *mixture*. The appropriate mathematical structure unifying such an investigation is the notion of an *abstract convex structure*, first introduced in 1949 by Marshall Harvey Stone. Note that the convexity has a theory in its own right, quite independent of the theory of linear spaces. According to the traditional view, however, the state spaces of classical mechanics and of quantum mechanics are linear spaces which acquire a convex structure with the usual convex combination of statistical states.

In the convex-state approach, the statistical states of a physical system are taken as undefined primitive elements. It is assumed that the set of statistical states permits the formation of statistical mixtures. The statistical states are taken as elements of an abstract convex structure in which no linear structure is presupposed but only the operation of forming *convex mixtures* has a meaning. A fundamental representation theorem by Stone (1949) says that every convex structure can be represented as a convex set.

Convex structures have been used by von Neumann and Morgenstern (1944, p.26) in their theory of games and economic behavior. For interesting applications of the abstract concept of convexity to the study of color vision and decision theory, compare Gudder (1977).

The convex-state approach to statistical physical theories uses the set of normalized statistical states as basic mathematical object. Observables are defined by means of operations on states. This formalism unifies the operational methods of axiomatic quantum mechanics, generalizes the traditional approach, and leaves open the possibility of nonlinear structures for generalized theories.

References on the convex-state approach

Axiomatic statistical physical theories based on the affine and topological structure of an abstract state space have been introduced by Ludwig (1964, 1967a-d, 1968, 1970, 1972), Gunson (1967), Pool (1968a-b) and Haag (1969). Developing a theory of quantum stochastic processes (compare Davies, 1969b, 1970a-b, 1971), Davies and Lewis (1970) introduced an abstract formulation of the operational approach to statistical physical systems where the set of states is represented by a generating cone in a real vector space. This approach has been further developed by Edwards (1970, 1971a-b, 1972a-b, 1974, 1975) and by Edwards and Gerzon (1970). A generalization of the operational approach by Davies and Lewis is due to Gudder (1973b, 1977a-b, 1979b) and Cornette and Gudder (1974); compare also the related work by Giles (1970), Guz (1977b-c), and Cook (1978). The relation between the structure of the convex-state space and the structure of transition probabilities has been investigated by Mielnik (1968, 1969, 1974), Krause (1974), Woiciechowski (1975), Cantoni (1975, 1976, 1977), Belinfante (1976) and Gudder (1978).

Important mathematical tools for the convex-state approach are reviewed in Alfsen (1971) and in the lecture notes edited by Hartkämper and Neumann (1974). From the newer mathematical literature, the characterization of the state spaces of orthomodular lattices by Shultz (1974), of Jordan algebras by Alfsen and Shultz 1978), and of C*-algebras by Alfsen, Hanche-Olsen and Shultz (1980) are of particular importance. Furthermore we would like to mention the important work by Alfsen and Shultz (1976, 1979) on a noncommutative spectral theory for affine functions.

The axiomatic approach has been further investigated by Dähn (1968, 1972a-b, 1973), Hellwig (1969), Hellwig and Kraus (1969, 1970a), Stolz (1969, 1971) Neumann (1971, 1972, 1974), Cycon and Hellwig (1977). A synoptic but elementary presentation of Ludwig's approach is contained in the text by Ludwig (1976). An elementary introduction in the operational approach of Davies and Lewis can be found in the text by Davies (1976a).

The mathematical structure of convex-state approach

In the convex-state approach to statistical physical theories the set S of statistical ensemble states is assumed to be σ-convex (i.e. countable convex combinations of elements in S exist and lie in S). The real vector space V generated by taking the σ-convex set S to be the base of a generating cone V_+ is a Banach space with respect to the natural base norm (Edwards and Gerzon, 1970). The cone V_+ defines a partial ordering on V if $u \le v$ $(u,v \in V)$ is defined to mean that $v-u \in V_+$. With this, V becomes a partially ordered real Banach space, called the *state space* of the system. A separable complete (V,S) is the axiomatic framework of the so-called *operational approach* to generalized quantum mechanics, independently proposed by Davies and Lewis (1970) and Ludwig (1970). The

Banach-space dual V^* of V is a complete order-unit space (V^*,e) with an order unit e (compare Alfsen, 1971, chapt.2). The order unit is a distinguished linear functional on V which on V_+ coincides with the norm in V so that the set S of statistical states is given by

$$S = \{\rho \,|\, \rho \in V_+,\ e(\rho)=1\} \quad .$$

The vectors in the cone V_+ are interpreted as describing statistical ensembles, and the order unit e as strength functional such that for each $\rho \in V_+$ the number $e(\rho)$ is proportional to the number of copies of the system in the statistical ensemble represented by ρ.

An *operation* τ on the system is represented by a positive norm-increasing linear operator $\tau: V \to V$. An operation is supposed to represent some kind of physical process effecting the ensemble of copies of the system in a given state. For each state $\rho \in V_+$, the expression $e\{\tau(\rho)\}/\tau(\rho)$ is interpreted as the probability that the new state of the system is $\tau(\rho)$ when before the operation the system was in the state ρ.

Each operation τ gives rise to an element $T \in V^*$, defined by $T(\rho)=e\{\tau(\rho)\}$ for every $\rho \in V_+$, and fulfilling (relative to the partial ordering in the dual cone V_+^*) the relation $0 \le T \le e$. Conversely, every functional in the dual order interval $[0,e]$ arises in this way. Such functionals are called *simple observables* by Edwards (1970), *effects* by Ludwig (1964), and *tests* by Giles (1970). The probability of measuring an effect T in a statistical ensemble state ρ is given by the value $T(\rho)$ of the functional T on ρ. The extreme points of the order interval $[0,e]$ are the *decision effects* of Ludwig, they correspond to the *propositions* of quantum logic. If some additional conditions are fulfilled, the set of decision effects is a complete orthomodular lattice with respect to the ordering induced by the ordering of V^*. In this formalism, *observables* can be introduced as *effect-valued measures* $A: \Sigma_{\mathbb{R}} \to [0,e]$, defined on the σ-algebra $\Sigma_{\mathbb{R}}$ of Borel sets of the real axis $\mathbb{R}$.

With every system characterized by the Banach space V there is associated a *classical system* which is represented by the Banach space $Z(V^*)_*$, the predual of the center $Z(V^*)$ of the dual space V^* of V (Edwards, 1974, 1975).

Since a *symmetry* of a system is a transformation leaving all significant features invariant, a symmetry ν acting on the statistical states, and commuting with the operation of mixing (Kadison, 1965):

$$\nu\{c\rho_1+(1-c)\rho_2\} = c\nu(\rho_1)+(1-c)\nu(\rho_2) \quad ,$$

where $0\leq c\leq 1$ and $\rho_1,\rho_2\in S$. That is, a symmetry is an affine bijection of the convex set S of all statistical states.

First example: pioneer quantum mechanics

As a first example for the convex-state approach we discuss the standard model of pioneer quantum mechanics where the vector space of state vectors is a separable infinite-dimensional complex Hilbert space H. The statistical states are represented by density operators, i.e. as self-adjoint nonnegative nuclear operators of trace one, so that

$$S = \{D|D\in B_1(H),\ D=D^*\geq 0,\ \mathrm{tr}(D)=1\} \quad ,$$

where $B_1(H)$ is the Banach space of all nuclear operators acting on H, with the trace norm $\|A\|_1=\mathrm{tr}\{(A^*A)^{1/2}\}$. The span of S is the real, partially ordered base-normed space of self-adjoint nuclear operators on H,

$$V = \{V|V\in B_1(H),\ V=V^*\} \quad ,$$

The norm of V is given by $\|V\|_1=\mathrm{tr}\{|V|\}$, while the ordering is defined by

$$V\geq 0 \quad \text{iff} \quad \langle\Psi|V\Psi\rangle\geq 0 \quad \text{for every} \quad \Psi\in H \quad ,$$

where $\langle\cdot|\cdot\rangle$ is the inner product in H. The dual V^* of V is order-isometrically identified with the Banach space of all bounded self-adjoint operators on H,

$$V^* = \{A|A\in B(H),\ A=A^*\} \quad ,$$

where $B(H)$ is the algebra of all bounded operators on H. The order unit e is identified with the identity operator 1 in $B(H)$. An effect is represented by an operator $F\in V^*$ fulfilling $0\leq F\leq 1$. The bilinear pairing of V and V^* is given by the trace, so that $\mathrm{tr}(DF)$ is the probability for measuring the effect $F\in V^*$ in an ensemble characterized by the density operator $D\in V$. Since the center of $B(H)$ equals the complex numbers, the classical system associated with pioneer quantum mechanics is trivial.

Second example: classical probability theory

The basic mathematical structure of classical probability theory is Kolmogorov's probability space (Ω,Σ,μ) where Ω is the set of elementary events, Σ is a σ-algebra of subsets of Ω, and μ is a countably additive probability measure. Let Ω be a metrizable compact Hausdorff space so that the Baire and Borel σ-algebras coincide and every Borel measure on Ω is regular. The generating cone V_+ equals the set of positive bounded Borel measures on Ω so that the state space V is given as the space of real bounded Borel measures on Ω. The strength functional e is given by

$$e(\nu) = \nu(\Omega) \quad , \quad \nu \in V \quad .$$

Instead of using the abstract dual space V^* it is more convenient to use the space $B(\Omega)$ of all real-valued bounded Borel functions on Ω. There is a natural embeddding of $B(\Omega)$ into V^* defined by

$$(f|\nu) = \int_\Omega f(\omega)\nu(d\omega) \quad , \quad \nu \in V \quad , \quad f \in B(\Omega) \quad .$$

This embedding takes positive functions f to positive elements of V^*, and takes the function $\nu \to f(\omega)=1$ to the order unit $e \in V^*$. The observables of classical probability theory are the bounded Borel functions (for more details compare Davies and Lewis, 1970).

Third example: W-systems*

If the generating cone V_+ is chosen to be the cone of positive normal linear functionals on a W*-algebra, then the theory of W*-systems is obtained. Let A be a W*-algebra and A_* its predual. Then the state space V is the space of real elements of A_*, while V_+ is the set of positive elements of A_*, and S equals the set of normal states on A. The norm in S is given by

$$\|\rho\| = \sup\{|\rho(A)| : A \in A \ , \ \|A\| \le 1\} \quad , \quad \rho \in S \quad .$$

With this, (V,S) is a complete base normed space with closed generating cone V_+. Let V^* be the set of self-adjoint elements of A, then (V^*,e) is the order-unit space dual to (V,S). The order unit interval $[0,e]$ equals the set of positive elements A of A such that $A \le 1$, where 1 is the identity of A. An effect A is a decision effect if and only if A is a projection.

In the C*-algebraic approach, a W*-system is just a special case of a C*-system. In the framework of the convex-state approach, however, a C*-system is a special case of a W*-system. Let C be a C*-algebra with identity, and consider C as an algebra of observables in the sense of

C*-algebraic quantum mechanics. With this, the state space S is given by the set $(C^*)^1_+$ of normalized positive linear functionals on C,

$$S = (C^*)^1_+ \quad ,$$

so that the generating cone V_+ of the state space V is given by $(C^*)_+$,

$$V_+ = (C^*)_+ \quad .$$

The real Banach space V is the space C^*_H of bounded hermitean linear functionals on C

$$V = C^*_H \quad \text{where} \quad C^* = C^*_H + iC^*_H \quad .$$

Since the dual C^* of the C*-algebra C can be identified with the predual A_* of the W*-envelope $A=C^{**}$, a C*-system with the C*-algebra C is a particular case of a W*-system obtained by choosing the W*-algebra A to be the W*-envelope C^{**} of C. As an example, take again pioneer quantum mechanics. Considered as a C*-system, pioneer quantum mechanics is characterized by the C*-algebra $B_\infty(H)$ of compact operators on some Hilbert space H, $C=B_\infty(H)$. Then the dual C^* of C can be identified with the Banach space $B_1(H)$ of nuclear operators acting on H, while the second dual C^{**} may be identified with the Banach space $B(H)$ of all bounded operators on H (Schatten, 1960, chapt.IV). With this, pioneer quantum mechanics can also be considered as a W*-system whose W*-algebra A of observables is the algebra of all bounded operators on some Hilbert space H, $A=B(H)$. References: Edwards (1970, 1971b).

Remark: Not every W-system is also a C*-system*

In the framework of the convex-state approach every C*-system is a W*-system, but not every W*-system is also a C*-system. The reason is that the second dual C^{**} of a C*-algebra C is always a W*-algebra containing atoms. Hence the predual A_* of an atom-free W*-algebra A cannot equal the dual C^* of a C*-algebra C. For example a W*-system whose algebra A of observables is a factor of type III is not a C*-system in the sense of the convex-state approach.

4.6 INTERRELATIONS AND SYNTHESIS

Generalized quantum mechanics

The approaches to axiomatic quantum mechanics we have reviewed do not exhaust the current range of extensions of pioneer quantum mechanics but they are sufficiently representative and diverse to show that both classical mechanics and pioneer quantum mechanics arise from an oversimplification of physics.

The various frameworks for studying generalized quantum theories are based on different primitive elements. Each approach stresses another aspect and with this adds new insight into the structure of the theory of matter. A common feature of all generalizations of quantum mechanics is that nowadays the adjective "quantum" no longer refers to Planck's quantum of action but to the existence of incompatible properties. Moreover, all these generalizations are no longer "mechanical", their basic concepts do not even refer to the notion of a three-dimensional physical space. Spatial notions can be introduced later, if necessary, by an appropriate kinematical symmetry group.

The three main frameworks are:

(i) *Quantum logic*, where the primitive elements are called propositions, and are postulated to be elements of a separable σ-complete orthomodular lattice. The logical symmetries are the lattice orthoisomorphisms.

(ii) *Algebraic quantum theory*, where the primitive elements are called bounded observables; they are postulated to be the self-adjoint elements of a C*-algebra. The logical symmetries are given by the Jordan *-automorphisms.

(iii) The *convex-state approach*, where the primitive elements are the epistemic states; they are postulated to be positive and normalized elements of a Banach space. The logical symmetries are given by the affine bijections on the set of all epistemic states.

References to papers dealing with the interconnections between the various generalizations of quantum theory

The relation between C*-algebraic quantum mechanics and Mackey's approach has been investigated by Gunson (1967), Plymen (1968a-b), Davies (1968), Deliyannis (1969, 1975), Gudder and Boyce (1970), Rüttimann (1970), and Shultz (1977).

The convex-state approach by Davies and Lewis (1970) has been compared with the quantum logic axiomatic scheme by Guz (1978b) and with the algebraic approach by Edwards (1970, 1971b). Chen (1971) has shown that a C*-algebra with identity satisfies Ludwig's (1964) axiom of sensitivity increase of effects if and only if it is a W*-algebra.

W*-systems fulfill the axioms of quantum logic, of algebraic quantum theory and of the convex-state approach; they have been investigated by Jauch (1960), Jauch and Misra (1961), Pool (1968a-b), Cirelli and Gallone (1973), Cirelli et al. (1975), Primas and Müller-Herold (1978), and Primas (1980).

W-systems*

The three main approaches to generalized quantum theories are not equivalent but they have a most interesting and rich intersection: the theory of W*-systems. The basic mathematical structure of W*-systems is a W*-algebra A with a separable predual A_*. W*-systems have a lattice-theoretical, an algebraic and a state-space characterization:

(i) A quantum logic is a W*-system if and only if the lattice of propositions is given by the projection lattice of a W*-algebra.

(ii) An algebraic quantum system is a W*-system if and only if the algebra of observables is a W*-algebra.

(iii) The Banach space generated by the epistemic states of a quantum system is the state space of a W*-system if and only if it is the predual of a W*-algebra.

W*-systems not only fulfill all proposed axiomatic schemes, they also provide the paradigmatic examples for physical systems. *Classical mechanics* is represented by an atom-free commutative W*-algebra, while *pioneer quantum mechanics* is represented by a factor of type I. *Quantum thermodynamics* is represented by a W*-algebra of type III whose center contains the typical observables of phenomenological thermodynamics (like temperature and chemical potential).

Because of the scarcity of projections in general C*-algebras, general C*-systems cannot be connected with quantum logic. On the other hand, every W*-algebra is generated by its projections; moreover, the set of all projections of every W*-algebra is a complete orthomodular and semimodular lattice (Kaplansky, 1951; Loomis, 1955; Topping, 1967). (Note that not every complete orthomodular and semimodular lattice is isomorphic to the projection lattice of a W*-algebra, there is not yet a purely lattice-theoretical characterization of the projection lattices among general orthomodular lattices; compare also sect.5 in Holland, 1964).

The decisive link between the logical and the W*-algebraic approach is given by an isomorphy relation due to Dye (1955): every lattice-orthoisomorphism of the projection lattice of a W*-algebra (without type I_2 direct summands) is implemented by a Jordan *-isomorphism on the W*-algebra. That is, *for a W*-system the symmetries of quantum logics are exactly the symmetries of algebraic quantum mechanics.*

The convex-state approach assumes that the state space is a Banach space; it coincides with the algebraic approach if the dual of the state space equals the C*-algebra of observables. According to Sakai's (1956) characterization of W*-algebras, this is the case if and only if the algebra of observables is a W*-algebra. The link between the convex-state and the algebraic approach is given by results due to Kadison (1965) which imply that the Jordan *-automorphisms of a W*-algebra A are in one-to-one correspondence to the affine bijections of the predual A_* which acts as a state space of the epistemic states (compare also Bratteli and Robinson, 1979, theorem 3.2.8). That is, *for a W*-system the symmetries of algebraic quantum mechanics are exactly the symmetries of the convex-state approach.*

The relation between the algebraic and the logical approach

The algebraic approach to generalized quantum theory has been frequently criticized on the grounds that it has a weak conceptual foundation. Quantum logic, on the other hand, is conceptually well-founded but it lacks the powerful mathematical tools of the algebraic approach Hence it is worthwhile to investigate under what conditions these two approaches can be related.

The self-adjoint and idempotent elements of a C*-algebra are called projections. In every C*-algebra one can partially order the projections by specifying that $F \leq G$ means $FG=F$ (or, equivalently, $GF=F$). However, it is known that a C*-algebra may possess no projections at all. Moreover, even if a C*-algebra possesses many projections they do as a rule not form a lattice. In order to connect the C*-algebraic approach with the lattice theoretical approach, the following two minimal requirements for the C*-algebra A of observables have to be fulfilled:

(i) each maximal commutative C*-subalgebra of A is generated by its projections,

(ii) in the partially ordered set of projections of A, every collection of mutually orthogonal projections has a supremum.

A C*-algebra fulfilling these two conditions is called an *AW*-algebra* (Kaplansky, 1951). Postulate (i) guarantees a large supply of projections such that every observable can be interpreted in terms of the elements of a Boolean algebra. Postulate (ii) is a necessary condition for the projections of A to form a complete lattice. In fact, Kaplansky (1951) proved that the projections in every AW*-algebra form a complete lattice. Moreover, the projection lattice of every AW*-algebra is orthomodular, whereby the orthocomplementation $F \to F^{\perp}$ is given as usual by $F^{\perp}=1-F$ (compare Maeda and Maeda, 1970; Berberian, 1972).

Every W*-algebra is an AW*-algebra but the converse is false. There exist commutative AW*-algebras which are not W*-algebras, and there exist type III AW*-factors which are not W*-algebras. However, every type I-AW*-factor is a W*-algebra.

In the commutative case the difference between AW*-algebras and W*-algebras is familiar from the theory of Boolean algebras. The projection lattice of an AW*-algebra A is a complete Boolean algebra which is a measure algebra if and only if A is a W*-algebra. This situation generalizes to noncommutative AW*-algebras. Pedersen (1972) has shown that an AW*-algebra with a separating family of completely additive states is a W*-algebra (compare also Pedersen, 1979, sect.3.9). An AW*-algebra is said to be *wild* if the only positive functional on A that is normal on projections is the zero functional. Wright (1979) has shown that every AW*-algebra can be decomposed (by a unique central projection) into a wild AW*-algebra and a W*-algebra. In particular, every AW*-factor is either wild or a W*-factor. If we insist on the possibility of a statistical interpretation of quantum logic, we need probability measures (in the sense of Kolmogorov), hence normal states. Accordingly, the only AW*-algebras that admit a statistical interpretation in terms of Kolmogorov's probability theory are the W*-algebras. In this sense, *the C*-algebraic approach coincides with the quantum-logical approach if and only if the C*-algebra is a W*-algebra.*

An appealing motivation for the use of orthomodular lattices in quantum logics is that it permits us to combine the compatible Boolean algebras of the various possible classical theories into a transitive partial Boolean algebra (compare the summary given in Primas, 1980, sect.4.2). This approach can be phrased in terms of the following characterization of a W*-algebra due to Pedersen (1972; 1979, sect. 2.8): *a C*-algebra A is a W*-algebra if and only if every maximal commutative C*-subalgebra of A is a W*-algebra.* In a given universe of discourse every set of mutually compatible propositions defines a *context* amenable to a classical description. A maximal context (given by a maximal set of mutually compatible propositions) is represented by a Boolean measure algebra. Every such Boolean measure algebra is orthoisomorphic to the projection lattice of a W*-algebra. The smallest W*-algebra containing all the commutative W*-algebras of the maximal contexts of the universe of discourse is the algebra of observables. Conversely, every minimal commutative W*-subalgebra of the W*-algebra of observables characterizes a feasible maximal context in the universe of discourse.

5. A FRAMEWORK FOR THEORETICAL CHEMISTRY

Truth is not that which is demonstrable
but that which is ineluctable.
Antoine de Saint-Exupéry, 1939

5.1 REEVALUATION OF THE PARADIGMS OF THEORETICAL CHEMISTRY

For decades, the presuppositions of pioneer quantum mechanics have served as paradigms for theoretical chemistry. Numerical quantum chemistry has become the natural way of looking at problems for most theoretical chemists. As stressed by Ludwik Fleck (1935) and Thomas S. Kuhn (1962), a paradigm on the one hand acts as a method by means of which facts are observed, interpreted and organized. On the other hand every paradigm acts also like a blinder. The snag is that the rules imposed by a paradigm are implicit and not recognized as tentative working hypotheses. In contemporary chemistry, the rules of pioneer quantum mechanics operate automatically, and a good theoretical chemist has become a person who feels that he is doing the obvious thing when using pioneer quantum mechanics to solve a chemical problem.

Since we are emotionally attached to things with which we are familiar and which give us pleasure, it is understandable that the repeated failure of a paradigm has as a rule little effect on the tenacity of our expectation that the failure is just apparent and caused by, say, obscure data or confused thinking. For example, not many contemporary theoreticians admit that we do not have a coherent theory of molecular structure. Most chemists just refuse to see that the Einstein-Podolsky-Rosen correlations predicted by pioneer quantum mechanics compellingly exclude any classical concept of molecular structure. The paradigms of contemporary quantum chemistry keep the researcher on a productive course by preventing him from wasting time with the hard problem of the omnipresence of Einstein-Podolsky-Rosen correlations. Nevertheless, these correlations do not cease to exist because the current paradigms play them down.

In spite of all the efforts, pioneer quantum mechanics fails to explain classical phenomena. This is one reason why there is much more to chemistry than pioneer quantum mechanics can explain. As a result of the rise of generalized quantum theories which have overcome the dualism between pioneer quantum mechanics and classical mechanics, we are provided with new tools permitting a new approach to theoretical chemistry. The various approaches to generalized quantum theories (reviewed in chapter 4) should make it clear that there is no unique formalism generalizing both pioneer quantum mechanics and classical mechanics. Every consistent generalization presumably can be used as an improved theoretical foundation for the natural sciences. There is no question of accepting "the correct theory", and rejecting "the false ones". Since no theory can be inferred from empirical data, there is an element of choice and invention involved. We certainly use, among other things, pragmatic and aesthetic criteria, so that human creativity plays an important role in choosing the basic theoretical structures.

Moreover, the generalized quantum theoretical formalisms which are at our disposal are not enough by themselves, they have to be interpreted. A generalized quantum theory without an interpretation is a theory without sense. We have to know what our theories are about. Technically speaking, we should know the *referent* of our theories. We should not forget that science is fundamentally related to philosophy. Not worrying about philosophical questions in science is too cheap a solution.

Naturally, a chemist would like to have a theory about substances and molecules, a theory which gives a correct account of typical phenomena in the realm of chemistry (among many other things such a theory would include chemical taxonomy, chemical thermodynamics, chemical kinetics and chemical system theory). The interpretation should provide a language which is appropriate to the phenomenology of chemical reality. Contemporary quantum chemistry is not such a theory for it is neither good chemistry nor good physics. The fact that contemporary quantum chemistry has an impressive empirical corroboration does not mean too much since every theory can be compared with experiments only by using the theory's own interpretative framework. The problem with contemporary quantum chemistry is not that it is false (nowadays, that would be no problem at all) but that often it is not appropriate. In theoretical science we are looking for explanations, and explanations are answers to questions. But *these questions must make sense within the chosen theoretical framework.* We have to respect the fact that chemistry has its own legitimate area of activity and that it has autochthonous concepts. It cannot be the task of theoretical chemistry to narrow down the scope of scientific inquiry by eliminating all genuine chemical concepts.

Contemporary quantum chemistry is based on pioneer quantum mechanics and reflects the dogmas of this approach. Many of the difficulties associated with the interpretations of pioneer quantum mechanics are created by philosophical preconceptions which we are free to discard as inconvenient ideologies. For example, *we are free to give up operationalism* (i.e. the view that the concepts used in scientific statements must be defined in terms of identifiable and repeatable operations), and *we are free to give up logical monism* (i.e. the view that there is only one kind of logic, and, more particularly, the assumption that the logical structure of a theory must be the same as that of its metatheory).

We are free to reject many preconceptions but we cannot avoid to adopt some *regulative principles* to fix the interpretation of generalized quantum theory. Regulative principles are normative rules of a metaphysical nature, hence not susceptible of experimental proof or disproof. However, the vital spark of a theory is closely linked with the regulative principle adopted. For example, there exist neither an empirical nor a logical necessity to adopt *realism*, i.e. the metaphysical view that some entities are real in their own right and exist independently of their being known. But are we really allowed to treat theories in isolation from the rest of science and our culture? If we give up realism, we get into trouble. There is no clear-cut transition from microphysics to macrophysics, so it would be inconvenient to use different regulative principles when speaking of electrons, atoms, molecules, macromolecules, enzymes, biological cells, bacteria or crystals. We all think that it is sensible to speak of chairs, books and animals as objectively existing entities, existing independently of our knowledge of them. Hence a reasonable and consistent unified language for theoretical chemistry will have to choose its regulative principles in such a manner that the fair requirements of a moderate realism are fulfilled. That is, everything which is practically real should appear as objectively real in the theory.

We shall therefore arrange our theoretical discussion in such a way that it is about objectively existing entities. In a more technical language this means that we adopt an *ontic interpretation* of generalized quantum theory. Many physicists and philosophers have claimed that a consistent ontic interpretation of quantum mechanics is impossible. *This view is wrong.* The main problem in adopting realism as a regulative principle for quantum mechanics was pointed out already by Einstein, Podolsky and Rosen in 1935 (compare sect.3.7). The existence of Einstein-Podolsky-Rosen correlations implies that nature is holistic so that the classical mode of investigating nature by compartmentalization is no longer defensible.

We accept the existence of Einstein-Podolsky-Rosen correlations as experimentally confirmed (compare sect.3.7). Hence we cannot anymore adhere to the old dream of a single frame of reference that permits us to eliminate the pluralism of physical, chemical and biological theories. We can break the holistic symmetry of the world only if we abstract *deliberately* from some Einstein-Podolsky-Rosen correlations. Without such an abstraction there are no phenomena. There are no entities in our world which have observable attributes independently of any abstraction. *Observable phenomena are created by abstracting from some Einstein-Podolsky-Rosen correlations.* Using different abstractions, one observes different phenomena. Each abstraction creates its own reality. This concept of reality is objective since it is intersubjective: everybody who makes the same abstractions gets the same observable phenomena. The idea that experienced reality is conditional and abstraction-dependent is hardly new but is in fact part of our culture. It is true that this idea has been tacitly suppressed by the theoretical framework of nineteenth century physics (which still dominates present scientific thought) so that its reappearance in a new quantum-theoretical dressing is perceived as a radical change in our attitude toward science, as a paradigm shift.

5.2 THE LOGIC OF PROPERTIES

Primitive concepts

Innocent theories of natural phenomena begin with entities whose existence is assumed to be indubitable since they are supposed to be given in immediate experience. This approach has been superseded, we challenge the alleged primacy of facts inasmuch as there are no bare facts. Moreover, we know that modern science presents us with a much richer reality than we are familiar with from everyday life or classical physics. We are looking for a theoretical framework which is suitable for any system of whatever size and complexity, which encompasses all genuine results of classical physics and pioneer quantum mechanics, and which can be conditioned by a variety of contexts.

Every theoretical framework must start with some undefined concepts which need no further explanation and can be taken as granted without proof, definition or analysis. Such concepts are called *primitive concepts*. Outside the theoretical framework considered, we may have an intuitive understanding of primitive concepts and we may discuss them in terms of a metalanguage. *Within* the framework itself, primitive concepts have no concrete a priori meaning, but we hope that in the course of the further development of the theory they will give rise to a meaningful structure and to operationalized concepts. This situation is difficult but unavoidable. *"Just go on, and faith will soon return"*, as Jean Le Rond d'Alembert (1717-1783) is supposed to have assured a hesitating friend.

We take the notion of *property* as a primitive concept. On a non-formal intuitive level, we imagine that properties represent inherent traits of the universe of discourse. Instead of "universe of discourse" we often use the linguistically more convenient (but philosophically dangerous) term *system*. In the following discussion, the notion "system" always includes a "definite way of looking" and *never* means a Kantian transcendental "system-in-itself". Nevertheless, we try to adopt a realistic and objectivistic view by relating the concept of a property to elements of an objective reality. That is, we associate properties to entities having (in an appropriate sense) real being. As a guiding rule, we require that whenever we are able to derive a *classical* theory from the basic theoretical framework, we use this subtheory in its almost universally accepted traditional realistic interpretation.

In the technical jargon, we say that we adopt an *ontic interpretation*. Such an interpretation may not satisfy the extreme operationalists or positivists, but we do not care. All we care about is that an ontic interpretation is a logically permissible albeit not necessary way of talking. No doubt other strategies are available. All we claim is that the ontic interpretation is a sound and viable option.

Propositions about properties

To every property f we associate a proposition F such that

if F ist true, we say the system *has* the property f,

if F is false, we say the system does not have the property f.

This normative rule will lead to a realist interpretation. What is decisive, is that *we do not assume that every proposition is either true or false*. That is, in a terminology introduced by Healey (1979), we *reject* a "*naive* realist interpretation".

If g is a generic term subsuming the property f, we say that the property f implies the property g (compare also Kanthack and Wegener, 1976). If a property f implies a property g, we write for the corresponding propositions $F \leq G$,

$$F \leq G \text{ iff } \text{"F is true" implies "G is true"} \quad .$$

Two properties which imply each other are called *equal*, and we write

$$F = G \text{ iff } F \leq G \text{ and } G \leq F \quad .$$

A property which is not implied by another property of the system is called an *atomic property*, and the corresponding proposition is called an *atom*.

To every proposition F we associate another proposition $F^{\perp}$ such that whenever F is true then $F^{\perp}$ is false, and if F is false then $F^{\perp}$ is true,

$$\text{"F is true" implies "}F^{\perp}\text{ is false".}$$

Logic and time

The feasible propositions of the universe of discourse constitute the raw material for the logic of properties. We have to distinguish two kinds of propositions: temporal propositions and timeless propositions. *Temporal propositions* stand for tensed statements, they refer to a particular time and their truth-values may vary with time. *Timeless propositions* are true or false without any reference to time. The timeless propositions characterize the *system*, whereas the truth-values of the temporal propositions characterize the *state* (at a particular time) of the system.

We cannot help but use an unreflected concept of *time*. The notion of time is mysterious and our understanding of it is inadequate. Moreover, in all Indo-European languages verbs are tensed hence we have the fundamental difficulty - already stressed by St.Augustine (354-430) - that we cannot discuss time without referring to time.

References: The nature of time

Many results of the research on the nature of time can be traced from the surveys edited by Fraser (1966, 1975), Gold (1967), Zeman (1971), Fraser et al. (1972, 1978), Fraser and Lawrence (1975). Compare also the monographs by Reichenbach (1928, 1956), and by Grünbaum (1964).

Salecker and Wigner (1958) claimed that no accurate clock can be of microscopic size. In the following, we naively *assume* the existence of macroscopic clocks outside our universe of discourse, and our notion "time" always refers to such external clocks. That is, we consider historical time not as an observable within our theoretical framework but as a fundamental trait of the external world. In order to formulate "if - then" relations, we need an "earlier - later" relationship. We *assume* that the relations "earlier than" and "later than" have a significance apart from any human conscious awareness of events coming into being, and define the "past" as the image of earlier time as remembered by a memory. That is, we *presuppose* the existence of memories so that we can give a mind-independent definition of the anisotropy of time as follows. If any memory M contains an event E_1 but not the event E_2, and if the memory M is contained in a memory M' that contains both events E_1 and E_2, then we *define* the event E_1 as being earlier than the event E_2.

Aside: Time and consciousness

The concepts "event" and "temporal order" are reciprocally related, each being essential to the other. The one-sideness of time is akin to the *coming into being of events*. Becoming is a change from the status of being potentially possible to the status of being factual, but becoming is also the passage from the future to the past. While Newton proposed the idea that "time" exists in its own right, Kant (in his *Critique of the Pure Reason*) came to the conclusion that our idea of time is merely a part of our mental apparatus for imagining the world. There exists a notion of time that is derived from introspection; this introspective notion of time exists on a subjective level, as a category of our conscious mind. Furthermore, the "*awareness of what we today call time seems to be dependent on the functioning of our conscious mind*" (Franz, 1966). It seems to be difficult not to agree with Denbigh (1972) who regards "*the criterion of before and after which is offered by consciousness as having primacy over any criterion offered by science*". Hermann Weyl has put forward the thesis that physical events are already there and do not happen; when we speak of "coming into being" we refer to something which depends on the presence of a conscious awareness. *"Only the consciousness that passes on in one portion of this world experiences the detached piece which comes to meet it, and passes behind it as history"* (Weyl, 1922, p.127).

The crucial issue which we have to face is as follows: is time *constitutive* in the sense that it regulates the rate of natural processes, or is time *relational*, i.e. are events prior to time so that time is a concept derived from events? In quantum mechanics, the answer to this question depends on the interpretation adopted. From an external point of view (e.g. in Bohr's interpretation of quantum mechanics), time has to be considered as constitutive. In a fundamental theory the relational view sounds quite alluring. As a non-Boolean theory, quantum theory makes a distinction between the potentially possible and the actual. The existential status of *future* events is that of the potentially possible, whereas past events are factual. In the Everett interpretation of pioneer quantum mechanics, an event is the same as the splitting of a state into classically inequivalent states; coming into existence means to be on one of several possible branches. In the Everett interpretation, coming into existence is the same as becoming aware of being on a certain branch in the Everett world. *Consciousness is a classical property*, coherent superpositions of observers' consciousness in different branches are impossible. *My consciousness labels my branch in the Everett world.* A conscious being travels along *his* branch in the Everett world. In this sense, consciousness and events imply time.

In the following we do *not* adopt the illuminating but exorbitant Everett interpretation. Yet the above discussion indicates that quantum events define a temporal order. There could be no historical time if they were no happenings. Adopting an external point of view we have to make assumptions that amount to a distinction between future and past since no external observation of a system is possible without introducing a time asymmetry. Every deep analysis of the "earlier - later" relationship has to be related to some fundamental theory of observations (such as the notorious "measurement process" of pioneer quantum mechanics).

Difficulties with Einsteinian relativity

In the following we will base the theoretical framework on the *temporal logic* of the propositions about the properties of a system. The philosophical merit of a temporal logic lies in the fact that it accounts for the incompatibility of potential properties of quantum systems in a fundamental way. In such an approach to physical theories, there is no longer any deep-seated connection between quantum logic and Newtonian mechanics. A priori, there is not even any connection between quantum logic and the three-dimensional physical space. Yet, the Galilean space-time structure can easily be introduced into temporal quantum logic by declaring the Galilei group to be a kinematical group which has to be represented by the symmetries of the theory.

Einsteinian space-time is conceptually much more difficult. While Galilei transformations only lead to a *dynamization of space*, Lorentz transformations lead in addition to a *spatialization of time*. If the space coordinates are represented by nonclassical observables (as is the case in pioneer quantum mechanics), then a spatialization of time turns time into a physical quantity which is not compatible with every other physical quantity. Such a time cannot be used as an ordering parameter in a temporal logic.

The Einsteinian relativistic viewpoint is in harmony with any classical temporal logic (represented by some Boolean lattice), but it is at odds with temporal quantum logic (represented by some non-distributive lattice). That is, the amalgamation of Einsteinian relativity with quantum theory is an unsolved *conceptual* problem. Contemporary quantum logics shows no holistic effects connecting events at different instants of time.

Is a global temporal logic of properties possible?

Recall that an *epistemic interpretation* of a theory refers to our knowledge. As opposed to that, an *ontic interpretation* refers to real being, it asserts that the propositions of the theory refer to properties of the object system itself. While an epistemic interpretation deals with the relationship between object and subject, an ontic interpretation agrees with the view that fundamental theories should deal with the objective world rather than with our knowledge of it.

The referent of a theory stands for that which a theory is about. If the referent of a theory is an individual object, we speak of an *individual interpretation*. If the referent is a fictitious ensemble of replicas, we speak of a *statistical interpretation*. One can always turn from an individual interpretation to a statistical interpretation by applying the individual interpretation to a Gibbsian ensemble.

If we want to represent the world, then a statistical and epistemic interpretation may be adequate. If we want to understand the world in terms of imaginative and intuitive concepts and finally to explain it, an individual and ontic interpretation of the basic theoretical framework seems to be indispensable.

Question: Is it possible to choose an individual and ontic interpretation of generalized quantum theory? *Answer:* Yes, provided we accept that the temporal logic of properties is non-Boolean.

Question: Is it possible to describe individual objects in spite of the unbroken wholeness of nature? *Answer:* Essentially yes, but the existence of Einstein-Podolsky-Rosen correlations gives us a hard nut to crack. We are forced to break with an old tradition: the idea that nature can be investigated by compartmentalization. The classical notion of the analyzability of nature into separately existing parts has to be replaced by the notion of the context-dependence of every description so that theory and experiment cannot any longer be separated.

Question: Is it possible to deduce the phenomenological theories of physics and chemistry from some fundamental theory? *Answer:* Yes and no. Certainly chemistry cannot be derived from physics. Yet, there is a more general theoretical scheme than physics in which all know nonrelativistic physical theories and all formalized chemical theories can be embedded in a natural way as subtheories. At present the precise embedding function can in most cases be only guessed at.

In the following sections we sketch a purely algebraic formulation of a universal theoretical framework for nonrelativistic physics and for chemistry which is completely independent of any Bohrian correspondence considerations. We will adopt an individual and ontic interpretation, and posit that every system is in a definite ontic state even if no information about the system is available.

5.3 ORTHOMODULAR TEMPORAL LOGIC AND ITS ONTIC INTERPRETATION

The partial Boolean structure of temporal logic

Quantum logic provides a unified and explicitly formulated framework for the scientific study of nature, it is a mode of description which takes the wholeness of nature into account in a natural way. It can be developed in an abstract way which owes little to the historical development of quantum theory.

As discussed in section 4.4, orthomodular logic generalizes the Boolean logic of classical theories and the Hilbert logic (given by the lattice of all closed subspaces of a Hilbert space) of pioneer quantum mechanics. Algebraically, an orthomodular logic is characterized by a *separable, complete and orthomodular lattice* L (compare also Primas, 1980, sect.4.2). In a temporal logic, the elements of L are interpreted as temporal propositions pertaining to the properties of the system. The partial order $\leq$ in L is interpreted as the *implication* (i.e. $F \leq G$ iff "F is true" implies "G is true"), while the orthocomplementation $F \rightarrow F^{\perp}$ in L is interpreted as *negation* (i.e. "F is true" iff "$F^{\perp}$ is false"). Recall that two elements F,G in L are called *orthogonal* if $F \leq G^{\perp}$. The lattice L is called *separable* if every subset of mutually orthogonal elements in L is countable.

A more detailed interpretation of an orthomodular temporal logic L is induced by the interpretation of the partial Boolean structure of L (compare also Primas, 1980, sect.4.3). Every Boolean sublattice of the orthomodular lattice L is considered as a feasible Boolean frame of reference which admits the usual ontic interpretation of classical theories.

In the ontic interpretation of a classical theory one stipulates that the system always *is* in a state with respect to which *every* temporal proposition is truth-definite (i.e. every proposition is either true or false). Clearly such a "principle of excluded middle" for the temporal propositions of an arbitrary system is no logical necessity. The view is possible for classical systems since the logic of temporal propositions is Boolean and all propositions are mutually compatible. For a general orthomodular logic one says that two propositions F,G are *compatible*, written $F \leftrightarrow G$, if the orthocomplemented sublattice generated by $\{F,F^{\perp},G,G^{\perp}\}$ is Boolean. Two propositions which are not compatible are called incompatible.

With any two propositions $F,G\in L$ of an orthomodular logic L we associate a proposition $C(F,G)\in L$ by

$$C(F,G) \overset{\text{def}}{=} (F\wedge G)\vee(F\wedge G^{\perp})\vee(F^{\perp}\wedge G)\vee(F^{\perp}\wedge G^{\perp}) \quad .$$

Piron (1964) has shown that the proposition $C(F,G)$ has the following remarkable properties:

(i) $C(F,G) = E$ if and only if $F\leftrightarrow G$,

(ii) $C(F,G) \leftrightarrow F$ and $C(F,G)\leftrightarrow G$,

(iii) $F_C \leftrightarrow G_C$ where $F_C \overset{\text{def}}{=} F\wedge C(F,G)$, $G_C=G\wedge C(F,G)$.

If $F\wedge G\neq 0$, then there exists a state ρ with $\rho(F\wedge G)=1$, so that $\rho(F)=\rho(G)=1$. That is, there exist states in which F and G are simultaneously truth-definite in spite of the fact that they are incompatible. In such states the proposition $C(F,G)$ necessarily has to be true and that is possible if and only if $C(F,G)\neq 0$, where 0 denotes the absurd proposition which is always false. With this, the proposition $C(F,G)$ measures the *degree of incompatibility* of the two propositions F,G (compare also Emch and Jauch, 1965). In particular, we call two propositions F,G *maximal incompatible* if $C(F,G)=0$, so that there is no state with respect to which F and G can be simultaneously truth-definite (compare also Hardegree, 1977). Hence, if F,G are maximally incompatible and if F is truth-definite with respect to some state, then G is neither true nor false with respect to this state. It easily follows that two propositions F,G are maximally incompatible if and only if they are "in position p" in the sense of Dixmier (1948), i.e. if

$$F\wedge G = F\wedge G^{\perp} = F^{\perp}\wedge G = F^{\perp}\wedge G^{\perp} = 0 \quad .$$

Note that in the terminology of pioneer quantum mechanics, compatible properties are *always* simultaneously measurable, maximal incompatible properties *never*, while incompatible (but not maximally incompatible) properties sometimes are and sometimes are not simultaneously measurable.

The crucial difference between classical and quantum theories is the existence of incompatible properties in quantum systems though the existence of incompatible properties is by no means a discovery of quantum physics. Incompatible properties have a priori nothing to do with Planck's constant of action, they are all about us if we only care to look. A system possessing incompatible properties exhibits a holistic behavior. On the other hand, the wholeness of nature implies that its temporal logic is not Boolean. Since the Hilbert logic of pioneer quantum mechanics admits no homomorphism into any Boolean algebra, there exists no single Boolean description of nature but only mutually complementary classical descriptions of some aspects of nature, each being given by a Boolean logic.

Our thinking is hardly independent from our biological evolution. The evolutionary development seems to have had a preference for clear-cut yes-no decisions so that we are in some way biased in favor of Boolean reasonings. If we want to understand nature properly, we must free ourselves from unfounded Booleanizations which lead to a simple-minded black-and-white account of nature. On the other hand, it would be unwise to believe that we can jump over our own shadow. Fortunately, a non-Boolean orthomodular logic still has a strong partial Boolean structure which allows us to transfer the familiar ontic interpretation of the classical theories to nonclassical theories with an orthomodular temporal logic.

Quantum logic provides a *global* description which in case of an orthomodular logic L can be represented by structured families of *local* Boolean descriptions, called Boolean atlases (Domotor, 1974). Any local Boolean description is given by a Boolean sublattice B of the temporal logic L. Two Boolean sublattices B_1, B_2 of L are said to be *compatible* if either $B_1 \subseteq B_2$ or $B_2 \subseteq B_1$, otherwise they are called *incompatible*. Two Boolean sublattices B_1, B_2 whose intersection is trivial (i.e. equals the Boolean algebra consisting of the minimal element O and the maximal element E only) are called *maximally incompatible*. A Boolean sublattice of L which is not properly contained in any other Boolean sublattice of L, is called *maximally Boolean*. Note that two different maximal Boolean sublattices of L are not necessarily maximally incompatible. An orthomodular logic is not a disjoint collection of Boolean logics but there is a strong coherence between its Boolean sublattices. The intersection of two incompatible, but not maximally incompatible Boolean sublattices B_1, B_2 is always a Boolean sublattice which is compatible both with B_1 and with B_2. With this the orthomodular logic L becomes a *partial Boolean algebra*. In a partial Boolean algebra the Boolean operations are not defined between every pair of propositions, but when they are defined, they have their usual classical interpretation.

Conversely, one can start to describe the universe of discourse locally, with the help of classical descriptions. Every classical description is characterized by a separable and complete Boolean algebra. Since such Boolean descriptions are incomplete, we need a whole family of complementary descriptions, given by mutually incompatible Boolean algebras. The various descriptions overlap so that a partial ordering can be introduced into the family of all feasible Boolean partial descriptions of the universe of discourse. If in this family the relations $F \leq G$, $G \leq H$ imply that $F \leftrightarrow G$ hence $F \leftrightarrow H$, then it is called a *transitive partial Boolean algebra* (a concept introduced by Kochen and Specker, 1965a). This assumption guarantees the needed coherence between the various Boolean atlases. In this case it can be shown that the family of all feasible Boolean partial descriptions of the universe of discourse forms in fact a separable complete orthomodular lattice (for more details and references, compare Primas, 1980, sect.4.2).

The ontic interpretation of temporal logic

In classical theories it is formally consistent to make the ontological claim that these theories are about objects of a certain kind, each of which has an autonomous status and can be described exhaustively in terms of its properties. An ontic interpretation of a classical theory with a Boolean temporal logic $\mathcal{B}$ stipulates that all propositions of $\mathcal{B}$ are truth-definite so that at every instant every proposition of $\mathcal{B}$ is either true or false, tertium non datur. With this, there exists at every instant of time t a truth-function τ_t which says for every proposition $F \in \mathcal{B}$ whether it is true or false. Using the convention

$$\tau_t(F) = 1 \quad \text{iff} \quad F \text{ is true at time } t \quad ,$$
$$\tau_t(F) = 0 \quad \text{iff} \quad F \text{ is false at time } t \quad ,$$

we call the truth-function $\tau_t : \mathcal{B} \rightarrow \{0,1\}$ *the ontic state of the classical system at time* t.

By definition, a truth-function complies with the rules of classical logic. That is, in a classical theory an ontic state is a Boolean homomorphism $\tau_t : \mathcal{B} \rightarrow \mathcal{B}_2$ from the Boolean algebra $\mathcal{B}$ of the temporal propositions of the system into the Boolean algebra $\mathcal{B}_2 = \{0,1\}$ consisting of the truth values 0 and 1. It is a mathematical fact that in general an orthomodular lattice L admits no homomorphism into the Boolean lattice B_2 of the two truth values. This circumstance makes the formulation of an ontic interpretation of non-Boolean theories more cumbersome but by no means impossible.

In holistic theories we can no longer assume that every potential property of the universe of discourse is truth-definite. Propositions which are true at time t correspond to *actualized* properties the system *has* at time t. Properties which are neither true nor false at time t correspond to potential properties not actualized at time t. Consequently, a truth-function is not defined for all propositions of a temporal logic L but only for those propositions of L which correspond to actualized properties.

A function on some subset $D \subseteq L$ is called a *partial functional* on L. A *partial truth function* on L assigns the truth-value 0 or 1 to propositions corresponding to actualized properties but assigns no value to a proposition associated with a potential but not actualized property. The set of all propositions to which a truth-function assigns values is called the *domain* D of this truth-function. Since a truth-function is an assignement of the truth-values 0,1 to the propositions in D such that the rules of classical logic are fulfilled, the domain D has to be at least an orthomodular lattice (but neither necessarily distributive nor σ-complete). Accordingly, a partial truth-function on an orthomodular logic L is a pair (D, τ) consisting of an orthomodular sublattice $D \subseteq L$ and an ortholattice homomorphism $\tau : D \rightarrow B_2$ from the domain D onto the Boolean algebra B_2 of the two truth-values 0 and 1.

Since every Boolean lattice admits an everywhere defined truth function, it seems to be reasonable to expect that the domain D of a partial truth-function on an orthomodular logic L always contains a maximal Boolean sublattice B of L, $B \subseteq L$. But there are no reasons to think that the maximal domain of a partial truth-function cannot be larger than a maximal Boolean sublattice. In fact, as a rule the domain of a partial truth-function can be extended to a nondistributive orthomodular lattice containing a maximal Boolean algebra. Let τ_1, τ_2 be partial truth-functions on L with the domains D_1, D_2. If $D_2 \supset D_1$, and if τ_2 coincides on D_1 with τ_1, then we say that τ_2 is an *extension* of τ_1. A partial truth-function is called *maximal* if there exists no proper extension of it.

The ontic interpretation of an arbitrary orthomodular temporal logic L posits that at every instant t the actualized propositions of L equal the domain D_t of a maximal partial truth-function τ_t on L. The ortholattice homomorphism $\tau_t : D_t \to B_2$ is called the *ontic state* of the system at time t. The domain D_t of the ontic state τ_t consists of those propositions which correspond to properties actualized at time t.

Remark: Unsolved mathematical problems

For every partial truth-function there exist (in virtue of Zorn's lemma) extensions to maximal domains, however the characterization of conditions warranting the uniqueness of an extension is very difficult. For the special case of a projection lattice L of a W*-algebra, it is known that if τ is a σ-lattice-orthoisomorphism from a maximal σ-complete Boolean sublattice $B \subset L$ to B_2, then there is a *unique* extension of τ to a maximal partial truth-function. If an ontic state τ_t is a σ-homomorphism, we speak of a *normal ontic state*, if τ_t is not a σ-homomorphism, we say that τ_t is a *singular ontic state*. The case of singular ontic states is conceptually important but mathematically far more difficult than the case of normal ontic states. These problems are presently under detailed investigation (Raggio, 1981).

The intersection of all Boolean sublattices of a σ-complete orthomodular lattice L is called the *center* $Z(L)$ of L; it is a σ-complete Boolean algebra. The elements of the center $Z(L)$ of a temporal logic L are called *classical propositions*, they correspond to the *classical properties* of the system. A system whose temporal logic L is identical with its center, $Z(L)=L$, is called a *classical system*. A system with trivial center (consisting of only the absurd and the trivial propositions) is called *purely quantal*. The center of any temporal logic is contained in the domain of every maximal partial truth-function, so that in the ontic interpretation classical propositions are truth-definite at every instant. Two ontic states on a temporal logic L are called *classically equivalent* if their restrictions on the center of L are equal. That is, the ontic states τ_t and τ'_t are classically equivalent if and only if $\tau_t(F)=\tau'_t(F)$ for every classical proposition $F \in Z(L)$.

The concept of an ontic state applies to quantum systems as well as to classical systems, it refers to individual systems (as opposed to an ensemble of systems) and is independent of any probabilistic notion. An ontic state is attributed to every *closed* system, and it is irrelevant whether we know this state or not. Ontic states relate to properties having real being, hence they have no a priori operational meaning. The states which refer to our knowledge are called *epistemic states*, they can be characterized by the information we obtain from a well-designed experiment intended to determine the probability that the system is in a given ontic state. The relations between ontic and epistemic states will be discussed in greater detail in section 5.4.

Symmetries

We have characterized the structure of temporal logic by a separable, complete orthomodular lattice L, together with its ontic interpretation. A *logical symmetry* (or *symmetry* for short) of a theory is a mapping of L onto L which preserves the logical structure of the temporal logic L together with its interpretation.

It is useful to consider symmetries as special cases of structure-preserving mappings from a temporal logic L to another temporal logic $\tilde{L}$. A mapping $\varphi: L \to \tilde{L}$ is called an *ortholattice homomorphism* from the orthomodular lattice L (with maximal element E) onto the orthomodular lattice $\tilde{L}$ (with maximal element $\tilde{E}$) if

(i) $\varphi(E) = \tilde{E}$,

(ii) $\varphi(F^{\perp}) = \varphi(F)^{\perp}$,

(iii) $F \leq G$ implies $\varphi(F) \leq \varphi(G)$.

If φ is a bijective homomorphism from L onto $\tilde{L}$, then it is called an *ortholattice isomorphism*, and L is said to be *isomorphic* to $\tilde{L}$. In the special case $L = \tilde{L}$ an ortholattice isomorphism from L onto L is called an *automorphism* of L. The set of all automorphisms of L forms a group, called the *automorphism group* Aut(L) of L. With this we can say that the logical symmetries of a temporal logic L are given by the elements of the automorphism group Aut(L).

The relation of an external observer to a system is given by a *kinematical group* G, usually a Lie group (such as, for example, the Galilei group). In order to introduce kinematical symmetries into a system with the temporal logic L, the kinematical group G has to be realized faithfully by a subgroup $\{\alpha_g | g \in G\}$ of automorphisms $\alpha_g \in \mathrm{Aut}(L)$. Since the physically interesting kinematical groups are *topological* groups with a physically relevant topology, we are looking for *continuous* realizations of arbitrary Lie groups by automorphisms of orthomodular lattices. In general, this is an unsolved mathematical problem; in what follows we therefore consider only a class of topologically well-behaved orthomodular lattices: the projection lattices of a W*-algebra. This class is very large and includes all known physically relevant systems.

5.4 W*-LOGIC FOR CHEMISTRY

Projection lattices of W-algebras*

There exist quite pathological orthomodular lattices so that it seems likely that the general orthomodular temporal logic overgeneralizes the logical systems useful in physics and chemistry. We therefore restrict our further discussion to logics whose lattice of propositions is given by the projection lattice of a W*-algebra. Recall that a *W*-algebra* A is an abstract C*-algebra which is the dual space of some Banach space, called the *predual* A_* of A. This special class of orthomodular logics will be called *W*-logics*.

In the following A denotes a *countably decomposable W*-algebra*, i.e every family of mutually orthogonal non-zero projections in A is at most countable. This is for example the case if A has a *separable* predual A_*, or if A is a weakly closed self-adjoint algebra of operators acting on a *separable* Hilbert space. The set of all projections in A will be denoted by $P(A)$,

$$P(A) \overset{\text{def}}{=} \{P \mid P \in A \ , \ P=P^*=P^2\} \qquad .$$

For every countably decomposable W*-algebra A the set $P(A)$ is a separable, complete, orthomodular and semimodular lattice. We call $P(A)$ the *projection lattice* of the W*-algebra A, and a temporal logic whose lattice of propositions equals $P(A)$ is said to be a W*-logic.

From a conceptual point of view all fundamental notions of a W*-logic have to be related directly to the lattice $P(A)$. We consider the embedding of the lattice $P(A)$ into the algebra A merely as a most convenient *technical* artifice of no conceptual import. Unfortunately, there is as yet no purely lattice-theoretical characterization of the projection lattices of W*-algebras so that we cannot avoid to characterize the basic mathematical structure $P(A)$ of W*-logic via the W*-algebra A.

W*-algebras have a rich topological structure which is of crucial importance to W*-logic. Firstly, every W*-algebra is a C*-algebra with its norm topology defined by the norm $\|\cdot\|$ on the C*-algebra A. Secondly, a W*-algebra A is the dual space of a Banach space A_* which induces the weak *-topology $\sigma(A,A_*)$ on A, called the *σ-topology* on A. The set of all norm-continuous linear functionals over A is called the *dual* A^* of A, while the set of all σ-weakly continuous linear functionals over A equals the predual A_*. Clearly we have $A_* \subseteq A^*$, and $A_* = A^*$ if and only if the algebra A is finite-dimensional. Both the norm topology and the σ-topology are of great conceptual significance for temporal logic.

The representation of properties by projections

We consider a temporal logic L whose lattice of propositions is given by the projection lattice $P(A)$ of a W*-algebra A, $L=P(A)$. Accordingly, the propositions in L are projection operators which are in a natural way embedded in the richer algebraic and topological structure of A. This is a technical advantage since some of the lattice-theoretical relations can be written in a simpler way as algebraic relations.

For every $F \in P(A)$, the *orthocomplement* $F^\perp \in P(A)$ is given by

$$F^\perp = 1 - F \quad ,$$

where 1 denotes the unit in A and equals the maximal element in the lattice $P(A)$. The partial ordering in $P(A)$ is characterized algebraically by

$$F \le G \quad \text{iff} \quad FG = F \quad , \quad F, G \in P(A) \quad .$$

Two propositions F,G in $P(A)$ are *compatible* if and only if they commute,

$$F \leftrightarrow G \quad \text{iff} \quad FG = GF \quad .$$

If F and G are compatible, the lattice operations have simple algebraic expressions:

$$\text{if} \quad F \leftrightarrow G \quad \text{then} \quad F \wedge G = FG \quad ,$$

$$\text{if} \quad F \leftrightarrow G \quad \text{then} \quad F \vee G = F + G - FG \quad .$$

Let $\|\cdot\|$ be the norm of the W*-algebra A, then the distance between two propositions $F, G \in P(A)$ can be defined by $\|F-G\|$, and it can be shown that

$$\|F-G\| \le 1 \quad \text{for every} \quad F, G \in P(A) \quad .$$

For $F \ne G$, the relation $\|F-G\| < 1$ implies that F and G are incompatible. Moreover, Maeda (1977) has shown that

$$\|F-G\| < 1 \quad \text{iff} \quad F \wedge G^\perp = F^\perp \wedge G = 0 \quad .$$

Since two propositions F,G are maximally incompatible if and only if $F \wedge G = F \wedge G^\perp = F^\perp \wedge G = F^\perp \wedge G^\perp = 0$ (compare sect.5.3), it follows that *two propositions* F,G *are maximally incompatible if and only if* $\|F-G\| < 1$ *and* $\|F-G^\perp\| < 1$.

Araki et al. (1971) have shown that two projections F,G with $\|F-G\|<1$ are not only unitarily equivalent but that they can even be continuously deformed into each other on a projection-valued norm-continuous path $\lambda\rightarrow P(\lambda)\in\mathcal{P}(\mathcal{A})$, $0\leq\lambda\leq 1$, such that $P(0)=F$ and $P(1)=G$. That is, there exists a self-adjoint operator $S\in\mathcal{A}$ such that

$$P(\lambda) = \exp(i\lambda S)F\exp(-i\lambda S) \quad , \quad 0\leq\lambda\leq 1 \quad .$$

This result implies that it makes sense to speak of *neighboring incompatible properties* and that the norm of the difference of their propositions is a measure of their closeness. We can therefore extend the interpretation of the projections in $\mathcal{A}$ as propositions about potential properties of the system by a continuity argument (Primas, 1975): we say that two maximally incompatible properties f,g are similar if the angle $\varphi(F,G)$ between the corresponding projections $F,G\in\mathcal{P}(\mathcal{A})$ is small, where

$$\sin\varphi(F,G) \stackrel{\text{def}}{=} \|F-G\| \quad .$$

With this, the norm topology acquires a conceptual meaning: it is that topology which controls the likeness of properties. The norm closure of the algebra generated by the elements of the temporal logic $\mathcal{P}(\mathcal{A})$ equals the W*-algebra $\mathcal{A}$ and is traditionally called the *algebra of observables*.

Observables in W-systems*

Every family of mutually compatible properties can be put together into a *feature*. Every feature can be represented by a *spectral measure*, say $E:\Sigma_{\mathbb{R}}\rightarrow\mathcal{P}(\mathcal{A})$, where $\Sigma_{\mathbb{R}}$ is the σ-algebra of the Borel sets of the real axis $\mathbb{R}$. Instead of the spectral family $\{E(B)|B\Sigma_{\mathbb{R}}\}$ one can introduce a particular observable A related to this feature by *spectral synthesis*

$$A = \int_{\mathbb{R}} a(\lambda)E(d\lambda) \quad ,$$

where the integral is a Radon-Stieltjes integral with respect to the strong topology on $\mathcal{A}$. The real-valued Borel function $a:\mathbb{R}\rightarrow\mathbb{R}$ gives a valuation of the feature. If a is an essentially bounded function then A is a self-adjoint element of $\mathcal{A}$ and is called a *bounded observable*. Unbounded observables (which are said to be *affiliated* to $\mathcal{A}$) can be introduced as unbounded functions of bounded observables in $\mathcal{A}$. Conversely, for every self-adjoint element in $\mathcal{A}$ there exists a unique spectral resolution with a spectral measure $E:\Sigma_{\mathbb{R}}\rightarrow\mathcal{P}(\mathcal{A})$ (compare Sakai, 1971, p.26). Accordingly, there is a one-to-one correspondence between valuated features and observables.

Remark: Observables in C- and in W*-systems*

The reader probably has noticed that quantum theory is cursed with the worst possible terminology, and so he will not be surprised to learn that an algebra of observables is conceptually something very different in the theory of C*- and of W*-systems.

In the theory of W*-systems, the algebra of observables is the smallest W*-algebra containing all the propositions of its temporal logic. The interpretation of observables of W*-systems is simply taken over from the basic interpretation of projections as propositions about properties.

This interpretation of observables as features is not possible in C*-systems which are not W*-systems, since in general a C*-algebra has far too few projections. For example, if we consider pioneer quantum mechanics as a W*-system, its W*-algebra of observables is given by the algebra $\mathcal{B}(\mathcal{H})$ of all bounded operators acting on the separable Hilbert space $\mathcal{H}$ of state vectors. If we consider the very same pioneer quantum mechanics as a C*-system, then the so-called C*-algebra of observables is given by the C*-algebra of the compact operators on $\mathcal{H}$. For convenience, we enlarge this C*-algebra to a unital C*-algebra by adjoining the identity $\mathbb{1}$. This C*-algebra is distinguished by the fact that every one of its irreducible *-representations is equivalent to the given one (Rosenberg, 1953). Hence the weak closure of every representation of this C*-algebra on a Hilbert space equals the algebra $\mathcal{B}(\mathcal{H})$, the algebra of observables in the sense of the theory of W*-systems. However, this example is not typical. In general, a C*-algebra (even if it is simple) admits very many inequivalent representations whose weak closures realize inequivalent W*-systems.

Experts in C*-algebraic quantum mechanics carefully avoid to give any conceptual characterization of the C*-algebras they use so that the following account should be taken with caution. Conceptually important is some *kinematical group* which describes our relation to the system. In the framework of C*-algebraic quantum mechanics one has to use a *minimal* and simple C*-algebra, defined as the smallest C*-algebra containing the set of unitary operators representing the kinematical group. This minimal C*-algebra generates (via GNS-constructions) all feasible W*-algebras of observables that are in harmony with the kinematical symmetry described by the unitary operators of the minimal C*-algebra. However, this minimal C*-algebra does not represent features or properties (it cannot, since it does not contain enough projections!). The attribution of properties to a system always amounts to the admission of ignorance on our part. As a rule, there are many possibilities, hence many feasible viewpoints. The fact that (as a rule) a minimal C*-algebra is not uniquely embeddable in a W*-algebra of observables (in the sense of the theory of W*-systems) corresponds to the fact that there is a multiplicity of viewpoints. Every W*-embedding (say by a GNS-construction via some reference state) reflects a particular point of view. It is remarkable that the generating minimal C*-algebra can be recovered from every particular W*-embedding. This implies that every particular viewpoint (represented by a particular W*-algebra) contains enough information to adopt a different viewpoint which is compatible with the same basic kinematical structure.

In order to avoid conceptual confusion, we shall call the minimal C*-algebra the *kinematical algebra*. In our terminology, the *algebra of observables* is a W*-algebra, it describes the properties and the features of the system which can be studied from a chosen point of view. It may be obtained by a particular W*-embedding of the kinematical algebra.

The kinematical algebra of a mechanical system over some symplectic phase space is given by the minimal Weyl-C*-algebra in the sense of Manuceau (1968) and Slawny (1972). The corresponding kinematical group is the Weyl group over the symplectic phase space, consisting of unitary operators fulfilling the canonical commutation relations.

In every ontic state τ in which a feature is truth-definite, every observable $A \in \mathcal{A}$ related to the spectral measure $E:\Sigma_{\mathbb{R}} \to \mathcal{P}(\mathcal{A})$ of this feature by $A=\int_{\mathbb{R}} a(\lambda)E(d\lambda)$ *has* the value $\tau(A) \overset{\text{def}}{=} \int_{\mathbb{R}} a(\lambda)\mu(d\lambda)$, where $\mu(B)=\tau\{E(B)\}$ for every Borel set $B \in \Sigma_{\mathbb{R}}$.

If a feature belongs to the classical part of the system, i.e. if $E(B) \in \mathcal{Z}\{\mathcal{P}(\mathcal{A})\}$, the associated observables are called *classical observables*. Bounded classical observables are self-adjoint elements of the center $\mathcal{Z}(\mathcal{A})$ of the algebra $\mathcal{A}$ of observables,

$$\mathcal{Z}(\mathcal{A}) \overset{\text{def}}{=} \{Z \,|\, Z \in \mathcal{A} \quad , \quad ZA=AZ \quad \text{for all} \quad A \in \mathcal{A}\} \quad .$$

The center of a W*-algebra is a commutative W*-algebra, and represents a classical sub-W*-system. A center is called trivial if it is just the complex numbers; a W*-algebra with trivial center is called a *factor*. A system whose algebra of observables is a factor has no classical observable and is called a *purely quantal W*-system*. If the algebra $\mathcal{A}$ is commutative, we have $\mathcal{Z}(\mathcal{A})=\mathcal{A}$, hence all observables are classical so that we speak of a *classical W*-system*. Classical features are truth-definite in every ontic state, hence classical observables *have* a value in any ontic state. Since the ontic interpretation stipulates that a classical system is always in some ontic state, we can say that *every classical observable always has a definite value*. It is logically consistent to assume that classical observables in principle are measurable to an arbitrary (though finite) accuracy without thereby introducing perturbations which would invalidate predictions based on this observation.

Ontic states in W-systems*

Recall that an ontic state τ_t is a maximal partial truth-function on the temporal logic. The domain $\mathcal{D}_t$ of τ_t is the smallest orthomodular lattice containing all propositions which correspond to properties actualized at time t. It is an important problem whether or not this concept of an ontic state can be extended to features, observables, or even to the W*-algebra $\mathcal{A}$ of all observables.

Unfortunately, the mathematicians have plagiarized the state concept of physics and presented it as a purely mathematical notion in the theory of *-algebras. According to the mathematicians, a state on a C*-algebra A simply is a normalized positive linear functional on the algebra A.

Memento: States on W-algebras*

A *state* ρ (in the sense of the mathematicians) on a W*-algebra A is an element of the dual A^* which is positive (i.e. $\rho(A^*A)\geq 0$ for every $A\in A$) and normalized (i.e. $\rho(1)=1$). The set of all states is convex, its extremal points are called *pure states*. A state ρ is said to be *normal* if it belongs to the predual A_* of A. Normal states on A are countably additive on the projection lattice $P(A)$ of A. A state ρ is called *singular* if for each $F\in P(A)$ with $\rho(F)>0$ there exists a $G\in P(A)$, with $G\neq 0$, $G\leq F$ and $\rho(G)=0$. Every state $\rho\in A^*$ has a unique decomposition into a normal state $\rho_n\in A_*$ and a singular state ρ_s, $\rho=c\rho_n+(1-c)\rho_s$, $0\leq c\leq 1$. Pure states are always either normal or singular.

A projection $F\in P(A)$ is called *truth-definite* relative to a state $\rho\in A^*$ if $\rho(F)=1$. If F is truth-definite relative to ρ, then we have $\rho(AF)=\rho(FA)=\rho(A)\rho(F)$ for every $A\in A$.

A state $\rho\in A^*$ is called *dispersion-free* on some W*-subalgebra M of A if $\rho(M^2)=\rho(M)^2$ for all $M=M^*\in M$. A dispersion-free state on $M\subseteq A$ is a pure state on M, and has a (not necessarily unique) extension to a pure state on A.

A state $\rho\in A^*$ is said to be a *factor state* if the weak closure of the GNS-representation $\pi_\rho(A)$ is a factor. For every factor state $\rho\in A^*$ we have $\omega(AZ)=\omega(A)\omega(Z)$ for every $Z\in Z(A)$ and every $A\in A$. Every pure state is a factor state. Factor states having different values for some elements of the center are called *disjoint*.

This terminology is so firmly established that we have to live with it despite the confusion it generates. The identification of physical states with mathematical states is premature since it is one of the important conceptual problems in the foundation of generalized quantum theories whether every ontic state can be extended uniquely to a pure state on the algebra of observables, and conversely, whether every pure state on the algebra of observables defines a unique ontic state. A priori, it is by no means clear that every state has an extension to a *linear* functional on A, and whether such an extension is unique.

For the special case of *normal* ontic states these questions have been settled by Raggio (1981). Every normal ontic state on $P(A)$ has a *unique* extension to a *normal pure state* (in the sense of the mathematicians) on the algebra of observables. On the other hand, every normal pure state on the W*-algebra of observables defines a unique ontic state on $P(A)$ (i.e. a σ-homomorphism from a maximal truth-definite σ-complete orthomodular sublattice of $P(A)$ onto the Boolean algebra B_2 of the two truth values).

It seems to be likely that similar conclusions can be drawn for singular ontic states. It is known that every ontic state on $P(A)$ has extensions to *linear* states on A but the conditions for uniqueness are not yet settled. In the case of Boolean temporal logics (i.e. for commutative W*-algebras), Raggio (1981) has shown that the extension of *every* ontic state to a pure state on the algebra of observables is unique.

Ontic states of temporal W*-logics are always represented by pure states on the W*-algebra A of observables, they refer to the actualized properties the system *has*. If $\rho \in A^*$ is a pure state and $F \in P(A)$ a proposition about the potential property f, then the relation $\rho(F)=1$ implies that in the ontic state ρ the system *has* the property f. If the ontic state ρ is dispersion-free for some observable $A \in A$, $\rho(A^2)=\rho(A)^2$, then we say that this observable A *has* the value $\rho(A)$ in the state ρ. Since every ontic state is pure, it is a factor state, hence dispersion-free on the center $Z(A)$. That is, every classical observable *has* at every instant a definite value.

The ontic states of a W*-system have a conceptually important continuity characteristic. Two properties f,g are similar if the norm-distance $\|F-G\|$ between the corresponding propositions F,G is small. Since states are norm-continuous, it is possible to attribute a certain degree of "f-ness" to an ontic state $\rho \in A^*$ for which $\rho(G)=1$ and $\|F-G\|$ is very small. For an arbitrary proposition $F \in P(A)$, we can consider the number $\rho(F)$, $0 \leq \rho(F) \leq 1$, as representing the grade of "f-ness" the system has in the ontic state ρ. The nearer the value of $\rho(F)$ is to unity, the higher is the grade of "f-ness" the system has; $\rho(F)=1$ implying that the system has the property f in the strict sense. Note that this concept of grade of "f-ness" is of *nonstatistical nature*, it fits precisely Zadeh's (1965) theory of fuzzy sets. Fuzziness has to be distinguished from randomness. Randomness involves an uncertainty about membership or nonmembership of an event in a Boolean yes-no classification. Fuzziness deals with the case where the object itself has an intrinsic indefiniteness, not allowing a Boolean yes-no classification.

Epistemic states in W-systems*

There is a sophisticated duality between observables and states. If we choose properties or features as primitive concepts, then the state space is defined as the topological dual A^* of the algebra A of observables. In the adopted ontic interpretation, A represents the features of the system while A^* carries the ontic states. If we turn to an operational view, we start from some space E of experimentally accessible states, called the *epistemic states*. The features are then inferred from the epistemic states so that from an operational view the feature space equals the topological dual E^* of the state space E of epistemic states. In the theory of W*-systems the feature space is a W*-algebra A which is the dual of the Banach space E of epistemic states, $A=E^*$, and we write for convenience $E=A_*$. Hence *ontic states are arbitrary pure states* in the dual A^*, while *epistemic states are arbitrary normal states* in the predual A_*. In general, the dual A^* and the predual A_* do not coincide; we have $A_* \subseteq A^*$, and $A_*=A^*$ if and only if the W*-algebra A is finite-dimensional (i.e. isomorphic to a matrix algebra).

While the norm-topology of A determines the proximity of properties, the operationally relevant topology is the σ-topology of A. This situation is well-known from classical mechanics and Kolmogorov's probability theory (for a review, compare Primas, 1980, chapters 2 and 3). The results of any experiment can be represented mathematically by a probability space (Ω,Σ,μ) consisting of a sample space Ω, a σ-algebra Σ, and a σ-additive probability measure μ on Σ. A state $\rho \in A^*$ on a W*-algebra A induces a σ-additive measure μ on every spectral measure with values in $P(A)$ if and only if ρ is normal, $\rho \in A_*$, or what is the same, if and only if ρ is continuous in the σ-topology of A.

Epistemic states refer to our *knowledge* about the true state of the system. An epistemic state represents maximal knowledge if and only if it is a pure state. If this maximal knowledge is in addition free of errors, then the corresponding pure epistemic state is an ontic state. However, as a rule our knowledge is not maximal, either because of practical reasons or because the ontic state of the system is singular, (i.e. not continuous, hence not approximable in the σ-topology).

The outcome of an experimental investigation can be summarized by specifying an epistemic state $\varphi \in A_*$. The *support* of a normal state φ is defined as the smallest true proposition S_φ which implies every proposition which is true in the state φ, i.e.

$$S_\varphi \overset{\text{def}}{=} \inf\{F \mid F \in P(A)\ ,\ \varphi(F)=1\} \quad .$$

We say that an epistemic state χ contains more information than an epistemic state φ if every proposition which is true relative to the state φ also is true relative to the state χ , that is if $S_\chi \le S_\varphi$. Equivalently, we can say that $\chi \in A_*$ contains more information than $\varphi \in A_*$ if $\chi(S_\varphi)=1$. Analogously, we say that a singular state $\rho \in A^*$ contains more information than a normal state $\varphi \in A_*$ if $\rho(S_\varphi)=1$.

Every good experiment provides some information about the ontic state of the system. Either the experiment attempts to determine the actual state without essentially disturbing the system, or the experiment attempts to prepare the system in the neighborhood of some ontic state. An *ideal* experiment may give only partial information, but information without any errors. We call such an experiment *state-determining*. The outcome of a state-determining experiment is an epistemic state $\varphi \in A_*$ such that the experimental knowledge is in harmony with the information given by the ontic state ρ, so that $\rho(S_\varphi)=1$ where S_φ is the support of the epistemic state φ. If S_φ is not an atom, then there exists a *finer experiment*, giving a more informative epistemic state, say $\chi \in A_*$ such that $S_\chi \le S_\varphi$ and $\rho(S_\chi)=1$. If the ontic state is normal, $\rho \in A_*$, then there exists a most informative experiment whose outcome determines a pure epistemic state which, in this case, coincides with the ontic state. This situation is possible if and only if the temporal logic $P(A)$ contains atoms. The paradigmatic example for such W*-systems is pioneer quantum mechanics whose algebra of observables is an atomic W*-algebra.

The pure states of atom-free W*-algebras are all singular. Examples for W*-systems with atom-free algebras of observables are classical mechanics and thermodynamic type III-W*-systems. In atom-free W*-systems every ontic state is singular, hence never coincides with an epistemic state. If an ontic state ρ is singular then it is not carried by a true proposition, $\inf\{F \mid F \in P(A), \rho(F)=1\}=0$, so that for *every* epistemic state $\varphi \in A_*$ with $\rho(S_\varphi)=1$ there exists a more informative state $\chi \in A_*$ with $\rho(S_\chi)=1$ and $S_\chi < S_\varphi$.

Singular ontic states are experimentally inaccessible. No actual experimental arrangement can determine or prepare a singular state. All we can do experimentally is to specify a distribution function that gives the probability that the system is in a given singular state. This probability can be said to be due to our ignorance, but it has to be added that we cannot eliminate this ignorance even if we can reduce it as much as we like (or can afford). Since pioneer quantum mechanics discusses only normal states (represented by state vectors in the Hilbert space of the irreducible representation of its W*-algebra of observables), the existence of singular ontic states may seem embarassing. However, it should come as no surprise since this situation is well-known in classical theories.

Example: Temperature is characterized by a singular ontic state
Consider a classical system in which the temperature is an observable and whose ontic state is characterized by the temperature $T \in \mathbb{R}^+$. In thermodynamics T is a classical observable and has at every instant a definite value, either a rational or an irrational positive number. Certainly, no experiment can determine whether the temperature is a rational or an irrational number, the best we can do is to specifiy a small but finite interval of $\mathbb{R}^+$ such that the temperature the system has lies in the interval. For every temperature measurement there exists a finer one, specifying a smaller interval in which the ontic state lies. However, there exists no finest experiment giving the true value of the temperature.

The fact that there are no pure epistemic states on some W*-algebras provides a strong argument in favor of the view that even strictly reversible and deterministic systems may require a probabilistic description (compare the example of the classical K-flow, e.g. in Primas, 1980, sect.3.6). The old question whether the probabilities of physics are irreducible or due to our ignorance is not well put. The probabilities arising in *classical* physics are due to the hidden character of the ontic states. In every atom-free W*-system we have to deal with ignorance probabilities which are irreducible. The irreducible probabilities of pioneer quantum mechanics have a different nature, they are not ignorance probabilities but arise from the non-Boolean structure of an atomic temporal logic (compare Primas and Müller-Herold, 1978, sect.IV). In both cases the epistemic states can be used as a statistical expectation functional, interpreted as relative frequencies of a fictitious Gibbsian ensemble.

Symmetries of W-systems*

Isomorphisms of W*-systems are defined by the isomorphisms of their temporal logics (compare sect.5.3). Let A and $\tilde{A}$ be the W*-algebras of observables of two W*-systems with temporal logics $P(A)$ and $P(\tilde{A})$, respectively. These two W*-systems are said to be *isomorphic* if there exists an ortholattice isomorphism from the orthomodular lattice $P(A)$ onto the orthomodular lattice $P(\tilde{A})$ (recall that an ortholattice isomorphism is a bijective mapping that preserves the partial ordering and the orthocomplementation). If A has no direct summands of type I_2, a fundamental theorem by Dye (1955) says that every ortholattice isomorphism from $P(A)$ onto $P(\tilde{A})$ is implemented by a Jordan-*-isomorphism from the W*-algebra A onto the W*-algebra $\tilde{A}$, which in turn can be represented by a direct sum of a *-isomorphism and a *-antiisomorphism from A onto $\tilde{A}$. That is, *the logical symmetries of a W*-system are given by *-isomorphisms and *-antiisomorphisms of the algebra of observables.*

Memento: Morphisms of W-algebras*

A mapping $\varphi: A_1 \to A_2$ from a W*-algebra A_1 into a W*-algebra A_2 is called a **-homomorphism* if it preserves the *algebraic* structure, that is for all $A B \in A_1$ and $\lambda \in \mathbb{C}$ we have: (i) $\varphi(\lambda A)=\lambda\varphi(A)$, (ii) $\varphi(A+B)=\varphi(A)+\varphi(B)$, (iii) $\varphi(AB)=\varphi(A)\varphi(B)$, (iv) $\varphi(A^*)=\varphi(A)^*$. If (iii) is replaced by (iii') $\varphi(AB)=\varphi(B)\varphi(A)$, we call φ an *anti-*-homomorphism*. A faithful *-homomorphism from A_1 onto A_2 is called a **-isomorphism*. A *-isomorphism from A onto A is called a **-automorphism*. A morphism $\varphi: A_1 \to A_2$ is called *isometric* if $\|\varphi(A)\|=\|A\|$ for all $A \in A_1$. A morphism $\varphi: A_1 \to A_2$ is called *normal*, if it is completely additive, i.e. if $\varphi(\sup_\alpha\{E_\alpha\})=\Sigma_\alpha\varphi(E_\alpha)$ for each orthogonal family $\{E_\alpha\}$ of projections $E_\alpha \in A_1$. Algebraic *-isomorphisms of W*-algebras exhibit rather strange strong continuity properties, in particular we have: Every *-homomorphism φ is a contraction, i.e. $\|\varphi(A)\| \le \|A\|$. Every *-isomorphism φ is isometric, i.e. $\|\varphi(A)\|=\|A\|$. Every normal *-homomorphism is σ-weakly continuous. Every *-isomorphism is normal. *Warning:* not every *-homomorphism of a W*-algebra into a W*-algebra is normal. However, the image of a normal homomorphism is a W*-algebra.

A Jordan-*-isomorphism of a W*-algebra A_1 onto a W*-algebra A_2 is a bijective mapping $\varphi: A_1 \to A_2$ satisfying (i) $\varphi(\lambda A)=\lambda\varphi(A)$, (ii) $(A+B)=\varphi(A)+\varphi(B)$, (iii) $\varphi(AB+BA)=\varphi(A)\varphi(B)+\varphi(B)\varphi(A)$, (iv) $\varphi(A^*)=\varphi(A)^*$. A Jordan-*-isomorphism of two W*-algebras is normal. Clearly, every *-isomorphism and every *-anti-isomorphism is a Jordan-*-isomorphism. A fundamental structure theorem by Kadison (1951) exhibits a Jordan-*-isomorphism as the direct sum of a *-isomorphism and a *-anti-isomorphism. In particular, every Jordan-*-isomorphism of a factor is either a *-isomorphism or a *-anti-isomorphism. In general, the decomposition of a Jordan-*-isomorphism into a direct sum of a *-isomorphism and *-anti-isomorphism is quite involved (note that the decomposition theorems given by Kruszynski, 1976, are wrong).

Theories which use Jordan-*-isomorphic W*-algebras are conceptually indistinguishable. Every W*-system can be realized in various ways by W*-isomorphic Hilbert space models. Different spatial realizations of the same theory may look quite different, may have different technical advantages and difficulties. Hence it is sensible to make good use of the possibility to choose unitarily inequivalent but W*-equivalent Hilbert-space realizations of a W*-system.

Example: Unitarily inequivalent representations of pioneer quantum mechanics

Pioneer quantum mechanics is a W*-system whose algebra A of observables is a type I factor. The traditional Schrödinger representation is irreducible, it represents A as the algebra of all bounded operators acting on some Hilbert space, in particular on the Hilbert space $L_2(\Lambda, dq)$ of Lebesgue square-integrable functions on the configuration space. A very different looking but convenient representation of A is the so-called standard representation of the factor of type I, which is highly reducible. The Hilbert space of the standard representation can be taken as

$L_2(\Lambda\times\Lambda, dpdq)$, the space of all Lebesgue square-integrable functions on the phase space $\Lambda\times\Lambda$. These two representations of pioneer quantum mechanics are unitarily inequivalent, have different technical advantages and disadvantages, but are nevertheless physically fully equivalent.

Dynamical W-systems*

The conceptually relevant time evolution is given by the time evolution of the states. A system-theoretic description requires that a state provide the minimal amount of information such that any state uniquely determines the temporally subsequent states that develop from it. This requirement leads to a *dynamical semigroup* describing the time evolution of the states. Experience shows that at the most fundamental level the dynamics is even given by a *group* of symmetries. This dynamical group can most conveniently be discussed on the algebra A of observables. An *invertible dynamical W*-system* is defined to be a pair $(A, t\rightarrow\alpha_t)$, consisting of a W*-algebra A having separable predual A_*, and a σ-weakly continuous one-parameter group $\{\alpha_t | t\in\mathbb{R}\}$ of automorphisms $\alpha_t\in \mathrm{Aut}(A)$, called the *dynamical group*. The dynamical automorphisms $\alpha_t : A\rightarrow A$ corresponds to the Heisenberg picture for the time evolution of the observables. The conceptually more relevant Schrödinger picture for the time evolution of the states is given by the preadjoint group $\eta_t : A_*\rightarrow A_*$, defined by $\rho_t(A) = \rho(A_t)$ for all $\rho\in A_*$ and $A\in A$, where $\rho_t \overset{\text{def}}{=} \eta_t(\rho)$ and $A_t \overset{\text{def}}{=} \alpha_t(A)$. Since $t\rightarrow\alpha_t$ has been postulated to be σ-weakly continuous, $t\rightarrow\eta_t$ is strongly continuous. By duality, the Schrödinger dynamics is also well-defined on the dual A^* but the time evolution of non-normal states is as a rule not blessed with nice continuity properties.

Remark: Why is the dynamics given by σ-weakly continuous groups of automorphisms?
There exist σ-weakly continuous groups $t\rightarrow\alpha_t$, $t\in\mathbb{R}$, of Jordan-*-automorphisms on W*-algebras where for each non-zero t, the mapping α_t is neither a *-automorphism nor a *-anti-automorphism (G.A.Raggio, private communication). The reason why the physical dynamics has to be a group of *-automorphisms $\alpha_t\in\mathrm{Aut}(A)$ is that the map α_t has to be completely positive for each $t\in\mathbb{R}$.

The assumption that the dynamical group $t\rightarrow\alpha_t$ is σ-weakly continuous is fulfilled for every reasonable system since it follows from the measurability of the map $t\rightarrow\rho\{\alpha_t(A)\}$ for every normal state ρ and for every $A\in A$. Moreover, $t\rightarrow\alpha_t$ is σ-weakly continuous. Any stronger continuity assumption is physically not defensible since a flow $t\rightarrow\alpha_t$ on a W*-algebra is weakly continuous if and only if it is uniformly continuous. Uniformly continuous flows have bounded generators; however, we know that most physically interesting systems have unbounded generators.

5.5 THEORY REDUCTION

The lattice of subtheories

In the ontic interpretation of W*-systems, the totality of all potential properties of a system are represented by the projection lattice of some W*-algebra A. The self-adjoint elements of A are called bounded observables, they represent the potential features of the system. Often not all potential features of a fundamental theory are really germane to a particular inquiry. The relevance of a feature is contingent on the abstraction we consider, and on how we choose to interact with the system. That is, the relevant algebra of observables is never an intrinsic property of the system but it reflects our particular point of view. The presuppositions which determine the *relevant algebra of observables* can be represented by a process of abstraction in which a number of observables of the original system are excluded. A minimal requirement for such an abstraction is that the retained observables form again a W*-algebra. Hence it is natural to relate to any W*-subalgebra A_α of A a subtheory which corresponds to the viewpoint adopted in choosing A_α as the algebra of observables.

We start with some basic theory T characterized by the W*-algebra A. A theory T_α having A_α as its algebra of observables is called a *subtheory* of T, written $T_\alpha \leq T$, if the W*-algebra A_α is a subalgebra of A, i.e.

$$T_\alpha \leq T \quad \text{iff} \quad A_\alpha \subseteq A \quad .$$

This relation $\leq$ is reflexive, antisymmetric and transitive, so that it defines a *partial order* on the set T of all subtheories of the basic theory T.

The set of all W*-subalgebras of some fixed W*-algebra A form a *lattice* where for $A_\alpha \subseteq A$ and $A_\beta \subseteq A$ the lattice operations $\vee$, $\wedge$ can be characterized as follows:

$A_\alpha \wedge A_\beta$ is defined as the largest W*-subalgebra of A which is contained both in A_α and A_β.

$A_\alpha \vee A_\beta$ is defined as the smallest W*-subalgebra of A which contains both A_α and A_β.

If we define $T_\alpha \wedge T_\beta$ and $T_\alpha \vee T_\beta$ as the subtheories whose algebras of observables are given by $A_\alpha \wedge A_\beta$ and $A_\alpha \vee A_\beta$, respectively, then the partially ordered set $(T,\leq)$ becomes a *lattice* with

$$T_\alpha \leq T_\beta \quad \text{iff} \quad T_\alpha \wedge T_\beta = T_\alpha \quad \text{iff} \quad T_\alpha \vee T_\beta = T_\beta$$

If the relation $T_\alpha \leq T_\beta$ is valid, we say that T_β *dominates* T_α, or equivalently, that T_α is a subtheory of T_β. To every pair (T_α, T_β) of subtheories $T_\alpha \leq T$, $T_\beta \leq T$, there exists a *smallest dominating theory* $T_\alpha \vee T_\beta$ and a *largest dominated theory* $T_\alpha \wedge T_\beta$,

$$T_\alpha \leq T_\alpha \vee T_\beta \quad , \quad T_\beta \leq T_\alpha \vee T_\beta \quad ,$$

$$T_\alpha \wedge T_\beta \leq T_\alpha \quad , \quad T_\alpha \wedge T_\beta \leq T_\beta \quad .$$

If $\cap$ denotes the set-theoretical intersection, and $\cup$ the set-theoretical union, then we have $A_\alpha \cap A_\beta = A_\alpha \wedge A_\beta$, and $A_\alpha \cup A_\beta \subseteq A_\alpha \vee A_\beta$ but in general $A_\alpha \cup A_\beta \neq A_\alpha \vee A_\beta$. That is, subtheories of a non-Boolean theory cannot be brought into the usual Boolean relations of inclusion, exclusion and overlap. Moreover, in a non-Boolean basic theory the hierarchically different descriptive levels are in general *not* connected by a homomorphic map (Primas, 1977). Two subtheories of one and the same basic theory may be structurally comparable or incomparable.

Comparable subtheories

Before we can give a precise definition of comparable subtheories we have to introduce the notion of a commutant of a subalgebra. For an arbitrary W*-subalgebra A_α of a W*-algebra A we define the *commutant* A_α^c of A_α (relative to A) by

$$A_\alpha^c \overset{\text{def}}{=} \{A \mid A \in A,\ AB=BA \text{ for all } B \in A_\alpha\} \quad .$$

It easily follows that A_α^c is a W*-subalgebra of A. The double commutant $(A_\alpha^c)^c$ of A_α is denoted by A_α^{cc}.

Some properties of commutants

Let A be a W*-algebra with the center Z, and let A_α, A_β be arbitrary W*-subalgebras of A. Then the following relations are easy consequences of the definition of a commutant:

(i) $A_\alpha \subseteq A_\alpha^{cc}$, $A_\alpha^c = A_\alpha^{ccc}$,

(ii) $A_\alpha \subseteq A_\beta$ implies $A_\beta^c \subseteq A_\alpha^c$ and $A_\alpha^{cc} \subseteq A_\beta^{cc}$,

(iii) $(A_\alpha^{cc} \cap A_\beta^{cc})^{cc} = A_\alpha^{cc} \cap A_\beta^{cc}$,

(iv) $A_\alpha^c \cap A_\beta^c = (A_\alpha^{cc} \cup A_\beta^{cc})^c \subseteq (A_\alpha \cup A_\beta)^c$,

(v) $(A_\alpha^{cc} \cap A_\beta^{cc})^c = (A_\alpha^c \cup A_\beta^c)^{cc} \subseteq (A_\alpha \cap A_\beta)^c$,

(vi) $A^c = Z$, $A^{cc} = A = Z^c$,

(vii) $Z \subseteq A_\alpha^c$, $Z \subseteq A_\alpha^{cc}$.

If T is the basic theory having A as its algebra of observables, and if T_α is a subtheory having $A_\alpha \subseteq A$ as its algebra of observables, then the subtheory with A_α^c as its algebra of observables will be denoted by T_α^c. The relation $T_\alpha \le T_\beta^c$ implies $T_\beta \le T_\alpha^c$, and means that all observables of the subtheory T_α commute with all observables of the subtheory T_β, and vice versa. In this case, the ontic states of the W*-system T_α cannot be entangled (in the sense of section 3.7) with the ontic states of the W*-system T_β. For that reason, we say that two subtheories T_α, T_β are *unentangled* if $T_\alpha \le T_\beta^c$ (hence $T_\beta \le T_\alpha^c$).

Unentangled theories and subtheories are two conceptually different examples of *comparable* theories but these concepts coincide in the classical case. A theory T_α is classical if and only if $T_\alpha \le T_\alpha^c$. If T_α and T_β are two *unentangled* classical theories, then the dominating theory $T_\alpha \vee T_\beta$ is also classical, so that T_α and T_β are classical subtheories of a classical theory. On the other hand, if T_α is a subtheory of a classical theory T_β, then we have $T_\alpha \le T_\beta$ and $T_\beta \le T_\beta^c$, hence $T_\alpha \le T_\beta^c$, so that T_α and T_β are classical unentangled theories.

In the general case subtheories are entangled. But even in the nonclassical case it is appropriate to call a subtheory T_α of T_β comparable with T_β since anything that is true in T_α is also true in T_β (but not the other way round). If T_α is a subtheory of T_β, we have $T_\alpha \le T_\beta$, or equivalently

$$T_\alpha = T_\alpha \wedge T_\beta \quad .$$

If T_α and T_β are unentangled, we have $T_\alpha \le T_\beta^c$, or equivalently

$$T_\alpha = T_\alpha \wedge T_\beta^c \quad .$$

Combining these two cases, we arrive at the condition

$$T_\alpha = (T_\alpha \wedge T_\beta) \vee (T_\alpha \wedge T_\beta^c) \quad .$$

Accordingly, we choose the following definition:

> Let T_α and T_β be subtheories of a basic theory T. The theory T_α is said to be *comparable* with the theory T_β, written $T_\alpha \leftarrow T_\beta$, if $T_\alpha = (T_\alpha \wedge T_\beta) \vee (T_\alpha \wedge T_\beta^c)$.

As an easy consequence of this definition, the comparability relation $\Leftarrow$ satisfies the following relations:

(i) $T_\alpha \Leftarrow T_\alpha$,

(ii) $T_\alpha \leq T_\beta$ implies $T_\alpha \Leftarrow T_\beta$,

(iii) $T_\alpha \leq T_\beta^c$ implies $T_\alpha \Leftarrow T_\beta$ and $T_\beta \Leftarrow T_\alpha$.

Note that in general the comparability relation $\Leftarrow$ is *not* symmetric, i.e. $T_\alpha \Leftarrow T_\beta$ does not imply $T_\beta \Leftarrow T_\alpha$. However, comparable classical theories are always mutually comparable: if T_α and T_β are classical $T_\alpha \leq T_\alpha^c, T_\beta \leq T_\beta^c$, then $T_\alpha \Leftarrow T_\beta$ implies $T_\beta \Leftarrow T_\alpha$.

Complementary subtheories

Let $\mathcal{T}$ be the lattice of all subtheories $T_\alpha \leq T$ of some *basic theory* T,

$$T = \bigvee_{T_\alpha \in \mathcal{T}} T_\alpha \quad .$$

The minimal element of $\mathcal{T}$ is called the *trivial theory* T_o,

$$T_o = \bigwedge_{T_\alpha \in \mathcal{T}} T_\alpha \quad ,$$

whose algebra $\mathcal{A}_o$ of observables equals the complex numbers, $\mathcal{A}_o = \mathbb{C}$.

Recall that a subtheory T_α is called comparable with T_β if $T_\alpha = (T_\alpha \wedge T_\beta) \vee (T_\alpha \wedge T_\beta^c)$. If T_α is nontrivial, $T_\alpha \neq T_o$, then this relation is violated in a maximal way if $T_\alpha \wedge T_\beta = T_o$ and $T_\alpha \wedge T_\beta^c = T_o$, so that we may say that T_α is *maximally incomparable* with T_β if $T_\alpha \wedge T_\beta = T_o$ and $T_\alpha \wedge T_\beta^c = T_o$. Note that if T_α is maximally incomparable with T_β, then T_α is also maximally incomparable with T_β^c. In the extreme case where both T_α and T_α^c are maximally incomparable with T_β (hence also with T_β^c), we call the two theories T_α and T_β complementary:

Two subtheories T_α and T_β of some basic theory T are called *complementary* if

$$T_\alpha \wedge T_\beta = T_\alpha \wedge T_\beta^c = T_\alpha^c \wedge T_\beta = T_\alpha^c \wedge T_\beta^c = T_o \quad .$$

Examples

Let the algebra $\mathcal{A}$ of the observables of the basic theory be a factor of type I_∞, and represent $\mathcal{A}$ by the bounded operators acting on the Hilbert space $\mathcal{H} = L_2(\mathbb{R},dq) \otimes L_2(\mathbb{R},dq)$, $\mathcal{A} = \mathcal{B}(\mathcal{H})$. Let (P,Q) be a Schrödinger pair acting irreducibly on $L_2(\mathbb{R},dq)$,

$$\{Q\Psi\}(q) = q\Psi(q)$$
$$\{P\Psi\}(q) = -id\Psi(q)/dq$$

for all Ψ of an appropriate domain in $L_2(\mathbb{R},dq)$. We define the following von Neumann subalgebras in $\mathcal{B}(H)$, where the double commutant with respect to $\mathcal{B}(H)$ is denoted by ":

$$A_1 \overset{\text{def}}{=} \{Q\otimes 1, P\otimes 1\}'' \quad ,$$
$$A_2 \overset{\text{def}}{=} \{1\otimes Q, 1\otimes P\}'' \quad ,$$
$$A_3 \overset{\text{def}}{=} \{Q\otimes 1+1\otimes Q \ , \ P\otimes 1+1\otimes P\}'' \quad ,$$
$$A_4 \overset{\text{def}}{=} \{Q\otimes 1-1\otimes Q \ , \ P\otimes 1-1\otimes P\}'' \quad ,$$
$$A_5 \overset{\text{def}}{=} \{Q\otimes 1, 1\otimes Q\}'' \quad ,$$
$$A_6 \overset{\text{def}}{=} \{P\otimes 1, 1\otimes P\}'' \quad ,$$
$$A_7 \overset{\text{def}}{=} \{Q\otimes 1+1\otimes Q \ , \ P\otimes 1-1\otimes P\}'' \quad ,$$
$$A_8 \overset{\text{def}}{=} \{Q\otimes 1-1\otimes Q \ , \ P\otimes 1+1\otimes P\}'' \quad .$$

The subalgebras A_1, A_2, A_3 and A_4 are factors of type I_∞, while the subalgebras A_5, A_6, A_7 and A_8 are maximal commutative W*-algebras. Clearly we have the following relations

$$A_1 = A_2^c \ , \quad A_2 = A_1^c \quad ,$$
$$A_3 = A_4^c \ , \quad A_4 = A_3^c \quad ,$$
$$A_5 = A_5^c \ , \quad A_6 = A_6^c \quad ,$$
$$A_7 = A_7^c \ , \quad A_8 = A_8^c \quad .$$

Let $\mathbb{C}$ be the trivial W*-algebra consisting of the multiples of the identity, then the two relations

$$A_j \wedge A_h = \mathbb{C} \quad \text{and} \quad A_j \vee A_h = A$$

are valid for the following pairs (j,h):

$$(j,h) = (1,2),\ (1,3),\ (1,4),\ (1,7),\ (1,8),\ (2,3),\ (2,4),\ (2,7),$$
$$(2,8),\ (3,4),\ (3,5),\ (3,6),\ (4,5),\ (4,6),\ (5,6),\ (7,8).$$

Let T be the basic theory having A as its algebra of observables and let T_0 be the trivial theory having $\mathbb{C}$ as its algebra of observables. Denote the subtheory with $A_\alpha (\alpha=1,..,8)$ as algebra of observables by T_α. Then the following relations are valid:

T_1 is comparable with T_2, and T_2 is comparable with T_1 ,
T_1 is not comparable with T_5 but T_5 is comparable with T_1 ,
T_1 is not comparable with T_6 but T_6 is comparable with T_1 ,
T_1 is complementary to T_3, to T_4, to T_7 and to T_8 .

The pairs (T_5,T_6) and (T_7,T_8) are examples for complementary classical theories. The classical theories T_5 and T_7 are examples of classical theories which are neither comparable nor complementary.

Aside: Coherent superpositions of theories

The superposition principle of quantum mechanics is closely related to the existence of complementary properties. Three properties are said to form a coherent triple if their representing projections F_1, F_2, F_3 fulfill the relations $F_1\wedge F_2=F_2\wedge F_3=F_3\wedge F_1=0$ and $F_1\vee F_2=F_2\vee F_3=F_3\vee F_1$. These relations cannot be fulfilled if any two of the three projections are compatible, hence there are no coherent triples of properties in classical theories. These concepts of quantum logic (which is represented by an orthomodular logic) can be applied in like manner to the lattice $\mathcal{T}$ of subtheories in spite of the fact that $\mathcal{T}$ is (in general) not orthomodular.

Three nontrivial subtheories T_1, T_2, T_3 of a common basic theory T are said to form a *coherent triple* if

$$T_1 \wedge T_2 = T_2 \wedge T_3 = T_3 \wedge T_1 = T_0$$

and

$$T_1 \vee T_2 = T_2 \vee T_3 = T_3 \vee T_1 \quad .$$

If the basic theory T is classical, there exists no coherent triple of subtheories of T.

The simplest example of a coherent triple of theories is generated by the Pauli (2×2)-matrices σ_1, σ_2, σ_3. Let A_j be the commutative W*-algebra generated by σ_j. Then we have $A_1 \wedge A_2 = A_2 \wedge A_3 = A_3 \wedge A_1 = \mathbb{C}$, and $A_1 \vee A_2 = A_2 \vee A_3 = A_3 \vee A_1 = A$, where A is a factor of type I_2. The three classical theories T_1, T_2, T_3 form a coherent triple of subtheories of T.

Another interesting triple of classical theories is generated by $A_1 = \{Q\}''$, $A_2 = \{P\}''$, $A_3 = \{P^2+Q^2\}$, where (P,Q) is a Schrödinger pair acting irreducibly on a Hilbert space H. Again we have $A_1 \wedge A_2 = A_2 \wedge A_3 = A_3 \wedge A_1 = \mathbb{C}$, and $A_1 \vee A_2 = A_2 \vee A_3 = A_3 \vee A_1 = A$, where $A = B(H)$ is a factor of type I_∞. The three classical theories T_1, T_2, T_3 form a coherent triple of subtheories of T.

In the previous example the triples (T_1,T_2,T_3), (T_1,T_2,T_4), (T_1,T_3,T_4), (T_2,T_3,T_4) are coherent triples consisting of three factors of type I_∞, the triples (T_1,T_2,T_7), (T_1,T_2,T_8), (T_3,T_4,T_5), (T_3,T_4,T_6) are coherent triples consisting of two factors and one classical theory, while the triples (T_1,T_7,T_8), (T_2,T_7,T_8), (T_3,T_5,T_6), (T_4,T_5,T_6) are coherent triples consisting of one factor and two classical theories.

Classification of W-subtheories*

The examples we have given should explain the concepts of comparable theories and complementary theories, but they cannot give an idea of the incredible richness of the family of all W*-theories. The classification of W*-theories is in one-to-one correspondence to the classification of W*-algebras so that every result in the theory of classification of W*-algebras has a counterpart in terms of classifications of W*-theories and their subtheories.

We begin by collecting some basic facts regarding the classification of W*-algebras. The most important algebraic characterization of an algebra refers to its center. The center $Z(A)$ of an algebra A is defined as the set of all those operators in A which commute with every operator in A,

$$Z(A) \stackrel{\text{def}}{=} \{Z \mid Z \in A \;,\; ZA = AZ \text{ for every } A \in A\} \quad .$$

The center $Z(A)$ of a W*-algebra A is a commutative W*-subalgebra of A, the corresponding subtheory is classical, it is called the *classical part* of the theory with A as the algebra of observables. If the W*-algebra A of observables is commutative, then it equals its center, $A = Z(A)$, and the corresponding theory is called a *classical theory*. A W*-algebra is called a *factor* if its center is trivial (i.e. if the center consists of the scalar multiples of the identity only), the corresponding W*-theory has no classical part and is therefore called a *pure quantum theory*. General W*-theories are mixed, they are nonclassical but have a classical part. There is a decomposition theory (cf. Sakai, 1971, sect.3.2) saying that every W*-algebra A with a separable predual A_* can be essentially uniquely expressed as a direct integral

$$A = \int_{\Lambda}^{\oplus} F_\lambda \, \mu(d\lambda)$$

of factors F_λ, such that the center $Z(A)$ of A is W*-isomorphic to the commutative W*-algebra $L_\infty(\Lambda,\mu)$. This decomposition of an arbitrary W*-algebra into factors is referred to as the *central decomposition*, it reduces the study of general W*-algebras to the study of factors and commutative W*-algebras. Hence we do not lose much generality by studying classical theories and pure quantum theories.

Examples of central decompositions

The simplest (and perhaps most important) example of a mixed W*-theory uses as algebra A of observables the W*-tensor product $A=F\bar{\otimes}C$, where F is a factor and C is a commutative W*-algebra. The classical part of the corresponding W*-theory is given by the center $Z(A)=\mathbb{C}\bar{\otimes}C$, where $\mathbb{C}$ denotes the trivial algebra consisting of the multiples of the identity.

Since the central direct-integral decomposition is technically rather awkward, it is important to know that the observable algebra $F\bar{\otimes}C$ has a most convenient L_∞-representation. Every commutative W*-algebra C with separable predual C_* is W*-isomorphic to the algebra $L_\infty(\Lambda,\mu)$ of complex-valued essentially bounded measurable functions on some probability space (Λ,μ). The predual of the W*-algebra $L_\infty(\Lambda,\mu)$ is given by $L_1(\Lambda,\mu)$. The W*-tensor product $A=B\bar{\otimes}C$ of a commutative W*-algebra C (isomorphic to $L_\infty(\Lambda,\mu)$) with another W*-algebra B (having a separable predual B_*) is W*-isomorphic to the algebra $L_\infty(\Lambda,\mu,B)$ of all weakly *-measurable B-valued functions on Λ, with pointwise multiplication. The predual of $L_\infty(\Lambda,\mu,B)$ is given by $L_1(\Lambda,\mu,B_*)$ (compare Sakai, 1971, sect.3.2).

Virtually all important W*-theories have W*-algebras A which can be represented by the W*-tensor product $A=F\bar{\otimes}C$, or more generally by a direct sum of such expressions, i.e. by $A=\oplus_n F_n\bar{\otimes}C_n$, where $\{F_n\}$ is a countable family of factors and $\{C_n\}$ is a family of commutative W*-algebras. However, this is not the most general situation: there exist truly global W*-algebras (with separable preduals) which cannot be expressed as a countable direct sum of nonisomorphic W*-algebras of the form $F_n\bar{\otimes}C_n$ (cf. Sakai, 1971, sect.4.5). To the best of our knowledge there are no physically relevant theories with such truly global W*-algebras of observables.

The central decomposition reduces the classification of W*-algebras to the classification of factors which is based on the properties of their projection lattices. In the projection lattice $P(A)$ of a W*-algebra A there is a natural equivalence relation: two projections F,G are called *equivalent* (written $F \sim G$) if there exists in A an operator T such that $T^*T=F$ and $TT^*=G$. A projection F is said to be *finite* if $G \sim F$ and $G \le F$ imply $G=F$. If F is not finite, it is called *infinite*. A W*-algebra with the unit 1 is called

(i) *finite* if for every $F \in P(A)$, $F \sim 1$ implies $F=1$,

(ii) *infinite* if A is not finite .

Recall that an element $F \in P(A)$ is called an *atom* if $F \neq 0$ and if for every $G \in P(A)$ the relation $G \le F$ implies either $G=0$ or $G=F$. The factors of W*-algebras are classified into three major groups, those of type I, II and III. A factor is said to be

(i) of type I if it contains an atom,

(ii) of type II if it is atomfree and if it contains some nonzero finite projection,

(iii) of type III if it does not contain any nonzero finite projection.

Clearly, the type of a factor is preserved under *-isomorphisms. A factor of type I is said to be of type I_n $(n=1,2,\ldots)$ if it is *-isomorphic to the W*-algebra of $n \times n$-matrices, and of type $I_{\aleph_0}$ (usually written as I_∞) if it is *-isomorphic to the algebra $B(H)$ of all bounded operators acting on some infinite-dimensional separable Hilbert space H. A factor of type II is said to be of type II_1 if all its projections are finite; it is called type II_∞ if it contains also infinite projections. Every type II_∞ factor can be realized as the W*-tensor product of a type II_1 factor and a type I_∞ factor.

With this, the classification of type I factors is complete. However, there are uncountably many nonisomorphic factors of type II_1, II_∞ and III whose *complete* classification is altogether out of reach.

Open questions: The role of injective W-algebras*

There are reasons to believe that the category of W*-algebras with separable preduals is much too large and not necessary for the theoretical description of nature. First of all, the known examples of physically relevant W*-systems all belong to a small subclass: the class of systems whose algebras of observables are approximately finite-dimensional W*-algebras. A W*-algebra A with a separable predual A_* is called *approximately finite-dimensional* (synonym: *hyperfinite*) when it is generated by an increasing sequence of finite-dimensional W*-subalgebras of A. The class of approximately finite-dimensional W*-algebras is closed under direct sums and direct integrals; it contains all commutative W*-algebras and all factors of type I. There exist uncountably many mutually nonisomorphic factors of type II and of type III which are not approximately finite-dimensional but none of them ever occurred in a physically relevant model. Therefore it is not unreasonable to conjecture that the W*-algebras occurring in the theory of W*-systems are approximately finite-dimensional.

Approximately finite-dimensional W*-algebras have many remarkable properties. For a von Neumann algebra $\mathcal{A}$ acting on a separable Hilbert space $\mathcal{H}$, the following properties are equivalent (Connes, 1976; Choi and Effros, 1976; Bunce and Paschke, 1978):

(i) $\mathcal{A}$ is approximately finite-dimensional,

(ii) $\mathcal{A}$ is injective,

(iii) there exists a quasi-expectation from $\mathcal{B}(\mathcal{H})$ on $\mathcal{A}$,

(iv) $\mathcal{A}$ is semidiscrete.

A von Neumann algebra $\mathcal{A}\subseteq\mathcal{B}(\mathcal{H})$ is called *injective* if there is a projection of norm one from the algebra $\mathcal{B}(\mathcal{H})$ of all bounded operators on $\mathcal{H}$ onto $\mathcal{A}$. A *quasi-expectation* from $\mathcal{B}(\mathcal{H})$ on $\mathcal{A}$ is a bounded linear projection $\varepsilon:\mathcal{B}(\mathcal{H})\to\mathcal{A}$ such that $\varepsilon(XAY)=X\varepsilon(A)Y$ for every $X,Y\in\mathcal{B}(\mathcal{H})$ and every $A\in\mathcal{A}$. A von Neumann algebra $\mathcal{A}$ is called *semidiscrete* if for arbitrary $A_n\in\mathcal{A}$ and $B_n\in\mathcal{A}'$ one has $\|\Sigma_n A_n B_n\|\le\|\Sigma_n A_n\otimes B_n\|$.

W*-algebras of observables are often generated by representations of Lie groups. For this situation the following characterization is relevant: Any representation of a connected locally compact separable group in a Hilbert space generates an approximately finite-dimensional von Neumann algebra (Connes, 1976).

The complete classification of approximately finite-dimensional W*-algebras is essentially known. Murray and von Neumann (1936, 1943) showed that there is (up to *-isomorphisms) exactly one approximately finite-dimensional factor of type II_1, and forty years later Connes (1976) proved that there is exactly one approximately finite-dimensional factor of type II_∞. The Tomita-Takesaki theory of modular automorphisms (reviewed in Takesaki, 1970) achieves a major breakthrough in the classification of type III. In this thesis Connes (1973) introduced a one-parameter classification of type III factors into factors of type III_λ with $0\le\lambda\le1$, and showed later (Connes, 1976) that there is (up to isomorphisms) only one approximately finite-dimensional factor of type III_λ for each λ with $0<\lambda<1$. Approximately finite-dimensional factors of type III_0 are classified exactly in terms of so-called Krieger factors. In case III_1, the only known approximately finite-dimensional factor is the factor R_∞ of Araki and Woods (1968) which is the relevant algebra of observables for an infinite free boson gas above the critical temperature (Araki and Woods, 1963). The question of the unicity of the approximately finite-dimensional factor of type III_1 is of great physical interest, however, this important mathematical problem is still open. Note that Golodec's (1977) proof of unicity is wrong! (compare Golodec's withdrawal added in the English translation by the American Mathematical Society).

The richness of the category of W-theories*

W*-theories having nonisomorphic W*-algebras are called *inequivalent*; they represent inequivalent viewpoints. The explanatory power and the general intellectual appeal of the category of W*-theories depends to a considerable extent on its capacity to describe inequivalent theories in a uniform way. If we start with some basic theory T, we may derive from T subtheories $T_\alpha, T_\beta, \ldots$ showing entirely different aspects. We say a theory T_α can be *reduced* to a more fundamental theory T_β if there exists a basic theory T such that T_α and T_β are subtheories of T, $T_\alpha\le T, T_\beta\le T$, and such that T_α is a subtheory of T_β, $T_\alpha\le T_\beta$. (It is assumed that the dynamics in T_α and in T_β is induced by the dynamics of the basic theory T).

Comment: Theory reduction and dynamics

The definition of theory reduction is essentially a kinematical condition. The dynamics is an intrinsic property of a system, it does not depend on the chosen point of view. A particular algebra of observables is determined by the adopted viewpoint so that there is no reason to expect that the dynamics of the system is given by a group of automorphisms. In general one chooses the basic theory T in such a way that it represents an autonomous W*-system. Nevertheless, the W*-system corresponding to some subtheory $T_\alpha \leq T$ will in general not be autonomous but will exhibit phenomena which have to be interpreted as interactions with the environment. When we speak of "relaxation" or of "noisy communication channels" we refer precisely to such interactions. A subtheory is particularly interesting if the whole prehistory of the system can be absorbed into the specification of the initial state of this W*-subsystem. In this special case a (deterministic or stochastic) system-theoretic description of the subsystem is possible.

Note that the concepts "atom", "pure state", "mixed state" always refer to a particular theory; these characteristics may change if we go to a subtheory. Whether a property is atomic, whether a state is ontic, and to what degree an epistemic state is mixed - all this depends on the chosen point of view. A more fundamental theory is not necessarily a truer theory than any of its subtheories for the subtheories may give a totally different description of the world. If the basic theory is non-Boolean, then its subtheories contain genuinely new phenomena which have no counterparts on the level of the basic theory. As a rule, a more fundamental theory does not provide any words for describing the things we see in a subtheory. In order to understand and give a meaning to what happens in the subtheories, it is necessary to create a new language, corresponding to the autochthonous languages of the various empirical sciences (for an example, compare sect.6.4). Each subtheory has its own language constituting the essence of what we can discuss within the chosen viewpoint.

The universal enveloping theory

All phenomenological and many fundamental theories contain parameters which should in principle be considered as particular values of classical observables. Examples for such parameters are the force constants or the chemical shift in phenomenological molecular Hamiltonians, the temperature in many macroscopic theories, or the mass and the electric charge in the Hamiltonians used in quantum chemistry. One may argue that a truly fundamental theory should not contain any parameters, and that classical observables arise always by a restriction of the universe of discourse. If we accept this as a plausible argument, then it is reasonable to require that the most fundamental theory T should have no classical observables so that its W*-algebra of observables is a factor. In this case we have $T^c = T_o$, $T_o^c = T$, hence $T = T^{cc}$ and $T_o = T_o^{cc}$, where T_o is the trivial theory having the complex numbers $\mathbb{C}$ as algebras of observables.

Past experience has shown that a factor of type I_∞ is the only reasonable candidate for the algebra of observables of a truly fundamental theory. In fact, *every* W*-algebra with a separable predual is a W*-subalgebra of the unique factor of type I_∞ with a separable predual. Accordingly we do not lose any generality if we assume that the algebra A of the observables of the basic theory T for all feasible chemically relevant theories is the factor of type $I_{\aleph_0}$. This choice has the following important consequences:

(i) every approximately finite-dimensional subtheory of T arises via a projection of norm one,

(ii) every subtheory $T_\alpha \leq T$ is normal, i.e. $T_\alpha^{cc} = T_\alpha$.

Remark: Normalcy of W-algebras*

Let A be a W*-algebra with center Z. A W*-subalgebra B of A whose double commutant B^{cc} relative to A coincides with B is said to be normal in A. A W*-algebra A is called *normal* if every W*-subalgebra $B \subseteq A$ which contains Z is normal in A, i.e. A is normal if and only if $B = B^{cc}$ for every $B \subseteq A$ with $Z \subseteq B$.

The question which W*-algebras are normal has a long history (compare Kadison, 1962); the final answer is very simple: A W*-algebra with separable predual is normal if and only if it is of type I (Willig, 1974). Hence every W*-subalgebra of a type I factor with separable predual enjoys the double commutant property.

An interesting characterization of the normalcy of a commutative W*-subalgebra is due to Kadison (1962): A commutative W*-subalgebra $B \subseteq A$ is normal in A if and only if B is the intersection of maximal commutative subalgebras of A.

If the algebra of observables of the basic theory T is a factor of type I (hence normal), the commutant operation $T_\alpha \to T_\alpha^c$ is an involution satisfying $T_\alpha^{cc} = T_\alpha$, $(T_\alpha \wedge T_\beta)^c = T_\alpha^c \vee T_\beta^c$, and $(T_\alpha \vee T_\beta)^c = T_\alpha^c \wedge T_\beta^c$ for all subtheories T_α, T_β of T, so that the family $\mathcal{T}$ of all subtheories of T becomes an *involutive lattice*. In this case, every subtheory T_α is comparable with its commutant T_α^c,

$$T_\alpha \Leftarrow T_\alpha^c \quad \text{and} \quad T_\alpha^c \Leftarrow T_\alpha \qquad \text{for every } T_\alpha \in \mathcal{T} \quad .$$

If A_α is a W*-algebra of type I, II or III, we call the corresponding theory T_α a theory of type I, II or III, respectively. If A_α is a factor, we call T_α a factor theory. If the basic theory is a factor theory of type I, then the commutants T_α^c of a type I, II or III subtheory T_α are themselves type I, II or III subtheories, respectively. Type I subtheories are well-known and easy to discuss (compare the following division). Subtheories of type III play a crucial role for the description of thermodynamic systems. All known examples of physically relevant thermodynamic systems (with nonzero finite temperature) have an approximately finite-dimensional W*-algebra of type III_1 as algebra of observables. The only known case of a type II theory is the somewhat pathological case of thermodynamic systems at infinite temperature.

The success of pioneer quantum mechanics (which is a factor theory of type I) can be understood if it is regarded as the universal enveloping theory which dominates all chemically relevant theories. The drawbacks of pioneer quantum mechanics are due to a confusion of the basic theory with its subtheories.

Theory reduction within the family of type I theories

Dirac (1930, §14) introduced into quantum mechanics the postulate of the existence of complete sets of commuting observables. In a mathematically rigorous form Jauch (1960) interpreted Dirac's postulate to mean that the algebra of observables contains at least one maximal commutative algebra. We therefore say that a subtheory T_α of a factor theory T of type I represents a *Dirac system* if its W*-algebra A_α of observables contains at least one commutative subalgebra which is maximal with respect to the basic type I factor A. This condition is fulfilled if and only if the center Z_α of A_α equals the commutant A_α^c with respect to A,

$$T_\alpha \text{ is a Dirac system iff } Z_\alpha = A_\alpha^c \quad .$$

Since A_α^c is commutative, the algebra A_α is of type I, so that *every Dirac system is a type I theory*. The relation $Z_\alpha = A_\alpha^c$ implies $A_\alpha \wedge A_\alpha^c = A_\alpha^c$, hence $T_\alpha^c \leq T_\alpha$. Since T_α is classical if and only if $T_\alpha \leq T_\alpha^c$, a classical Dirac system is characterized by the condition $T_\alpha = T_\alpha^c$, that is by a maximal commutative W*-algebra. Dirac systems together with classical subtheories exemplify theories of type I.

The traditional concept of theory reduction can be described as follows. The basic theory is taken to be a classical theory T_{cl}, $T_{cl} = T_{cl}^c$, so that all subtheories of T_{cl} are again classical. If T_α, T_β

are subtheories of the classical basic theory T_{cl} with $T_\alpha \leq T_\beta$, then T_β is simply an enlargement of the theory T_α such that T_α gives a cruder description than T_β which has a richer theoretical structure and more factual properties. This situation corresponds to the traditional case of homogeneous theory reduction as discussed by Nagel (1961). Within this Boolean framework the growth of scientific knowledge can be described by the cumulative process $T_\alpha \rightarrow T_\beta \rightarrow T_\gamma \rightarrow \ldots$ with $T_\alpha \leq T_\beta \leq T_\gamma \leq \ldots \leq T_{cl}$. This process eventually will be complete when we reach the basic classical theory T_{cl}. If we adopt a non-Boolean framework, take as basic theory T a factor theory of type I, and choose T_{cl} as one of its maximal commutative subtheories, $T_{cl} = T_{cl}^c \leq T$, then the relations $T_\alpha \leq T_\beta \leq T_\gamma \leq \ldots \leq T_{cl}$ describe the cumulative growth of knowledge *within a fixed point of view* which is given by the maximal classical theory $T_{cl} \leq T$. However, in the basic factor theory T there exist many complementary Boolean frames of reference (each described by a maximal classical subtheory) so that the growth of knowledge is no longer cumulative if the frame of reference is changed.

Theory reduction within a classical dominating theory is quite trivial. A nontrivial and diametrically opposite behavior under theory reduction is displayed by the Dirac subtheories of a basic factor theory of type I. If T_α and T_β describe Dirac systems, $T_\alpha^c \leq T_\alpha \leq T$, $T_\beta^c \leq T_\beta \leq T$, such that T_α is a subtheory of T_β, $T_\alpha \leq T_\beta$, then we have $T_\beta^c \leq T_\alpha^c$ or $Z_\beta \subseteq Z_\alpha$. Since the center describes the directly perceptible features of a system, we have the following important (though at first sight counterintuitive) result: if a Dirac theory T_α can be reduced to a more fundamental Dirac theory T_β, then *the less fundamental theory* T_α *is phenomenologically richer than the dominating theory* T_β. Within the family of Dirac theories, abstraction in the theoretical domain means creation of new phenomena in the factual domain. As opposed to classical theories, the reduction of a Dirac theory T_α to a more fundamental theory T_β does not allow us to consider the dominated theory T_α as superseded by the dominating theory T_β since there are factual features in T_α which are not factual but only potential in T_β (compare also Primas, 1977; Primas und Gans, 1979).

5.6 OBJECTS IN A QUANTUM WORLD

Some preliminary remarks: What are objects?

Are there molecules? This seems to be a silly question. For a contemporary chemist, nothing is more obvious than the existence of molecules. Likewise, quantum chemistry regards the existence of molecules as being beyond any doubt and does not touch the question whether or not this assumption emerges from the principles of our most fundamental theories. Clearly, we can speak of a molecule only if we can distinguish it from its environment. It is one of the basic tenets of chemistry that molecular systems can be separated from their environments. The doctrine of the existence of molecules is so widely received and so respectable that we may have some difficulty to discuss it seriously. However, if our intention is to arrive at a proper understanding of the situation then we certainly cannot take the notion of an "object" for granted. The omnipresence of Einstein-Podolsky-Rosen correlations (cf. sect.3.7), makes it far from clear how to define an object in quantum theories, meaning something with an individuality and intrinsic properties.

As it is usually rather difficult to produce good definitions for very general concepts, we have to let examples guide us on our way. Thus we should choose the definition in such a way that planets, stones, crystals and dust particles are objects. The question whether we can find a theoretically sound definition that turns also single molecules and electrons into objects, is much more delicate. To set the stage for our discussion, we quote a dictionary definition of the word "object" as "*something of which the mind by any of its activities takes cognizance, whether a thing external in space and time or a conception formed by the mind itself*" (Webster, 1961, p.1555, 3b2). For the time being we shall accept this as a heuristically useful characterization but we shall be more precise and more technical later on.

Note that we do not restrict the notion of an "object" to concrete things (as some philosophers like Franz Brentano have proposed to do) or to spatiotemporal entities. Objects may be quite abstract individuals. We consider "object" as the most general ontological term of something having *individuality* and *properties*. In the framework of quantum theory, objects will be described by certain W*-systems but we do not assume that these systems correspond to isolated systems.

The only essential point is that objects do not lose their identity in the course of time.

Remark: Systems localized in space and time

Many of the physically interesting systems are (at least approximately) localized in space-time. If such a localized system is arbitrarily far from another localized system, the two systems may still be entangled by Einstein-Podolsky-Rosen correlations. In this case the localized systems cannot be considered as objects.

A definition of an "object" can be given in a very general framework (such as the category of W*-systems), no concept of space is necessary for this. The concepts "object" and "localization in space" are therefore independent of each other. There are localized systems which are not objects, and there are objects which are not localized in space.

The question of localization in time is conceptually much more difficult. If quantum logic is considered as a temporal logic, then time is a parameter specifying the order of occurence of states. If time is regarded as a variable, then we have a fundamental superselection rule with respect to time. We restrict our discussion to this case, hence assume that objects are strictly localized in time. We do so not because we are convinced that it is correct but because it is traditional. But it would be interesting to break the superselection rule with respect to time and to discuss coherent superpositions of states having different times. Such states seem to be unavoidable in a consistent Einstein-relativistic quantum theory.

We consider objects to be mind-independent. That is not to say that objects exist in an absolute sense, independently of any abstraction. The existence of objects in an absolute sense is a chimaera. No affront to common sense is implied but an explanation is certainly needed. The crucial point is the triviality that *nothing can be said about nature unless some abstractions have been made*. Objects exist only by virtue of abstractions. *The notion "object" is abstraction-dependent but it can be taken as being mind-independent.*

Example

According to quantum mechanics the electrons of the moon are entangled with their radiation field. If we are not willing to abstract from the quantum mechanical structure of this radiation field on the grounds that it is irrelevant for the problem under discussion, then the moon becomes entangled with the sun etc. and cannot be said to possess an individuality. So without abstracting from the quantum structure of the radiation field, the moon cannot be an object.

The nonseparability of nature

Tacit premises are the basis of mental understanding between the members of any social group. One of the most fundamental preconceptions of science is the assumption that we need not consider the whole universe at once but that in a reasonable approximation we can go ahead by compartmentalization. The view of classical science that "the analysis of nature in terms of approximately independent parts is in accordance with nature" is an example of making an unsubstantiated assertion and then treating it as an established fact. The old objection against science that it destroys the universal interdependence by separating parts which should remain united has vanished from the sight of the classical scientists.

It is a mathematical property of *classical mechanics* that the states of subsystems of a system determine the state of the whole system. This property is called *separability*. If a system does not possess this property, we call it *nonspearable*. Pioneer quantum mechanics is the first mathematically formalized theory of nonseparable systems, it exhibits *holistic effects* which can be described by the so-called Einstein-Podolsky-Rosen correlations (compare sect.3.7). Adopting the terminology introduced by Schrödinger (1935a,b; 1936), we call noninteracting systems having Einstein-Podolksy-Rosen correlations *entangled systems* (in German: "verschränkte Systeme"). Entangled systems have holistic features, they have properties which are absent from any of their subsystems.

The existence of Einstein-Podolsky-Rosen correlations implies that nature is holistic. In contrast to classical theories, pioneer quantum mechanics predicts an entanglement of every system with its environment under the influence of even extremely weak interactions. This whole-part relationship of pioneer quantum mechanics casts severe doubts on the existence of isolated systems.

The idea of the nonseparability of nature has been a shock to classical scientists. Even today some deep thinkers claim that a violation of the separability principle is impossible "*unless we are prepared to give up extremely general ideas that are all deeply engrained within our minds and that nobody has hitherto suggested dropping in this context*" (d'Espagnat, 1977). This of course, is, a matter of tradition and the adopted paradigm. Classical scientists are so accustomed to accept the separability of nature as something self-evident and so they find it hard to realize how artificial this doctrine really is. Outside the Newtonian world view, separability has hardly played a role in the development of human thought. Moreover, the compartmentalization of the analytical scientist has been condemned at all times, for example by Goethe in his famous passage of Faust: "*The parts in his hand he may hold and class / But the spiritual link is lost, alas!*".

The nonseparability implied by quantum mechanics is anything but an unsatisfactory feature: it is a triumph of theoretical science that it can grasp such an utterly counterintuitive phenomenon in terms of a conceptually sound and mathematically well-formulated theory. The Einstein-Podolsky-Rosen correlations predicted by quantum mechanics are *genuine characteristics of nature and there is no valid reason for a universal validity of the principle of separability*. This view implies a shift in paradigms and in our understanding of physical reality but it does not create any logical difficulties.

The nonseparability of nature compels us to attach much more attention to the background than is necessary in classical theories. In a holistic theory like quantum mechanics an object cannot be interpreted as a subsystem determinate in itself but has to fulfill stringent self-consistency requirements concerning the interactions with its environment. Since every object (except the whole cosmological universe which we do not consider) interacts with its environment (e.g. via the electromagnetic radiation field), we always have to discuss the embedding of the object system into this environment. In classical theories, weak interactions lead to weak correlations so that it may be permissible to neglect the influence of weak background effects. The possibility of Einstein-Podolsky-Rosen correlations in quantal systems does not permit us to draw such a conclusion: as a rule, the smallness of the energetic interactions of a subsystem with its environment does not imply that the subsystem is even in an approximately ontic state. Since the concept of an object entails individuality, and individuality necessitates the existence of ontic states, we can characterize objects as (possibly open) subsystems that are at every instant in an ontic state. Equivalently, *we define an object as a W*-subsystem that cannot have Einstein-Podolsky-Rosen correlations with its environment.*

Every description of nature depends on the division of the world into a part whose effects are to be considered, and another part whose effects can be ignored. *Objects are created by abstracting from the Einstein-Podolsky-Rosen correlations with their environments.* The ontic state of an object depends on the state of the environment in which it is embedded. The dynamics of an object is determined uniquely by a requirement of self-consistency: an object has to be a time-independent notion in a holistic theory of nature, objects have no longer an absolute status, they are created by a particular point of view. But given the viewpoint, objects have well-defined properties which are possessed independently of our knowledge of them.

In the following we give a precise definition of an object in the framework of the theory of W*-systems. It will be shown that non-classical objects exist if and only if their environment is classical. This result has far-reaching implications which will be illustrated by examples.

Factorizations of W-systems*

Before proceeding to the subject proper, we discuss the important problem under what conditions a closed system can be decomposed into two (possibly open) subsystems. Recall that a system is called *closed* if all phenomena which can influence the system can be taken into account in the initial specification. A system is called *open* if the state of the system at time $t>t_0$ is not a function of the state t_0 alone.

Given a W*-theory T describing a closed system, can we factorize T into well-defined subtheories T_1 and T_2 such that the corresponding subsystems have sufficient individuality to consider the whole system as being composed of these two subsystems? We are very interested in the case where T represents the universe of discourse, while $T_1 \leq T$ represents a subsystem of particular interest and $T_2 \leq T$ represents its environment.

The following two postulates are *minimal* requirements which have to be fulfilled by any sensible factorization of a theory T into two subtheories T_1, T_2:

(i) *Completeness*
No part of the environment of the subtheory T_1 should be neglected, so that T_1 and T_2 generate T, i.e. $T=T_1 \vee T_2$.

(ii) *Unentangledness*
Every property of the subsystem T_1 has to be compatible with every property of the environment T_2, so that T_1 and T_2 have to be unentangled theories (in the sense of sect.5.5), i.e. $T_1 \leq T_2^c$ and $T_2 \leq T_1^c$.

For technical reasons we add a third postulate demanding kinematical independence of the properties of the subsystem from the properties of the environment. For this we require the existence of a full set of experimentally independent states. A state ρ on the algebra A of the observables of T is said to be independent relative to (T_1,T_2) if the knowledge of its restriction ρ_1 to the algebra A_1 of observables of T_1 does not give any information about its restriction ρ_2 to the algebra A_2 of the observables T_2. That is we postulate:

(iii) *Kinematical independence*
For every normal state ρ_1 on A_1, and every normal state ρ_2 on A_2, there exists a normal state ρ on A which extends both ρ_1 and ρ_2, i.e. $\rho_1=\rho|_{A_1}$ and $\rho_2=\rho|_{A_2}$.

The postulates (i), (ii) and (iii) characterize the factorization (T_1,T_2) as a *tensor-product factorization* in the sense that the W*-algebra A is *-isomorphic to the W*-tensor product $A_1\bar{\otimes}A_2$ (Bures, 1968; Tomiyama, 1969). Without restricting the generality we can put $A=B\bar{\otimes}C$ where B and C are W*-algebras such that $A_2=\mathbb{C}\bar{\otimes}C$ represents the environment. The trivial W*-algebra (consisting of the multiples of the identity only) is here denoted by $\mathbb{C}$. This result says that the rule for *decomposing* a W*-system into kinematically independent and unentangled subsystems is the same as the traditional rule of quantum mechanics for *constructing* a composite system from simpler systems. While the composition rule leads to a *unique* joint system, there is no reason to expect that the tensor-product factorization should be unique.

Comment

Note that postulate (iii) does not imply postulate (ii) nor is it implied by (i) (deFacio and Taylor, 1973). However, condition (iii) implies that $A_1\wedge A_2=\mathbb{C}$, so that the greatest dominated theory $T_1\wedge T_2$ equals the trivial theory T_0, $T_1\wedge T_2=T_0$. If A is a W*-algebra of type I, then $T=T_1\vee T_2$, $T_1\leq T_C$, and $T_1\wedge T_2=T_0$ imply that A is *-isomorphic to $A_1\bar{\otimes}A_2$ (Murray and von Neumann, 1936), so that postulate (iii) is automatically fulfilled. However, the postulate (iii) cannot in general be replaced by the requirement $T_1\wedge T_2=T_0$. This situation is related to the fact that a W*-algebra with separable predual is normal if and only if it is of type I (Willig, 1974).

In the following, we adopt postulates (i), (ii) and (iii), i.e. the tensor-product factorization for technical convenience. However, it should not be forgotten that from a conceptual viewpoint postulate (iii) is not compelling for object factorizations. One should try to base object factorizations on postulates $T=T_1\vee T_2$, $T_0=T_1\wedge T_2$ and $T_1\leq T_2^c$, but this problem has not been investigated. Such a factorization is of course only more general in the cases of type II and type III theories. In particular, thermodynamic W*-systems should be investigated from this point of view.

For the discussion of a composite W*-system whose algebra A of observables is a W*-tensor product $A=B\bar{\otimes}C$ of two arbitrary W*-algebras

$\mathcal{B}$ and $\mathcal{C}$, the notion of product states is of crucial importance. Let ρ be a state (i.e. a normalized, positive linear functional) on $\mathcal{A}=\mathcal{B}\bar{\otimes}\mathcal{C}$, and define the associated *partial states* $\rho_{\mathcal{B}}\in\mathcal{B}^*$ and $\rho_{\mathcal{C}}\in\mathcal{C}^*$ by

$$\rho_{\mathcal{B}}(B) \overset{\text{def}}{=} \rho(B\otimes 1) \quad , \quad B\in\mathcal{B} \quad ,$$

and

$$\rho_{\mathcal{C}}(C) \overset{\text{def}}{=} \rho(1\otimes C) \quad , \quad C\in\mathcal{C} \quad .$$

If ρ is normal, then both $\rho_{\mathcal{B}}$ and $\rho_{\mathcal{C}}$ are normal. A state ρ on a W*-tensor product $\mathcal{B}\bar{\otimes}\mathcal{C}$ is said to be a *product state* if $\rho(B\otimes C)=\rho_{\mathcal{B}}(B)\rho_{\mathcal{C}}(C)$ for every $B\in\mathcal{B}$ and every $C\in\mathcal{C}$. Such a product state ρ will be denoted by $\rho_{\mathcal{B}}\otimes\rho_{\mathcal{C}}$. Most states on a tensor product are *not* product states. This fact is of fundamental conceptual importance and the crux of the problem of nonseparability in quantum theories.

Object factorizations

We describe the universe of discourse by a basic theory T representing a closed W*-system having an arbitrary W*-algebra $\mathcal{A}$ of observables. If $\mathcal{A}$ can be written as a W*-tensor product $\mathcal{A}=\mathcal{B}\bar{\otimes}\mathcal{C}$ of two W*-algebras $\mathcal{B}$, $\mathcal{C}$ where neither $\mathcal{B}$ nor $\mathcal{C}$ is the trivial factor $\mathbb{C}$, then we speak of a *factorization* of the basic theory T into two subtheories T_1,T_2 and write $T=T_1\bar{\otimes}T_2$. Here T_α is the subtheory having $\mathcal{A}_\alpha$ as algebra of observables, where $\mathcal{A}_1=\mathcal{B}\bar{\otimes}\mathbb{C}$ and $\mathcal{A}_2=\mathbb{C}\bar{\otimes}\mathcal{C}$. It is *not* assumed that T_α represents a closed system.

If the two subsystems of the joint theory $T_1\bar{\otimes}T_2$ are noninteracting, then the dynamics $t\to\alpha_t\in\mathrm{Aut}(\mathcal{A})$ of the joint system is given by $\alpha_t=\beta_t\otimes\gamma_t$, with $\beta_t\in\mathrm{Aut}(\mathcal{B})$ and $\gamma_t\in\mathrm{Aut}(\mathcal{C})$, so that the time evolution transforms product states into product states. If the product states are not the only pure states on $\mathcal{A}$, then the slightest interaction between the two subsystems changes the situation dramatically. In such a case an automorphic dynamics still transforms pure states into pure states but not product states into product states. That is, even if the two subsystems are initially uncorrelated, the evolution of Einstein-Podolsky-Rosen correlations is inevitable in the course of time. Such a factorization cannot be used to define an object system since the corresponding subsystems have (in general) no ontic states.

It is remarkable that there is a distinguished class of factorizations of the universe of discourse not showing any Einstein-Podolsky-

Rosen correlations *between* the subsystems. We call such factorizations *object factorizations*.

> *Definition: Object factorization*
> An object factorization of a theory T is a pair (T_1,T_2) of subtheories of T with $T_1 \vee T_2 = T$ and $T_1 \wedge T_2 = T_0$ such that the restriction of *every* ontic state of T to T_α is an ontic state of T_α, $\alpha=1,2$.

In an object factorization (T_1,T_2) we call one subsystem, say T_1 the *object system* and the other subsystem T_2 its *environment*.

If a theory allows no object factorizations at all, we call it a *holistic theory*. If every factorization of a theory is an object factorization, then we speak of a *separable theory*. Holistic and separable theories are extreme cases; if a theory is neither holistic nor separable we call it a *partially holistic theory*. A separable theory has no Einstein-Podolsky-Rosen correlations, a partially holistic theory has Einstein-Podolsky-Rosen correlations although there are still factorizations which are free from these correlations. In a holistic system, every factorization is marred by Einstein-Podolsky-Rosen correlations.

For tensor-product factorizations $T=T_1\bar{\otimes}T_2$, Raggio (1978, 1981) has proved that $T_1\bar{\otimes}T_2$ is an object factorization only if one of the subtheories T_1,T_2 is classical (i.e. has a commutative algebra of observables). Accordingly, there exist only three kinds of objects:

(i) classical objects embedded in a classical environment,
(ii) classical objects embedded in a nonclassical environment,
(iii) nonclassical objects embedded in a classical environment.

Note that there exists no nonclassical object having a nonclassical environment. If T is a theory of type I, II or III, then every object associated with T is given by a type I, II or III subtheory T_1, respectively. If we assume that the universal enveloping theory is of type I, then no object factorization of this theory can give a type II or type III object (like thermodynamics).

Remark: Open problems

Is every W*-tensor product $\mathcal{A}=\mathcal{A}_1\bar{\otimes}\mathcal{A}_2$, where $\mathcal{A}_1$ or $\mathcal{A}_2$ is a commutative W*-algebra, an object factorization? We do not know but conjecture yes. If $\mathcal{A}$ is commutative, or if all ontic states of $\mathcal{A}$ are normal, we know that the answer is yes. We know that every normal ontic state is represented uniquely by a normal pure state, and that the restriction of every pure normal state on $A_1\bar{\otimes}A_2$ to $A_1\bar{\otimes}\mathbb{C}$ or to $\mathbb{C}\bar{\otimes}A_2$ is again a pure normal state (Raggio, 1981). The situation is much more difficult for *singular* states on *noncommutative* W*-algebras. For C*-algebras, *every* pure state ρ on the injective C*-tensor product $A_1\otimes_{min}A_2$, where either A_1 or A_2 is a commutative W*-algebra, is of the form $\rho=\rho_1\otimes\rho_2$ for some pure states $\rho_1\in A_1^*$, $\rho_2\in A_2^*$ (Takesaki, 1979, p.211, theorem 4.14). But we know neither whether this theorem generalizes from the injective C*-tensor products to W*-tensor products, nor what the precise relationship is between singular ontic states and singular states.

Every W*-theory with no classical part (i.e. the algebra of observables is a *factor*) is holistic. There are extraordinary holistic theories whose algebras of observables do not allow any tensor product decomposition. Examples for such prime algebras are the factors of type I_n with n a prime number. In the molecular domain prime theories do not seem to play any important role and we will not consider them in detail. The factors of type I_∞, II_∞ and III are not prime, but the corresponding W*-theories are still holistic.

Every classical W*-theory (characterized by a *commutative* algebra of observables) is separable, hence neither holistic nor partially holistic. The simplest class of partially holistic theories is given by theories T whose classical part T_c appears by a factorization of T, $T=T_f\bar{\otimes}T_c$, so that T_f is a factor theory (whose algebra of observables is a nontrivial factor). Examples are given by the two-level hierarchical quantum systems (compare sect.6.2). A representative illustration of a type I two-level hierarchical theory $T=T_f\bar{\otimes}T_c$ is the adiabatic description of molecules in terms of an electronic part (described by T_f) and the hierarchically higher classical molecular structure (described by T_c) (compare sect.6.4).

Elementary objects

Elementarity is always defined with respect to some group, such as the kinematical group. In particular, the idea of an elementary "particle" is inherently related to the structure of space-time. In classical mechanics the oldfashioned notion of a "hard, solid, impenetrable particle" has been replaced by the idealization of a "mass point", which nowadays is defined via a *transitive* action of the Galilei (or Lorentz) group. In pioneer quantum mechanics, the so-called "elementary particles" are defined by *irreducible* projective unitary representations of the Galilei (or Lorentz) group. The measure-theoretic analogue of a transitive action is an ergodic action. In a Hilbert-space-free formulation of pioneer quantum mechanics in terms of abstract W*-systems, irreducible projective unitary representations have to be replaced by ergodic automorphic representations. We therefore adopt the proposal by Amann (1978) to define *elementary W*-systems quite generally as ergodic automorphic actions on the algebra of observables*. An object which is an elementary W*-system is called an *elementary object*.

Elementary W-systems*

More precisely, a G-elementary system is a W*-system $(A, G, g \to \alpha_g)$ consisting of a W*-algebra A, a Lie group G, and a faithful σ-weakly continuous representation $g \to \alpha_g$ of G by automorphisms $\alpha_g \in \mathrm{Aut}(A)$, $g \in G$, which acts ergodically on A. Recall that an automorphism group $\{\alpha_g | g \in G\}$ is said to act ergodically on A if the only elements $A \in A$ which satisfy $\alpha_g(A) = A$ for all $g \in G$ are the multiples of the identity.

Quantum logic is prior to mechanics. Four-dimensional space-time is just one of the ways of describing our experience. We can do so by choosing the Galilei group (or the Lorentz group) as the relevant kinematical group. In the theory of W*-systems, the Galilei group creates both Newtonian mechanics and pioneer quantum mechanics. Every *elementary Galilean system* of mechanics is characterized by a pair (m,s), $m \in \mathbb{R}$, $s \geq 0$, where the parameter m is called the *mass*, and the parameter s the *spin* of the elementary Galilean system (Bargmann, 1954; Loinger, 1962; Pauri and Prosperi, 1968). There exists a logical symmetry (given by an anti-automorphism of the algebra of observables) transforming the elementary system (m,s) into the elementary system $(-m,s)$, so that these two elementary systems are logically equivalent (that is, they cannot be distinguished in isolation). By convention, we call an elementary Galilean W*-system with positive mass a *Galileon*. For every Galileon (m,s) there exists a twin system $(-m,s)$, called an *anti-Galileon*. If the algebra of observables is chosen to be an atomfree commutative W*-algebra, then the Galileon $(m,0)$ is punctually localized in the physical 3-dimensional space and has all properties of a *Newtonian point particle*. If the algebra of observables is a factor of type I_∞, then the Galileons describe the elementary systems discussed in pioneer quantum mechanics (for example, if $m = 9{,}109 \ldots \cdot 10^{-31}$ kg and $s = \frac{1}{2}$, we speak of an electron). The Galileons of pioneer quantum mechanics and the Galileons of classical mechanics have utterly different properties so that the usual name "elementary particles" for quantum Galileons is bad and altogether misleading. Galileons are defined as substantive entities neither in classical nor in quantum mechanics; they are defined as structures which reflect in a simple way the geometrical aspects of space-time, and which help us to understand reality. This modern view is quite in accordance with the ideas of Ludwig Boltzmann (1844-1906), the brilliant defender of the atomic theory. Boltzmann never attributed to atoms an exceptional physical reality but treated them as "entities of imagination" ("Vorstellungseinheiten", compare Boltzmann, 1896).

Do elementary objects really exist?

Elementary systems such as Galileons are defined by their transformation properties under the actions of the kinematical group. We may say that elementary objects exist if we are able to isolate them in a reasonable approximation. Clearly, no system can be completely isolated from the rest of the world and we are therefore obliged to discuss the correlations and the weak long-range interactions of an elementary system with its environment. Galileons without long-range interactions seem to be feasible from a purely logical point of view but it is hard to imagine how an elementary system without long-range interactions could be subjected to a direct experimental investigation. It is a most remarkable fact that all known mechanical Galileons with nonvanishing mass interact among themselves by long-range forces. The only two long-range forces known are related to the electromagnetic and to the gravitational field. Electromagnetic and gravitational interactions can neither be turned off nor be completely screened and it is noteworthy that these two fields allow measurements to be made upon objects without the objects losing their identity. As a matter of fact, *every* measurement whatsoever uses either the electromagnetic or the gravitational field as an informational channel.

Both electromagnetic field and the gravitational field can be characterized as representations of the inhomogeneous Lorentz group corresponding to "particles" of mass zero, and spin 1 for the photons and spin 2 for the gravitons. Maxwell's theory of electromagnetism and Einstein's theory of gravitation are the essentially unique Lorentz-invariant theories of massless particles with spin 1 and 2, respectively, but there are no Lorentz-invariant theories of massless particles of spin ≥ 3, hence no further theories which yield an inverse-square-law macroscopic force (Weinberg, 1965). In the following we pass over gravitation since so little is known about this extremely weak interaction.

The electromagnetic field has two peculiar properties which are essential both for the *existence* of Galileons and for the possibility of *measurements* on Galileons with nonvanishing mass. First of all, the photon mass is *zero* which implies that the electromagnetic forces are long-range and that the electromagnetic field has a so-called infrared singularity. Secondly, the free electromagnetic field is governed by *linear* equations of motion so that its description by quantum mechanics and by stochastic classical mechanics gives in many cases identical or similar results.

The quantization of the electromagnetic field is of special character. In spite of the vector nature of the photon (spin 1), the vanishing mass of the photon implies that the electromagnetic field has only two pairs of canonical fields so that a representation by four potentials (or six field components) is redundant. The redundant components have to be eliminated by supplementary equations and gauge conditions, neither of which make sense as operator equations.

All these facts have to be considered carefully when we ask: *are there molecules?* Contemporary textbooks are far from being careful. We have to investigate whether we can isolate Galileons, atoms and molecules from their residual radiation field. This is an exceedingly difficult mathematical problem since it has to be *proved* by a stability analysis that the weak coupling of these systems with their environment (which is a system having *infinitely* many degrees of freedom!) does not change their *qualitative* behavior. From an engineering point of view, the most important property of long-wave-length radiation is that it *cannot be screened*. Photons with wave numbers $k \to 0$ find their way through any screen or container, and couple every molecular system with any other molecular system. Every molecular system is influenced by low frequency photons and we must investigate whether this interaction has a negligible effect or not.

If we can find a factorization of the entire system (including the radiation field) such that a stability analysis of the factorized objects validates the behavior expected by ignoring the environment, then we speak of a *robust object*. Nonrobust objects are *structurally unstable*, they are not adapted to their environment and may exhibit qualitatively new properties when even very weak interactions with the environment are included. Structurally unstable systems have to be stabilized by including some parts of their environment into the object system itself. If we ignore the long-range forces in the definition of elementary objects we get as a rule dynamically unstable systems. One also speaks of "bare particles" since they can be "dressed" by including an appropriate long-range part of the electromagnetic field. Such a dressing transformation can change a structurally unstable object into a robust one, and may have far-reaching consequences since the dressed object may acquire altogether new properties. If a properly dressed system represents a robust object, the corresponding factorization of the entire system is referred to as a *robust factorization*. The question whether there are molecules has therefore be rephrased as: is there a robust factorization for systems consisting of Galileons and the electromagnetic field such that molecules are robust objects?

Molecules as robust objects

A molecule is not a thing-in-itself whose ontic state is independent of the background. Every molecular object has to achieve compatibility with its environment which means that a molecular object *adapts* its state to its environment. Since this environment always contains the radiation field, *every molecular object coexists with its electrodynamic radiation field*. The interaction between charged Galileons (like electrons and nuclei) is mediated by the electromagnetic field.

The formulation of the interaction of matter and the electromagnetic radiation field is affected by ambiguities since the theory is invariant under gauge transformations. Traditional quantum chemistry works in the *Coulomb gauge* where the total electromagnetic field is split into two parts: the static Coulomb field and the transverse radiation field. The crucial point is that *the Coulomb field is assigned to the molecular object* while the transverse radiation field is regarded as environment. We will call this procedure the *Coulomb dressing* and the corresponding factorization of the universe of discourse will be referred to as the *Coulomb factorization*. In a fully quantum mechanical description using the Coulomb gauge, the longitudinal part of the magnetic vector potential is a c-number (that is, a classical observable), and only the transverse fields are quantized.

The reason for the popularity of the Coulomb factorization is the belief that the residual interactions between Coulomb-dressed Galileons and the transverse radiation field is small. However, as we all know, the smallness of an interaction Hamiltonian is not at all a measure of the actual strength of interaction. What is relevant is the perturbation matrix element divided by the energy difference between the two corresponding unperturbed eigenstates. That is, if we have almost degenerate molecular eigenstates, even a very weak interaction with the environment may lead to a significant change in the molecular system. Since the radiation field is a system with infinitely many degrees of freedom, a cavalier treatment of this problem in terms of perturbation theory is not possible. The problem goes beyond the scope of pioneer quantum mechanics and has to be discussed in terms of modern algebraic quantum mechanics. The result is that in the Coulomb factorization a molecular object is robust only if the difference between the energy of the ground state and the first excited state of the unperturbed object is larger than a certain critical value (which is fully determined by the free Hamiltonian of the object in question). If the free object system has an almost degenerate ground state then it is structurally unstable and does not represent an object which can be observed experimentally.

First example: Macroscopic objects

Zeh (1970, 1971) and Baumann (1970, 1972) have given arguments to the effect that macroscopic systems can never be considered as isolated quantum systems. They argue that macroscopic systems possess extremely dense energy spectra so that already exorbitantly weak interactions with the environment lead to an entanglement with the environment. This remark is important and sound but Zeh's and Baumann's conclusions that macroscopic systems cannot be assumed to be in definite pure states is wrong. In order to discuss such structurally unstable systems a much deeper analysis is required. The simplest case is discussed in the following example.

Second example: Chiral molecules

Pfeifer (1980) has shown that a two-fold near-degeneracy of low-lying eigenstates of Coulomb-type molecules can be turned into a strict degeneracy by the interaction with the molecular transverse radiation field. The transverse radiation field (which is described by a quasi-local algebra of observables) is here split into two mutually inequivalent non-Fock representations (soft-photon coherent-state representations). This phenomenon is due to the infrared singularity of the radiation field, and can be described by a molecular superselection rule which separates two chiral molecular states.

Depending on whether the difference between the two energy levels of the Coulomb-dressed Galileons is above or below a critical level (determined by the Coulomb-type Hamiltonian), the combined system molecule plus radiation field possesses one or two ground states which are of product form and whose field parts are generalized coherent states. If there are two ground states, then the associated superselection rule explains the existence of chiral molecules (like alanine). If there is only one true ground state, it represents an achiral molecule (like NHDT).

Inherited classical observables

There are molecules - this is not a commonplace but a highly nontrivial theoretical result. A robust molecular object is dynamically adapted to its environment, it inherits from the rest of the world some of the characteristics which distinguish inequivalent environmental states. Phenomenologically such inherited properties are represented by some parameters, which from a theoretical viewpoint are particular values of classical observables. Strictly speaking, such classical observables belong to the algebra of observables of the combined system consisting of object and environment. In an object factorization the total ontic state factorizes into an object and environment state. We can even disregard the environment *provided* we transfer the classical observables of the entire system to the object system. (Note that this is what is done in the Coulomb gauge when the static Coulomb field is attributed to the molecular object). Classical observables of the object system which emerge in this way are called *inherited classical observables*.

A nontechnical example for such a relational property is the *nationality* of a person: it is not an intrinsic property of a person but it is inherited from the environment and makes sense only in relation to other inequivalent environments. Examples for inherited molecular classical observables are the chirality of molecules, the electronic dipole moment of molecules, the temperature and the chemical potential of molecular systems. Larger molecular systems have as a rule more almost-degenerate energy eigenvalues than small systems so that there is a tendency for complex molecular systems to have more inherited classical properties than simple ones. It seems that *all* classical observables of molecular systems are inherited from the environment. *It is the wealth of inequivalent representations of the environment which makes the molecular world so rich and interesting.*

Why may the Hartree approximation be better than an allegedly exact treatment?

How can we find an appropriate object factorization of the universe of discourse? This question is difficult and has no straightforward answer. We must have a preliminary idea of what degrees of freedom of the object system could lead to a factorization $\mathcal{A}=\mathcal{A}_1\otimes\mathcal{A}_2$, where $\mathcal{A}_1$ is intended to describe the object system. Most probably the combined system represents a single system rather than two coupled systems, hence the factorization $\mathcal{A}_1\otimes\mathcal{A}_2$ cannot be expected to be an object factorization (i.e. neither $\mathcal{A}_1$ nor $\mathcal{A}_2$ is commutative). But we can *enforce* an object factorization by a Hartree-type approximation.

The Hartree method invokes the use of product states $\rho_1(t)\otimes\rho_2(t)\in(\mathcal{A}_1\otimes\mathcal{A}_2)_*$ for the states of combined and interacting systems at all times $t\geq 0$. It is well-known that this amounts replacing the original dynamical group $t \to \alpha_t \in Aut(A_1 \otimes A_2)$ by the approximate but self-consistent Hartree dynamics $t \to \alpha_t^H = \alpha_{1t} \otimes \alpha_{2t}$, where α_{jt} maps A_j into A_j but is not necessarily an one-parameter group of automorphisms (compare also sect. 6.3). In several well-investigated cases (e.g. a molecule in its radiation field, or the adiabatic description of molecules, cf. sect.6.4), the Hartree approximation is equivalent to a *semiclassical analysis*. In this case, the partial states of one (say the second) system do not reach all of $(A_2)_*$ but lie within a Choquet simplex $C_* \subset (A_2)_*$, so that the effective algebra of observables is the dual of C_*, hence a commutative W*-algebra C. With this we have found an *object factorization* $A = B \otimes C$, where B is essentially A_1, possibly enlarged by some dressing observables.

If A_2 represents a system with infinitely many degrees of freedom, the Hartree-type approach is an interesting method for the calculation of classically inequivalent environment states. If one takes very many, but only finitely many degrees of freedom into account, one can apply the usual methods of pioneer quantum mechanics. But it turns out that the problem is as a rule ill-conditioned in the sense that it is not robust if larger and larger parts of the environment are taken into account. The Hartree approach proves to be a regularization of this ill-conditioned Schrödinger equation by mimicking inequivalent representations (which rigorously exist only for infinitely many degrees of freedom) already for finitely many degrees of freedom; if we have a problem which only generalized algebraic quantum mechanics can solve, but nevertheless use pioneer quantum mechanics, then an ingenious application of a Hartree factorization may give useful results while the exact solution of pioneer quantum mechanics fails even qualitatively.

In a Hartree-type object factorization the discussion can be reduced to the object system by an elimination of the environment variables. Such a reduction has the consequence that the resulting Schrödinger equation for the state vector of the object system becomes (as a rule) *nonlinear* (compare also Davies, 1979b; Pfeifer, 1981).

If one applies the Hartree factorization to a molecular system interacting with its own radiation field, then the Hartree time evolution transforms coherent radiation states into coherent states: if the electromagnetic radiation field is initially in a coherent state, then it remains coherent for all times so that all radiation states are contained in the Choquet simplex of coherent states which is isomorphic to the Lebesgue space L_1 over the phase space of the coherent states. The algebra of the environment observables is then the commutative W*-algebra $C = (L_1)^* = L_\infty$. The radiation states we get from the Hartree approximation are normal (i.e. elements of $L_1 = (L_\infty)_*$), while the ontic states of L_∞ are singular (i.e. elements of L_∞^* but not of L_1). The situation finds its explanation in the fact that the radiation field, of the Hartree approach, is a Gaussian stochastic classical field, that fulfills the classical Maxwell equations but does not vanish when the molecular sources vanish. The normal classical states represent a probability distribution while the true ontic states are not observable. Inequivalent coherent states are represented by classical stochastic fields having mutually singular probability measures. In the Hartree approximation, the emergence of new classical observables is again due to the singular behavior of the electromagnetic radiation field of extremely long wavelength (Pfeifer, 1981).

6. REDUCTIONISM, HOLISM AND COMPLEMENTARITY

"Eine Hauptursache der Armut in den Wissenschaften ist meist eingebildeter Reichtum."
Bertolt Brecht (Leben des Galilei)

6.1 THE CONTROVERSY REDUCTIONISM VS. HOLISM

Reductionism

In spite of the obvious plurality of scientific explanations on various levels of description, there still exists a bias toward theoretical monism. Neglecting the possibility to view nature from different perspectives and ignoring the fact that the decomposition of nature into parts is not God-given, traditional reductionism treats the various theories and models as, to be sure, incompletely articulated but ultimately reducible to an all-embracing fundamental theory. In the context of biology, reductionism is the view that all phenomena of life can be ultimately reduced to the laws of physics and chemistry. There are many variants of reductionism differing in the explanation of what "reduced to" should mean; for example, it may be defined as "accounted for", "described by", or "deduced from". In its extreme form, reductionism denies that a concept is scientifically meaningful unless it is unambiguously defined in terms of fundamental physics.

Reductionism, if accepted, is usually accepted on faith and without any logical evidence or plausible reasons. Many scientists and philosophers maintain that a reduction of "higher sciences" (like biology) to "lower" ones (like physics) is in principle possible. For example, Bertrand Russell (1948, p.50) claims: *"There is no reason to suppose living matter subject to any laws other than those to which inanimate matter is subject, and considerable reason to think that everything in the behaviour of living matter is theoretically explicable in terms of physics and chemistry"*. A strong claim which would deserve a most careful discussion; however, Russell's amplification of this thesis is just meaningless chatter. Armchair philosophers (who evidently did not go to the trouble of learning some chemistry and physics) even claim that chemistry already *has* been reduced to physics (e.g. Kemeny and Oppenheim, 1956; Oppenheim and Putnam, 1958). In contrast to the unsubstantiated claims of science fiction writers it has to be stressed that *most theoretical concepts of chemistry have not yet been successfully reduced to quantum mechanics and it is an open question whether such a reduction can always be achieved.* Moreover, we know from recent work in algebraic statistical mechanics that successful reductions require often new and very sophisticated mathematical tools. Certainly, the elimination of theoretical pluralism is not sheer routine.

The reductionist's hypothesis is simple: since organisms are built of matter, every biological question must be sought in terms of the fundamental theory of matter - there is no room for autochthonous biological laws. On the other hand, vitalists, holists and emergentists deny that the laws of physics are sufficient to explain the phenomena of life. In their arguments, antireductionist biologists emphasize the inherently global properties like complexity, organization and organic wholeness. Physiologists and physicians have always been stressing that they have to acknowledge a living organism as a pre-established whole, as an "*ensemble harmonique*" , as Claude Bernard (1865, p.150) put it in his classic on medical science.

The vitalists require organizing principles that do not have a *local* character but refer to holistic properties of the organisms. However that may be, experimental biologists have often deprecated vitalism not on philosophical grounds but because such a working hypothesis has greatly hampered the progress of science (compare e.g. Bernard, 1865, p.154). While vitalism tends to restrict science, reductionism has always been a productive heuristic principle. In fact, many difficulties of explaining living systems in terms of chemistry and physics, as stressed by the classical vitalists like Driesch (1908) have disappeared one by one. Yet, as P.W. Anderson (1972) has warned us "*the reductionist hypothesis does not by any means imply a "constructionist" one: The ability to reduce everything to simple fundamental laws does not imply the ability to start from those laws and reconstruct the universe*". And as Kenneth F. Schaffner said "*the organismic point of view may well prove heuristically valuable at certain stages of biological inquiry*" and "*there may be good heuristic reasons for not attempting in all possible areas to develop physiochemical explanations of biological phenomena, and good reasons for attempting to formulate specific biological theories*".

Reduction of theories

To say that "biology can be reduced to the laws of physics and chemistry" has a dubious meaning unless it is clearly explained what we should understand by "reduction of theories" and by the "laws of physics and chemistry".

The following traditional concept of reduction of theories has been widely accepted as a working hypothesis of scientific methodology. According to this view one says that a theory T_α has been reduced to a theory T_β if one can infer T_α from T_β, or from T_β and some additional assumptions. Nagel (1961) distinguishes two types of theory reduction, the *homogenous reduction* when T_α and T_β are expressed in the same language, and the *heterogenous reduction* when the languages are different so that we have in addition translation problems. It is said that the reducing theory T_β *explains* the reduced theory T_α so that theory reduction is often considered as a special case of deductive-nomological explanation in which the explanandum T_α is the logical consequence of the explanans T_β (cf. Hempel and Oppenheim, 1948).

This traditional concept of theory reduction rests on much too simple a view about the structure of scientific theories. Nearly all of the philosophical studies of theory reduction have their origins in an analysis of *classical* physics, and, therefore, reflect the dogmas and limitations of that approach. Traditional theory reduction presupposes that subtheories can always be totally ordered. This is true for classical theories but wrong for quantum theories. The subtheories of a non-Boolean theory may be incomparable and form a directed set which cannot be totally ordered (Primas, 1977).

Since there is no doubt that at least some phenomena have to be described by quantum theory, it is not permissible to describe chemical and biological phenomena in the conceptual framework of classical physics which presupposes a Boolean propositional logic. Accordingly, theory reduction has to be discussed on the methodological level of non-Boolean theories. In the context of W*-systems, we have given in section 5.5 a precise definition of theory reduction which agrees with the defintion by Nagel, Hempel and Oppenheim in the special case of classical theories but which leads to utterly different consequences in the case of quantum theories. Since all known mathematically well-formulated nonrelativistic theories can be rephrased in terms of W*-systems, it is tempting to try the following working hypothesis: *every consistent phenomenological theory is a W*-subtheory of some universal W*-theory.* While the universal theory refers to the undivided wholeness of nature and has no subject-object separation and therewith no observable phenomena, every subtheory derived from the global universal theory reflects some specific inquiry. Of course, we do not know in what domains the W*-reductionist thesis may be successful. In any case, it has a well-defined meaning, includes all successful reductions so far, and can account for the emergence of properties of wholes not possessed by their parts (compare Primas, 1977; Primas and Gans, 1979).

Besides physics, there is history

In every physical theory, the description of natural phenomena is divided into two categories: the equations of motion and the initial conditions. It is assumed that the initial conditions can be chosen arbitrarily, or to put it differently: if we disregard cosmology, *there is no physical theory of initial conditions.*

The phenomenon of the evolution of new species makes biology a different kind of field from pyhsics. An electron has no interesting history but every cell represents a history of more than a billion years of evolution.

Even if evolution obeys the laws of physics, can physics explain evolution? Polanyi (1967) holds that no machine or machinelike feature of an organism can be represented in terms of physics. "*If all men were exterminated, this would not affect the laws of inanimate nature. But the production of machines would stop, and not until men arose again could machines be formed once more*" (Polanyi, 1968). Every machine relies for its operation on the laws of physics and chemistry, but the machine's design is a higher-order principle; in a physical description it is relegated to the innocent-looking initial conditions. In this sense *machines are irreducible to physics.*

The separation of a description into laws and initial conditions depends on the chosen point of view and is to some extent arbitrary. Biology studies life as it happens to have evolved, so that biology is much more than physics the study of the historical process of evolution.

Hence reductionism can be a sensible thesis only if the laws of physics are combined with historical facts.

Holism

In the context of biology, holism is the thesis that the parts of an organic whole exhibit patterns that they do not show outside these wholes. More specifically it is often held that even if we possess a comprehensive theory of the parts, we still could not derive from that theory a description of the whole. An organism is considered to be prior to its constituent parts in the sense that any understanding of their function presupposes an understanding of the whole organism.

An important argument put forward by holists relates to *emergent properties*. The theory of emergence assumes that there are different levels of existence such that entities on a higher level, are characterized by specific properties that do not occur on lower levels, and such that it is impossible to deduce the characteristics of a higher level from those of a lower level. Hence we have the possibility of *essential novelty* of laws in a higher level description. An *emergent law* involves the emergence of new variables describing new qualities. The question of a functional relationship of these new variables with the lower variables is controversial.

The emergent evolutionists hold that the nature of emergent characters can only be learnt by experience of their occurence (Morgan, 1923, p.5), so that we must wait till we meet an actual instance of an object of higher order before we can come to an emergent law (Broad, 1929, p.79). Hence emergent laws are considered as a priori unpredictable.

It goes without saying that such ideas are beyond the comprehension of logical positivists. Indeed, Hempel and Oppenheim (1948) maintain that this argument contains a logical mistake and that knowledge of wholes can be taken into account in defining concepts concerning their parts. While the ideas of the holists may be vague, the statement by Hempel and Oppenheim is clear but wrong. Their statement is refuted by explicit examples even in the realm of modern physics. There *are* facts which are not ultimately accounted for by the intrinsic properties of elementary objects. For example, substances may have a temperature, single molecules do not. In the framework of W*-systems, thermodynamics can be considered as a W*-subtheory of pioneer quantum mechanics, and it can be shown that a change from a molecular viewpoint to a thermodynamic viewpoint implies the emergence of new quantities such as the temperature (Primas, 1977).

Holism is often considered as the opposite of reductionism but this view must be rejected as naive. Quantum logic is a perfectly well-defined holistic theory and non-Boolean theory reduction represents a sophisticated variant of reductionism. In this framework reductionism is in harmony with holism, and the emergence of essential novelty in a higher level description is a compelling consequence of the theory. This kind of emergence is possible only in virtue of the existence of incompatible properties.

6.2 COMPLEX SYSTEMS

Von Neumann and Chaitin systems

Empirical science nourishes the idea that the value of a scientific theory is that it enables us to replace a lengthy listing of incoherent empirical facts by a deductive scheme based on a few theoretical hypotheses. Following up this idea we are led to grave problems. For one thing, in order to reconstruct empirical data from a short theoretical scheme, the computations that have to be made can be very long and complicated so that we have to spend an enormous amount of time to predict from the few axioms the empirically accessible data. For another thing, there is no reason to believe that every phenomenon has a theoretical description which is simpler or shorter than the listing of the corresponding empirical data.

In really complex systems, both dilemmas are unavoidable. Using the concept of information-theoretical complexity, Chaitin (1974) has shown that in every formal description of a complex system one has either few bits of axioms and incredibly long proofs, or short proofs and an incredibly great number of bits of axioms. Here, "incredibly great" means that these quantities increase more quickly than any computable function of the computational complexity. (For an introduction to the concept of complexity of computations, compare e.g. the text of Brainerd and Landweber, 1974).

In the context of automata theory, there is a suggestion by John von Neumann (1966, pp.47-56, p.78) saying that when an automaton is not very complicated, the description of its *function* is simpler than the description of the automaton, but that the situation is reversed with respect to very complex automata. That is, for objects above a critical complexity ("von Neumann systems", for short) it is simpler to say how they are made rather than what they do. Perhaps the most extreme case has been discussed by Chaitin (1977): *a Chaitin system is a classical system for which the computational complexity of the whole is less than the sum of its parts*. Chaitin (1977, 1979) proposed such systems as a new criterion for life.

Whether such systems play in fact a role in biology is an open question. We have mentioned here the von Neumann and the Chaitin systems only to point at an important limitation of the physico-chemical methodology which presupposes that natural phenomena can be effectively described by a comparatively short description in terms of basic building blocks (like the elementary objects associated with the space-time symmetry group). Von Neumann and Chaitin systems are classical system of ultimate complexity, they represent an organic whole whose description in terms of its parts is logically feasible but utterly absurd.

Self-referring systems

The idea that there may exist a theory having truly universal validity gives rise to nontrivial logical problems. For example, in such a theory it would be possible to have a self-describing system which can explain itself in an autonomous manner.

Self-reference can lead to contradictions. For example, the self-referential sentence "this statement is false" is true if false and false if true, and so it is both true and false, hence contradictory.

We do *not* claim that self-reference can play no role in biological or sociological systems. But in order to avoid logical circles and paradoxes of the Richard type, we do not assume that the theories we consider can also explain our own actions. Otherwise we would be forced to look at our brains as systems that produce the very theories we have about brains.

Remark

Self-reference is a deep problem, related to Gödel's incompleteness theórem. A simple but competent *and* imaginative discussion of some problems of self-reference and incompleteness can be found in the extraordinary book "Gödel, Escher, Bach" by Douglas R. Hofstadter (1979).

Complexity is not an intrinsic property

Matter admits many modes of description, all equally valid and real. The plurality of ways of seeing and describing systems is related to their complexity. As emphasized by Robert Rosen (1977), "*complexity is not an intrinsic property of systems, but rather arises from the number of ways in which we are able (or desire) to interact with a system. ... A system like a stone typically is regarded as simple, because we interact with it only in a few ways. For a geologist, who multiplies the number of ways in which he interacts with a stone, such a system can appear infinitely complex*".

Accordingly we consider complexity as associated with *descriptions* (Löfgren, 1977) and define a complex system as one for which we have at our disposal a variety of different modes of descriptions, referring to differently chosen subsystems, having a high degree of coherence and autonomy, and being engaged simultaneously in several distinguishable activities.

A simple but important class of complex systems arises when there exist subsystems which exercise control over others while being controlled by others. In this case we have a new order, called *hierarchical order*.

Hierarchical systems

Many complex systems we know from physics, chemistry, biology and sociology are hierarchically organized and have the property of dynamical near-decomposability (Simon, 1962). This hierarchical structure permits an analysis by a comparatively simple description so that the complexity of most hierarchical systems is below the critical complexity suggested by von Neumann. This may explain the fact that hierarchically organized complex systems often represent the thermodynamically preferred situation.

A hierarchical system is defined as a complex system which can be decomposed into an ascending family of successively more encompassing subsystems such that every lower level system is subordinated by an authority relation to the next higher level. Accordingly, a hierarchical system exhibits a series of "levels of organization" but a particular level does not exist independently of the entire system. In particular, a higher level cannot be understood if considered in isolation from the lower levels, it exists only in virtue of some specific lower-level mechanism. The authority relation constrains the behavior on the lower level so that a lower level system shows some coherent activity. On the other hand, all processes at a higher level act in conformity with the laws of the lower levels. Accordingly, adjacent hierarchical levels are connected by a feedback loop.

Example: Biosystems

Biological systems have always been described by a variety of hierarchical structures such as: molecules, cells, tissues, organisms, breeding populations, species, social systems, ecosystems. To describe the various activities of a biosystem, each of these levels requires its own kind of state description as an open dynamical system.

If we focus our interest on the coherent activities of a hierarchical feedback system, we can get a simple description provided we use for every level the appropriate language. As a rule, the description of a hierarchical level with a language appropriate for a lower level is possible and may be very complex and almost incomprehensible, and the relevant patterns of the higher level are not put in evidence. On every hierarchical level new entities ("collective modes") emerge as new natural variables in terms of which a simple description of the coherent activities is possible. These new variables allow a new language such that a high-level view of the system may contain explanatory power which is absent in a lower-level description. The fact that some phenomena can be explained on a high level quite easily, but not at all on a lower level, shows that a reduction to a more fundamental level may mask essential characteristics of complex systems. Since we are not interested in incomprehensible descriptions, we have to invent new viewpoints which render the description the simplest possible.

First example: Zoology

A molecular description in all its detailed complexity of honey bees would be bad, not because our computers are hopelessly too slow but because such a description would be entirely irrelevant to the understanding of, say, how honeybees signal the location of a rich source of nectar (as described by Frisch, 1965).

Second example: Chemistry

In the last hundred years, chemistry has been dominated by the idea that the description of chemical systems in terms of molecular constituents is the correct mode of analysis. Yet chemistry is not only the science of molecules, it is also the science of substances. The qualities characterizing substances and molecules are not the same. For example, in the conceptual framework of chemical thermodynamics there is no room for the notion of a molecule. It is well-known that the concept of a "pure substance" should not (and usually cannot) be given in molecular terms (compare: van der Waals, 1927, §92). To cite an instance: we know that water is a pure substance in spite of the fact that its molecular structure is still highly controversial.

Third example: Water

The established dogma of theoretical chemistry says: "*In order to interpret the properties of steam, ice, and liquid water, we must understand the water molecule*" (Eisenberg and Kauzmann, 1969, p.1). But this program does not work: "*In spite of the convincing nature of the general qualitative picture and the extensive range of physico-chemical measurements which have been made on water, including in recent years numerous spectroscopic studies, it remains unfortunately true that no satisfactory detailed molecular picture has yet emerged. The present situation is too confused to be briefly summarized, and the same statement is even more true of efforts to give a molecular interpretation of the role of the solvent in solvent-solute interactions in water and other hydrogen-bonded solvent systems*" (Bailar et al. 1973, p.137). In the latest volume of a comprehensive treatise on water, Richards (1979, p.123) summarizes the recent "advances" by the statement: "*Calculations based on a brute force approach have not contributed anything of outstanding significance ... In fact, even a system such as* $(H_2O)_3$ *has so many possible variables in terms of the position of the constituent nuclei that it is not possible to do sufficient calculations to satisfy a statistical mechanician without using grotesque amounts of computer time*". Even though the current paradigm commands it, it is pure science fiction that substances can be described in terms of molecules. The molecular language may be the wrong language for describing chemical substances.

Fourth example: Superconductivity

The transition of a metal from the normal to the superconductive state is a phase transition involving a change in the order of the system. The essentials of this phase transition are nowadays well understood in terms of so-called Cooper bound electron pairs (Cooper, 1956; Bardeen, Cooper and Schrieffer, 1957).

In a modernized version of this theory the Cooper pair is replaced by a boson-type quasiparticle which describes the collective modes responsible for the macroscopic order in superconductors (Leplae et al. 1974, Matsumoto and Umezawa, 1976). These bosons behave under Galilei transformations as elementary particles but they appear on a hierarchically higher level than the electrons of the microscopic description of the metal. The microscopic theory of superconductivity explicitly explains how the boson quasiparticles are made out of electrons. Once created, the boson quasiparticles constitute a distinct entity and in a reasonable approximation it is permissible to neglect the dynamical factors on which their existence depends. The equations which determine the macroscopic electromagnetic field, the current and the energy in a superconductor under given boundary conditions are much simpler on the boson level than the microscopic description on the hierarchically lower electron level. For example, the Josephson effect can be described simply but rigorously in terms of quasi-boson tunneling.

It is a thesis of organismic biology that the laws determining the dynamics at a given hierarchical level have nothing to say on the dynamics at a higher level. From examples of hierarchical systems we completely understand in physical terms (like superconductivity), we know that such a claim cannot be true for general hierarchical systems. But it is true that a derivation of a higher-level dynamics from a lower-level theory is a highly nontrivial business which requires sophisticated mathematical tools.

The time scales in hierarchical systems

A hierarchical order is called stable if the hierarchical relationship is preserved during the operation of the system. Sudden changes in organization and structure are among the characteristic features of hierarchical instabilities. Well-known examples of hierarchical instabilities are social revolutions and the formation of larger integrated systems. In every stable social system the government has to be conservative and to act slowly such that the subordinates can follow. This property seems to be typical for *any* hierarchical system and can be used to discriminate hierarchical levels according to their reaction time.

A higher level in a hierarchy has always a much longer reaction time than a level classified as lower. This condition is necessary in order that a lower level is controlled ("slaved", in the terminology of Haken, 1977) by the dynamical variables ("order parameters") of the next higher level, and that the lower level variables adapt themselves immediately. Accordingly, we can expect an hierarchical order only if the system possesses modes grossly differing in reaction time. In such a case, we can use asymptotic expansions and contract a low-level description by introducing a few secular variables (having a slow time variation) which establish a higher hierarchical level.

Example: Biological time scale and reductionism

In biology one distinguishes the following time scales (cf. Goodwin, 1963):

(i) the biochemical scale (a fraction of a second or less),
(ii) the metabolic scale (in the order of a minute),
(iii) the epigenetic scale (several hours),
(iv) the development scale (days or years),
(v) the evolutionary scale (thousands to millions of years).

An inveterate reductionist who likes to reduce biology to pioneer quantum mechanics should not forget that the accuracy of the molecular Hamiltonian he needs to describe biological phenomena is of the order h/T, where h is Planck's constant and T the relevant biological reaction time. In order to get this accuracy in a science-fiction-free way, he would have to include a large part of our world into his theoretical universe of discourse. That is, a discussion of the long-time behavior of dynamical systems and long-range predictions on the basis of a fundamental theory is practically impossible.

A reductionist is obliged to attempt a description of a hierarchical system in terms of a single fundamental theory. If such a description is possible then it necessarily behaves *nonuniformly* with respect to time if a parameter in the fundamental equation of motion for the entire system tends toward some limiting value of interest. A direct expansion in powers of these parameters fails badly in the domain of main interest, and one speaks of a *singular perturbation problem*. In chemistry, well-known examples of singular perturbation problems are the time-dependent Born-Oppenheimer approximation (where the ratio between electron and nuclear mass is the characteristic expansion parameter), and the steady state approximation in chemical kinetics.

The nonuniformity in a fundamental description of a hierarchical system is related to a multiscale dependence of the solution. In order to get a *uniformly valid*, hence easily interpretable asymptotic expansion to the nonuniformly behaving solution one has to introduce explicitly the *intrinsic time scales* of the various hierarchical levels.

The use of time scales which are pertinent to the physics of the problem at hand has been emphasized by Bogoliubov (1946). For example, a sensible description of a gas requires at least two time scales. Phenomena occurring during collisions have to be described by a time variable t_0 which is read on a fast clock. Phenomena occurring during the relaxation of the gas toward thermodynamic equilibrium have to be described by a time variable t_1 which is read on a slow clock.

The method of multiple time scales

The multiscale behavior of hierarchic systems can be discussed by the "method of multiple time scales" (for reviews, compare e.g. Nayfeh, 1973, chapt.6, and Smith, 1975). Consider for example a real vector differential equation of the form

$$\dot{x} = f(t,x,\varepsilon) \quad ,$$

where $x(t)\in\mathbb{R}^n$, $f\in\mathbb{R}^n$, $t\geq 0$ and $\varepsilon\in\mathbb{R}$ is assumed to be a small parameter. The method of multiple time scales makes explicit the distinction between physical time scales and assumes that the exact time dependence can be considered to be the resultant of independent time behavior on several distinct time scales. With this, the solution of the equation of motion can be found in the form

$$x = x_0(t_0,t_1,\ldots) + \varepsilon x_1(t_0,t_1,\ldots) + \ldots$$

where $t_0=t$, $t_1=\varepsilon t,\ldots$ are treated as independent variables. When the solution is obtained in terms of $t_0,t_1,\ldots$, they can be converted to expressions in terms of the physical time t by $t_n=\varepsilon^n t$. The otherwise unspecified functional dependence of x_n on $t_0,t_1,\ldots$ is given by the requirement that no secular behavior should exist.

The simplest molecular hierarchic system is a *molecule*. The usual adiabatic description of a molecule considers the electronic motions as fast while the nuclear motions (oscillations, rotations) are considered as slow. If t_0 denotes the fast time variable (describing the electronic motions), and if t_1 denotes the slow time variable (describing the nuclear motions), then we have $t_1=\varepsilon t_0$ where $\varepsilon^2=m/M$ denotes the ratio between the electron mass m and a typical nuclear mass M. The nuclear government controls the electronic rank and file by actions which can be measured with the slow clock t_1. The electronic rank and file obey at once, i.e. the electronic motion follows the motion of nuclei with a fast reaction time which can be measured using the fast clock t_0. Since there is no government without rank and file, the dynamics of the nuclear control level is not given a priori, it is created via a feedback interaction by the state of the electronic rank and file. As in any feedback system, the stability of this molecular hierarchy is not automatically warranted by the dynamical laws. A sudden change in the nuclear variables leads (as a rule) to a molecular revolution. The possibility of a breakdown of the adiabatic description of a molecule exemplifies the fact that the hierarchical structure of a system is never an *intrinsic* property of a system, it rather arises from a particular way we interact with the system. The hierarchical description of molecules is appropriate if we interact with molecules in a chemically relevant way.

Hierarchical quantum systems

The adiabatic description of a molecule gives an example of a hierarchic *quantum system*. But there is a delicate point in such an approach, since pioneer quantum mechanics is not the proper framework to discuss hierarchical quantum systems. If a higher hierarchical level is to give perspicuous and unambiguous orders, then the control variables must be *classical* variables. The fact that pioneer quantum mechanics has no classical observables implies that in pioneer quantum mechanics a description of hierarchical systems is possible only with the aid of difficult classical limits. The traditional way to handle such systems in pioneer quantum mechanics is to introduce adiabatic approximations like the Born-Oppenheimer description of molecules.

The time-honored and rather accurate way to discuss molecular spectra goes back to a classical paper by Born and Oppenheimer (1927). In this approach, one solves the eigenvalue problem for the fast electrons by holding the slow nuclear variables fixed. These eigenvalues (one for each electronic state) are then used as an effective potential for the eigenvalue problem of the slow nuclei. The Born-Oppenheimer separation of nuclear and electronic motions turned out to be the key for the interpretation of molecular spectra, the Born-Oppenheimer potential energy surfaces underlie the whole field of inealstic and reactive molecular collisions. The perturbation expansion in terms of the ratio $(m/M)^{1/4}$ of the electron mass m to the molecular mass M as perturbation parameter, as used by Born and Oppenheimer, is singular since a small parameter multiplies the Laplacian for the nuclear coordinates so that the nuclear energy is *not* a small perturbation. However, this problem can be cast into the framework of modern asymptotic perturbation theory (cf. Kato, 1966, chapt.8), and in some cases the Born-Oppenheimer method for stationary states can be rigorously justified (Combes, 1975, 1977; Combes and Seiler, 1978).

In spite of the practical success of the Born-Oppenheimer approximation in molecular spectroscopy, and in spite of its mathematical justification by the modern theory of singular perturbations, this approach is conceptually not satisfactory for chemistry. The concept of *molecular structure* (as used by chemists) is a *classical* concept, hence extraneous to pioneer quantum mechanics (for a review of the interpretative problems of molecular structure in quantum theory, compare Claverie and Diner, 1980). Since the algebra of observables of pioneer quantum mechanics has no classical observables, it remains a mystery why the Born-Oppenheimer description should lead to a notion of molecular structure which can be interpreted classically. In fact, the time-independent Born-Oppenheimer formalism misses the typical features of a hierarchical description. Clearly it is not sufficient to expand eigenvalues and eigenvectors of the molecular Schrödinger equation in terms of some small parameter like $(m/M)^{1/4}$, and to prove the existence of such an expansion. Even an asymptotic expansion of the equations of motion is not enough since the equations of motion do not yet determine a full theory. For a chemically relevant description of behavior of molecules we have to know what the new observables are in the asymptotic limit. As a rule, every asymptotic limit changes the algebra of observables (compare also sect.6.3). The Born-Oppenheimer limit of large nuclear masses is very similar to the limit of vanishing Planck's constant, hence highly singular and delicate. A careful discussion shows that in the adiabatic asymptotic limit the algebra of nuclear observables contracts to a commutative W*-algebra. With this, the proper framework for a chemical description of a molecule is a W*-system whose algebra $\mathcal{A}$ of observables is given by the tensor product $\mathcal{A}=\mathcal{B}\bar{\otimes}\mathcal{C}$, where $\mathcal{B}$ is a factor of type I_∞ describing the electronic degress of freedom, while $\mathcal{C}$ is an atomfree commutative W*-algebra describing the molecular structure and the stochastic classical motions of the nuclei (for more details, compare sect.6.4).

Example: 2-level hierarchical type I W-system*

The simplest example of a hierarchical quantum system is a W*-system $(\mathcal{A}, t\to\alpha_t)$ with two hierarchical levels. If $\mathcal{A}$ is of type I, then the W*-algebra $\mathcal{A}$ of the total system can be represented as a W*-tensor product $\mathcal{A}=\mathcal{B}\bar{\otimes}\mathcal{C}$, where the W*-algebra $\mathcal{B}$ refers to the lower level, and the W*-algebra $\mathcal{C}$ refers to the higher level. Since the control observables are as a necessity classical, the algebra $\mathcal{C}$ must possess a nontrivial center $\mathcal{Z}$. To simplify matters, we put $\mathcal{C}=\mathcal{Z}$, and choose for $\mathcal{C}$ an atomfree commutative W*-algebra. Since we have assumed that the subordinate system corresponds to the lowest hierarchical level, the W*-algebra $\mathcal{B}$ need not have a center, for the sake of simplicity we take $\mathcal{B}$ to be a factor of type I. In this case, the tensor-product representation $\mathcal{A}=\mathcal{B}\bar{\otimes}\mathcal{C}$ is an object factorization so that no Einstein-Podolsky-Rosen correlation exist between the two hierarchical levels. In spite of their mutual interaction, both the lower and the higher level have separate individuality.

For an object factorization with an ontic product state as initial state, the time evolution $t \to \alpha_t$ of $\mathcal{B} \bar{\otimes} \mathcal{C}$ can be described in terms of effective (Hartree-type) time evolutions $t \to \beta_t$ and $t \to \gamma_t$ of the two hierarchical levels $\mathcal{B}$ and $\mathcal{C}$, respectively. The time evolution $t \to \beta_t$ of the subordinate W*-system $(\mathcal{B}, t \to \beta_t)$ is controlled by the governing W*-system $(\mathcal{C}, t \to \gamma_t)$ via some classical control observables $C_n \in \mathcal{C}$. Since $\mathcal{B}$ is a factor of type I, the mapping $\beta_t : \mathcal{B} \to \mathcal{B}$ can be unitarily implemented in $\mathcal{B}$, so that for every $B_t \stackrel{\text{def}}{=} \beta_t(B)$, $B \in \mathcal{B}$, we have

$$\dot{B}_t = i[H_B(t), B_t]_- \quad ,$$

where the time-dependent Hamiltonian $H_B(t)$ is affiliated to $\mathcal{B}$, and can be written in the form

$$H_B(t) = H_B^0 + \Sigma_n c_n(t) B_n \quad \text{with} \quad B_n \in \mathcal{B} \quad ,$$

where H_B^0 is the Hamiltonian in absence of a hierarchical coupling, and $c_n(t)$ is the value of the classical observable $C_n \in \mathcal{C}$ at time t.

The feedback of the subordinate system $(\mathcal{B}, t \to \beta_t)$ to the control system $(\mathcal{C}, t \to \gamma_t)$ can be represented by an external time-dependent force acting on the classical control system. If, for example, $(\mathcal{C}, t \to \gamma_t)$ is a classical Hamiltonian system, the equations of motion for a classical observable $C_t \stackrel{\text{def}}{=} \gamma_t(C)$, $C \in \mathcal{C}$, can be written as

$$\dot{C}_t = \{C_t, H_C(t)\} \quad ,$$

where $\{,\}$ denotes the Poisson bracket, and $H_C(t)$ is the classical Hamiltonian function, which may have the form

$$H_C(t) = H_C^0 + \Sigma_n b_n(t) C_n \quad , \quad C_n \in \mathcal{C} \quad ,$$

where H_C^0 is the classical Hamiltonian function in absence of hierarchical coupling , and the real numbers $b_n(t)$ are the expectation values of the coupling operators $B_n \in \mathcal{B}$ at time t.

The dynamical W*-system appropriate for a chemical description of a molecule has exactly this structure.

6.3 PATTERNS IN HOLISTIC SYSTEMS

Breaking the holistic symmetry

Recall that a system is called *holistic* if it cannot be decomposed into nontrivial subsystems in such a way that the states of the subsystems determine the state of the whole system. The opposites are *separable systems*. The states of the subsystems of *every* decomposition of a separable system determine the state of the composite system. Holistic and separable systems are extreme cases. A system which is neither holistic nor separable is called *partially holistic*, it has both holistic and separable features.

Every classical system is separable, hence shows no holistic effects. Pioneer quantum mechanics is an example for a holistic theory, it has no separable features. In the framework of the theory of W*-systems, separable systems are characterized by a *commutative* W*-algebra of observables. The algebra $\mathcal{A}$ of observables of a partially holistic system can be written as a W*-tensor product $\mathcal{A}=\mathcal{B}\bar{\otimes}\mathcal{C}$ where $\mathcal{B}$ is a non-commutative W*-algebra and $\mathcal{C}$ is a commutative W*-algebra. If the algebra of observables is a factor (i.e. has a trivial center), then the corresponding W*-system is holistic (compare also sect.5.6). If we regard W*-quantum logic as a unified theory of reality, then the universal enveloping theory (in the sense of non-Boolean theory reduction, cf. sect.5.5) is a type I_∞-factor W*-system, hence a holistic theory.

Most theories of chemical, biological and social systems are still limited by the classical paradigm of explanation presupposing in an unreflected way the separability of these systems. This paradigm leads to the belief in the existence of a single frame of reference for the description of reality. That is, according to the classical view one might expect an ordered sequence of theories, each of them including the preceding ones, and describing reality better and better. No reasonable scientist claims that the presently known theories are more than approximately valid. However, it is an open question whether such a theory is an approximation of a perfect presently unknown description of nature. If reality is nonseparable, this premise is wrong.

A holistic theory describes an unbroken wholeness. How do we come to recognize objects in a holistic world? Apart from the unbroken wholeness there are no absolute objects or absolute patterns. Consequently there is no such thing as an unprejudiced description of nature. In the science of art, this is well-known: "*There is no reality without interpretation; just as there is no innocent eye, there is no innocent ear*" (Gombrich, 1960, p.307). An indispensable prerequisit of every observation is a splitting of the world into *essential* and *accidental* components. What is essential is not an intrinsic property but depends on the particular viewpoint adapted. *A point of view is characterized by a deliberate lack of interest which breaks the holistic unity of nature.* Every description of nature presupposes a particular point of view, which may be largely implicit but nevertheless defines the relevant universe of discourse. Observable phenomena come into being by breaking the holistic symmetry. Any alternative viewpoint with a different emphasis leads to an inequivalent description. *There is only one reality but there are many points of view.*

Contexts as frames of reference

In every observation and in every experiment one has to leave out of consideration a myriad of detail. In every experiment there is a stage of data processing, *direct observations without data processing do not exist.* Accordingly, *every fact is conditional.*

Since there are no bare facts we need some frames of reference which constitute necessary and sufficient conditions for the existence of facts. A point of view relative to which observations and measurements can be made and relative to which a partial description of the world can be given will be called a context. More precisely, *we define a context as a part of the world which is singled out by a well-defined set of prior conceptions whose ontological structure is amenable to the application of classical two-valued logic.* As a consequence, a context is a *Boolean* frame of reference so that within one and the same context all properties are compatible and all experimental questions are simultaneously decidable. Incompatible features are intentionally dropped so that a context precisely specifies our deliberate lack of interest. In experimental science, a context is characterized by a particular apparatus used for an experimental investigation, together with instructions laying down what counts as relevant and what is irrelevant.

Contexts constitute complex sets of rules that enable us to interact with nature, they define a point of view, a paradigm. Since we always preceive reality filtered through a Boolean frame of reference, these rules may be implicit and are often not recognized as presuppositions. Nonetheless, the observational language we use is never absolute but *always* determined by a context, and it may change when the context changes. Changing contexts means changing paradigms and changing the nature of theorizing.

There are always factors outside the context in question. If they cannot be ignored, engineers refer to them as "noise", and include some of their effects by discussing *stochastic* systems. Again, the notions "noise" and "signal" are context-dependent. What counts as noise and what as a signal depends on the question we ask.

Concept formation by abstraction

In order to describe an aspect of the holistic reality we have to ignore certain factors such that the remainder separates into facts. Inevitably, such a description is true only within the adopted partition of the world, that is, within the chosen context. It would be very narrow-mined to use only *one* context: *we have to learn to be able imagining different points of view.* While classical science encourages discoveries within a single given context, modern quantum theories encourage the invention of new contexts, complementary to those already known. We accept that there are many possible forms of truth which are mutually incompatible but not contradictory since they can be interconnected by dialectical thinking. Even before the advent of quantum theory, scientific thought always has had a strongly dialectical character.

Contexts are created by deliberately chosen abstractions. Every isolation of a subsystem from a quantum system requires a *process of abstraction* which discards potential properties present in the original system. In science abstractions are indispensable. Every actual event is unique and is not in the domain of science. Science deals with *equivalence classes* of events. These classes are generated by grouping actual events, ignoring some unique properties of the actual events. For example, the repetition of any experiment requires some assumption like "other things being equal". Of course, the "other things" are not equal but they belong to the same equivalence class which is generated by the abstraction used.

Every phenomenon is created by abstractions alone and does not otherwise exist. The objects we recognize with our senses are created by abstractions performed by our sense organs and the associated physiological processes. An analysis of the perception process shows that some features of the world of sense objects are due to our mental organization. We must remind ourselves that man's world picture has its roots in man's primitive past, and that we know very little about our own faculty to recognize pattern in nature. If we would know in detail the criteria we use to cognize patterns and use them as reference data for future recognition, we could apply them in a fundamental holistic theory of reality and discuss whether or not this theory can reproduce these patterns.

We should not underrate the overwhelming role of biological evolution. We decipher the outside world on the basis of techniques developed by evolutionary selection over hundreds of millions of years. The ability of animals to recognize pattern is a marvellously developed faculty. In particular, to learn from paradigms is an outstanding gift of man. It is astonishing to see how few paradigmatic samples of the equivalence classes "cats" and "dogs" a say one year old child needs to learn whether a new example of an animal is a cat, a dog, or a "non-cat-non-dog". Learning from paradigms is an inductive process, and it is well known (at least since David Hume) that an inductive process cannot be a purely logical process. Accordingly, the problem of partitioning a given set of objects into classes in such a way that the members of the same class are similar to each other, and the elements in different classes are dissimilar, cannot be solved by logic and the given data alone. In addition we have to weight the importance of the various predicates (compare also Watanabe, 1969a,c). One extralogical factor in human pattern recognition is historical and given by the pattern recognition mechanisms that have evolved in the biological evolution of homo sapiens. Every pattern recognition mechanism effectuates an abstraction, an abstraction which involves the emphasis of some aspects.

Besides the built-in biological pattern recognition mechanisms, we all have preferred abstractions since we are members of an organized society and since we are educated to have particular viewpoints. Clearly, there is no logical necessity for such abstractions but we cannot help being prejudiced by the mental climate. We can neither expect nor would it be desirable that such biologically and sociologically preferred abstractions are incorporated in a truly fundamental theory of matter. If we would like to recover the familiar constructs of everyday human observation and reasoning from such a fundamental theory, we have to learn what criteria we use to recognize "patterns" in our world. We have to understand how human intelligence interacts with the outside world and we have to study the nature of decision process in humans.

The problem of how we know the nature of qualities and things has played a central role in the development of Western thought (for a short review, compare Weinberg, 1968). There are many thoughtful ideas but few attempts to construct a mathematical framework within which Aristotle's view that objects are the results of abstractions can be discussed. One difficulty is that there may be abstractions which cannot be made verbally at all (Langer, 1957).

Pattern recognition is of considerable importance in engineering science, and has become a large interdisciplinary field (compare e.g. Andrews, 1972; Watanabe, 1972; Sklansky, 1973; Tou and Gonzalez, 1974; Young and Calvert, 1974). Pattern recognition technologies have been applied in such fields as radar detection, weather prediction, character recognition, speech recognition, fingerprint identification, analysis of bubble chamber photographs, medical diagnosis, etc. In spite of much research efforts, most pattern recognition problems are unsolved. We possess a bag of useful tricks and heuristic methods but no comprehensive theory of pattern recognition is available at present. In particular, the basic mechanisms of the extraordinary pattern recognition capabilities of animals and men are poorly understood.

Human intelligence is able to reason successfully in non-quantitative and imprecise terms. We all know what a cat is, we have learnt it by paradigms; nevertheless we are probably not capable to define the "equivalence class of cats" either by *intension* (i.e. connotation, given by the set of characteristics that determine to which objects the name applies) or by *extension* (i.e. the collection of referents to which the name applies). Quite generally, *most equivalence classes used in science are exemplified by paradigms and not by definitions which are understandable in terms of a fundamental theory*. One reason why only a tiny part of chemistry has been reduced to fundamental molecular quantum mechanics is that most chemical concepts are not (or, maybe, not yet) defined in a language which can be transferred to the fundamental theory. Quantum chemists used to say that "we can compute everything" but by "everything" they tacitly refer to things that make sense in terms of mechanical degrees of freedom. They cannot "calculate a cat", not because their computers are too small but because they have no idea what a cat is (in terms of mechanical degrees of freedom). Nevertheless, there are cats. The situation is not very different for much simpler system. For example, keto groups exist in chemical taxonomy, but they have no natural place in the framework of pioneer quantum mechanics. There is nothing in quantum theory which tells us we ought to invent a notion of "ketons". Furthermore, without smuggling an extralogical value judment for the importance of predicates, the traditional chemical concept "keto group" cannot be derived from pioneer quantum mechanics and some paradigmatic examples.

A pattern is important if it leads to a classification which is *useful* for some purpose. Usefulness is a value judment, hence of extralogical nature. Since pioneer quantum mechanics has no built-in value structure it cannot explain natural kind of objects: all atomic properties are equivalent, all observables are equally important. In order to classify objects according to common patterns, we have to introduce *preferred observables* via some kinematical group, or more generally, by some pattern group.

Patterns and pattern groups

While it is true that some pattern come into being by taking something away from our sense impressions, science knows of patterns represented by purely abstract structures which do not arise in this way. For some patterns there is nothing concrete to start with and to abstract from. In such cases, the *concrete* comes into being by an abstraction process. Hence we should not try to define patterns by their primitive components and their composition.

Well, what is then a pattern? A survey by Verhagen (1975) reveals that the term "pattern" has a different meaning in different areas and with different authors. Usually no abstract definition is given. We may agree with Meisel's (1972, p.1) remark "*in their widest sense, patterns are the means by which we interpret the world*", but for our purpose a more exact even though a more abstract characterization is necessary. A conceptually convincing definition of pattern has to be in terms of their invariant properties. Such a group-theoretical defintion has been given by Grenander (1969): "*By a pattern we shall mean a set of images which is invariant under similarity transformation*". Two images (or objects) arising from each other by some transformation are regarded as equivalent if they are indistinguishable in a given context when each is considered by itself and not in their mutual relation. The class of equivalent images is called a *pattern*. The set of all similarity transformations which connect equivalent images form a group which we call the *pattern group*.

On the other hand, if G is any abstract group, M a set, and f a mapping from $G\times M$ into M, then we can define a pattern group as *transformation group* by putting $g(x) \overset{\text{def}}{=} f(g,x) \in M$ for every $g \in G$ and every $x \in M$, and requiring that (i) for the identity element e of G we have $e(x)=x$, and (ii) $\{gh\}(x)=g\{h(x)\}$ for every $x \in M$ and any $g,h \in G$. If G is a topological group and M a topological space, and if the mapping $x \to g(x)$ of M onto itself is continuous, then G is called a *topological transformation group* of M. A transformation group G is said to act *transitively* on M if for every pair of points $x,y \in M$ there exists a group element $g \in G$ such that $g(x)=y$. A *pattern* is defined as an *orbit* of a transformation group G, i.e. by a set $\{y \mid y=g(x),\ g \in G\}$, where x is some fixed element of M. Accordingly, the patterns of M with respect to the pattern group G are given by a family of disjoint and G-invariant subsets of M. If the pattern group G acts transitively on M, then we call M an *elementary pattern*.

In the context of the theory of W*-systems, the concept of pattern groups is a generalization of the concept of kinematical groups, as introduced by Hermann Weyl (1927). The kinematical group determines the patterns of feasible abstract motions of mechanical systems apart from considerations of mass and force. The elementary patterns of Galilean space-time are given by the ergodic automorphic realizations of the Galilei group on W*-algebras, and are called Galileons (like the Newtonian point particle in classical mechanics, or the electron in pioneer quantum mechanics). While the kinematical groups are related to space-time patterns, general pattern groups refer to any patterns whatsoever, so that in principle any group can be applied as a pattern group (Primas, 1978). Whether we regard the corresponding patterns as interesting or not does not only depend on the structure of the system alone but also on our mode of interacting with the system.

A particular viewpoint characterizes a particular W*-subtheory of a holistic W*-theory. It is tempting to assume that it is the pattern group which determines the W*-algebra together with its properly transforming distinguished observables of the subtheory corresponding to the chosen viewpoint. Today we know only very few pattern groups so that this conjecture can be studied only in a few cases. The Newtonian concepts like momentum, energy, force and power indeed emerge from the Galilei group. The recent developments in W*-algebraic thermodynamics makes it plausible that the patterns "temperature" and "chemical potential" are directly related with the automorphic action of a thermodynamic pattern group (the affine and the torus groups, respectively). Whether concepts like entropy, organization, specificty, memory, adaptiveness, growth, function and purpose - which are all important for the scientific inquiry but extraneous to quantum theory - are related to pattern groups is a difficult open question. The most promising tools we have to investigate such questions are the theory of Lie group contractions and theory of asymptotic expansions. From a conceptual point of view, group contractions would be the ideal tool to generate new pattern groups. Yet, only for very few patterns the relationships with group contractions are known explicitly (compare for example Dubin, 1974, p.78 and p.81; Cattaneo and Wreszinski, 1979).

Asymptotic patterns

According to Josiah Willard Gibbs (1839-1903) "*one of the principal objects of theoretical research in any department of knowledge is to find the point of view from which the subject appears in its greatest simplicity*". If we find it impossible to discover the line which separates one pattern from another it is because no such line of demarcation exists. A good description is a good caricature, exaggerating some aspects by deliberate simplification and by permitting extravagance. To ask whether one caricature is "better" than another cannot be answered without considering "better for what purpose". A caricature is neither a replica of something seen nor does it rely on pre-existing forms. In art, a caricature is a true creation by the artist, a creation which enables us to see reality from a new perspective.

In exact science, the basic theoretical tool for creating caricatures is the study of a system in limiting approximations. As discussed by Hammer (1964, 1971), "*the result of an approximating process is the substi-*

tution of one entity for another, with the intention that the former shall play the same role in some regards as the latter".

First example: Shadows

The following résumé is taken from a most interesting discussion of asymptotic phenomena by K.O. Friederichs (1955).

The propagation of light is governed by Maxwell equations, that is by partial differential equations which have *continuous* solutions. A shadow appears when a light wave passes an object. In the exact theory there is a continuous transition from light to dark. The shadow boundary is narrow but not discontinuous. The pattern of a sharp shadow boundary is a Stokes phenomenon of the asymptotic expansion of Maxwell's theory leading to geometrical optics. A sharp shadow is not only a most convenient description but has a basic conceptual significance since it leads to a description of macroscopic objects possessing sharp outlines.

Asymptotic approximations simplify and schematize complex systems so as to disclose typical patterns. Different limiting processes represent different viewpoints and create different patterns which may be appropriate for different aims. Inevitably, a price has to be paid for making asymptotic approximations: we may say things about nature which are not strictly true. Yet it would be stupid to object that we look at nature through half-shut eyes. The aim of a caricature is not to represent the truth, the whole truth, and nothing but the truth. It is the essence of a caricature that new patterns come into evidence. "*One grievous error in interpreting approximations is to allow only good approximations*" (Hammer, 1971).

In recent years considerable progress has been made in establishing precise links between microscopic dynamical laws and the dynamics of higher hierarchical levels. In the thermodynamic description of irreversible processes one encounters many situations where some of the variables vary on a much slower time-scale than others. A rigorous asymptotic description of such systems is possible by rescaling of the variables such that even holistic systems factorize asymptotically into object systems though the actual limit cannot be reached exactly. The asymptotically induced patterns are often an excellent caricature of actual systems.

Second example: The van Hove limit

The weak-coupling long-time limit has been introduced by van Hove (1955) in order to derive an irreducible dynamics from a Hamiltonian dynamics without ad hoc assumptions. He proposed to split the Hamiltonian of a large system into an unperturbed part H and a small perturbation εV, and he showed heuristically that an irreversible behavior can be expected in terms of a rescaled time $\tau=\varepsilon^2 t$ in the limit $\varepsilon\to 0$. The asymptotic limit

$$\varepsilon\to 0\ ,\quad t\to\infty \quad \text{such that } \varepsilon^2 t \overset{\text{def}}{=} \tau\neq 0 \ \text{ and finite,}$$

is referred to as the van Hove limit. The rescaling of the time accounts for the fact that the decay of the system becomes slower with decreasing coupling constant.

By tracing out the degrees of freedom corresponding to the environment, the evolution of the reduced density operator D of the open system is governed by a non-Markovian master equation,

$$\frac{dD(t)}{dt} = i[H,D(t)]_- + \varepsilon^2 \int_0^t K_\varepsilon(s)D(t-s)ds \quad .$$

Introducing the rescaled time $\tau=\varepsilon^2 t$, and going to the interaction representation $\bar{D}(\varepsilon^2 t)=\exp(-iHt)D(t)\exp(iHt)$, we get

$$\frac{d\bar{D}(\tau)}{d\tau} = \int_0^{\varepsilon^{-2}\tau} \bar{K}_\varepsilon(s)\bar{D}(\tau-\varepsilon^2 s)ds \quad ,$$

so that in the limit $\varepsilon\to 0$ we get formally an ordinary differential equation of semigroup type in the rescaled time variable τ,

$$\frac{d}{d\tau}\bar{D}(\tau) = G\bar{D}(\tau) \quad \text{with} \quad G \overset{\text{def}}{=} \int_0^\infty \bar{K}_0(s)ds \quad .$$

The mathematical justification of these steps is by no means trivial.

A precise mathematical formulation of the van Hove limit by Davies (1974, 1975, 1976b, 1977a) allows rigorous derivations of Markovian master equations for open quantum systems. Accordingly we know in principle how to deduce rigorously contracted descriptions by open W*-systems with a semigroup dynamics from a basic W*-system with an automorphic time evolution. The use of the van Hove limit is meaningful if the relaxation time T_S of the open system is much larger than the relaxation time T_R of the environment (often called "reservoir"), $T_S \gg T_R$.

For a review of various applications of the van Hove limit, and additional literature references compare Davies (1976a, 1977b), Gorini et al. (1978), Spohn and Lebowitz (1978), Spohn (1979).

Third example: The Boltzmann-Grad limit

Consider a gas with a low density ρ of scatterers. The mean free path and the mean free time are of the order $1/\rho$, so that in a meaningful limit $\rho\to 0$ the space variable q and the time variable t have to be rescaled such that the mean free path and the mean free time remain finite. That is we have to discuss the following asymptotic limit $\varepsilon\to 0$:

$$\begin{aligned} &\rho\to 0 \text{ such that } \rho/\varepsilon \neq 0 \text{ is finite,}\\ &q\to\infty \text{ such that } \varepsilon q \neq 0 \text{ is finite,}\\ &t\to\infty \text{ such that } \varepsilon t \neq 0 \text{ is finite.}\end{aligned}$$

In this limit the Boltzmann equation goes over into the *linear* Boltzmann transport equation for a dilute gas, the cornerstone of the classical kinetic theory of gases (for a review of this limit, compare Spohn, 1979).

Fourth example: The Brownian-motion limit

Consider the Brownian motion of particle with a large mass M suspended in a fluid. The position q of the Brownian particle changes much more slowly than its velocity $v=dq/dt=p/M$. According to Langevin's idea, in the limit of large mass, the motion of a Brownian particle should be governed by a Markov process. This is indeed rigorously the case if we rescale mass, time, position and momentum by a parameter ε, and consider the asymptotic limit

$$
\begin{array}{llcl}
M\to\infty & \text{such that} & \varepsilon^2 M \overset{\text{def}}{=} \bar{M} \neq 0 & \text{and finite,} \\
t\to\infty & \text{such that} & \varepsilon^2 t = \bar{t} \neq 0 & \text{and finite,} \\
q\to\infty & \text{such that} & \varepsilon q = \bar{q} \neq 0 & \text{and finite,} \\
p\to\infty & \text{such that} & \varepsilon p = \bar{p} \neq 0 & \text{and finite.}
\end{array}
$$

With this rescaling, the velocity is given by $\bar{v}=d\bar{q}/d\bar{t}$, so that $\bar{p}=\bar{M}\bar{v}$. (For a review with references to modern investigations compare Spohn, 1979).

Fifth example: The Hartree limit

The Hartree-Fock method is a well-known variational method for approximating the many-body problems of quantum chemistry. Originally, Hartree considered his approach not as a variational method but as a mean-field theory whose dynamics is given by an effective Hamiltonian which depends linearly on the one-density operator so that the equation of motion for the state becomes nonlinear.

It is a remarkable fact that the Hartree method can become exact in the limit of infinitely many particles. As an example, consider a system of N distinguishable particles described by a bounded Hamiltonian H_N,

$$H_N = \sum_{j=1}^{N} K_j + \frac{1}{N} \sum_{j<k} V_{jh} \quad ,$$

acting on the Hilbert space $\mathcal{H}\otimes\mathcal{H}\otimes..\otimes\mathcal{H}$ (N times). The single particle operator K_j and the two-particle operator V_{jh} are defined as usual by

$$K_j \bigotimes_{n=1}^{N} \varphi_n = (K\varphi_j)\otimes(\bigotimes_{n\neq j}\varphi_n) \quad , \quad K\in\mathcal{B}(\mathcal{H}) \quad ,$$

$$V_{jk} \bigotimes_{n=1}^{N} \varphi_n = (V\varphi_j\otimes\varphi_k)\otimes(\bigotimes_{n\neq j,k}\varphi_n) \quad , \quad V\in\mathcal{B}(\mathcal{H}\otimes\mathcal{H}) \quad ,$$

for all $\varphi_n\in\mathcal{H}$.

The evolution equation for the density operator $D_N(t)$ is given by

$$i\frac{d}{dt} D_N(t) = [H_N, D_N(t)]_- \quad .$$

Let $tr_{n,N}$ denote the partial trace over the Hilbert space $\bigotimes_{j=n+1}^{N}\mathcal{H}$, and denote the corresponding reduced density operator by $D_{n,N}(t)$,

$$D_{n,N}(t) \overset{\text{def}}{=} tr_{n,N}\{D_N(t)\} \quad .$$

If the initial state is a product state, $D_N(t=0)=D\otimes D\otimes..\otimes D$, $D\in\mathcal{B}_1(\mathcal{H})$, then in the asymptotic limit $N\to\infty$, the reduced density operator $(n<\infty)$ is still a product state,

$$\lim_{N\to\infty} D_{n,N}(t) = D(t)\otimes D(t)\otimes..\otimes D(t) \quad \text{(n times)} \quad .$$

Here $D(t)\in\mathcal{B}_1(\mathcal{H})$ is the solution of the Hartree equation

$$i\frac{d}{dt}D(t) = [K,D(t)]_- + \tfrac{1}{2}\, tr_2\{[V_{12}+V_{21}, D(t)\otimes D(t)]\} \quad ,$$

where tr_2 denotes the partial trace over the second Hilbert space (Spohn, 1979).

Unfortunately, rigorous results for the Hartree limit for particles fulfilling the Bose or Fermi statistics are not available at present. Nevertheless, one would expect that some Hartree or Hartree-Fock approximation is asymptotically exact for large molecular systems (for more details, examples and references, compare Spohn, 1979).

6.4 A NEW LOOK AT MOLECULAR PATTERNS

Molecular structure as an asymptotic pattern

One of the chemically most relevant pattern is the *molecular structure*. In pioneer quantum mechanics a molecular structure is defined via the Born-Oppenheimer approximation by a minimum energy configuration on the Born-Oppenheimer potential surface. In spite of the obvious merits of this definition, it is ad hoc. In pioneer quantum mechanics molecular structure does not represent a feature of the system since by definition features are represented by elements of the algebra of observables. According to the general use of this concept by chemists, a molecular structure is something a molecule *has*, and which can be discussed by using an unrestricted Boolean language. That is, molecular structure is a *classical* concept, and should be represented by classical observables. However, in pioneer quantum mechanics there are no classical observables.

Since a conceptually acceptable discussion of the pattern of molecular structure does not seem to be available in the published literature, we give a somewhat detailed (although only heuristic) exposition of the main points. In the framework of pioneer quantum mechanics, a molecule has no molecular structure in the sense of the chemist. Nevertheless, we use pioneer quantum mechanics as a starting point, and *derive* the molecular structure as an *asymptotic pattern* in the singular limit of infinite molecular masses. Here, attention must be paid to the fact that the limiting situation goes beyond the scope of pioneer quantum mechanics. However, the limit is well-defined as a dynamical W*-system having two hierarchical levels.

We consider a molecule consisting of r electrons and s nuclei, and denote the position operators of the electrons by $q=(q_1,..,q_{3r})$ and the position operators of the nuclei by $Q=(Q_1,..Q_{3s})$. First we formulate the molecular Hamiltonian H in terms of pioneer quantum mechanics so that the electronic momentum $p=(p_1,..,p_{3r})$ and the nuclear momentum operators $P=(P_1,..,P_{3s})$ fulfill Heisenberg's canonical commutation relations

$$[q_j,p_{j'}]_- = i\hbar\delta_{jj'} , \quad j,j' = 1,..,3r ,$$

$$[Q_k,P_{k'}]_- = i\hbar\delta_{kk'} , \quad k,k' = 1,..,3s .$$

on appropriate domains. The Hamiltonian is taken as

$$H = \frac{1}{2m}\sum_{j=1}^{3r} p_j^2 + \sum_{k=1}^{3s} \frac{1}{2M_k} P_k^2 + V_e(q) + V_n(Q) + U(q,Q) \quad ,$$

where m is the electronic mass and M_k is a nuclear mass. The Coulomb interaction between the electrons is given by the potential V_e, the Coulomb interaction between the nuclei is denoted by V_n, while U is the Coulomb interaction between electrons and nuclei.

In the Heisenberg picture of pioneer quantum mechanics, the equations of motion are given by (j=1,..,r; k=1,..,s):

$$\frac{d}{dt} q_j(t) = \frac{1}{m} p_j(t) \quad ,$$

$$\frac{d}{dt} p_j(t) = -\frac{\partial}{\partial q_j(t)} \Big(V_e\{q(t)\} + U\{q(t),Q(t)\} \Big) \quad ,$$

$$\frac{d}{dt} Q_k(t) = \frac{1}{M_k} P_k(t) \quad ,$$

$$\frac{d}{dt} P_k(t) = -\frac{\partial}{\partial Q_k(t)} \Big(V_n\{Q(t)\} + U\{q(t),Q(t)\} \Big) \quad .$$

This description of a molecule in terms of pioneer quantum mechanics represents a W*-system $(A, t\to\alpha_t)$, where A is given by the W*-tensor product $A=A_e\bar{\otimes}A_n$ of two factors A_e and A_n of type I_∞. The W*-algebra A_e is generated by the canonical operators p(o), q(o) of the electrons, while A_n is the W*-algebra generated by the canonical operators P(o), Q(o) of the nuclei. The time evolution $t\to\alpha_t \in Aut(A)$ is given by the equations of motion for p(t), q(t), P(t) and Q(t). Note that for t>o the operators p(t) and q(t) are no longer affiliated to A_e so that the restriction of α_t to A_e is *not* an automorphism. That is, in this description the electronic subsystem cannot be regarded as an autonomous dynamical system.

The ratio of the intrinsic time scales of electronic and nuclear motions is given by $\sqrt{m/M} \overset{def}{=} \varepsilon$, where M is a typical nuclear mass, say $M=\min\{M_1,..,M_{3s}\}$. In order to investigate the asymptotic limit $\varepsilon\to o$, we rescale the nuclear variables as follows:

$$\tau \overset{def}{=} \varepsilon t \quad , \qquad m_k \overset{def}{=} \varepsilon^2 M_k \quad ,$$

$$X_k(\tau) \overset{def}{=} Q_k(\tau/\varepsilon) \quad , \qquad Y_k(\tau) \overset{def}{=} \varepsilon P_k(\tau/\varepsilon) \quad ,$$

so that

$$[X_k(\tau), Y_{k'}(\tau)]_- = i\varepsilon\hbar\delta_{kk'} \quad .$$

In the rescaled variables, the equations of motion are given by

$$\frac{d}{dt} q_j(t) = \frac{1}{m} p_j(t) \quad ,$$

$$\frac{d}{dt} p_j(t) = - \frac{\partial}{\partial q_j(t)} \Big(V_e\{q(t)\} + U\{q(t), X(\varepsilon t)\} \Big) \quad ,$$

$$\frac{d}{d\tau} X_k(\tau) = \frac{1}{m_k} Y_k(\tau) \quad ,$$

$$\frac{d}{d\tau} Y_k(\tau) = - \frac{\partial}{\partial X_k(\tau)} \Big(V_n\{X(\tau)\} + U\{q(\tau/\varepsilon), X(\tau)\} \Big) \quad .$$

Note that for $\varepsilon > o$, these equations still represent the motion as given by pioneer quantum mechanics. The limiting situation $\varepsilon = o$ changes the situation radically and has to be discussed with great care. Here we content ourselves with a heuristic discussion.

In the infinite nuclear mass limit, $\varepsilon \to o$, the electronic equations of motion become autonomous, so that

$$\frac{d}{dt} q_j(t) = \frac{1}{m} p_j(t) \quad ,$$

$$\frac{d}{dt} p_j(t) = - \frac{\partial}{\partial q_j(t)} \Big(V_e\{q(t)\} + U\{q(t), X(o)\} \Big) \quad ,$$

define a one parameter group $t \to \beta_t^o$ of automorphisms on the factor $\mathcal{A}_e$, which still is generated by the canonical observables $p(o)$ and $q(o)$. The vector $X(o)$ acts as a fixed parameter. The Hamiltonian which generates this automorphism is given by $H_e\{X(o)\}$ where

$$H_e^o(X) \overset{\text{def}}{=} \sum_{j=1}^{3r} \frac{1}{2m} p_j^2 + V_e(q) + U(q,X) \quad , \quad X \in \mathbb{R}^{3s}$$

and coincides with the traditional Born-Oppenheimer electronic Hamiltonian.

The equations of motion for the nuclear variables X and Y represent an intrinsically slow dynamical system which is in addition subjected to a very rapidly fluctuating external force

$$\partial U\{q(\tau/\varepsilon), X(\tau)\}/\partial X(\tau) \quad .$$

The standard way of treating such dynamical systems is by means of the method of averaging, as discussed by Bogoliubov and Mitropolsky (1961, chapt.6). Instead of Bogoliubov's time average, we take the average over the electronic degrees of freedom with respect to some *stationary* normal state $\rho_e \in (A_e)_*$, and define

$$X(\tau) \overset{\text{def}}{=} \overline{X}(\tau) + x(\tau) \quad , \text{ where } \quad \overline{X}(\tau) \overset{\text{def}}{=} \rho_e\{X(\tau)\} \quad ,$$

$$Y(\tau) \overset{\text{def}}{=} \overline{Y}(\tau) + y(\tau) \quad , \text{ where } \quad \overline{Y}(\tau) \overset{\text{def}}{=} \rho_e\{Y(\tau)\} \quad .$$

Using the usual tools of stochastic analysis, one can write down the equations of motion for the averages $\overline{X}$ and $\overline{Y}$ of the nuclear canonical variables X and Y, respectively (compare e.g. Primas, 1961; Hakim, 1968). These equations simplify a great deal after the limit $\varepsilon \to o$ is taken:

$$\frac{d}{d\tau} \overline{X}_k(\tau) = \frac{1}{m_k} \overline{Y}_k(\tau) \quad ,$$

$$\frac{d}{d\tau} \overline{Y}_k(\tau) = - \frac{\partial}{\partial \overline{X}_k(\tau)} \Big(V_n\{\overline{X}(\tau)\} + \Phi\{\overline{X}(\tau)\} \Big) \quad ,$$

where the effective potential Φ is given by

$$\Phi\{\overline{X}(\tau)\} = \rho_e\big(U\{q, X(\tau)\}\big) \quad .$$

Since ρ_e is stationary, $\rho_e\{\beta_t^o(A)\} = \rho_e(A)$ for every $A \in A_e$, the potential Φ is explicitly time-independent and does not depend on ε.

The relation of this adiabatic limit to the traditional Born-Oppenheimer approach can be seen by introducing the Born-Oppenheimer potential Φ_{BO} by

$$\Phi_{BO}(X) \overset{\text{def}}{=} \rho_e\{H_e^o(X)\} \quad , \quad \rho_e \in (A_e)_* \quad , \quad X \in \mathbb{R}^{3s} \quad .$$

Since for a stationary electronic state ρ_e we have

$$\frac{\partial \Phi_{BO}(X)}{\partial X_k} = \frac{\partial \Phi(X)}{\partial X_k} \quad ,$$

we can write the equations of motion in the adiabatic limit $\varepsilon \to o$ also as

$$\frac{d}{d\tau} \overline{X}_k(\tau) = \frac{1}{m_k} \overline{Y}_k(\tau) \quad ,$$

$$\frac{d}{d\tau}\,\bar{Y}_k(\tau) = -\frac{\partial}{\partial\bar{X}_k}\left(V_n\{\bar{X}(\tau)\} + \Phi_{BO}\{\bar{X}(\tau)\}\right) \quad .$$

These equations are of the very same forms as the quantum mechanical equations of motion which arise from the traditional Born-Oppenheimer Hamiltonian H_{BO},

$$H_{BO} \stackrel{def}{=} \sum_{k=1}^{3s} \frac{1}{2M_k} P_k^2 + V_n(Q) + \Phi_{BO}(Q) \quad .$$

Despite the *formal* similarity of the equations of motion, *the Born-Oppenheimer approach and the time-dependent adiabatic limit are basically different.* In the scheme of Born and Oppenheimer, the nuclei are described by pioneer quantum mechanics so that the algebra of all nuclear observables is a factor of type I_∞. In the time-dependent rescaled asymptotic description, the nuclear observables $X(\tau)$ and $Y(\tau')$ commute for all τ,τ', and generate a *commutative* atomfree W*-algebra C, so that the algebra of observables for the adiabatic caricature of a molecule is given by

$$A_{ad} = A_e \bar{\otimes} C \quad ,$$

where the factor A_e of type I_∞ describes the electrons (as in pioneer quantum mechanics), while the commutative W*-algebra C describes the nuclei. As opposed to the Born-Oppenheimer description with the algebra $A_e\bar{\otimes}A_n$, the adiabatic description with the algebra $A_e\bar{\otimes}C$ is an *object factorization*: only in this description electrons and nuclei exist as individual entities. There is a new pattern - the molecular structure - which is described by the classical observables in A_{ad}. Moreover, the adiabatic caricature is a two-level hierarchical quantum system (as discussed in sect.6.2).

The infinite nuclear mass limit is highly singular, it changes pioneer quantum mechanics into a mixed quantum-classical W*-system where the electrons behave quantum mechanically while the nuclei are described classically. Of course, molecules have *finite* nuclear masses so that the infinite mass limit can be used only as a starting point for an asymptotic expansion in powers of ε around the singular point ε=o. Such an expansion does not exist in a uniform way within the original W*-system ($A_e\bar{\otimes}A_n$, $t\to\alpha_t$) but it can be represented by a 2-level hierarchical quantum system ($A_e\bar{\otimes}C$, $t\to\beta_t\otimes\gamma_{\varepsilon t}$). The electronic system represents the lower hierarchical level, it is an open W*-system $(A_e, t\to\beta_t)$ where A_e is a factor of type I_∞, and the time evolution $t\to\beta_t \in Aut(A_e)$ is generated by the *time-dependent* Hamiltonian

$$H_e(t) = \sum_{j=1}^{3r} \frac{1}{2m} p_j^2 + V_e(q) + U\{q, X(\varepsilon t)\} \quad ,$$

where X is an external time-dependent *classical* control variable. The nuclear system acts as a control system $(C, \tau\to\gamma_\tau)$. It is a *classical stochastic system* whose time evolution $\tau\to\gamma_\tau \in Aut(C)$ generates the equation of motion of the classical observables $X(\tau)=\gamma_\tau\{X\}$ and $Y(\tau)=\gamma_\tau\{Y\}$, which can be written in Hamiltonian form as

$$\frac{d}{d\tau} X_k(\tau) = \{X_k(\tau), H_n(\tau)\} \quad ,$$

$$\frac{d}{d\tau} Y_k(\tau) = \{Y_k(\tau), H_n(\tau)\} \quad ,$$

where {,} denotes the Poisson bracket with respect to X, Y. The time-dependent Hamiltonian function H_n is affiliated to C and given by $H_n(\tau)=H_n^0+\varepsilon H_n^1(\tau)+\ldots$, where H_n^0 is explicitly time-independent and given by

$$H_n^0 = \sum_{k=1}^{3s} \frac{1}{2m_k} Y_k^2 + V_n(X) + \Phi_{BO}(X) \quad .$$

The next correction term $H_n^1(\tau)$ contains corrections to the Born-Oppenheimer potential Φ_{BO} and a rapidly fluctuating stationary stochastic process. The main term H_n^0 is formally analogous to the Born-Oppenheimer Hamiltonian H_{BO} of pioneer quantum mechanics, but in H_n^0 all variables are c-numbers (i.e. classical observables).

In the infinite nuclear mass limit, the nuclei behave like Newtonian point particles: they move on definite trajectories $\tau\to\bar{X}(\tau)$. In the next higher approximation, we have to consider two things: a stochastic feedback from the electronic system (which is quite small and often not very important), and the nature of the initial conditions. The initial conditions are not given by the characterization of the dynamical system $(C,\tau\to\gamma_\tau)$, nevertheless we are not free to choose them ad libitum. The initial conditions have to take into account the now hidden quantum nature of the nuclei. (In a more rigorous development of the adiabatic limit, this hidden aspect is *automatically* taken into account by an additional but nonobservable classical random field). An admissible initial state for the nuclear system has to fulfill the stochastic inequality $\Delta X_k \Delta Y_k \geq \frac{1}{2}\varepsilon\hbar$, where ΔX_k and ΔY_k denote the variance of the *commuting* observables X_k and Y_k, respectively. In the first approximation ($\varepsilon=0$), we can take the initial state as an ontic state which has sharp values both for X and for Y. In the second approximation an admissible initial state necessarily is an epistemic state fulfilling the stochastic Heisenberg-type inequality $\Delta X_k \Delta Y_k \geq \frac{1}{2}\varepsilon\hbar$.

The ontic interpretation of the W*-system $(\mathcal{C}, t\rightarrow\gamma_t)$ implies, that in the adiabatic caricature of molecules the nuclei are Newtonian point particles with sharp position and sharp momentum at every instant, but they fluctuate very rapidly (in a time scale much more rapid than the collective nuclear motions like translation, vibration or rotation) so that the trajectories are only stochastically knowable. Their mean distribution corresponds to the description given by pioneer quantum mechanics.

Remark: On the rigorous formulation of the adiabatic limit

Every nonuniform singular limiting operation is difficult mathematically so that a mathematically rigorous formulation of the adiabatic limit is a nontrivial undertaking. Using the formalism of pioneer quantum mechanics, one realizes $\mathcal{A}_n$ as the algebra of bounded observables acting on some Hilbert space $\mathcal{H}_n$ such that the canonical Weyl operators act irreducibly on $\mathcal{H}_n$. One can then use the Weyl operators in order to construct a family $\{P_\varepsilon | \varepsilon>0\}$ of positive linear mappings from the predual $(\mathcal{A}_n)_*$ (realized as the Banach space of nuclear operators acting on $\mathcal{H}_n$) to the Banach space $L_1(\mathbb{R}^{6s})$, where $\mathbb{R}^{6s}$ is the classical phase space of nuclei. For $\varepsilon\rightarrow 0$, the positive linear mapping P_0 describes the adiabatic limit, whereby the commutative algebra $\mathcal{C}$ is realized as $L_\infty(\mathbb{R}^{6s})$. In roughly this way, analogous limits (e.g. for the case $\hbar\rightarrow 0$) have been discussed by Hepp (1974), Davies (1976c), Ingólfsson (1976), Ginibre and Velo (1979), and Hagedorn (1980a,b). From a mathematically rigorous point of view, the time-dependent adiabatic limit has recently been discussed by Hagedorn (1980c).

The traditional representation of pioneer quantum mechanics is rather inconvenient for the discussion of the adiabatic limit $(\mathcal{A}_n)_* \rightarrow (\mathcal{C})_*$ since the Hilbert space used for an irreducible representation of Weyl's canonical commutation relations is not large enough to allow a convenient representation of the normal states in $(\mathcal{A}_n)_*$ and in $(\mathcal{C})_*$ as vector states. Since W*-isomorphic W*-systems are physically equivalent, we are free to choose *any* faithful Hilbert space representation. It is not yet widely recognized that the so-called *standard representation* of a W*-system (in the sense of Haagerup, 1975) is a most convenient representation for the discussion of asymptotic limits. For the adiabatic limit an appropriate standard representation of $\mathcal{A}_n$ (acting reducibly on some Hilbert space $\mathcal{K}$) is generated by the (highly reducible) *regular representation* of the canonical Weyl operators on the Hilbert space $\mathcal{K}=L_2(\mathbb{R}^{6s})$, where $\mathbb{R}^{6s}$ is the phase space of nuclei. The classical limit ($\hbar\rightarrow 0$) or the adiabatic limit ($\varepsilon\rightarrow 0$) can then be performed quite easily without ever leaving the Hilbert space $\mathcal{K}$. For this, one constructs a family $\{\mathcal{S}_\varepsilon | \varepsilon>0\}$ of regular representations of Weyl's canonical commutation relations such that for every $\varepsilon>0$ the W*-algebra $\mathcal{S}_\varepsilon \subset \mathcal{B}(\mathcal{K})$ is a standard representation of $\mathcal{A}_n$, and such that $\mathcal{S}_0=L_\infty(\mathbb{R}^{3s})$ is a standard representation of $\mathcal{C}$. Although $\mathcal{S}_0$ is not a subalgebra of any $\mathcal{S}_\varepsilon$ with $\varepsilon>0$, it is a subalgebra of $\mathcal{B}(\mathcal{K})$ and this fact simplifies proofs in a pleasant way. (These investigations have been carried out some years ago together with Guido A. Raggio, but we have been too lazy to publish them).

Open problems

A description (correct in first order of ε) of the vibrational and rotational spectra in terms of the stochastic classical system $(\mathcal{C}, t \rightarrow \gamma_t)$ is feasible but has not yet been worked out in detail. Likewise, the details of a description of the dynamics of molecular structure in terms of stochastic W*-classical mechanics with nonlinear phase space have not yet been investigated. On a heuristic level, some of these problems have been studied by Miller (1978).

Neoclassical stochastic mechanics

It may be surprising for the uninitiate that a classical description of molecular spectra can be reasonable at all. Furthermore, many textbooks still state that classical electrodynamics fails in the microscopic domain. Such claims are very misleading since it is not at all plain what "classical electrodynamics" is. In *traditional* classical electrodynamics one adopts usually Lorentz's boundary condition implying that there is no radiation without matter. This is an ad hoc assumption, not an irrevocable decision. Neoclassical stochastic electrodynamics takes the same equations of motion but a different boundary condition: it postulates the existence of a universal random background field (even in the absence of matter) and arrives at very different results. In fact, many phenomena (like spontaneous emission or the Lamb shift) which are often considered to be inherently quantum mechanical effects can be explained at a completely classical level.

In order to understand whether a description in terms of some neoclassical theory is possible or not, we have to understand the difference between classical and quantum theories. Whether a theory is classical or not depends neither on the appearance of Planck's constant, nor on quantum jumps, nor on the validity of a Heisenberg-type uncertainty relation. It is true that every quantum theory shows stochastic effects but this is *not* its typical characteristic. *Both stochastic neoclassical mechanics and deterministic quantum mechanics exhibit stochastic phenomena but they crucially differ with respect to the superposition principle.* More precisely: the convex set of all epistemic states of a stochastic neoclassical theory is a simplex, while the state space of any quantum theory never is a simplex. (For W*-systems an equivalent statement is that the algebra of the observables of a neoclassical theory is commutative, while the algebra of observables of a quantum theory is noncommutative).

Consider as an example quantum electrodynamics versus neoclassical electrodynamics. It is possible to reproduce many phenomena of the interaction between matter and the electromagnetic field by applying pioneer quantum mechanics only to matter and treating the electromagentic field by *classical* stochastic electrodynamics (compare the review of Milonni, 1976). Dirac's (1927) successful treatment of the spontaneous emission of radiation by quantizing the radiation field marked the beginning of quantum electrodynamics but also led many to believe that spontaneous emission is a typical quantum effect. However, as we know from neoclassical electrodynamics, an explanation of spontaneous emission does by no means require a quantization of the radiation field. A stochastization of the classical electromagnetic field reproduces the effects of the quantum mechanical zero-point radiation, while the fluctuations enforce the emission of radiation from excited states, and furthermore can account for the Lamb shift (Welton, 1948). However, any neoclassical electrodynamics is bound to fail for phenomena involving Einstein-Podolsky-Rosen correlations. Indeed, the only known cases where a neoclassical description of the radiation field breaks down are experiments of the type first performed by Kocher and Commins (1967) which depend on the existence of Einstein-Podolsky-Rosen correlations (compare also Clauser, 1972).

If Einstein-Podolsky-Rosen correlations are important, quantum electrodynamics is the only valid theory. *Hence we do not consider neoclassical electrodynamics as a fundamental theory but as a derived theory operating on a hierarchically higher level.* The fact that many spectroscopically important radiation effects (like spontaneous emission, line broadening, line shifts) can be well understood in the framework of stochastic neoclassical theories shows that *these effects are not due to Einstein-Podolsky-Rosen correlations.* It is well known that the only possible way of describing classical external fields within quantum electrodynamics (without the artificial introduction of c-number fields) is via coherent states (Beck and Thellung, 1969). The convex set of all coherent states of quantum electrodynamics forms a simplex which can be taken as the state space for a neoclassical theory. Roughly we can say that stochastic neoclassical electrodynamics results from quantum electrodynamics if the state space is restricted to coherent states. To every coherent state of quantum electrodynamics there is a corresponding state in neoclassical electrodynamics, but superpositions of coherent states have no analogues in the neoclassical theory.

The equations of motion for neoclassical electrodynamics are given by Maxwell's equations. The difference to traditional classical electrodynamics is the assumption of a universal Gaussian random field, called background field. This background field is the classical counterpart of the vacuum zero-point field of quantum electrodynamics. Up to a constant, it is uniquely fixed by the requirement of Lorentz invariance, and can be characterized by the spectral energy density

$$\rho(\omega) = \frac{\hbar|\omega|^3}{2\pi^2 c^3} \quad ,$$

through which Planck's constant $\hbar$ enters this classical theory. Strange to say, stochastic classical electrodynamics can even reproduce some quantization effects of molecular matter. In a remarkable paper, Marshall (1963) has shown that a *classical* harmonic oscillator (in whose equation of motion the radiation reaction force $(2e^2/3c^3)\dddot{q}$ has been included) which is subjected to classical stochastic background radiation field behaves like a quantum mechanical harmonic oscillator (without radiation damping) in a stationary state.

Additional references:

Marshall's model has been investigated in more details by Boyer (1975a,b) and by de la Peña and Cetto (1979). For reviews of the present status of semiclassical and neoclassical radiation theories, compare Mandel and Wolf (1973, pp.35-119, 273-317), (1978, pp.469-509), Claverie and Diner (1977, 1980), Barut (1980).

For linear symplectic systems (like the harmonic oscillator, the rigid rotator, or the electromagnetic field) there is a very close correspondence between pioneer quantum mechanics and stochastic neoclassical mechanics. That is, though the adiabatic caricature of molecules describes the nuclei in a classical way, it is possible to interprete vibrational and rotational spectra by neoclassical harmonic oscillators and rigid rotators. From a practical point of view, the discussion of the response of a molecule to high-power infrared laser radiation is even much simpler than in pioneer quantum mechanics. On the other hand, nonlinear problems (like the anharmonic oscillator or the nonrigid oscillator) are difficult, a sound neoclassical description of such systems is not yet available.

The tale of the best theory

From the viewpoint of pioneer quantum mechanics it is tempting to say that molecules should always be "*discussed in a unified way using the best theory available, namely quantum mechanics*" (Woolley, 1976b). I strongly disagree. *There is no best theory.* If pioneer quantum mechanics would be the best theory, then the best description of a barking dog would use the observables of pioneer quantum mechanics. Admittedly, there is a difference between a barking dog and a rotating molecule. But if pioneer quantum mechanics is not appropriate for the description of a barking dog, why is it so certain that it is appropriate for a rotating molecule?

Pioneer quantum mechanics is not the *best* theory but it is a *fundamental* theory. That is not the same. Fundamental theories are never good for the description of higher hierarchical levels. There are even more fundamental theories than pioneer quantum mechanics (say relativistic quantum field theory), but these theories are even less appropriate for a description of molecules or dogs. Phenomena involving positrons cannot be described conveniently on the level of Galilei covariant pioneer quantum mechanics, and it may be true that some high resolution experiments on small molecules cannot be grasped conveniently at the hierarchical level of the adiabatic description of molecules. On the other hand, pioneer quantum mechanics cannot give a conceptually consistent *and* chemically relevant description of molecular structure. But molecular structure is too important a concept to be thrown overboard only because it is foreign to the edifice of ideas belonging to some fundamental theory.

Molecules can be studied on many hierarchical levels and it would be unwise to assume that there is a best description independent of our goals. The most fundamental theories we have hardly ever give a good description of higher-level phenomena. Nevertheless, fundamental theories are important because they allow the derivation of higher-level theories which may give an appropriate description of observable phenomena. The derivation of a higher hierarchical level from a fundamental theory *always* is a very difficult task. It is mathematically so delicate that it should not be left as a matter of routine which every good experimentalist can perform.

Pioneer quantum mechanics is a reasonable starting point for the *theoretician*, but chemistry deserves better theories, theories which can describe quantum properties and classical properties at the same time, and which operate on a chemically relevant level. A rigorous derivation of good higher level theories from pioneer quantum mechanics remains an outstanding conceptual and mathematical problem which is central to a more comprehensive understanding of the complex behavior of molecular matter. The dynamical W*-system created by the adiabatic limit is an example for such a higher-level theory. Neither the description by pioneer quantum mechanics nor the adiabatic caricature is of higher quality than the other, they are two incomensurable descriptions of the same reality, they should not be thought as contradictory in any sense. They serve different purposes.

6.5 ONE WAY OF TELLING IS NOT ENOUGH

The death of natural laws

Classical physics has believed in the existence of a real world behaving in accordance with objective, immutable and universal laws, revealing an omnipotent reason which rules nature. For example, Max Planck (1858-1947) argued that the laws of nature are not invented in the minds of men but that external factors force us to recognize these laws. This view is still quite popular. For example, Richard Feynman said in his BBC television lectures: "*The age in which we live is the age in which we are discovering the fundamental laws of nature, and that day will never come again. It is very exciting, it is marvellous, but this excitement will have to go*" (Feynman, 1965, p.172). In some cases, this excitement has already gone since we are beginning to see through the myth of universal laws.

Consider for example the atomic idea. Feynman thinks that the most informative fact of scientific knowledge is "*that all things are made of atoms - little particles that move around in perpetual motion, attracting each other when they are a little distance apart, but repelling upon being squeezed into another*" (Feynman et al., 1963, p.1-2). Many modern scientists still think that this atomic idea is a context-independent fact. They think so because they tacitly accept some forms of interaction with reality as natural. The modern formulation of the atomic idea is in terms of elementary particles, where elementarity is defined in a group-theoretical sense with respect to an appropriate kinematical space-time group. For example, in the nonrelativistic theory of matter, an elementary particle is defined as an ergodic representation of the Galilei group. The question "do atoms really exist?" is much too naive to allow a clear-cut answer. *If* we interact with matter in a way that emphasizes the space-time structure and its description via the Galilei group, then we have a starting point for a sensible answer. However, a priori there is no reason which would force us to adopt this viewpoint. That is, an elementary particle does not exist as a thing-in-itself but the atomic idea is *enforced* by the adopted viewpoint.

Werner Heisenberg was once thinking of the possibility that physics might one day be finished by the discovery of the ultimate world equation which describes everything in terms of one single spinor equation. If physics should ever come to such an equation then we will probably recognize it as a truism. A universal world equation is never a deep divine law but a group theoretical triviality: it does not represent nature but our presuppositions.

Example: The triviality of the Schrödinger equation

Given the Hilbert-space formalism of pioneer quantum mechanics, the Schrödinger equations $i\dot{\phi}=H\phi$ and $H\Psi=E\Psi$ are trivialities which follow via Stone's theorem from the *presupposed* existence of a dynamical group. Schrödinger's equations are said to be valid for any closed system, where a closed system is characterized by the fact that its time evolution is governed by a one-parameter group.

More interesting than the form of Schrödinger's equations is the specific form of the Hamiltonian H. For noninteracting elementary objects, even the form of the Hamiltonian is a truism in the sense that it follows from the *presupposed* behavior of the system under Galilei transformations.

Modern theoretical research has recognized many of the historical "great natural laws" as *truisms*. They are not really descriptions of nature but consequences of the physical approach. The law of conservation of energy is a simple consequence of the presupposed homogenity of time, the laws of conservation of momentum and angular momentum follow from the presupposed homogenity and isotropy of space, Feynman's dictum that all things are made of particles says nothing but that every representation of the kinematical group can be constructed out of ergodic representations, the fundamental laws of thermodynamics do not describe nature but are prescriptions to scientists to act in certain ways rather than in others.

Natural laws are not discovered, they are human creations enforced by a particular point of view. We do no longer believe in "great natural laws" since truisms are not the kind of statements to believe in. *Natural laws are compelling consequences of some presupposed group theoretical structure*. Or in the words of Sir Arthur Eddington: "*We have found a strange foot-print on the shores of the unknown. We have devised profound theories, one after another, to account for its origin. At last, we have succeeded in reconstructing the creature that made the foot-print. And Lo! it is our own*" (Eddington, 1920, p.201).

When Friedrich Nietzsche proclaimed the "Death of God" he meant that the existence of God in the modern world was no longer a natural fact as it was for men in earlier centuries. By the same token we may speak of the "Death of Natural Laws" in modern science. This development implies a farewell to the idea that science rests on facts and natural laws.

Once scientists believed in fundamental laws, hence they had imagination and found laws. Science was fascinating, it was in touch with the soul. Modern group theory allowed us to see through the alleged fundamental laws so that the magical validity of natural laws disappeared and science has lost its conviction. Theories have no longer to be true but just useful. We approach nature with such methods that we can only find what we do find. With this, the classical scientific language is dead since it does no longer carry soul. A new style in science began with quantum mechanics when Niels Bohr initiated a spiritual renewal by introducing his concept of complementarity. It turned out that complementarity is far more important than quantum mechanics, it has led to a development of science that encourages a holistic vision.

Complementarity in modern science

Bohr's conception of complementarity marks a singular turning point in the development of physics. It was in September 1927, during the commemoration of the hundredth anniversary of Volta's death in his native city of Como, that Niels Bohr introduced the term "complementarity" to account for situations where two different conditions of observations yield conclusions that are conceptually incompatible. For physical situations, a more precise characterization has been given by Pauli: "*Wenn ... die Benutzbarkeit eines klassischen Begriffes in einem ausschliessenden Verhältnis zu der eines anderen steht, nennen wir diese beiden Begriffe ... komplementär*" (Pauli, 1933, Ziff.1). Bohr made it very clear that the notion of complementarity is not restricted to quantum physics. He stressed that the idea of complementarity is related "*to the general difficulty in the formation of human ideas, inherent in the distinction between subject and object*" (Bohr, 1928).

Additional references

A short discussion of the history of the concept of complementarity can be found in Jammer (1974, sect.4.4). For a profound analysis of Bohr's ideas compare Meyer-Abich (1965). In a little known book, Pascual Jordan (1947) has discussed the relations between quantum physics and the psychology of the unconscious, in particular the deep analogy between the physical concept of the complementarity and the psychological notion of repression.

While Bohr's concept of complementarity has a unique position in the history of physics, the general idea is as old as religious and philosophical thought. For example, as we can learn from Carl Gustav Jung (Coll. Works, vol.12), the union of opposites has played a decisive role in the alchemical process. In his studies of psychological types of 1921, Jung (Coll. Works, vol.6, par.854-855) made the following modern sounding remarks: "*For anyone who thinks there is only one true explanation of a psychic process, this vitality of psychic contents, which necessitates two contradictory theories, is a matter for despair, espcecially if he is enarmoured of simple and uncomplicated thruths, incapable maybe of thinking both at the same time. ... I believe that other equally 'true' explanations of the pyschic process can still be put forward, just as many in fact as there are types.*"

Complementary descriptions refer to the same reality but from different points of view, and such that no experiment can be devised that could demonstrate complementary aspects in a single observational context. Neither mode of two complementary descriptions can be subsumed into the other. A particular description is a *projection* of reality into a particular context. Complementary descriptions are not independent of each other since they refer to the same reality, they are not contradictionary since they refer to different contexts. "*Truths may clash without contradicting each other*" (Saint-Exupéry, 1948).

Examples

a) *Mechanics vs. thermodynamics*
The description of a macroscopic system in terms of the dynamics of molecular motion is complementary to the thermodynamic description (Bohr, 1932, compare in particular p.376). The conditions allowing a complete mechanical description exclude the possibility of applying thermodynamic concepts. On the other hand, to assign to a system a definite temperature requires conditions of observation under which the particle dynamics escapes control.

b) *Newtonian vs. Einsteinian mechanics*
Newtonian mechanics and Einstein's special theory of relativity are *not* complementary descriptions since there exist experiments which can decide between the predictions of these two theories.

c) *Physics vs. chemistry*
The description of a molecule in terms of pioneer quantum mechanics is complementary to the chemist's description. Both describe the same system but they have different viewpoints, hence they use different W*-systems. The physicist uses pioneer quantum mechanics, that is he chooses a factor of type I_∞ as his algebra of observables so that his theoretical framework does not allow the description of classical features. The chemist's description is different, he describes molecules as having a classical molecular structure and uses a type I_∞-W*-algebra with a large center as his algebra of observables.

In the physicist's description electrons and nuclei are correlated by Einstein-Podolsky-Rosen correlations so that neither electrons nor nuclei exist as individual objects. In the chemist's description electrons and nuclei are not correlated in the sense of Einstein, Podolsky and Rosen (in spite of their strong electromagnetic interaction), so that nuclei exist as classically describable individual objects.

These two descriptions of a molecule are complementary, they correspond to the complementary ways physicists and chemists interact with nature. Nevertheless, the higher-level chemical view follows in a mathematically rigorous way by an asymptotic caricature. As discussed in sect.6.4, this limit is given by $\varepsilon \to 0$, $t \to \infty$, with $\varepsilon t =$ constant, where $\varepsilon^2 = m/M$ is the ratio between the electron mass m and the molecular mass M.

d) *Structure vs. function*
The dialectical nature of the structure-function relationship marks the complementarity between the physical and system-theoretic approach. The complementarity between the structural properties of a system and its behavioral activities is characteristic for organized systems and is particularly important in the biological realm. The structural properties of a system and its behavioral activities is how the mechanisms work, the function of a system refers to what it is made *for*. But structural descriptions do not refer to functional relations, functional descriptions do not refer to structural relations. In general, a given structure can put into effect a variety of different functions, and a given function can be realized by a variety of different structures.

For most phenomena Boolean logic is far too simple a language. Nature manifests herself to a scientist in many complementary modes. Even within the narrow compass of exact science we need mutually exclusive viewpoints and antithetical concepts. No particular description of nature is complete, no single abstraction is appropriate for all contexts. A full account requires the simultaneous articulation of complementary modes of descriptions, it can be achieved only by enlarging the universe of discourse to include all complementary descriptions of reality. Non-Boolean quantum logic provides such a common frame for the exact natural sciences. According to the ontic interpretation of quantum logic, the existence of incompatible potential properties is the root of the existence of complementary descriptions and the basic condition for the emergence of novelty.

Complementarity is by no means a mysterious or vague concept but is open to the penetrating mind. The category of W*-systems allows a precise characterization of complementarity in description of a large class of natural systems. Within an encompassing basic W*-theory, the concept of non-Boolean theory reduction (compare sect.5.5) shows how complementary subtheories are related. Every W*-subtheory corresponds to a particular view, complementary subtheories correspond to complementary views. We are free to choose between mutually exclusive viewpoints (Pauli, 1948), but every choice has to be paid for by the loss of complementary knowledge.

On the objectivity of the many viewpoints

If we speak of the reality of an entity we refer without exception to a concept relative to a given context. Is such a notion objective? Yes, provided we agree that the criterion "all observers having the same point of view do agree" is sufficient to define objectivity. The adopted viewpoint reflects a subjective aspect in our description of the world. Subjective aspects are not mere illusions but necessary conditions for objectivity. Or in the words of Georg Picht: "*Nur wo es gelingt, alles was 'nicht dazugehört', als störende Faktoren zu eliminieren, gewinnt man Erkenntnisse, die nachprüfbar sind. Man nennt diese Erkenntnis objektiv. Die Objektivität der modernen Wissenschaft ist geradezu dadurch definiert, dass sie der Wirklichkeit nicht entspricht, sondern auf solche Erkenntisse eingeschränkt ist, die sich durch eine eindeutig zu kontrollierende Versuchsanordnung gewinnen lassen*" (Picht, 1969, p.90). That is, objectivity can never mean anything else but *conditional intersubjective agreement*, conditioned by a jointly accepted context. The criterion of objectivity is that the perceptions can be shared.

Invention is prior to discovery

The classical scientist believes in *facts* which are discovered through observation of natural reality. For example, Garrett Birkhoff (1950, p.179) thinks that "*there is hidden order in Nature, to be found only by patient search*". However, whether there *is* order in nature is a hotly debated question. Order in nature has been frequently asserted because her creator is wise, reasonable and no lover of chaos. On the other hand, one can argue that we stand in relation to nature more as inventors than discoverers. However that may be, the world does not present itself to us neatly divided into facts but, as observed by Susanne Langer (1942, p.273) "*our world 'divides into facts' because we so divided it*".

Our obstinate concern for hard facts deprived us of the insight that *every fact is conditional and context-dependent*. What a person or society means by a 'fact' cannot be understood in isolation from the whole conceptual structure. The cognition of a "fact" is a strongly paradigm-conditioned activity of man. This social nature of scientific knowledge has been discussed by Ludwik Fleck (1935), in particular he stressed that *scientific facts are invented, not discovered.*

For every problem, for every observation, for every experiment we have to choose the appropriate context. If we want to explore nature, we have to *invent* a suitable context. Invention is logically prior to discovery. Only if we have chosen a context, we can discover. Even if we are not entirely free, we have a choice of contexts. The creation of a new context enlist's a man's utmost technical skill in the service of his imagination. Newton's remark that apples are attracted by the earth was not a new empirical fact or a discovery but a great idea. Similarly, the general theory of relativity did not start with a new empirical fact but, in Albert Einstein's own words, with "*the happiest thought of my life*", namely: "*if one considers an observer in free fall, e.g. from the roof of a house, there exists for him during his fall no gravitational field - at least in his immediate vicinity*" (Einstein, 1919). It requires a reorientation to acknowledge that *science starts with ideas, not with facts*. New ideas are viable if they lead to new viewpoints which can be used to define new contexts that allow interesting divisions of the holistic world.

Classical physics has believed in a universal frame of reference, a universal context that permits independent variations of the elements. This doctrine has encouraged the search of basic building blocks of matter through which one hoped to understand nature. Such an approach has the flavor of a purely empirical undertaking in which *discoveries* play the fundamental role. However, quantum mechanics taught us that the hunt for a universal context is in vain. Nature is holistic and her separability into facts a matter of context. In a holistic world *objects have to be invented;* in a theoretical description *objects are created by abstracting from the Einstein-Podoslky-Rosen-correlation between the object and its environment.* The properties of the invented objects can be discovered experimentally, by experiments which neglect some existing correlations as irrelevant. The exploration of nature leads to inventions and discoveries. *Discoverers have a fixed point of view, inventors create new contexts.*

In retrospect: Knowledge without wisdom

Science deals with ideas and its freedom is its essence. But at what cost? There are no scientific facts untouched by our inquiry, every scientific experiment is an essential intervention with the wholeness of nature. Furthermore, we should not underrate the reality of new ideas and symbolic forms. Ideas transform the meaning of the world.

It is probable that in future scientists may have a rough time to defend their freedom to develop new ideas. We praise the joy of science and the sheer delight of invention and discovery. Yet we are free to invent and to discover only insofar as we can take the *full* responsibility for our creations. It is a fact that a large part of our scientific manpower is associated with the development of most potent means of destruction. We are proud of the international traditions in science and of the fraternity of scientists, but we hardly ever mention that about half of all scientists work on military programs. Still worse, most scientists working in weapon technology are *fascinated* by problems like binary chemical weapons, laser guided smart bombs and automated battlefields. Science is by no means socially and politically neutral but serves more and more sinister ends. It would be utterly infantile to assume that scientists have no responsibility for their ideas and creations. One has to be naive indeed to believe in the intellectual superiority of scientists and in the unquestionable value of scientific knowledge for humanity. The excitement and fascination of science cannot any longer be the basis for the activities of responsible men. Science has an uncontrollable impact on society, why should society not be allowed to control science?

Mediocre science should not be supported and good science is too dangerous for intelligent fools who are not aware of its limitation. If our interest is restricted to science it becomes an obsessive and destructive ideology. We only see what we look for.

Contemporary science has a tendency to propagate a narrow intellectualism, to slander insights found elsewhere, and to take over the role of the one and only one redeeming church. The success of science should not blind us to see the fact that the world view of science abstracts from vital aspects of reality so that scientific knowledge represents only a tiny fraction of human knowledge. Scientific progress has been made by artificial isolation and compartmentalization. It would be downright dishonest not to examine most painstakingly whether or not it is wise to further develop science with its pre-eminence given to extraverted thinking rather than to explore other types of orientations which can give complementary knowledge about the relations between man and his world. "*A good teacher will not just make people accept a form of life, he will also provide them with means of seeing it in perspective and perhaps of even rejecting it*" (Feyerabend, 1979, p.86).

There is little to be gained in rejecting science but we should know about the myth that scientists are living. We can accept science without accepting that science is the only valid way of understanding the world. Science itself is not in the position to explain what is outside its boundaries. For seeing science in perspective we have to understand it, but we can only understand the scientific world view if we have antithetical viewpoints available. In order to grasp the many aspects of reality we have to look - like the Roman god Janus - into several opposite directions at the same time. There is a basic harmony between the rational and the irrational, the regular and the irregular, the civilized and the primitive, and the logic and the mythic.

Even though the prevailing scientific doctrine still glorifies the classical ideal of explanation, the legendary idea that wholes are explicable in terms of their parts is dead. But new symbolic languages are once again in the making. Nowadays, complementarity speaks with a living tongue. The multiplicity of complementary views corresponds to the psychic reality of modern man with his multiplicity of souls and voices. In modern quantum theory we accept the necessity of complementary viewpoints. None of them is more authentic than the other, none can replace the others, all are necessary, none is sufficient. Despite of the precision of each language, the paradoxical aspect of reality is not lost.

A hopeful prospect is that man will limit science by supplementing it with complementary modes of orientations. There is a strong tendency in contemporary culture favoring holistic views. Since holism and reductionism are complementary viewpoints, both can be accepted, both are necessary, none of them is sufficient. We are beginning to realize that the hidden part of nature is still the largest and maybe the most difficult one to see. Perhaps one day we will come to recognize the wisdom of the paradoxical truth and again cultivate complementary viewpoints and alternative instruments for probing reality.

BIBLIOGRAPHY AND AUTHOR INDEX

References are by the author's name, followed by the year of publication. Whenever possible, the year refers to the first published version. For the convenience of the reader, republished or translated versions are occasionally added. Author's names transcribed from other alphabets are standardized, the originally published version of the name is added for cross reference. Section numbers for quotations of the reference are listed to permit use of this bibliography as an author index.

Aarnes, J.F. (*1969*): Physical states on a C*-algebra. Acta Math. 122, 161-172.
Quoted in sect.4.5.

Aarnes, J.F. (*1970*): Quasi-states on C*-algebras. Trans.Amer.Math.Soc. 149, 601-625.
Quoted in sect.4.5.

Abrikosov, A.A., *Gorkov*, L.P. *and Dzyaloshinski*, I.E. (*1962*): Methods of quantum field theory in statistical physics. Prentice-Hall, Englewood Cliffs, N.J., 1963. (Russian original: Fizmatgiz, Moscow, 1962).
Quoted in sect.4.3.

Accardi, L. (*1975*): The noncommutative Markov property. Functional Anal.Appl. 9, 1-7. (Russian original: Funkcional.Anal.i Priložen. 9, 1-8).
Quoted in sect.4.5.

Accardi, L. (*1976*): Nonrelativistic quantum mechanics as a noncommutative Markov process. Advances in Mathematics 20, 329-366.
Quoted in sect.4.5.

Accardi, L. (*1978*): Noncommutative Markov chains associated to a preassigned evolution: an application to the quantum theory of measurement. Advances in Mathematics 29, 226-243.
Quoted in sect.4.5.

Accardi, L. (*1979*): Local perturbations of conditional expectations. J.Math.Anal.Appl. 72, 34-69.
Quoted in sect.4.5.

Aerts, D. *and Daubechies*, I. (*1978a*): About structure-preserving maps of a quantum mechanical propositional system. Helv.Phys.Acta 51, 637-660.
Quoted in sect.4.4.

Aerts, D. *and Daubechies*, I. (*1978b*): Physical justification for using the tensor product to describe two quantum systems as a joint system. Helv.Phys.Acta 51, 661-675.
Quoted in sect.4.4.

Aerts, D. *and Daubechies*, I. (*1979*): A connection between propositional systems in Hilbert spaces and von Neumann algebras. Helv.Phys.Acta 52, 184-199.
Quoted in sect.4.4.

Aharonov, Y. *and Petersen*, A. (*1967*): Some quantum aspects of interference experiments. IEEE Trans.Antennas Propagat. AP 15, 186-187.
Quoted in sect.3.7.

Akemann, C.A. *and Newberger*, S.M. (*1973*): Physical states on a C*-algebra. Proc.Amer. Math.Soc. 40, 500.
Quoted in sect.4.5.

Albert, A.A. (*1934*): On a certain algebra of quantum mechanics. Annals of Mathematics 35, 65-73.
Quoted in sect.4.2.

Albertson,W. (*1963*): Quantum-mechanical measurement operator. Phys.Rev. 129, 940-943.
Quoted in sect.3.5.

Alfsen,E.M. (*1971*): Compact convex sets and boundary integrals. Springer, Berlin.
Quoted in sect.4.5.

Alfsen,E.M. and Shultz,F.W. (*1976*): Non-commutative spectral theory for affine function spaces on convex sets. Memoirs Amer.Math.Soc., vol.172, Providence, R.I.
Quoted in sect.4.5.

Alfsen,E.M. and Shultz,F.W. (*1978*): State spaces of Jordan algebras. Acta Math. 140, 155-190.
Quoted in sect.4.5.

Alfsen,E.M. and Shultz,F.W. (*1979*): On non-commutative spectral theory and Jordan algebras. Proc.London Math.Soc. 38, 497-516.
Quoted in sect.4.5.

Alfsen,E.M., Hanche-Olsen,H. and Shultz,F.W. (*1980*): State spaces of C*-algebras. Acta Math. 144, 267-305.
Quoted in sect.4.5.

Alfsen,E.M., Shultz,F.W. and Størmer,E. (*1978*): A Gelfand-Neumark theorem for Jordan algebras. Advances in Math. 28, 11-56.
Quoted in sect.4.2.

Ali,S.T. and Emch,G.G. (*1974*): Fuzzy observables in quantum mechanics. J.Math.Phys. (N.Y.) 15, 176-182.
Quoted in sect.3.5.

Amai,S. (*1962*): On observation in quantum mechanics. Progr.Theor.Phys. 28, 401-402.
Quoted in sect.3.5.

Amai,S. (*1963*): Theory of measurement in quantum mechanics. Destruction of interference. Progr.Theor.Phys. 30, 550-562.
Quoted in sect.3.5.

Amai,S. (*1964*): On the entropy problem in the theory of measurement in quantum mechanics. Progr.Theor.Phys. 31, 931-933.
Quoted in sect.3.5.

Amann,A. (*1978*): Eine Verallgemeinerung des quantenmechanischen Elementaritätsbegriffs. Diplomarbeit an der Abteilung für Chemie der ETH Zürich. Unpublished.
Quoted in sect.5.6.

Amemiya,I. and Araki,H. (*1966*): A remark on Piron's paper. Publ.Res.Inst.Math.Sci. (Kyoto),Ser.A2, 423-427.
Quoted in sect.4.4.

Anderson,P.W. (*1963*): Plasmons, gauge invariance, and mass. Phys.Rev. 130, 439-442.
Quoted in sect.3.3.

Anderson,P.W. (*1972*): More is different. Science 177, 393-396.
Quoted in sect.6.1.

Andrews,H.C. (*1972*): Introductions to mathematical techniques in pattern recognition. Wiley-Interscience, New York.
Quoted in sect.6.3.

Antoine,J.P. and Gleit,A. (*1971*): Space-time structure and measurement theory. Int.J.Theor.Phys. 4, 197-216.
Quoted in sects.3.5, 3.7.

Araki,H. (*1968*): Multiple time analyticity of a quantum statistical state satisfying the KMS boundary condition. Publ.Res.Inst.Math.Sci.(Kyoto),Ser.A4, 361-371.
Quoted in sect.4.3.

Araki,H. (*1973*): Relative Hamiltonian for faithful normal states of a von Neumann algebra. Publ.Res.Inst.Math.Sci.(Kyoto),Ser.A9, 165-209.
Quoted in sect.4.3.

Araki,H. (*1974*): Some properties of modular conjugation operator of von Neumann algebras and a non-commutative Radon-Nikodym theorem with a chain rule. Pacific J.Math. 50, 309-354.
Quoted in sect.4.5.

Araki,H. *and Ion*,P.D.F. (*1974*): On the equivalence of KMS and Gibbs conditions for states of quantum lattice systems. Commun.Math.Phys. 35, 1-12.
Quoted in sect.4.3.

Araki,H. *and Kishimoto*,A. (*1977*): Symmetry and equilibrium states. Commun.Math.Phys. 52, 211-232.
Quoted in sect.4.3.

Araki,H. *and Miyata*,H. (*1968*): On KMS boundary condition. Publ.Res.Inst.Math.Sci. (Kyoto),Ser.A4, 373-385.
Quoted in sect.4.3.

Araki,H. *and Woods*,E.J. (*1963*): Representations of the canonical commutation relations describing a nonrelativistic infinite free Bose gas. J.Math.Phys. (N.Y.) 4, 637-662.
Quoted in sects.4.3, 5.5.

Araki,H. *and Woods*,E.J. (*1968*): A classification of factors. Publ.Res.Inst.Math.Sci. (Kyoto),Ser.A4, 51-130.
Quoted in sect.5.5.

Araki,H. *and Yanase*,M.M. (*1960*): Measurement of quantum mechanical operators. Phys.Rev. 120, 622-626.
Quoted in sect.3.5.

Araki,H., *Kastler*,D. *and Takesaki*,M. (*1977*): Extension of KMS states and chemical potential. Commun.Math.Phys. 53, 97-134.
Quoted in sect.4.3.

Araki,H., *Smith*,M.B. *and Smith*,L. (*1971*): On the homotopical significance of the type of von Neumann algebra factors. Commun.Math.Phys. 22, 71-88.
Quoted in sect.5.4.

Arthurs,E. *and Kelly*,J.L. (*1965*): On the simultaneous measurement of a pair of conjugate observables. Bell Syst.Tech.J. 44, 725-729.
Quoted in sect.4.5.

Auger,P. (*1961*): Current trends in scientific research. United Nations, New York.
Quoted in sect.1.1.

Bacry,H. *and Lévy-Leblond*,J.M. (*1968*): Possible kinematics. J.Math.Phys.(N.Y.) 9, 1605-1614.
Quoted in sect.3.3.

Bader,R.F.W. (*1975*): Molecular fragments or chemical bonds? Acc.Chem.Res. 8, 34-40.
Quoted in sect.1.3.

Bader,R.F.W. *and Preston*,H.T.J. (*1969*): The kinetic energy of molecular charge distributions and molecular stability. Int.J.Quantum Chem. 3, 327-347.
Quoted in sect.1.3.

Bader,R.F.W., *Anderson*,S.G. *and Duke*,A.J. (*1979a*): Quantum topology of molecular charge distributions.I. J.Amer.Chem.Soc. 101, 1389-1395.
Quoted in sect.1.3.

Bader,R.F.W., Nguyen-Dang,T.T. and Tal,Y. (1979b): Quantum topology of molecular charge distributions.II. Molecular structure and its change. J.Chem.Phys. 70, 4316-4329.
Quoted in sect.1.3.

Bailar,J.C., Emeléus,H.J., Nylholm,R. and Trotman-Dickenson,A.F. (1973): Comprehensive inorganic chemistry. Volume I. Pergamon Press, Oxford.
Quoted in sect.6.2.

Ballentine,L.E. (1970): The statistical interpretation of quantum mechanics. Rev.Mod. Phys. 42, 358-381.
Quoted in sect.3.4.

Bardeen,J., Cooper,L.N. and Schrieffer,J.R. (1957): Theory of superconductivity. Phys.Rev. 108, 1175-1204.
Quoted in sect.6.2.

Bargmann,V. (1954): On unitary ray representations of continuous groups. Annals of Mathematics 59, 1-46.
Quoted in sects.3.2, 3.3, 5.4.

Bargmann,V. (1964): Note on Wigner's theorem on symmetry operations. J.Math.Phys. (N.Y.) 5, 862-868.
Quoted in sect.3.3.

Bargmann,V. and Wigner,E.P. (1948): Group theoretical discussion of relativistic wave equations. Proc.Nat.Acad.Sci.(USA) 34, 211-223.
Quoted in sect.3.3.

Barnes,J.L. (1964): Information theoretic aspects of feedback control systems. IEEE Int.Conv.Rec. 1964, Part 1, 142-153. Revised version: Automatica 4, 165-185 (1968).
Quoted in sect.3.8.

Barnes,J.L. (1970): Laplace-Fourier transformation, the foundation for quantum information theory and linear physics. In: "Problems in analysis. A symposium in honor of Salomon Bochner"; ed. by R.C.Gunning; Princeton University Press, Princeton, New Jersey; pp. 157-173.
Quoted in sect.3.8.

Barut,A.O. (editor) (1980): Foundations of radiation theory and quantum electrodynamics. Plenum Press, New York.
Quoted in sect.6.4.

Batty,C.J.K. (1979): The strong law of large numbers for states and traces of a W*-algebra. Z.Wahrscheinlichkeitstheorie verw. Geb. 48, 177-191.
Quoted in sect.4.5.

Baumann,K. (1970): Quantenmechanik und Objektivierbarkeit. Z.Naturforsch.A 25, 1954-1956.
Quoted in sects.3.8, 5.6.

Baumann,K. (1972): Die objektiv-reale Wirklichkeit und die Quantenmechanik. Acta Phys. Austriaca 36, 1-6.
Quoted in sects.3.8, 5.6.

Beck,H. and Thellung,A. (1969): Representation of external fields by means of coherent states. Helv.Phys.Acta 42, 678-684.
Quoted in sect.6.4.

Becker,E. (1973): The denial of death. The Free Press, New York.
Quoted in sect.1.1.

Belinfante, J.G.F. (1973): A survey of hidden-variable theories. Pergamon Press, Oxford.
Quoted in sect.3.4.

Belinfante,J.G.F. *(1976)*: Transition probability spaces. J.Math.Phys.(N.Y.) 17, 285-291.
Quoted in sect.4.5.

Bell,J.S. *(1964)*: On the Einstein Podolsky Rosen paradox. Physics (Long Island City, N.Y.) 1, 195-200.
Quoted in sects.3.4, 3.7.

Bell,J.S. *(1966)*: On the problem of hidden variables in quantum mechanics. Rev.Mod. Phys. 38, 447-452.
Quoted in sect.3.4.

Bell,J.S. *(1971a)*: Introduction to the hidden-variable question. In: "Proceedings of the international school of physics 'Enrico Fermi'. Course 49: Foundations of quantum mechanics"; ed. by B.d'Espagnat; Academic Press, New York; pp. 171-181.
Quoted in sect.3.4.

Bell,J.S. *(1971b)*: On the hypothesis that the Schrödinger equation is exact. CERN-preprint TH.1424.
Quoted in sect.3.6.

Bell,J.S. *(1972)*: The measurement theory of Everett and de Broglie's pilot wave. CERN-preprint TH.1599.
Quoted in sect.3.6.

Bell,J.S. *and Nauenberg*,M. *(1966)*: The moral aspect of quantum mechanics. In: Preludes in theoretical physics. In honor of V.F.Weisskopf"; ed. by A.De-Shalit, H.Feshbach and L.van Hove; North Holland, Amsterdam; pp. 279-286.
Quoted in sect.3.5.

Beltrametti,E.G. *and Cassinelli*,G. *(1972)*: Quantum mechanics and p-adic numbers. Foundations of Physics 2, 1-7.
Quoted in sect.4.4.

Beltrametti,E.G. *and Cassinelli*,G. *(1973)*: On the logic of quantum mechanics. Z.Naturforsch.A 28, 1516-1530.
Quoted in sect.4.4.

Beltrametti,E.G. *and Cassinelli*,G. *(1976)*: Logical and mathematical structures of quantum mechanics. Rivista del Nuovo Cimento 6, 321-404.
Quoted in sect.4.4.

Benioff,P.A. *(1972a)*: Operator valued measures in quantum mechanics: finite and infinite processes. J.Math.Phys.(N.Y.) 13, 231-242.
Quoted in sect.3.5.

Benioff,P.A. *(1972b)*: Decision procedures in quantum mechanics. J.Math.Phys.(N.Y.) 13, 908-915.
Quoted in sect.3.5.

Benioff,P.A. *(1972c)*: Procedures in quantum mechanics without von Neumann's projection axiom. J.Math.Phys.(N.Y.) 13, 1347-1355.
Quoted in sect.3.5.

Benioff,P.A. *(1973a)*: On definitions of validity applied to quantum theories. Foundations of Physics 3, 359-379.
Quoted in sect.3.5.

Benioff,P.A. *(1973b)*: Possible strengthening of the interpretative rules of quantum mechanics. Phys.Rev.D 7, 3603-3609.
Quoted in sect.3.5.

Benioff,P.A. *(1974)*: Some consequences of the strengthened interpretative rules of quantum mechanics. J.Math.Phys.(N.Y). 15, 552-559.
Quoted in sect.3.5.

Bennet,M.K. (*1970*): A finite orthomodular lattice which does not admit a full set of states. SIAM Rev. 12, 267-271.
Quoted in sect.4.4.

Benoist,R.W. and Marchand,J.P. (*1979*): Statistical inference in coupled quantum systems. Letters in Mathematical Physics 3, 93-96.
Quoted in sect.4.5.

Benoist,R.W., Marchand,J.P. and Yourgrau,W. (*1977*): Statistical inference and quantum mechanical measurement. Foundations of Physics 7, 827-833.
Quoted in sect.4.5.

Berberian,S.K. (*1972*): Baer *-rings. Springer, Berlin.
Quoted in sect.4.6.

Bernard,C. (*1865*): Introduction à l'étude de la médicine expérimentale. Ballière et Fils, Paris. (English translation: "An introduction to the study of experimental medicine", Dover, New York, 1957).
Quoted in sect.6.1.

Bernstein,J. (*1974*): Spontaneous symmetry breaking, gauge theories, the Higgs mechanism and all that. Rev.Mod.Phys. 46, 7-48. Errata: ib. 46, 855 and 47, 259 (1975).
Quoted in sect.3.3.

Bertolini,G., Bettoni,M and Lazzarini,E. (*1955*): Angular correlation of scattered annihilation radiation. Nuovo Cimento 2, 661-662.
Quoted in sect.3.7.

Berzi,V. and Gorini,V. (*1969*): Reciprocity principle and the Lorentz transformations. J.Math.Phys.(N.Y.) 10, 1518-1524.
Quoted in sect.3.3.

Bethe,H.A. and Salpeter,E.E. (*1957*): Quantum mechanics of one-and-two-electron atoms. Handbuch der Physik, ed. by S.Flügge, Vol.35. Springer, Berlin.
Quoted in sect.3.3.

Birkhoff,G. (*1940*): Lattice theory. American Mathematical Society, Providence, Rhode Island; second revised edition, 1948; third new edition, 1967.
Quoted in sect.4.4.

Birkhoff,G. (*1950*): Hydrodynamics. A study in logic, fact, and similitude. Princeton University Press.
Quoted in sect.6.5.

Birkhoff,G. (*1961*): Lattices in applied mathematics. In: "Proceedings in Pure Mathematics". Vol.2; American Mathematical Society, Providence, R.I.; pp. 155-184.
Quoted in sects.3.2, 4.4.

Birkhoff,G. and Neumann,J.von (*1936*): The logic of quantum mechanics. Annals of Mathematics 37, 823-843.
Quoted in sect.4.4.

Blau,U. (*1973*): Zur 3-wertigen Logik der natürlichen Sprache. Papiere zur Linguistik 4, 20-96.
Quoted in sect.4.4.

Blau,U. (*1978*): Die dreiwertige Logik der Sprache. De Gruyter, Berlin.
Quoted in sect.4.4.

Bloch,F. (*1946*): Nuclear induction. Phys.Rev. 70, 460-474.
Quoted in sects.3.8, 4.5.

Bloch,F., Hansen,W.W. and Packard,M. (*1946*): The nuclear induction experiment. Phys.Rev. 70, 474-485.
Quoted in sect.4.5.

Blochinzew,D.I., see *Blokhintsev,D.I.*

Blokhintsev,D.I. (1953): Kritik der philosophischen Anschauungen der sogenannten 'Kopenhagener Schule' in der Physik. Sowjetwissenschaft, Naturwissenschaftliche Abteilung 6, 545-574.
Quoted in sect.3.4.

Blokhintsev,D.I. (1966): Fundamental problems in quantum mechanics. (In Russian). Nauka. (English translation with the title: "The philosophy of quantum mechanics"; Reidel, Dordrecht-Holland, 1968).
Quoted in sects.3.4, 3.5.

Blokhintsev,D.I. (1968): Interaction of a microsystem with a measuring instrument. Sov.Phys.Usp. 11, 320-327 (1968). (Russian original: Usp.Fiz.Nauk 95, 75-89, 1968).
Quoted in sects.3.4, 3.5.

Bocchieri,P. and Valz-Gris,F. (1972): Dynamical study of an anharmonic crystal interacting with an ideal gas. Lett.Nuovo Cimento 4, 685-689.
Quoted in sect.3.2.

Bocchieri,P. and Valz-Gris,F. (1974): Ergodic properties of an anharmonic two-dimensional crystal. Phys.Rev. A9, 1252-1256.
Quoted in sect.3.2.

Bocchieri,P. and Valz-Gris,F. (1975): The radiation of a charged oscillator under the action of a random force. Lett.Nuovo Cimento 12, 485-486.
Quoted in sect.3.2.

Bocchieri,P., Loinger,A. and Valz-Gris,F. (1974): Classical electrodynamics of a one-dimensional Hohlraum. Invalidity of the equipartition law. Nuovo Cimento B 19, 1-14.
Quoted in sect.3.2.

Bogoliubov,N.N. (1946): Problemy dinamičeskoĭ teorii v statističeskoĭ fizike. OGIZ, Moscow. (English translation: "Problems of dynamical theory in statistical physics", in: "Studies in statistical mechanics", ed. by J.de Boer and G.E.Uhlenbeck, North Holland, Amsterdam, 1962; pp. 1-118).
Quoted in sect.6.2.

Bogoliubov,N.N. and Mitropolsky,Y.A. (1961): Asymptotic methods in the theory of non-linear oscillators. Hindustan Publishing Corp., Delhi, and Gordon & Breach, New York. (Russian original: second edition, Fizmatgiz, Moscow, 1958).
Quoted in sect.6.4.

Bogolyubov,N. see *Bogoliubov,N.N.*

Bohm,D. (1951): Quantum theory. Prentice-Hall, Englewood Cliffs, New Jersey.
Quoted in sects.3.1, 3.5, 3.7.

Bohm,D. (1957): Causality and chance in modern physics. Routledge and Kegan Paul, London.
Quoted in sect.3.4.

Bohm,D. (1962): Hidden variables in the quantum theory. In: "Quantum theory, vol.3. Radiation and high energy physics"; ed. by D.R.Bates; Academic Press, New York; pp. 345-387.
Quoted in sect.3.4.

Bohm,D. and Aharonov,Y. (1957): Discussion of experimental proof for the paradox of Einstein, Rosen, and Podolsky. Phys.Rev. 108, 1070-1076.
Quoted in sect.3.7.

Bohm,D. and Aharonov,Y. (1960): Further discussion of possible experimental tests for the paradox of Einstein, Podolsky and Rosen. Nuovo Cimento 17, 964-976.
Quoted in sect.3.7.

Bohm, D. and Bub, J. (*1966a*): **A proposed solution of the measurement problem in quantum mechanics by a hidden variable theory.** Rev.Mod.Phys. 38, 453-469.
Quoted in sects.3.4, 3.5.

Bohm, D. and Bub, J. (*1966b*): **A refutation of the proof by Jauch and Piron that hidden variables can be excluded in quantum mechanics.** Rev.Mod.Phys. 38, 470-475.
Quoted in sects.3.4, 4.4.

Bohr, H. (*1967*): **My father.** In: "Niels Bohr", ed. by S.Rozental, North Holland. Amsterdam; pp. 325-339.
Quoted in sect.2.6.

Bohr, N. (*1913*): **On the constitution of atoms and molecules.** Phil.Mag. 26, 1-25, 476-502, 857-875.
Quoted in sect.3.2.

Bohr, N. (*1928*): **The quantum postulate and the recent development of atomic theory.** Nature (London) 121, 580-590.
Quoted in sects.3,4, 6.5.

Bohr, N. (*1931*): **Atomtheorie und Naturbeschreibung.** Springer, Berlin. (English translation: "Atomic theory and the description of nature"; Cambridge University Press, London, 1934).
Quoted in sects.3.1, 3.5.

Bohr, N. (*1932*): **Chemistry and the quantum theory of atomic constitution. Faraday Lecture, delivered before the Fellows of the Chemical Society at the Salter's Hall on May 8th, 1930.** J.Chem.Soc. 134, 349-384.
Quoted in sects.2.4, 6.5.

Bohr, N. (*1935*): **Can quantum-mechanical description of physical reality be considered complete?** Phys.Rev. 48, 696-702.
Quoted in sects.3.4, 3.7.

Bohr, N. (*1936*): **Kausalität und Komplementarität.** Erkenntnis 6, 293-303. (English version: "Causality and complementarity", Philosophy of Science 4, 289-298, 1937).
Quoted in sect.3.4.

Bohr, N. (*1948*): **On the notions of causality and complementarity.** Dialectica 2, 312-319. (Reprinted in: Science 111, 51-54, 1950).
Quoted in sects.2.4., 3.4.

Bohr, N. (*1949*): **Discussion with Einstein on epistemological problems in atomic physics.** In: "Albert Einstein: Philosopher-Scientist", ed. by P.A.Schilpp; Library of living Philosophers, Evanston, Illinois, pp. 199-241.
Quoted in sects.2.4, 3.4, 3.7.

Bohr, N. *1958*): **Quantum physics and philosophy, causality and complementarity.** In: "Philosophy in the mid-century; a survey"; ed. by R.Klibansky; La Nuova Italia Editrice, Florence, pp. 308-314. (Essentially the same version appeared in German: "Ueber Erkenntnisfragen der Quantenphysik", in: "Max Planck Festschrift 1958", hg. von B.Kockel, W.Macke, A.Papapetrou; Deutscher Verlag der Wissenschaften, Berlin, 1959, pp. 169-175. Reprinted in: Naturwissenschaftliche Rundschau 13, 252-255, 1960. Russian translation: "Kvantovaya fizika i filosofiya", Usp.Fiz.Nauk 67, 37, 1959).
Quoted in sects.1.5, 3.4.

Bohr, N. (*1963*): **Essays 1958-1962 on atomic physics and human knowledge.** Wiley-Interscience, New York.
Quoted in sect.3.5.

Boltzmann,L. (1896): **Ueber die Unentbehrlichkeit der Atomistik in der Naturwissenschaft.** Wiener Berichte 105, 907-922. Reprinted in: Annalen der Physik und Chemie 60, 231-247 (1897), and in: "Populäre Schriften", Barth, Leipzig, 1905, pp. 141-157.
Quoted in sect.5.6.

Bolyai,W. (1899): **Briefwechsel zwischen C.F.Gauss und W.Bolyai.** Herausgegeben von F.Schmidt und P.Stäckel. Leipzig.
Quoted in sect.2.6.

Bonch-Bruevich,V.L. and Tyablikov,S.V. (1961): **The Green function method in statistical mechanics.** North-Holland, Amsterdam, 1962. (Russian original: Fizmatgiz, Moscow, 1961).
Quoted in sect.4.3.

Boolos,G.S. and Jeffrey,R.C. (1974): **Computability and logic.** Cambridge University Press, London.
Quoted in sects.2.2, 4.5.

Born,M. (1925): **Vorlesungen über Atommechanik. Erster Band.** Springer, Berlin.
Quoted in sect.3.2.

Born,M. (1926a): **Zur Quantenmechanik der Stossvorgänge.** Z.Phys. 37, 863-867.
Quoted in sect.3.2.

Born,M. (1926b): **Zur Quantenmechanik der Stossvorgänge.** Z.Phys. 38, 803-827.
Quoted in sect.3.2.

Born,M. (1926c): **Zur Wellenmechanik der Stossvorgänge.** Nachr.Ges.Wiss. Göttingen, Math.-Phys.Kl., 146-160.
Quoted in sect.3.2.

Born,M. (1926d): **Das Adiabatenprinzip in der Quantenmechanik.** Z.Phys. 40, 167-192.
Quoted in sect.3.2.

Born,M. (1927): **Quantenmechanik und Statistik.** Naturwissenschaften 15, 238-242.
Quoted in sect.3.2.

Born,M. (1953): **Physical reality.** Philosophical Quarterly 3, 139-149. (German translation: "Physikalische Wirklichkeit"; Physikalische Blätter 10, 49-61, 1954).
Quoted in sect.3.4.

Born,M. (1955): **Die statistische Deutung der Quantenmechanik.** In: "Les prix Nobel en 1954". Stockholm, pp. 79-90. (Reprinted in: Physikalische Blätter 11, 193-202, 1955).
Quoted in sect.3.2.

Born,M. (1961): **Bemerkungen zur statistischen Deutung der Quantenmechanik.** In: "Werner Heisenberg und die Physik unserer Zeit", hg. von F.Bopp, Vieweg, Braunschweig, 1961, pp. 103-118.
Quoted in sect.3.2.

Born,M. (1969): **Albert Einstein, Hedwig und Max Born. Briefwechsel 1916-1955.** Nymphenburger Verlagshandlung, München. (English translation: "The Born-Einstein letters"; Walker and Company, 1971).
Quoted in sects.1.4, 3.2.

Born,M. and Jordan,P. (1925): **Zur Quantenmechanik.** Z.Phys. 34, 858-888.
Quoted in sect.3.2.

Born,M. and Oppenheimer,R. (1927): **Zur Quantentheorie der Molekeln.** Annalen der Physik 84, 457-484.
Quoted in sect.6.2.

Born,M., Heisenberg,W. and Jordan,P. (*1926*): Zur Quantenmechanik II. Z.Phys. 35, 557-615.
Quoted in sect.3.2.

Bose,S.N. (*1924*): Plancks Gesetz und Lichtquantenhypothese. Z.Phys. 26, 178-181.
Quoted in sect.3.2.

Boyer,T.H. (*1969*): Classical statistical thermodynamics and electromagnetic zero-point radiation. Phys.Rev. 186, 1304-1318.
Quoted in sect.3.2.

Boyer,T.H. (*1975a*): Random electrodynamics: The theory of classical electrodynamics with classical electromagnetic zero-point radiation. Phys.Rev. D 11, 790-808.
Quoted in sect.6.4.

Boyer,T.H. (*1975b*): General connection between random electrodynamics and quantum electrodynamics for free electromagnetic fields and for dipole oscillator systems. Phys.Rev. D 11, 809-830.
Quoted in sect.6.4.

Brainerd,W.S. and Landweber,L.H. (*1974*): Theory of computation. Wiley-Interscience, New York.
Quoted in sects.4.5, 6.2.

Bratteli,O. and Kastler,D. (*1976*): Relaxing the clustering condition in the derivation of the KMS property. Commun.Math.Phys. 46, 37-42.
Quoted in sect.4.3.

Bratteli,O. and Robinson,D.W. (*1976*): Greens function, Hamiltonians and modular automorphism. Commun.Math.Phys. 50, 133-156.
Quoted in sect.4.2.

Bratteli,O. and Robinson,D.W. (*1979*): Operator algebras and quantum statistical mechanics 1.Springer, New York.
Quoted in sects.4.2, 4.3, 4.6.

Bratteli,O., Kishimoto,A. and Robinson,D.W. (*1978*): Stability properties and KMS condition. Commun.Math.Phys. 61, 209-238.
Quoted in sect.4.3.

Braun,H. and Koecher,M. (*1966*): Jordan-Algebren. Springer, Berlin.
Quoted in sect.4.2.

Brecht,B. (*1967*): Turandot oder Der Kongress der Weisswäscher. In: Bertolt Brecht, Gesammelte Werke, Suhrkamp Verlag, Frankfurt am Main; Band 5, pp. 2211-2212.
Quoted in sect.3.8.

Breitenberger,E. (*1965*): On the so-called paradox of Einstein, Podolsky and Rosen. Nuovo Cimento 38, 356-360.
Quoted in sect.3.7.

Brennich,R.H. (*1970*): The irreducible ray representations of the full inhomogeneous Galilei group. Ann.Inst.Henri Poincaré A 13, 137-161.
Quoted in sect.3.3.

Bridgman,P.W. (*1927*): The logic of modern physics. Macmillan, New York.
Quoted in sect.3.8.

Bridgman,P.W. (*1959*): P.W. Bridgman's 'The Logic of Modern Physics' after thirty years. Daedalus 88, 518-526.
Quoted in sect.3.8.

Broad,C.D. (*1929*): The mind and its place in nature. London.
Quoted in sect.6.1.

Brodsky,S.J. and Primack,J.R. (*1968*): The electromagnetic interactions of loosely bound composite systems. Phys.Rev. 174, 2071-2073.
Quoted in sect.3.3.

Broglie,L. de (1923): **Ondes et quanta.** C.R.Acad.Sci.(Paris) 177, 507-510.
Quoted in sect.3.2.

Broglie,L. de (1971): **A new interpretation concerning the coexistence of waves and particles.** In: "Perspectives in quantum theory. Essays in honor of Alfred Landé". Ed. by W.Yourgrau and A. van der Merve; MIT Press, Cambridge, Massachusetts; pp. 5.16.
Quoted in sect.3.2.

Bub,J. (1968a): **The Daneri-Loinger-Prosperi quantum theory of measurement.** Nuovo Cimento B 57, 503-520.
Quoted in sect.3.5.

Bub,J. (1968b): **Hidden variables and the Copenhagen interpretation - a reconciliation.** British J. for the Philosophy of Science 19, 185-210.
Quoted in sect.3.4.

Bub,J. (1969): **What is a hidden variable theory of quantum phenomena?** Int.J.Theor. Phys. 2, 101-123.
Quoted in sect.3.4.

Bub,J. (1971): **Comment of the Daneri-Loinger-Prosperi quantum theory of measurement.** In: "Quantum theory and beyond"; ed. by T.Bastin; Cambridge University Press, London; pp. 65-70.
Quoted in sect.3.5.

Bub,J. (1973): **On the completeness of quantum mechanics.** In: "Contemporary research in the foundations and philosophy of quantum theory"; ed. by C.A.Hooker; Reidel, Dordrecht-Holland; pp. 1-65.
Quoted in sects.3.4, 4.4.

Bub,J. (1974): **The interpretation of quantum mechanics.** Reidel, Dordrecht-Holland.
Quoted in sects.3.4, 4.4.

Bub,J. (1976): **The statistics of non-Boolean structures.** In: "Foundations of probability theory, statistical inference, and statistical theories of science"; ed. by W.L.Harper and C.A.Hooker; Reidel, Dordrecht-Holland; pp. 1-16.
Quoted in sect.3.4.

Bugajska,K. (1974): **On the representation theorem for quantum logic.** Int.J.Theor. Phys. 9, 93-99.
Quoted in sect.4.4.

Bugajska,K. and Bugajski,S. (1972a): **On the axioms of quantum mechanics.** Bull.Acad. Pol.Sci.Ser.Sci.Math.Astron.Phys. 20, 231-234.
Quoted in sect.4.4.

Bugajska,K. and Bugajski,S. (1972b): **Hidden variables and 2-dimensional Hilbert space.** Ann.Inst.Henri Poincaré A 16, 93-102.
Quoted in sect.3.4.

Bugajska,K. and Bugajski,S. (1973a): **Description of physical systems.** Reports on Mathemaitcal Physics 4, 1-20.
Quoted in sect.4.4.

Bugajska,K. and Bugajski,S. (1973b): **The projection postulate in quantum logic.** Bull.Acad.Pol.Sci.Ser.Sci.Math.Astron.Phys. 21, 873-877.
Quoted in sect.4.4.

Bugajska,K. and Bugajski,S. (1973c): **The lattice structure of quantum logics.** Ann. Inst.Henri Poincaré A 19, 333-340.
Quoted in sect.4.4.

Bugajski,S. (1978): **Probability implication in the logics of classical and quantum mechanics.** J.Philosophical Logic 7, 95-106.
Quoted in sect.4.4.

Bühler,K. (*1918*): Die geistige Entwicklung des Kindes. Jena
Quoted in sect.2.5.

Bunce,J.W. *and Paschke*,W.L. (*1978*): Quasi-expectation and amenable von Neumann algebras. Proc.Amer.Math.Soc. 71, 232-236.
Quoted in sect.5.5.

Bunge,M. (*1967a*): A ghost-free axiomatization of quantum mechanics. In: "Quantum theory and reality"; ed. by M.Bunge; Springer, Berlin; pp. 105-117.
Quoted in sect.3.5.

Bunge,M. (*1967b*): Foundation of Physics. Springer, Berlin.
Quoted in sects.3.5, 4.4.

Bunge,M. (*1973*): Philosophy of Physics. Reidel, Dordrecht-Holland.
Quoted in sect.3.8.

Bures,D. (*1968*): Tensor products of W*-algebras. Pacific J.Math. 27, 13-37.
Quoted in sect.5.6.

Cameron,R.H. *and Martin*,W.T. (*1947*): The behavior of measure and measurability under change of scale in Wiener space. Bull.Amer.Math.Soc. 53, 130-137.
Quoted in sect.4.2.

Cannon,J.T. (*1973*): Infinite volume limits of the canonical free Bose gas states on the Weyl algebra. Commun.Math.Phys. 29, 89-104.
Quoted in sect.4.3.

Cantoni,V. (*1975*): Generalized 'transition probability'. Commun.Math.Phys. 44, 125-128.
Quoted in sect.4.5.

Cantoni,V. (*1976*): Enveloping subspaces and superposition of states. Commun.Math. Phys. 50, 241-244.
Quoted in sect.4.5.

Cantoni,V. (*1977*): The Riemannian structure on the states of quantum-like systems. Commun.Math.Phys. 56, 189-193.
Quoted in sect.4.5.

Capasso,V., *Fortunato*,D. *and Selleri*,F. (*1970*): Von Neumann's theorem and hidden variable models. Rivista del Nuovo Cimento 2, 149-199.
Quoted in sect.3.4.

Carathéodory,C. (*1909*): Untersuchungen über die Grundlagen der Thermodynamik. Mathematische Annalen 67, 355-386.
Quoted in sect.3.8.

Casher,A., *Frieder*,G., *Glück*,M. *and Peres*,A. (*1965*): Reducibility of parastatistics representations. Nucl.Phys. 66, 632-634.
Quoted in sect.3.3.

Casinelli,G. *and Beltrametti*,E.G. (*1975*): Ideal, first-kind measurements in a proposition-state structure. Commun.Math.Phys. 40, 7-13.
Quoted in sect.4.2.

Cattaneo,U. *and Wreszinski*,W. (*1979*): Trotter limits of Lie algebra representations and coherent states. Helv.Phys.Acta 52, 313-327.
Quoted in sect.6.3.

Celeghini,E., *Lasanna*,L. *and Sorace*,E. (*1976*): Galilean invariance, gauge invariance and spin-dependent Hamiltonians. Nuovo Cimento A 31, 89-99.
Quoted in sect.3.3.

Chaitin,G. (*1966*): On the length of programs for computing finite binary sequences. J.Assoc.Comp.Mach. 13, 547-569.
Quoted in sect.4.5.

Chaitin,G.J. (1974): Information-theoretic computational complexity. IEEE Transact. Inform.Theory IT-20, 10-15.
Quoted in sect.6.2.

Chaitin,G.J. (1977): Algorithmic information theory. IBM J.Res.Develop. 21, 350-359.
Quoted in sects.2.2, 6.2.

Chaitin,G.J. (1979): Toward a mathematical definition of "life". In: "The maximum entropy formalism", ed. by R.D.Levine and M.Tribus, MIT Press, Cambridge, Mass.; pp. 477-498.
Quoted in sect.6.2.

Chen,E. (1971): Operator algebras and axioms of measurements. J.Math.Phys.(N.Y.) 12, 2364-2371.
Quoted in sect.4.6.

Chen,E. (1973): Facial aspect of superposition principle in algebraic quantum theory. J.Math.Phys.(N.Y.) 14, 1462-1465.
Quoted in sect.4.4.

Chen,E. (1976): Markovian subdynamics in quantum dynamical systems. J.Math.Phys. (N.Y.) 17, 1785-1789.
Quoted in sect.4.5.

Chen,E. (1977): Entropy and superposition principle. Reports on Mathematical Physics 11, 189-195.
Quoted in sect.4.5.

Choi,M.D. (1972): Positive linear maps on C*-algebras. Canad.J.Math 24, 520-529.
Quoted in sect.4.5.

Choi,M.D. *and Effros*,E.G. (1976): Separable nuclear C*-algebras and injectivity. Duke Math.J. 43, 309-322.
Quoted in sect.5.5.

Christou,E. (1963): The logos of the soul. Spring Publications, New York and Zürich.
Quoted in sect.2.5.

Church,A. (1940): On the concept of a random sequence. Bull.Amer.Math.Soc. 46, 130-135.
Quoted in sect.4.5.

Cirelli,R. *and Cotta-Ramusino*,P. (1973): On the isomorphism of a 'quantal logic' with the logic of the projections in a Hilbert space. Int.J.Theor.Phys. 8, 11-29.
Quoted in sect.4.4.

Cirelli,R. *and Gallone*,F. (1973): Algebra of observables and quantum logic. Ann.Inst. Henri Poincaré A19, 297-331.
Quoted in sect.4.6.

Cirelli,R. *Gallone*,F. *and Gubbay*,B. (1975): An algebraic representation of continuous superselection rules. J.Math.Phys.(N.Y.) 16, 201-213.
Quoted in sect.4.6.

Clark,I.D. (1973): An axiomatisation of quantum logic. J.Symbolic Logic 38, 389-392.
Quoted in sect.4.4.

Clauser,J.F. (1972): Experimental limitations to the validity of semiclassical radiation theories. Phys.Rev.A 6, 49-54.
Quoted in sects.3.4, 6.4.

Clauser,J.F. (1974): Experimental distinction between the quantum and classical field-theoretic predictions for the photoelectric effect. Phys.Rev.D 9, 853-860.
Quoted in sect.3.4.

Clauser,J.F. (1976): Experimental investigation of a polarization correlation anomaly. Phys.Rev.Lett. 36, 1223-1226.
Quoted in sects,3.4, 3.7.

Clauser,J.F. *and Shimony*,A. (*1978*): Bell's theorem: experimental tests and implications. Rep.Prog.Phys. 41, 1881-1927.
Quoted in sect.3.7.

Clauser,J.F., *Horne*,M.A., *Shimony*,A. *and Holt*,R.A. (*1969*): Proposed experiment to test local hidden-variable theories. Phys.Rev.Lett. 23, 880-884.
Quoted in sects.3.4, 3.7.

Claverie,P. *and Diner*,S. (*1977*): Stochastic electrodynamics and quantum theory. Int.J.Quantum Chem. 12, Suppl. 1, 41-82.
Quoted in sects.3.4, 6.4.

Claverie,P. *and Diner*,S. (*1980*): The concept of molecular structure in quantum theory: interpretation problems. Israel J.Chemistry 19, 54-81.
Quoted in sects.6.2, 6.4.

Clementi,E. (*1973*): In: Proceedings of the Robert A.Welch Foundation Conference on Chemical Research. XVI. Theoretical Chemistry. Robert A.Welch Foundation, Houston; p.117.
Quoted in sect.1.3.

Cockcroft,A.M. *and Hudson*,R.L. (*1977*): Quantum mechanical Wiener processes. J.Multivariate Analysis 7, 107-124.
Quoted in sect.4.5.

Combes,J.M. (*1975*): On the Born-Oppenheimer approximation. In: "Lecture Notes in Physics", vol.39, ed. by H.Araki, Springer, Berlin; pp.467-471.
Quoted in sect.6.2.

Combes,J.M. (*1977*): The Born-Oppenheimer approximation. Acta Phys. Austriaca, Suppl. 17, 139-159.
Quoted in sect.6.2.

Combes,J.M. *and Seiler*,R. (*1978*): Regularity and asymptotic properties of the discrete spectrum of electronic Hamiltonians. Int.J. Quantum Chem. 14, 213-299.
Quoted in sect.6.2.

Compton,A.H. *and Simon*,A.W. (*1925*): Directed quanta of scattered X-rays. Phys.Rev. 26, 289-299.
Quoted in sect.3.5.

Connes,A. (*1973*): Une classification des facteurs de type III. Annales Scientifiques de l'École Normale Supérieure 6, 133-252.
Quoted in sect.5.5.

Connes,A. (*1976*): Classification of injective factors. Annals of Mathematics 104, 73-115.
Quoted in sect.5.5.

Cook,J.M. (*1953*): The mathematics of second quantization. Trans.Amer.Math.Soc. 74, 222-245.
Quoted in sect.4.2.

Cook,T.A. (*1978*): The geometry of generalized quantum logics. Int.J.Theor.Phys. 17, 941-955.
Quoted in sect.4.5.

Cooper,J.L.B. (*1950*): The paradox of separated systems in quantum theory. Proc.Cambridge Phil.Soc. 46, 620-625.
Quoted in sect.3.7.

Cooper,L.N. (*1956*): Bound electron pairs in degenerate Fermi gas. Phys.Rev. 104, 1189-1190.
Quoted in sect.6.2.

Cooper,L.N. and Vechten,D.van (1969): On the interpretation of measurement within quantum theory. Amer.J.Phys. 37, 1212-1220.
Quoted in sect.3.6.

Copernicus,N. (1543): De Revolutionibus Orbium Coelestium Libri IV. With an anonymous introduction (due to Andreas Osiander). (German translation: "Ueber die Kreisbewegungen der Weltkörper", Thorn 1879).
Quoted in sect.2.1.

Cornette,W.M. and Gudder,S.P. (1974): The mixture of quantum states. J.Math.Phys.(N.Y.) 15, 842-850.
Quoted in sect.4.5.

Csonka,P.L. (1971): Prediction in quantum physics. Phys.Rev.D 4, 1607-1611.
Quoted in sect.3.5.

Cycon,H. and Hellwig,K.E. (1977): Conditional expectations in generalized probability theory. J.Math.Phys.(N.Y.) 18, 1154-1161.
Quoted in sect.4.5.

Czelakowski,J. (1974): Logics based on partial Boolean σ-algebras. 1. Studia Logica 33, 371-396.
Quoted in sect.4.4.

Czelakowski,J. (1975): Logics based on partial Boolean σ-algebras. 2. Studia Logica 34, 69-86.
Quoted in sect.4.4.

Daele,A.van (1975): A Radon Nikodym theorem for weights on von Neumann algebras. Pacific J.Math. 61, 527-542.
Quoted in sect.4.5.

Dähn,G. (1968): Attempt of an axiomatic foundation of quantum mechanics and more general theories IV. Commun.Math.Phys. 9, 192-211.
Quoted in sect. 4.5.

Dähn,G. (1972a): The algebra generated by physical filters. Commun.Math.Phys. 28, 109-122.
Quoted in sect.4.5.

Dähn,G. (1972b): Symmetry of the physical probability function implies modularity of the lattice of decision effects. Commun.Math.Phys. 28, 123-132.
Quoted in sect.4.5.

Dähn,G. (1973): Two equivalent criteria for modularity of the lattice of all physical decision effects. Commun.Math.Phys. 30, 79-78.
Quoted in sect.4.5.

Dalenius,T., Karlsson,G. and Malmquist,S. (1970): Scientists at work. Festschrift in honor of Herman Wold. Almquist and Wiksell, Stockholm.
Quoted in sect.2.5.

Dalla Chiara,M.L. (1977): Quantum logic and physical modalities. J.Philosophical Logic 6, 391-404.
Quoted in sect.4.4.

Daneri,A., Loinger,A. and Prosperi,G.M. (1962): Quantum theory of measurement and ergodicity conditions. Nucl.Phys. 33, 297-319.
Quoted in sect.3.5.

Daneri,A., Loinger,A. and Prosperi,G.M. (1966): Further remarks on the relations between statistical mechanics and quantum theory of measurement. Nuovo Cimento 44, 119-128.
Quoted in sect.3.5.

Davies,E.B. (1968): On the Borel structure of C*-algebras. (With an appendix by R.V.Kadison). Commun.Math.Phys. 8, 147-163.
Quoted in sects.4.5, 4.6.

Davies,E.B. (*1969a*): The structure of Σ*-algebras. Quart.J.Math.(Oxford) 20, 351-366.
Quoted in sect.4.5.

Davies,E.B. (*1969b*): Quantum stochastic processes. Commun.Math.Phys. 15, 277-304.
Quoted in sects.3.5, 4.5.

Davies,E.B. (*1970a*): Quantum stochastic processes II. Commun.Math.Phys. 19, 83-105.
Quoted in sects.3.5, 4.5.

Davies,E.B. (*1970b*): On the repeated measurement of continuous observables in quantum mechanics. J.Functional Analysis 6, 318-346.
Quoted in sect.3.5, 4.5.

Davies,E.B. (*1971*): Quantum stochastic processes. III. Commun.Math.Phys. 22, 51-70.
Quoted in sects.3.5, 4.5.

Davies,E.B. (*1972a*): Some contraction semigroups in quantum probability. Z.Wahrscheinlichkeitstheorie verw.Geb. 23, 261-273.
Quoted in sects.3.5, 4.5.

Davies,E.B. (*1972b*): Example related to the foundation of quantum theory. J.Math.Phys. (N.Y.) 13, 39-41.
Quoted in sect.4.4.

Davies,E.B. (*1973*): The ideal boson gas in an external scalar potential. Commun.Math. Phys. 30, 229-247.
Quoted in sect.4.3.

Davies,E.B. (*1974*): Markovian master equations. Commun.Math.Phys. 39, 91-110.
Quoted in sect.6.3.

Davies,E.B. (*1975*): Markovian master equations. III. Ann.Inst. Henri Poincaré B 11, 265-273.
Quoted in sect.6.3.

Davies,E.B. (*1976a*): Quantum theory of open systems. Academic Press, London.
Quoted in sects.3.5, 4.5, 6.3.

Davies,E.B. (*1976b*): Markovian master equations. II. Math.Ann. 219, 147-158.
Quoted in sect.6.3.

Davies,E.B. (*1976c*): The classical limit for quantum dynamical semigroups. Commun. Math.Phys. 49, 113-129.
Quoted in sect.6.4.

Davies,E.B. (*1977a*): Asymptotic analysis of some abstract evolution equations. J.Functional Analysis 25, 81-101.
Quotes in sect.6.3.

Davies,E.B. (*1977b*): Master equations: a survey of rigorous results. Ren.Sem.Math. Fis.Milano 47, 165-173.
Quoted in sect.6.3.

Davies,E.B. (*1979a*): Generators of dynamical semigroups. J.Functional Analysis 34, 421-432.
Quoted in sect.4.5.

Davies,E.B. (*1979b*): Symmetry breaking for a non-linear Schrödinger equation. Commun. Math.Phys. 64, 191-210.
Quoted in sect.5.6.

Davies,E.B. *and Lewis*,J.T. (*1970*): An operational approach to quantum probability. Commun.Math.Phys. 17, 239-260.
Quoted in sects.3.5, 4.5, 4.6.

Davis,C. (*1961*): Operator valued entropy of quantum mechanical measurements. Proc. Japan Acad. 37, 533-538.
Quoted in sect.4.5.

Davis,M. (*editor*) (*1965*): The undecidable. Basic papers on undecidable propositions, unsolvable problems and computable functions. Raven Press, New York.
Quoted in sect.2.2.

Davisson,C.J. and Germer,L.H. (*1927*): Diffraction of electrons by a crystal of nickel. Phys.Rev. 30, 705-740.
Quoted in sect,3,2.

Day,T.B. (*1961*): Demonstration of quantum mechanics in the large. Phys.Rev. 121, 1204-1206.
Quoted in sect.3.7.

Deliyannis,P.C. (*1969*): Theory of observables. J.Math.Phys.(N.Y.) 10, 2114-2127.
Quoted in sect.4.6.

Deliyannis,P.C. (*1971a*): Density of states. J.Math.Phys.(N.Y.) 12, 860-862.
Quoted in sect.4.4.

Deliyannis,P.C. (*1971b*): Generalized hidden variable theorem. J.Math.Phys.(N.Y.) 12, 1013-1017.
Quoted in sect.3.4.

Deliyannis,P.C. (*1973*): Vector space models of abstract quantum logics. J.Math.Phys. (N.Y.) 14, 249-253.
Quoted in sect.4.4.

Deliyannis,P.C. (*1975*): Imbedding of Segal systems. J.Math.Phys.(N.Y.) 16, 163-170.
Quoted in sect.4.6.

Deliyannis,P.C. (*1976*): Superposition of states and the structure of quantum logics. J.Math.Phys.(N.Y.) 17, 248-254.
Quoted in sect.4.4.

Dell'Antonio,G.F. (*1967*): On the limits of sequences of normal states. Commun.Pure Appl.Math. 20, 413-429.
Quoted in sect.4.3.

Dell'Antonio,G.F., Doplicher,S. and Ruelle,D. (*1966*): A theorem on canonical commutation and anticommutation relations. Commun.Math.Phys. 2, 223-230.
Quoted in sect.4.3.

Demopoulos,W. (*1976*): The possibility structure of physical systems. In: Foundations of probability theory, statistical inference, and statistical theories of science; ed. by W.L. Harper and C.A.Hooker; Reidel, Dordrecht-Holland, pp.55-80.
Quoted in sect.3.4.

Denbigh,K.G. (*1972*): In defence of *the* direction of time. In: "The study of time"; ed. by J.T.Fraser, F.C.Haber, and G.H.Müller; Springer, Berlin; pp. 148-158.
Quoted in sect.5.2.

Destouches-Février,P. (*1951*): La structures des théories physiques. Presses Universitaires de France, Paris.
Quoted in sects.3.3, 4.4.

DeWitt,B.S. (*1967*): Quantum theory of gravity.I. The canonical theory. Phys.Rev. 160, 1113-1148.
Quoted in sect.3.6.

DeWitt,B.S. (*1968*): The Everett-Wheeler interpretation of quantum mechanics. In: "Battelle Rencontres, 1967 lectures in mathematics and physics"; ed. by C.M.DeWitt and J.A.Wheeler; Benjamin, New York; pp.318-332.
Quoted in sect.3.6.

DeWitt,B.S. (*1970*): Quantum mechanics and reality. Physics Today 23 (No.9, September 1970), 30-35.
Quoted in sect.3.6.

DeWitt,B.S. (*1971a*): The many-universes interpretation of quantum mechanics. In: "Proceedings of the international school of physics 'Enrico Fermi'. Course 49: Foundations of quantum mechanics"; ed. by B.d'Espagnat; Academic Press, New York; pp.211-262.
Quoted in sect.3.6.

DeWitt,B.S. (*1971b*): Quantum mechanics debate. DeWitt replies. Phys.Today 24 (No.4, April 1971), 41-44.
Quoted in sect.3.6.

Dieudonné,J. (*1970*): The work of Nicholas Bourbaki. Amer.Math.Monthly 77, 134-145.
Quoted in sect.2.5.

Dingler,H. (*1907*): Grundlinien einer Kritik und exakten Theorie der Wissenschaften, insbesondere der Mathematik. Ackermann, München.
Quoted in sect.3.8.

Dirac,P.A.M. (*1925*): The fundamental equations of quantum mechanics. Proc.Roy.Soc. London A 109, 642-653.
Quoted in sect.3.2.

Dirac,P.A.M. (*1926a*): Quantum mechanics and preliminary investigation of the hydrogen atom. Proc.Roy.Soc. London A 110, 561-579.
Quoted in sect.3.2.

Dirac,P.A.M. (*1926b*): On quantum algebra. Proc.Cambr.Phil.Soc. 23, 412-418.
Quoted in sect.3.2.

Dirac,P.A.M. (*1926c*): On the theory of quantum mechanics. Proc.Roy.Soc.London A 112, 661-677.
Quoted in sect.3.2.

Dirac,P.A.M. (*1926d*): The physical interpretation of quantum dynamics. Proc.Roy.Soc. A 113, 621-641.
Quoted in sect.3.2.

Dirac,P.A.M. (*1927*): The quantum theory of the emission and absorption of radiation. Proc.Roy.Soc.London A 114, 243-265.
Quoted in sect.6.4.

Dirac,P.A.M. (*1929a*): Quantum mechanics of many-electron systems. Proc.Roy.Soc. London A 123, 713-733.
Quoted in sect.1.3.

Dirac,P.A.M. (*1929b*): The basis of statistical quantum mechanics. Proc.Cambridge Phil.Soc. 25, 62-66.
Quoted in sect.3.3.

Dirac,P.A.M. (*1930*): The principles of quantum mechanics. Clarendon Press, Oxford; 1st ed. 1930, 4th ed. 1958.
Quoted in sects.3.1, 4.5, 5.5.

Dirac,P.A.M. (*1964*): Hamiltonian methods and quantum mechanics. Proc.Roy.Irish Acad. 63A, 49-59.
Quoted in sect.2.5.

Dishkant,H. (*1977*): Imbedding of the quantum logic in the modal system of Brower. J.Symbolic Logic 42, 321-328.
Quoted in sect.4.4.

Dishkant,H. (*1978*): An extension of the Lukasiewicz logic to the modal logic of quantum mechanics. Studia Logica 37, 149-155.
Quoted in sect.4.4.

Dixmier,J. (*1948*): Position relative de deux variétés linéaires fermées dans un espace de Hilbert. Revue Scientifique 86, 387-399.
Quoted in sect.5.3.

Dixmier,J. (1957): Les algèbres d'opérateurs dans l'espace Hilbertien. (Algèbres de von Neumann). Gauthier-Villars, Paris; première édition, 1957; deuxième édition, revue et augmentée, 1969.
Quoted in sect.4.2.

Dixmier,J. (1964): Les C*-algèbres et leurs représentations. Gauthier-Villars, Paris; première édition 1964; deuxième édition 1969. (English translation: "C*-algebras, North-Holland, Amsterdam, 1977).
Quoted in sects.3.3, 4.2.

Dolph,C.L. (1963): Positive real resolvents and linear passive Hilbert systems. Ann.Acad.Sci.Fennicae A I 336/9, 1-39.
Quoted in sect.4.3.

Dombrowski,H.D. and Horneffer,K. (1964): Der Begriff des physikalischen Systems in mathematischer Sicht. Nachr.Ges.Wiss.Göttingen, Math.-Phys.Kl. 1964, 67-100.
Quoted in sect.3.4.

Domotor,Z. (1974): The probability structure of quantum-mechanical systems. Synthese 29, 155-185.
Quoted in sect.5.3.

Doplicher,S. and Kastler,D. (1968): Ergodic states in noncommutative ergodic theory. Commun.Math.Phys. 7, 1-20.
Quoted in sect.4.3.

Doplicher,S., Kadison,R.V., Kastler,D. and Robinson,D.W. (1967): Asymptotically Abelian systems. Commun.Math.Phys. 6, 101-120.
Quoted in sect.4.3.

Driesch,H. (1908): The science and philosophy of the organism. 2 volumes. Black, London. (German translation: "Philosophie des Organischen", Engelmann, Leipzig, 1909, rev. ed. 1921).
Quoted in sect.6.1.

Drieschner,M. (1979): Voraussage - Wahrscheinlichkeit - Objekt. Ueber die begrifflichen Grundlagen der Quantenmechanik. Lecture Notes in Physics, vol.99, Springer, Berlin.
Quoted in sect.4.4.

Driessler,W. and Wilde,I.F. (1979): Stochastic independence in non-commutative probability theory. Math.Proc.Camb.Phil.Soc. 86, 103-114.
Quoted in sect.4.5.

Drisch,T. (1979): Generalization of Gleason's theorem. Int.J.Theor.Phys. 18, 239-243.
Quoted in sect.4.5.

Dubin,D.A. (1973): Bosons in thermal contact: a C*-algebraic model. Commun.Math.Phys. 32, 1-17.
Quoted in sect.4.3.

Dubin,D.A. (1974): Solvable models in algebraic statistical mechanics. Oxford University Press, London.
Quoted in sects.4.3, 6.3.

Dubin,D.A. and Sewell,G.L. (1970): Time translations in the algebraic formulation of statistical mechanics. J.Math.Phys.(N.Y.) 11, 2990-2998.
Quoted in sect.4.2.

Dugas,R. (1950): Histoire de la Mécanique. Éditions du Griffon, Neuchâtel. (English translation: "A history of mechanics", Editions du Griffon, Neuchâtel, 1955).
Quoted in sect.3.3.

Duhem,P. *(1908)*: **La théorie physique, son objet et sa structure.** (German translation: "Ziel und Struktur der pyhsikalischen Theorien"; Barth, Leipzig, 1908; Nachdruck: Felix Meiner, Hamburg, 1978. English translation: "The aim and structure of physical theory"; Princeton University Press, Princeton. 1954).
Quoted in sects.2.3, 2.4.

Durand,L. *(1960)*: **On the theory of measurement in quantum mechanical systems.** Philosophy of Science 27, 115-133.
Quoted in sect.3.5.

Dye,H.A. *(1952)*: **The Radon-Nikodým theorem for finite rings of operators.** Trans.Amer. Math.Soc. 72, 243-280.
Quoted in sect.4.5.

Dye,H.A. *(1955)*: **On the geometry of projections in certain operator algebras.** Annals of Mathematics 61, 73-89.
Quoted in sects.4.6, 5.4.

Earman,J. and Shimony,A. *(1968)*: **A note on measurement.** Nuovo Cimento B 54, 332-334.
Quoted in sect.3.5.

Echigo(-Choda),M. and Nakamura,M. *(1962)*: **A remark on the concept of channels.** Proc. Japan Acad. 38, 307-309.
Quoted in sect.4.5.

Eckmann,J.P. and Zabey,P.C. *(1969)*: **Impossibility of quantum mechanics in a Hilbert space over a finite field.** Helv.Phys.Acta 42, 420-424.
Quoted in sect.4.4.

Eddington,A. *(1920)*: **Space, time and gravitation.** Cambridge University Press, London.
Quoted in sect.6.5.

Edwards,C.M. *(1970)*: **The operational approach to algebraic quantum theory I.** Commun. Math.Phys. 16, 207-230.
Quoted in sects.4.5, 4.6.

Edwards,C.M. *(1971a)*: **Classes of operations in quantum theory.** Commun.Math.Phys. 20, 26-56.
Quoted in sect.4.5.

Edwards,C.M. *(1971b)*: **Sets of simple observables in the operational approach to quantum theory.** Ann.Inst. Henri Poincaré A 15, 1-14.
Quoted in sects.4.5, 4.6.

Edwards,C.M. *(1972a)*: **The theory of pure operations.** Commun.Math.Phys. 24, 260-288.
Quoted in sect.4.5.

Edwards,C.M. *(1972b)*: **Pure operations in statistical physical theories.** In: "Statistical mechanics and field theory"; ed. by R.N.Sen and C.Weil; Halsted Press-Wiley, New York and Israel University Press, Jerusalem; pp. 81-89.
Quoted in sect.4.5.

Edwards,C.M. *(1974)*: **The center of a physical system.** In: "Foundations of quantum mechanics and ordered linear spaces"; ed. by A.Hartkämper and H.Neumann; Springer, Berlin; pp. 199-205.
Quoted in sect.4.5.

Edwards,C.M. *(1975)*: **Alternative axioms for statistical physical theories.** Ann.Inst. Henri Poincaré A 22, 81-95.
Quoted in sect.4.5.

Edwards,C.M. and Gerzon,M.A. *(1970)*: **Monotone convergence in partially ordered vector spaces.** Ann.Inst. Henri Poincaré A 12, 323-328.
Quoted in sect.4.5.

Eilers,M. *and Horst*,E. *(1975)*: **The theorem of Gleason for nonseparable Hilbert spaces.** Int.J.Theor.Phys. 13, 419-424.
Quoted in sect.4.5.

Einstein,A. *(1905a)*: **Ueber einen die Ergänzung und Verwandlung des Lichtes betreffenden heuristischen Gesichtspunkt.** Annalen der Physik 17, 132-148.
Quoted in sect.3.2.

Einstein,A. *(1905b)*: **Zur Elektrodynamik bewegter Körper.** Annalen der Physik 17, 891-921. (English translation in: "The principle of relativity", by H.A.Lorentz, A.Einstein, H.Minkowsky and H.Weyl, Methuen, London 1923; reprinted by Dover, New York).
Quoted in sect.3.3.

Einstein,A. *(1916)*: **Ueber die spezielle und allgemeine Relativitätstheorie.** Vieweg, Braunschweig.
Quoted in sect.3.3.

Einstein,A. *(1919)*: **The fundamental idea of general relativity in its original form.** Unpublished manuscript in German, written in Einstein's own hand in approximately 1919. Einstein Archives, Institute for Advanced Study, Princeton, N.J. The quoted excerpt (translated by G.Holton) has been taken from: G.Holton, "Thematic origins of scientific thought", Harvard University Press, Cambridge, Mass., 1973; p. 364.
Quoted in sect.6.5.

Einstein,A. *(1922)*: **The meaning of relativity.** Sixth edition, revised, 1956; Science paperback, London 1967.
Quoted in sect.3.3.

Einstein,A. *(1924)*: **Quantentheorie des einatomigen Gases.** Sitzungsberichte der Preussischen Akad.d.Wiss., Phys.-Math.Kl., 1924, 261-267.
Quoted in sects.3.2, 4.3.

Einstein,A. *(1925)*: **Quantentheorie des einatomigen Gases. Zweite Abhandlung.** Sitzungsberichte der Preussischen Akad.d.Wiss., Phys.-Math.Kl., 1925, 3-14.
Quoted in sects.3.2, 4.3.

Einstein,A. *(1936)*: **Physik und Realität.** J.Franklin Inst. 221, 313-347. (English translation: "Physics and relativity", ib., pp. 349-382.)
Quoted in sects.2.4, 3.4, 3.7.

Einstein,A. *(1944)*: **Remarks on Bertrand Russell's theory of Knowledge.** In: "The philosophy of Bertrand Russell"; ed. by P.A.Schilp; Open Court, La Salle, Illinois; pp. 279-291.
Quoted in sect.1.5.

Einstein,A. *(1948)*: **Quanten-Mechanik und Wirklichkeit.** Dialectica 2, 320-324.
Quoted in sects.3.4, 3.7.

Einstein,A. *(1949)*: **Autobiographisches and Reply to criticisms.** In: "Albert Einstein: Philosophischer-Scientist"; ed. by P.A.Schilp; Library of Living Philosophers, Evanstone, Illinois; pp. 2-95 and pp. 663-693.
Quoted in sects.2.1, 3.4, 3.7.

Einstein,A. *(1953a)*: **Einleitende Bemerkungen über Grundbegriffe.** In: "Louis de Broglie, Physicien et Penseur"; Albin Michel, Paris.
Quoted in sects.3.4, 3.7.

Einstein,A. *(1953b)*: **Elementare Ueberlegungen zur Interpretation der Quanten-Mechanik.** In: "Scientific Papers, presented to Max Born"; Oliver Boyd, Edinburgh; pp. 33-40.
Quoted in sects.1.4, 3.4.

Einstein,A., Podolsky,B. and Rosen,N. (1935): Can quantum-mechanical description of physical reality be considered complete? Phys.Rev. 47, 777-780.
Quoted in sects.1.4, 3.4, 3.7, 5.1.

Eisenberg,D. and Kauzmann,W. (1969): The structure and properties of water. Clarendon Press, Oxford.
Quoted in 6.2.

Ekahguere,G.O.S. (1979): Markov fields in noncommutative probability theory on W*-algebras. J.Math.Phys.(N.Y.) 20, 1679-1683.
Quoted in sect.4.5.

Elsasser,W.M. (1937): On quantum measurements and the role of the uncertainty relations in statistical mechanics. Phys.Rev. 52, 987-999.
Quoted in sect.3.5.

Elsasser,W.M. (1951): Quantum mechanics, amplifying processes, and living matter. Philosophy of Science 18, 300-326. (Reprinted from Proc. of the Utah Academy of Sciences, Arts and Letters 26, 89, 1951).
Quoted in sect.3.5.

Elsasser,W.M. (1953): Les mesures et la réalité en mécanique quantique. In: "Louis de Broglie. Physicien et Penseur"; Albin Michel, Paris; pp. 87-108.
Quoted in sect.3.5.

Elsasser,W.M. (1961): Quanta and the concept of organismic law. J.Theor.Biol. 1, 27-58.
Quoted in sect.1.4.

Elsasser,W.M. (1962): Physical aspects of non-mechanistic biological theory. J.Theor. Biol. 3, 164-191.
Quoted in sect.1.4.

Elsasser,W.M. (1968): Theory of quantum-mechanical description. Proc.Nat.Acad. Sci.(USA) 59, 738-744.
Quoted in sect.3.4.

Elsasser,W.M. (1969a): The mathematical expression of generalized complementarity. J.Theor.Biol. 25, 276-296.
Quoted in sect.4.3.

Elsasser,W.M. (1969b): A causal phenomena in physics and biology: a case for reconstruction. American Scientist 57, 502-516.
Quoted in sect.1.4.

Elsasser,W.M. (1971): Philosophical dissonances in quantum mechanics. In: "Perspectives in quantum theory"; ed. by W.Yourgrau and A.van der Merwe; M.I.T. Press, Cambridge, Massachusetts; pp. 199-218.
Quoted in sect.3.4.

Elsasser,W.M. (1973): A natural philosophy of quantum mechanics based on induction. Foundations of Physics 3, 117-137.
Quoted in sect.3.4.

Emch,G.G. (1963): Mécanique quantique quaternionienne et relativité restreinte. Helv. Phys.Acta 36, 739-769, 770-788.
Quoted in sect.4.4.

Emch,G.G. (1972a): Algebraic methods in statistical mechanics and quantum field theory. Wiley-Interscience, New York.
Quoted in sect.4.3.

Emch,G.G. (1972b): On quantum measurement processes. Helv.Phys.Acta 45, 1049-1056.
Quoted in sect.3.5.

Emch,G.G. (*1972c*): **The C*-algebraic approach to phase transitions.** In: "Phase transitions and critical phenomena. Volum 1. Exact results"; ed. by C.Domb and M.S.Green; Academic Press, London; pp. 137-175.
Quoted in sect.4.3.

Emch,G.G. (*1975*): **Nonabelian special K-flows.** J.Functional Analysis 19, 1-12.
Quoted in sect.4.5.

Emch,G.G. (*1976*): **Generalized K-flows.** Commun.Math.Phys. 49, 191-215.
Quoted in sect.4.5.

Emch,G.G. and Jauch,J.M. (*1965*): **Structures logiques et matématiques en physique quantique.** Dialectica 19, 259-279.
Quoted in sects.4.4, 5.3.

Emch,G.G. and Knops,H.J.F. (*1970*): **Pure thermodynamical phases as extremal KMS states.** J.Math.Phys.(N.Y.) 11, 3008-3018.
Quoted in sect.4.3.

Emch,G.G. and Piron,C. (*1963*): **Symmetry in quantum theory.** J.Math.Phys.(N.Y.) 4, 469-473.
Quoted in sect.4.4.

Emch,G.G. and Varilly,J.C. (*1979*): **On the standard form of the Bloch equation.** Letters in Mathematical Physics 3, 113-116.
Quoted in sect.4.5.

Emch,G.G., Albeverio,S. and Eckmann,J.P. (*1978*): **Quasi-free generalized K-flows.** Reports on Mathematical Physics 13, 73-85.
Quoted in sect.4.5.

Emch,G.G., Knops,H.J.F. and Verboven,E.J. (*1970*): **Breaking of euclidean symmetry with an application to the theory of crystallization.** J.Math.Phys.(N.Y.) 11, 1655-1668.
Quoted in sect.4.3.

England,W., Salmon,L.S. and Ruedenberg,K. (*1971*): **Localized molecular orbitals: a bridge between chemical intuition and molecular mechanics.** Fortschr.Chem. Forsch. 23, 31-123.
Quoted in sect.1.3.

Erlichson,H. (*1972*): **The Einstein-Podolsky-Rosen paradox.** Philosophy of Science 39, 83-85.
Quoted in sect.3.7.

Espagnat,B. de (*1965*): **Conceptions de la physique contemporaine.** Hermann, Paris. (German translation: "Grundprobleme der gegenwärtigen Physik"; Vieweg, Braunschweig, 1971).
Quoted in sects.3.5, 3.7.

Espagnat,B. de (*1966a*): **An elementary note about mixtures.** In: "Preludes in theoretical physics. In honor of V.F.Weisskopf"; ed. by A.De-Shalit, H.Feshbach and L.van Hove; North-Holland, Amsterdam; pp. 185-191.
Quoted in sect.3.5.

Espagnat,B. de (*1966b*): **Two remarks on the theory of measurement.** Nuovo Cimento Suppl. 4, 828-838.
Quoted in sect.3.5.

Espagnat,B. de (*1971*): **Conceptual foundations of quantum mechanics.** Bejamin, Menlo Park, California. Second revised and enlarged edition 1976, Benjamin, Reading, Massachusetts.
Quoted in sects.3.5, 3.7.

Espagnat,B. *de* (1974): On measurement in general quantum theory. Nuovo Cimento B 21, 233-257.
Quoted in sect.3.5.

Espagnat,B. *de* (1977): Nonseparability and quantum logic. Ann.Japan Assoc.Philos.Sci. 5, 57-62
Quoted in sect.5.6.

Evans,D.E. *and Lewis*,J.T. (1976): Dilations of dynamical semi-groups. Commun.Math. Phys. 50, 219-227.
Quoted in sect.4.5.

Evans,D.E. *and Lewis*,J.T. (1977a): Some semigroups of completely positive maps on the CCR algebra. J.Functional Analysis 26, 369-377.
Quoted in sect.4.5.

Evans,D.E. *and Lewis*,J.T. (1977b): Dilations of irreversible evolutions in algebraic quantum theory. Commun. Dublin Inst. Advanced Studies, ser.A, no.24.
Quoted in sect.4.5.

Everett,H. (1957): 'Relative state' formulation of quantum mechanics. Rev.Mod.Phys. 29, 454-462.
Quoted in sect.3.6.

Everett,H. (1973): The theory of the universal wave function. Undated manuscript, published in: "The many-worlds interpretation of quantum mechnaics"; ed. by B.S.DeWitt and N.Graham; Princeton University Press, Princeton, New Jersey; pp. 1-140.

Facio,B.*de and Taylor*,D.C. (1973): Commutativity and causal independence. Phys.Rev. D8, 2719-2731.
Quoted in sect.5.6.

Fannes,M. *and Verbeure*,A. (1974): On the time evolution automorphisms of the CCR-algebra for quantum mechanics. Commun.Math.Phys. 35, 257-264.
Quoted in sect.4.2.

Fáy,G. (1970): Phenomenological foundation of quantum logic. Acta Phys.Acad.Sci. Hungar. 29, 27-33.
Quoted in sect.4.4.

Fehrs,M.H. *and Shimony*,A. (1974): Approximate measurement in quantum mechanics. Phys.Rev.D 9, 2317-2320.
Quoted in sect.3.5.

Feyerabend,P.K. (1957): Zur Quantentheorie der Messung. Z.Phys. 148, 551-559. (English version: "On the quantum theory of measurement", in: "Observation and interpretation"; ed. by S.Körner; Butterworth, London, 1957; reprinted by Dover, New York, 1962; pp. 121-130.)
Quoted in sect.3.5.

Feyerabend,P.K. (1958a): Complementarity. Supplementary Proceedings of the Aristotelian Society 32, 75-104.
Quoted in sect.3.4.

Feyerabend,P.K. (1958b): An attempt at a realistic interpretation of experience. Proceedings of the Aristotelian Society 58, 143-170.
Quoted in sect.2.4.

Feyerabend,P.K. (1962): Problems of microphysics. In: "Frontiers of science and philosophy"; ed. by R.G.Colodny:, University of Pittsburgh Press; pp. 189-283.
Quoted in sects.3.4, 3.5.

Feyerabend,P.K. (1968): On a recent critique of complementarity. Philosophy of Science 35, 309-331 (1968), 36, 82-105 (1969).
Quoted in sect.3.4.

Feyerabend,P.K. (*1970a*): **Problems of empiricism. Part II.** In: "The nature and function of scientific theories"; ed. by R.G.Colodny, University Pittsburgh Press; pp. 275-353.
Quoted in sect.2.4.

Feyerabend,P.K. (*1970b*): **Against method.** In: "Minnesota Studies in the Philosophy of Science, vol.IV"; ed. by M.Radner and S.Winokur, Minneapolis; pp. 17-130.
Quoted in sect.2.4.

Feyerabend,P.K. (*1970c*): **Consolations for the specialist.** In: "Criticism and the growth of knowledge"; ed. by I.Lakatos and A.Musgrave, Cambridge University Press, London; pp. 197-230.
Quoted in sect.2.4.

Feyerabend,P.K. (*1979*): **Dialogue on method.** In: "The structure and development of science"; ed. by G.Radnitzky and G.Andersson, Reidel, Dordrecht; pp. 63-131.
Quoted in sect.6.5.

Feynman,R. (*1965*): **The character of physical law.** British Broadcasting Corporation, London.
Quoted in sects.3.3, 6.5.

Feynman,R.P., Leighton,R.B. and Sands,M. (*1963*): **The Feynman lectures on physics. Volume I.** Addison-Wesley, Reading, Mass.
Quoted in sect.6.5.

Finch,P.D. (*1969a*): **Sasaki projections on orthocomplemented posets.** Bull.Austral. Math.Soc. 1, 319-324.
Quoted in sect.4.4.

Finch,P.D. (*1969b*): **On the lattice structure of quantum logic.** Bull.Austral.Math. Soc. 1, 333-340.
Quoted in sect.4.4.

Finch,P.D. (*1969c*): **On the structure of quantum logic.** J.Symbolic Logic 34, 275-282.
Quoted in sect.4.4.

Finch,P.D. (*1970*): **Quantum logic as an implication algebra.** Bull.Austral.Math.Soc. 2, 101-106.
Quoted in sect.4.4.

Fine,A.I. (*1969*): **On the general quantum theory of measurement.** Proc. Cambridge Phil.Soc. 65, 111-122.
Quoted in sect.3.5.

Fine,A. (*1970*): **Insolubility of the quantum measurement problem.** Phys.Rev. D 2, 2783-2787.
Quoted in sect.3.5.

Fine,A. (*1972*): **Some conceptual problems of quantum theory.** In: "Paradigms and paradoxes"; ed. by R.G.Colodny; University of Pittsburgh Press; pp. 3-31.
Quoted in scet.4.4.

Finkelstein,D. (*1962*): **The logic of quantum physics.** Trans.N.Y.Acad.Sci. 25, 621-637 (1962/63).
Quoted in sect.4.4.

Finkelstein, D. (*1969*): **Matter, space, and logic.** In: "Boston studies in the philosophy of science. Vol.5"; ed. by R.S.Cohen and M.W.Wartofsky, Reidel, Dordrecht-Holland; pp. 199-215.
Quoted in sect.4.4.

Finkelstein,D. (*1972*): **The physics of logic.** In: "Paradigms and paradoxes"; ed. by R.G.Colodny; University of Pittsburgh Press; pp. 47-66.
Quoted in sect.4.4.

Finkelstein,D., Jauch,J.M. and Speiser,D. (1963): **Quaternionic representations of compact groups.** J.Math.Phys.(N.Y.) 4, 136-140.
Quoted in sect.4.4.

Finkelstein,D., Jauch,J.M., Schiminovich,S. and Speiser,D. (1962): **Foundations of quaternion quantum mechanics.** J.Math.Phys.(N.Y.) 3, 207-220.
Quoted in sect.4.4.

Fischer,E. (1907): **Sur la convergence en moyenne.** C.R.Acad.Sci.(Paris) 144, 1022-1024.
Quoted in sect.3.2

Fischer,H.R. and Rüttimann,G.T. (1978): **The geometry of the state space.** In: "The mathematical foundations of quantum theory", ed. by A.R.Marlow, Academic Press, New York; pp. 153-176.
Quoted in sect.4.4.

Fleck,L. (1935): **Entstehung und Entwicklung einer naturwissenschaftlichen Tatsache.** Benno Schwabe, Basel. (Reprinted as: Suhrkamp Taschenbuch Wissenschaft 312, Suhrkamp, Frankfurt, 1980. English translation: "Genesis and development of a scientific fact", University of Chicago Press, Chicago, 1979).
Quoted in sects.5.1, 6.5.

Fock,V.A. (1932): **Konfigurationsraum und zweite Quantelung.** Z.Phys. 75, 622-647, 76, 852.
Quoted in sect.4.2.

Fock,V.A. (1951): **Kritik der Anschauungen Bohrs über die Quantenmechanik.** (In Russian). Usp.Fiz.Nauk 45, 3-14. (Reprinted in: Czech.J.Phys. 5, 436-448(1955). German translation in: Sowjetwissenschaften, Naturwissenschaftliche Abteilung 5, 123-131, 1952).
Quoted in sect.3.4.

Fock,V.A. (1952): **On the so-called ensembles in quantum mechanics.** Vestnik Leningrad Univ. 1952, No.6, 67-73, (in Russian).
Quoted in sect.3.4.

Fock,V.A. (1957): **On the interpretation of quantum mechanics.** Usp.Fiz.Nauk 62, 461. (In Russian). (English translation in: Czech.J.Phys. 7, 643-656(1957). Essentially the same version in German: "Ueber die Interpretation der Quantenmechanik", in: "Philosophische Probleme der modernen Naturwissenschaft. Materialien der Allunionskonferenz zu den philosophischen Fragen der Naturwissenschaft, Moskau, 1958", Akademie-Verlag, Berlin, 1962, pp. 189-212 und 503-505. Again the same paper: "Ueber die Deutung der Quantenmechanik", in: "Max Planck Festschrift 1958", hg. von B.Kockel, W.Macke, A.Papapetrou; Deutscher Verlag der Wissenschaften, Berlin, 1959, pp. 177-195).
Quoted in sects.1.5, 3.2, 3.4.

Fock,V.A. (1958): **Remarks on Bohr's article on his discussions with Einstein.** Usp. Fiz.Nauk 66, 599-602. (In Russian). (English translation: Sov.Phys.Usp. 66 (1), 208-210, 1958).
Quoted in sect.3.4.

Fock,V.A. (1959): **The theory of space, time and gravitation.** Pergamon, London.
Quoted in sect.3.3.

Fock,V.A. (1965): **La physique quantique et les idéalisations classiques.** Dialectica 19, 223-245. (Essentially the same paper in German: "Quantenphysik und Struktur der Materie", in: "Struktur und Formen der Materie; Deutscher Verlag der Wissenschaften, Berlin, 1969, pp. 147-174).
Quoted in sect.3.4.

Fock,V.A. (1969): **Quantum physics and philosophical problems.** Foundations of Physics 1, 293-306(1971). (First published in Russian in: "Lenin and modern natural sciences", Mysl.Moscow, 1969).
Quoted in sect.3.4.

Fock,V.A. (*1971*): **Personal communication.** Letter from October 12, 1971.
Quoted in sect.3.4.

Fok,V.A. see: *Fock,V.A.*

Foulis,D.J. and Randall,C.H. (*1972*): **Operational statistics.I. Basic concepts.** J.Math.Phys.(N.Y.) 13, 1667-1675.
Quoted in sect.4.4.

Foulis,D.J. and Randall,C.H. (*1974a*): **The empirical logic approach to the physical sciences.** In: "Foundations of quantum mechanics and ordered linear spaces"; ed. by A.Hartkämper and H.Neumann; Springer, Berlin; pp. 230-249.
Quoted in sect.4.4.

Foulis,D.J. and Randall,C.H. (*1974b*): **Empirical logic and quantum mechanics.** Synthese 29, 81-111.
Quoted in sect.4.4.

Foulis,D.J. and Randall,C.H. (*1978*): **Manuals, morphisms and quantum mechanics.** In: "Mathematical foundations of quantum theory", ed. by A.R.Marlow, Academic Press, New York; pp. 105-126.
Quoted in sect.4.4.

Fox,R. (*1971*): **Low-energy proton-proton scattering as a test of local hidden-variable theory.** Lett.Nuovo Cimento 2, 565-567.
Quoted in sect.3.4.

Fox,R. and Rosner,B. (*1971*): **Proposed experiment to test local hidden-variable theories.** Phys.Rev.D 4, 1243-1244.
Quoted in sects.3.4, 3.7.

Fraassen,B.C.van (*1972*): **A formal approach to the philosophy of science.** In: "Paradigm and paradoxes", ed. by R.G.Colodny, University of Pittsburgh Press; pp. 303-366.
Quoted in sect.4.4.

Fraassen,B.C.van (*1973*): **Semantic analysis of quantum logic.** In: "Contemporary research in the foundations and philosophy of quantum theory"; ed. by C.A.Hooker; Reidel, Dordrecht-Holland; pp. 80-113.
Quoted in sect.4.4.

Fraassen,B.C.van (*1974*): **The labyrinth of quantum logic.** In: "Logical and epistemological studies in contemporary physics"; ed. by R.S.Cohen and M.W.Wartofsky, Reidel, Dordrecht; pp. 224-254.
Quoted in sect.4.4.

Fraassen,B.C.van (*1979*): **Foundations of probability: A model frequency interpretation.** In: "Problems of the foundations of physics"; ed. by G.Toraldo di Francia, Proceedings of the International School of Physics "Enrico Fermi", course 72, North-Holland, Amsterdam; pp. 344-394.
Quoted in sect.4.4.

Frank,P. (*1909*): **Die Stellung des Relativitätsprinzips im System der Mechanik und Elektrodynamik.** Sitzungsberichte d. math.-naturwiss.Kl.Akad.Wiss. Wien 118, 373-446.
Quoted in sect.3.3.

Franz,M.L.von (*1966*): **Time and synchronicity in analytic psychology.** In: "The voices of time"; ed. by J.T.Fraser; Braziller, New York; pp. 218-232 and 633-642.
Quoted in sect.5.2.

Fraser,J.T. (*editor*) (*1966*): **The voices of time.** Braziller, New York.
Quoted in sect.3.0.

Fraser,J.T. (*1975*): **Of time, passion and knowledge.** Braziller, New York.
Quoted in sect.5.2.

Fraser,J.T. and Lawrence,N. (1975): The study of time II. Springer, Berlin.
Quoted in sect.5.2.

Fraser,J.T., Haber,F.C. and Müller,G.H. (editors) (1972): The study of time. Springer, Berlin.
Quoted in sect.3.0.

Fraser,J.T., Lawrence,N. and Park,D. (1978): The study of time III. Springer, New York.
Quoted in sect.5.2.

Freedman,S.J. and Clauser,J.F. (1972): Experimental test of local hidden-variable theories. Phys.Rev.Lett. 28, 938-941.
Quoted in sects.3.4, 3.7.

Freistadt,H. (1957): The causal formulation of quantum mechanics of particles (the theory of de Broglie, Bohm and Takabayasi). Nuovo Cimento Suppl. 5, 1-70.
Quoted in sect.3.4.

Freundlich,Y. (1972): Mind, matter and physicists. Foundations of Physics 2, 129-148.
Quoted in sect.3.5.

Friedman,C.N. (1972): Semigroup product formulas, compressions, and continual observations in quantum mechanics. Indiana University Mathematics J. 21, 1001-1011.
Quoted in sect.3.5.

Friedrichs,K.O. (1951): Mathematical aspects of the quantum theory of fields. Commun. Pure Appl.Math. 4, 161-224(1951), 5, 1-56, 349-411(1952), 6, 1-72(1953). (Collectively reissued: Interscience, New York, 1953).
Quoted in sect.4.2.

Friedrichs,K.O. (1955): Asymptotic phenomena in mathematical physics. Bull.Amer. Math.Soc. 61, 485-504.
Quoted in sect.6.3.

Frigerio,A. (1974): Quasi-local observables and the porblems of measurement in quantum mechanics. Ann.Inst. Henri Poincaré A21, 259-270.
Quoted in sect.3.5.

Frisch,K.von (1965): Tanzsprache und Orientierung der Bienen. Springer, Berlin. (English translation: Harvard University Press, Cambridge, Mass. 1967).
Quoted in sect.6.2.

Fry,E.S. and Thompson,R.C. (1976): Experimental test of local hidden-variable theories. Phys.Rev.Lett. 37, 465-468.
Quoted in sects.3.4, 3.7.

Fujiwara,I. (1972): Quantum theory of state reduction and measurement. Foundations of Physics 2, 83-110.
Quoted in sect.3.5.

Furry,W.H. (1936a): Note on the quantum-mechanical theory of measurement. Phys.Rev. 49, 393-399.
Quoted in sect.3.7.

Furry,W.H. (1936b): Remarks on measurements in quantum theory. Phys.Rev. 49, 476.
Quoted in sect.3.7.

Furry,W.H. (1966): Some aspects of the quantum theory of measurement. In: "Lectures in theoretical physics", Vol.8A, "Statistical physics and solid state physics"; ed. by W.E.Brittin; University of Colorado Press, Boulder; 1-64.
Quoted in sect.3.5.

Galilei,G. *(1623)*: **Il saggiatore.** Rome 1623. (English translation in: "Controversy on the comets of 1618"; ed. by C.D.O'Malley and S.Drake, Philadelphia, 1960).
Quoted in sect.2.2.

Galilei,G. *(1632)*: **Dialogo sopra i due massimi sistemi del mondo.** Florence. (Translated by S.Drake as "Dialogue concerning the two chief world systems"; Berkeley, 1953).
Quoted in sects.2.5, 3.3.

Galindo,A., *Morales*,A. *and Nuñez-Lagos*,R. *(1962)*: **Superselection principle and pure states of n identical particles.** J.Math.Phys.(N.Y.) 3, 324-328.
Quoted in sect.3.3.

Gallone,F. *and Mania*,A. *(1971)*: **Group representations by automorphisms of a proposition system.** Ann.Inst Henri Poincaré A15, 37-59.
Quoted in sect.4.4.

Gallone,F. *and Zecca*,A. *(1973)*: **Quantum logic axioms and the proposition-state structure.** Int.J.Theor.Phys. 8, 51-64.
Quoted in sect.4.4.

Gårding,L. *and Wightman*,A. *(1954a)*: **Representations of the anticommutation relations.** Proc.Nat.Acad.Sci.(USA) 40, 617-621.
Quoted in sect.4.2.

Gårding,L. *and Wightman*,A. *(1954b)*: **Representations of the commutation relations.** Proc.Nat.Acad.Sci.(USA) 40, 622-626.
Quoted in sect.4.2.

Gardner,M.R. *(1971)*: **Is quantum logic really logic?** Philosophy of Science 38, 508-529.
Quoted in sect.4.4.

Gauss,C.F. *(1805)*: **Letter of Gauss to Olbers, dated: Braunschweig, 1805, September 3.** Reprinted in: "Carl Friedrich Gauss, Werke", Vol.X,1; Teubner, Leipzig, 1917, p. 24-25.
Quoted in sect.2.5.

Gel'fand,I.M. *(1939)*: **Ueber normierte Algebren.** Dokl.Akad.Nauk SSSR 23, 430-432.
Quoted in sect.4.2.

Gel'fand,I.M. *(1941)*: **Normierte Ringe.** Mat.Sbornik 9, 3-24.
Quoted in sect.4.2.

Gel'fand,I.M. *and Naǐmark*,M.A. *(1943)*: **On the embedding of normed rings into the ring of operators in Hilbert space.** Mat.Sbornik 12, 197-213.
Quoted in sect.4.2.

Gel'fand,I.M. *and Vilenkin*,N.Ja. *(1961)*: **Generalized functions. Volume 4. Applications of harmonic analysis.** Academic Press, New York, 1964. (Russian original: Gosudarstv.Izdat.Fiz.-Mat.Lit., Moscow, 1961).
Quoted in sect.3.3.

George,C., *Prigogine*,I. *and Rosenfeld*,L. *(1972)*: **The macroscopic level of quantum mechanics.** Danske Vid.Selsk.Mat.-Fys.Medd. 38, No.12, 1-44.
Quoted in sect.3.5.

Gerber,J. *(1969)*: **Geschichte der Wellenmechanik.** Arch.History Exact Sci. 5, 349-416.
Quoted in sect.3.1.

Ghirardi,G.C., *Omero*,C., *Rimini*,A. *and Weber*,T. *(1978)*: **The stochastic interpretation of quantum mechanics: a critical review.** Rivista del Nuovo Cimento (3) 1, 1-34.
Quoted in sect.3.4.

Giles,R. *(1970)*: **Foundations for quantum mechanics.** J.Math.Phys.(N.Y.) 11, 2139-2160.
Quoted in sect.4.5.

Ginibre,J. and Velo,G. (1979): The classical field limit of scattering theory for non-relativistic many-boson systems.I. Commun.Math.Phys. 66, 37-76.
Quoted in sect.6.4.

Gleason,A.M. (1957): Measures on the closed subspaces of a Hilbert space. J.Mathematics abd Mechanics 6, 885-893.
Quoted in sects.3.4, 4.4, 4.5.

Glimm,J.G. and Kadison,R.V. (1960): Unitary operators in C*-algebras. Pacific J.Math. 10, 547-556.
Quoted in sect.4.2.

Gödel,K. (1944): Russell's mathematical logic. In: "The philosophy of Bertrand Russell"; ed. by P.A.Schilpp. Open Court, La Salle, Illinois; pp. 123-153.
Quoted in sect.2.5.

Gold,T. (editor) (1967): The nature of time. Cornell University Press, Ithaca, New York.
Quoted in sect.5.2.

Goldberger,M.L. and Watson,K.M. (1964): Measurement of time correlation for quantum-mechanical systems. Phys.Rev. 134, B919-B929.
Quoted in sect.3.5.

Goldblatt,R.I. (1974): Semantic analysis of orthologic. J.Philosophical Logic 3, 19-35.
Quoted in sect.4.4.

Golodec,V.Ja. (1977): On the automorphisms of von Neumann algebras. Dokl.Akad.Nauk SSSR 237, 770-772. (English translation incorporating corrections made by the author: Sov.Math.Dokl. 18, 1477-1580, 1977).
Quoted in sect.5.5.

Golodets,V.Ya. see *Golodec,V.Ja.*

Gombrich,E.H. (1960): Art and illusion. A study in the psychology of pictorial representation. Phaidon Press, London.
Quoted in sect.6.3.

Goodwin,B.C. (1963): Temporal organization in cells. Academic Press, New York.
Quoted in sect.6.2.

Gootman,E.C. and Kannan,D. (1976): Zero-one laws in finite W*-algebras. J.Math. Analysis and Applications 55, 743-756.
Quoted in sect.4.5.

Gordon,J.P. and Louisell,W.H. (1966): Simultaneous measurement of noncommuting observables. In: "Physics of quantum electronics"; ed. by P.L.Kelley, B.Lax, P.E.Tannenwald; McGraw-Hill, New York; pp. 833-840.
Quoted in sect.4.5.

Gorini,V. (1971): Linear kinematical groups. Commun.MathPhys. 21, 150-163.
Quoted in sect.3.3.

Gorini,V., Kossakowski,A. and Sudarshan,E.C.G. (1976): Completely positive dynamical semigroups of N-level systems. J.Math.Phys.(N.Y.) 17, 821-825.
Quoted in sect.4.5.

Gorini,V., Frigerio,A., Verri,M., Kossakowski,A. and Sudarshan,E.C.G. (1978): Properties of quantum Markovian master equations. Reports on Mathematical Physics 13, 149-173.
Quoted in sect.6.3.

Graham,N. (1970): The Everett interpretation of quantum mechanics. Ph.D. thesis, University of North Carolina at Chapel Hill; (unpublished), 97 pp.
Quoted in sect.3.6.

Graham,N. (*1973*): The measurement of relative frequency. In: "The many-worlds interpretation of quantum mechanics"; ed. by B.S.DeWitt and N.Graham; Princeton University Press, Princeton, New Jersey; pp. 229-253.
Quoted in sect.3.6.

Grattan-Guiness,I. (*1977*): Dear Russell - Dear Jourdain. Duckworth, London.
Quoted in sect.4.4.

Grawert,G. (*1953*): Eine Theorie der physikalischen Aussagen. Z.Phys. 136, 206-220.
Quoted in sect.4.4.

Greechie,R.J. (*1969*): An orthomodular poset with a full set of states not embeddable in Hilbert space. Caribbean J.Science and Mathematics 1, 15-26.
Quoted in sect.4.4.

Greechie,R.J. (*1971*): Orthomodular lattices admitting no states. J.Combinatorial Theory, Ser.A 10, 119-132.
Quoted in sect.4.4.

Greechie,R.J. *and Gudder*,S.P. (*1971*): Is a quantum logic a logic? Helv.Phys. Acta 44, 238-240.
Quoted in sect.4.4.

Greechie,R.J. *and Gudder*,S.P. (*1973*): Quantum logics. In: "Contemporary research in the foundations and philosophy of quantum theory"; ed. by C.A.Hooker; Reidel, Dordrecht-Holland; pp. 143-173.
Quoted in sect.4.4.

Green,H.S. (*1958*): Observation in quantum mechanics. Nuovo Cimento 9, 880-889.
Quoted in sect.3.5.

Greenstein,C.H. (*1978*): Dictionary of logical terms and symbols. Van Nostrand Reinhold, New York.
Quoted in sect.4.4.

Grenander,U. (*1969*): Foundations of pattern analysis. Quart.Appl.Math. 27, 1-55.
Quoted in sect.6.3.

Grib,A.A. (*1968*): Non-equivalent representations in quantum field theory and measurement theory. (in Russian). Vestnik Leningrad Univ., Seria Fizika i Khimiya No.16, 16-23.
Quoted in sect.3.5.

Groenewold,H.J. (*1946*): On the principles of elementary quantum mechanics. Physica (Utrecht) 12, 405-460.
Quoted in sect.3.5.

Groenewold,H.J. (*1952*): Information in quantum mechanics. Proc.Kon.Ned.Akad.Wetens. Ser.B 55, 219-227.
Quoted in sect.3.5.

Groenewold,H.J. (*1957*): Objective and subjective aspects of statistics in quantum description. In: "Observation and interpretation"; ed. by S.Körner; Butterworth, London, 1957 (reprinted by Dover, New York, 1962); pp. 197-203.
Quoted in sect.3.5.

Groenewold,H.J. (*1964*): Micro- and macromeasurements. Nucl.Phys. 57, 112-133.
Quoted in sects.3.4, 3.5.

Groenewold,H.J. (*1971*): Quantal observation in statistical interpretation. In: "Quantum theory and beyond"; ed. by T.Bastin; Cambridge University Press, London; pp. 43-54.
Quoted in sects.3.4, 3.5.

Grünbaum,A. (*1964*): Philosophical problems of space and time. Knopf, New York. Second, enlarged edition: Reidel, Dordrecht-Holland, 1973-
Quoted in sect.5.2.

Gudder,S.P. (1965): Spectral methods for a generalized probability theory. Trans.Amer.Math.Soc. 119, 428-442.
Quoted in sect.4.4.

Gudder,S.P. (1966): Uniqueness and existence properties of bounded observables. Pacific J.Math. 19, 81-93.
Quoted in sect.4.4.

Gudder,S.P. (1967a): Coordinate and momentum observables in axiomatic quantum mechanics. J.Math.Phys.(N.Y.) 8, 1848-1858.
Quoted in sect.4.4.

Gudder,S.P. (1967b): System of observables in axiomatic quantum mechanics. J.Math. Phys.(.N.Y.) 8, 2109-2113.
Quoted in sect.4.4.

Gudder,S.P. (1967c): Hilbert space, independence, and generalized probability. J.Math.Analaysis and Applications 20, 48-61.
Quoted in sect.4.4.

Gudder,S.P. (1968a): Hidden variables in quantum mechanics reconsidered. Rev.Mod. Phys. 40, 229-231.
Quoted in sect.3.4.

Gudder,S.P. (1968b): Dispersion-free states and the exclusion of hidden variables. Proc.Amer.Math.Soc. 19, 319-324.
Quoted in sect.3.4.

Gudder,S.P. (1968c): Complete sets of observables and pure states. Canad.J.Math. 20, 1276-1280.
Quoted in sect.4.4.

Gudder,S.P. (1969a): On the quantum logic approach to quantum mechanics. Commun.Math. Phys. 12, 1-15.
Quoted in sects.3.4, 4.4.

Gudder,S.P. (1969b): Quantum probability spaces. Proc.Amer.Math.Soc. 21, 296-302.
Quoted in sects.3.4, 4.4.

Gudder,S.P. (1970a): On hidden-variable theories. J.Math.Phys.(N.Y.) 11, 431-436.
Quoted in sect.3.4.

Gudder,S.P. (1970b): A superposition principle in physics. J.Math.Phys.(N.Y.) 11, 1037-1040.
Quoted in sect.4.4.

Gudder,S.P. (1970c): Axiomatic quantum mechanics and generalized probability theory. In: "Probabilistic methods in applied mathematics", Vol.2; ed. by A.T.Bharucha-Reid; Academic Press, New York.
Quoted in sect.4.4.

Gudder,S.P. (1970d): Projective representations of quantum logics. Int.J.Theor.Phys. 3, 99-108.
Quoted in sect.4.4.

Gudder,S.P. (1971): Representations of groups as automorphisms on orthomodular lattices and posets. Canad.J.Math. 23, 659-673.
Quoted in sect.4.4.

Gudder,S.P. (1972a): Hidden variable models for quantum mechanics. Nuovo Cimento B 10, 518-522.
Quoted in sect.3.4.

Gudder,S.P. (1972b): Plane frame functions and pure states in Hilbert space. Int.J.Theor.Phys. 6, 369-375.
Quoted in sect.4.4.

Gudder,S.P. (*1973a*): Quantum logics, physical space, position observables, and symmetry. Reports on Mathematical Physics 4, 193-202.
Quoted in sect.4.4.

Gudder,S.P. (*1973b*): Convex structures and operational quantum mechanics. Commun. Math.Phys. 29, 249-264.
Quoted in sect.4.5.

Gudder,S.P. (*1977a*): Convexity and mixtures. SIAM Review 19, 221-240. Erratum ib.20, 837 (1978).
Quoted in sect.4.5.

Gudder,S.P. (*1977b*): Four approaches to axiomatic quantum mechanics. In: "The uncertainty principle and foundations of quantum mechanics"; ed. by W.C.Price and S.S.Chissick, Wiley, London; pp. 247-276.
Quoted in sects.4.4, 4.5.

Gudder,S.P. (*1978*): Cantoni's generalized transition probability. Commun.Math.Phys. 63, 265-267.
Quoted in sect.4.5.

Gudder,S.P. (*1979a*): A Radon-Nikodym theorem for *-algebras. Pacific J.Math. 80, 141-149.
Quoted in sect.4.5.

Gudder,S.P. (*1979b*): Axiomatic operational quantum mechanics. Reports on Mathematical Physics 16, 147-166.
Quoted in sect.4.5.

Gudder,S.P. and Boyce,S. (*1970*): A Comparison of the Mackey and Segal models for quantum mechanics. Int.J.Theor.Phys. 3, 7-21.
Quoted in sect.4.6.

Gudder,S.P. and Hudson,R.L. (*1978*): A noncommutative probability theory. Trans.Amer. Math.Soc. 245, 1-41.
Quoted in sect.4.5.

Gudder,S.P. and Marchand,J.P. (*1972*): Non-commutative probability on von Neumann algebras. J.Math.Phys.(N.Y.) 13, 799-806.
Quoted in sect.4.5.

Gudder,S.P. and Michel,J.R. (*1979*): Embedding quantum logics in Hilbert space. Letters in Mathematical Physics 3, 379-386.
Quoted in sect.4.4.

Gudder,S.P. and Mullikin,H.C. (*1973*): Measure theoretic convergence of observables and operators. J.Math.Phys.(N.Y.) 14, 234-242.
Quoted in sect.4.4.

Gudder,S.P. and Piron,C. (*1971*): Observables and the field in quantum mechanics. J.Math.Phys.(N.Y.) 12, 1583-1588.
Quoted in sect.4.4.

Gudder,S.P., Marchand,J.P. and Wyss,W. (*1979*): Bures distance and relative entropy. J.Math.Phys.(N.Y.) 20, 1963-1966.
Quoted in sect.4.5.

Guenin,M. (*1966*): Axiomatic foundations of quantum theories. J.Math.Phys.(N.Y.) 7, 271-282.
Quoted in sect.4.4.

Gunson,J. (*1967*): On the algebraic structure of quantum mechanics. Commun.Math.Phys. 6, 262-285.
Quoted in sects.4.4, 4.5, 4.6.

Gunson,J. (*1972*): Physical states on quantum logics. Ann.Inst. Henri Poincaré, A 17, 295-311.
Quoted in sects.4.4, 4.5.

Gustafson,K. and Misra,B. (*1976*): Canonical commutation relations of quantum mechanics and stochastic regularity. Letters in Mathematical Physics 1, 275-280.
Quoted in sect.3.8.

Guz,W. (*1971*): Quantum logic and a theorem on commensurability. Reports on Mathematical Physics 2, 53-61.
Quoted in sect.4.4.

Guz,W. (*1974*): On the axiom system for non-relativistic quantum mechanics. Reports on Mathematical Physics 6, 445-454.
Quoted in sect.4.4.

Guz,W. (*1975*): A modification of the axiom system of quantum mechanics. Reports on Mathematical Physics 7, 313-320.
Quoted in sect.4.4.

Guz,W. (*1977a*): Axioms for nonrelativistic quantum mechanics. Int.J.Theor.Phys. 16, 299-306.
Quoted in sect.4.4.

Guz,W. (*1977b*): Axioms for statistical physical theories and GL-spaces. Reports on Mathematical Physics 12, 151-167.
Quoted in sect.4.5.

Guz,W. (*1977c*): Some spaces of the type GM and GL, basic properties. Reports on Mathematical Physics 12, 285-299.
Quoted in sect.4.5.

Guz,W. (*1978a*): On the lattice structure of quantum logics. Ann.Inst. Henri Poincaré A 28, 1-7.
Quoted in sect.4.4.

Guz,W. (*1978b*): Filter theory and covering law. Ann.Inst. Henri Poincaré A 29, 357-378.
Quoted in sects.4.4, 4.6.

Guz,W. (*1979*): On the simultaneous verifiability of yes-no measurements. Int.J.Theor. Phys. 17, 543-548.
Quoted in sect.4.4.

Guz,W. (*1980*): Event-phase-space structure: an alternative to quantum logic. J.Phys.A 13, 881-899.
Quoted in sect.4.4.

Haack,S. (*1974*): Deviant logic. Cambridge University Press, London.
Quoted in sect.4.4.

Haag,R. (*1955*): On quantum field theories. Danske Vid.Selsk.Mat.-Fys.Medd. 29, No.12, 1-37.
Quoted in sect.4.2.

Haag,R. (*1962*): The mathematical structure of the Bardeen-Cooper-Schrieffer model. Nuovo Cimento 25, 287-299.
Quoted in sect.4.2.

Haag,R. (*1969*): Bemerkungen zum Begriffsbild der Quantenphysik. Z.Phys. 229, 384-391.
Quoted in sect.4.5.

Haag,R. and Kastler,D. (*1964*): An algebraic approach to quantum field theory. J.Math.Phys.(N.Y.) 5, 848-861.
Quoted in sect.4.2.

Haag,R. and Trych-Pohlmeyer,E. (1977): Stability properties of equilibrium states. Commun.Math.Phys. 56, 213-224.
Quoted in sect.4.3.

Haag,R., Hugenholtz,N.M. and Winnink,M. (1967): On the equilibrium states in quantum statistical mechanics. Commun.Math.Phys. 5, 215-236.
Quoted in sects.4.2, 4.3.

Haag,R., Kadison,R.V. and Kastler,D. (1970): Nets of C*-algebras and classification of states. Commun.Math.Phys. 16, 81-104.
Quoted in sects.4.2, 4.3.

Haag,R., Kastler,D. and Trych-Pohlmeyer,E.B. (1974): Stability and equilibrium states. Commun.Math.Phys. 38, 173-193.
Quoted in sect.4.3.

Haagerup,U. (1975): The standard form of von Neumann algebras. Math.Scand. 37, 271-283.
Quoted in sect.6.4.

Haake,F. and Weidlich,W. (1968): A model for the measuring process in quantum theory. Z.Phys. 213, 451-465.
Quoted in sect.3.5.

Hack,M.N. (1964): Quantum theory of measurement: comment on a paper of Shimony. Amer.J.Phys. 32, 890-892.
Quoted in sect.3.5.

Hackenbroch,W. (1976): Non commutative integration in spectral theory. In: "Measure theory"; ed. by A.Bellow and D.Kölzow, Lectures Notes in Mathematics, Vol.541, Springer, Berlin; pp. 309-321.
Quoted in sect.4.5.

Hadamard,J. (1945): The psychology of invention in the mathematical field. Princeton University Press, Princeton, N.J.
Quoted in sect.2.5.

Hagedorn,G.A. (1980a): Semiclassical quantum mechanics for coherent states. In: "Mathematical problems in theoretical physics"; ed. by K.Osterwalder, Lecture Notes in Physics, Vol.116, Springer, Berlin; pp. 78-82.
Quoted in sect.6.4.

Hagedorn,G.A. (1980b): Semiclassical quantum mechanics.I. The $\hbar\to 0$ limit for coherent states. Commun.Math.Phys. 71, 77-93.
Quoted in sect.6.4.

Hagedorn,G.A. (1980c): A time dependent Born-Oppenheimer approximation. Commun.Math. Phys. 77, 1-19.
Quoted in sect.6.4.

Hagedorn,R. (1959): Note on symmetry operations in quantum mechanics. Nuovo Cimento Suppl. 12, 73-86.
Quoted in sect.3.3.

Haken,H. (1977): Synergetics. Springer, Berlin.
Quoted in sect.6.2.

Hakim,R. (1968): Relativistic stochastic processes. J.Math.Phys.(N.Y.) 9, 1805-1818.
Quoted in sect.6.4.

Hall,D. and Wightman.A.S. (1957): A theorem on invariant analytic functions with applications to relativistic quantum field theory. Danske Vid.Selsk.Mat.-Fys.Medd. 31, No.5, 1-41.
Quoted in sect.4.2.

Hammer,P.C. (1964): Topologies of approximation. J.SIAM Numer.Anal. B1, 69-75.
Quoted in sect.6.3.

Hammer,P.C. (*1971*): Mind pollution. Cybernetics 14, 5-19.
Quoted in sects.3.8, 6.3.

Hanner,O. (*1950*): Deterministic and non-deterministic stationary random processes. Arkiv för Matematik 1, 161-177.
Quoted in sect.3.8.

Hanson,N.R. (*1958*): Patterns of discovery. Cambridge University Press, Cambridge, England.
Quoted in sects.2.3, 2.4.

Hardegree,G.M. (*1974*): The conditional in quantum logic. Synthese 29, 63-80.
Quoted in sect.4.4.

Hardegree,G.M. (*1975a*): Stalnaker conditionals and quantum logic. J.Philosophical Logic 4, 399-421.
Quoted in sect.4.4.

Hardegree,G.M. (*1975b*): Quasi-implicative lattices and the logic of quantum mechanics. Z.Naturforsch. 30a, 1347-1360.
Quoted in sect.4.4.

Hardegree,G.M. (*1977*): Relative compatibility in conventional quantum mechanics. Foundations of Physics 7, 495-510.
Quoted in sect.5.3.

Hardy,G.H. (*1967*): A mathematician's apology. Cambridge University Press, London; first edition 1940, reprinted with a foreword by C.P.Snow, 1967.
Quoted in sect.2.5.

Hartkämper,A. *and Neumann*,H. (*1974*): Foundations of quantum mechanics and ordered linear spaces. Advanced study institute, Marburg, 1973. Lecture Notes in Physics, Vol.29, Springer, Berlin.
Quoted in sect.4.5.

Hartle,J.B. (*1968*): Quantum mechanics of individual systems. Amer.J.Phys. 36, 704-712.
Quoted in sect.3.6.

Hartmann,H. (*1964*): Die Bedeutung quantenmechanischer Modelle für die Chemie. Experimentia Supplement (Geburtstagsschrift für G.Schwarzenbach) 9, 94-97.
Quoted in sect.1.4.

Healey,R. (*1979*): Quantum realism: naiveté is no excuse. Synthese 42, 121-144.
Quoted in sect.5.2.

Heelan,P.A. (*1965*): Quantum mechanics and objectivity. A study of the physical philosophy of Werner Heisenberg. Nijhoff, The Hague.
Quoted in sect.3.4.

Heelan,P.A. (*1970a*): Quantum and classical logic: their respective roles. Synthese 21, 2-33.
Quoted in sect.4.4.

Heelan,P.A. (*1970b*): Complementarity, context dependence, and quantum logic. Foundations of Physics 1, 95-110.
Quoted in sect.4.4.

Heisenberg,W. (*1925*): Ueber quantentheoretische Umdeutung kinematischer und mechanischer Beziehungen. Z.Phys. 33, 879-893.
Quoted in sect.3.2.

Heisenberg,W. (*1927*): Ueber den anschaulichen Inhalt der quantentheoretischen Kinematik und Mechanik. Z.Phys. 43, 172-198.
Quoted in sect.3.4.

Heisenberg,W. (*1930*): **Die physikalischen Prinzipien der Quantentheorie.** Hirzel, Leipzig.
Quoted in sect.3.4.

Heisenberg,W. (*1955*): **The development of the interpretation of the quantum theory.** In: "Niels Bohr and the development of physics"; ed. by W.Pauli; Pergamon Press, London; pp. 12-29.
Quoted in sect.3.4.

Heisenberg,W. (*1956a*): **Die Entwicklung der Deutung der Quantenmechanik.** Physikalische Blätter 12, 289-304.
Quoted in sect.3.2.

Heisenberg,W. (*1956b*): **The representation of nature in contemporary physics.** Daedalus 87, 95-108, 1958. (German original: "Die Künste im technischen Zeitalter"; Wissenschaftliche Buchgesellschaft, Darmstadt; pp. 31-47.)
Quoted in sect.3.4.

Heisenberg,W. (*1958*): **Physics and Philosophy.** Harper and Brothers, New York. (German version: "Physik und Philosophie". Hirzel, Stuttgart, 1959.)
Quoted in sect.3.4.

Heisenberg,W. (*1959*): **Wolfgang Paulis philosophische Auffassungen.** Naturwissenschaften 46, 661-663.
Quoted in sect.1.1.

Heisenberg,W. (*1969*): **Der Teil und das Ganze.** Piper, München.
Quoted in sect.2.6.

Hellwig,K.E. (*1969*): **Coexistent effects in quantum mechanics.** Int.J.Theor.Phys. 2, 147-155.
Quoted in sect.4.5.

Hellwig,K.E. *and Kraus*,K. (*1969*): **Pure operations and measurements.** Commun.Math.Phys. 11, 214-220.
Quoted in sect.4.5.

Hellwig,K.E. *and Kraus*,K. (*1970a*): **Operations and measurements II.** Commun.Math.Phys. 16, 142-147.
Quoted in sect.4.5.

Hellwig,K.E. *and Kraus*,K. (*1970b*): **Formal descriptions of measurements in local quantum field theory.** Phys.Rev. D 1, 566-571.
Quoted in sects.3.5, 3.7.

Hellwig,K.E. *and Krausser*,D. (*1974a*): **Propositional systems and measurements.I.** Int.J.Theor.Phys. 9, 277-289.
Quoted in sect.4.4.

Hellwig,K.E. *and Krausser*,D. (*1974b*): **Propositional systems and measurements.II.** Int.J.Theor.Phys. 10, 261-272.
Quoted in sect.4.4.

Hellwig,K.E. *and Krausser*,D. (*1977*): **Propositional systems and measurements.III.** Int.J.Theor.Phys. 16, 775-793.
Quoted in sect.4.4.

Helstrom,C.W. (*1967*): **Detection theory and quantum mechanics.** Information and Control 10, 254-291.
Quoted in sect.4.5.

Helstrom,C.W. (*1968*): **Detection theory and quantum mechanics (II).** Information and Control 13, 156-171.
Quoted in sect.4.5.

Helstrom,C.W. (*1969*): Quantum detection and estimation theory. J.Statist.Phys. 1, 231-252.
Quoted in sect.4.5.

Helstrom,C.W. (*1973*): Cramér-Rao inequalities for operator-valued measures in quantum mechanics. Int.J.Theor.Phys. 8, 361-376.
Quoted in sect.4.5.

Helstrom,C.W. (*1974a*): Quantum Bayes estimation of the amplitude of a coherent signal. IEEE Trans.Inform.Theory IT-20, 374-376.
Quoted in sect.4.5.

Helstrom,C.W. (*1974b*): 'Simultaneous measurement' from the standpoint of a quantum measurement theory. Foundations of Physics 4, 453-463.
Quoted in sect.4.5.

Helstrom,C.W. (*1974c*): Estimation of a displacement parameter of a quantum system. Int.J.Theor.Phys. 11, 357-378.
Quoted in sect.4.5.

Helstrom,C.W. (*1976*): Quantum detection and estimation theory. Academic Press, New York.
Quoted in sect.4.5.

Helstrom,C.W. *and Kennedy*,R.S. (*1974*): Noncommuting observables in quantum detection and estimation theory. IEEE Trans.Inform.Theory IT-20, 16-24.
Quoted in sect.4.5.

Helstrom,C.W., *Liu*,J.W.S. *and Gordon*,J.P. (*1970*): Quantum mechanical communication theory. Proc.IEEE 58, 1578-1598.
Quoted in sect.4.5.

Hempel,C.G. (*1945*): Review of: Philosophic foundations of quantum mechanics by Hans Reichenbach. J.Symbolic Logic 10, 97-100.
Quoted in sect.4.4.

Hempel,C.G. (*1965*): Aspects of scientific explanation. Free Press, New York.
Quoted in sect.3.8.

Hempel,C.G. (*1966*): Philosophy of nature. Prentice-Hall, Englewood Cliffs, N.J.
Quoted in sect.3.8.

Hempel,C.G. *and Oppenheim*,P. (*1948*): Studies in the logic of explanation. Philosophy of Science 15, 135-175.
Quoted in sect.6.1.

Hepp,K. (*1972*): Quantum theory of measurement and macroscopic observables. Helv.Phys.Acta 45, 237-248.
Quoted in sects.3.5, 4.2, 4.3.

Hepp,K. (*1974*): The classical limit for quantum mechanical correlation functions. Commun.Math.Phys. 35, 265-277.
Quoted in sect.6.4.

Herbut,F. (*1969*): Derivation of the change of state in measurement from the concept of minimal measurement. Ann.Phys.(New York) 55, 271-300.
Quoted in sect.3.5.

Herbut,F. (*1974*): Minimal disturbance measurement as a specification in von Neumann's quantal theory of measurement. Int.J.Theor.Phys. 11, 193-204.
Quoted in sect.3.5.

Herman,L. *and Piziak*,R. (*1974*): Modal propositional logic on an orthomodular basis.I. J.Symbolic Logic 39, 478-488.
Quoted in sect.4.4.

Herman,L., Marsden,E.L. and Piziak,R. (*1975*): Implication connectives in orthomodular lattices. Notre Dame J.Formal Logic 16, 305-328.
Quoted in sect.4.4.

Herman,R.H. and Takesaki,M. (*1970*): States and automorphism groups of operator algebras. Commun.Math.Phys. 19, 142-160.
Quoted in sect.4.3.

Hermann,A. (*1963*): Dokumente der Naturwissenschaft. Abteilung Physik. Band 4. Die Kopenhagener Deutung der Quantentheorie. Battenberg Verlag, Stuttgart.
Quoted in sect.3.2.

Hermann,A. (*1969*): Frühgeschichte der Quantentheorie. Physik Verlag, Mosbach, (Baden).
Quoted in sect.3.1.

Herzberg,G. (*1970*): The dissociation energy of the hydrogen molecule. J.Mo.Spectrosc. 33, 147-169.
Quoted in sect.2.3.

Herzberg,G. and Monfils,A. (*1960*): The dissociation energies of the H_2, HD, and D_2 molecules. J.Mol.Spectrosc. 5, 482-498.
Quoted in sect.2.3.

Hiai,F., Ohya,M. and Tsukada,M. (*1979*): Sufficiency, KMS condition and relative entropy in von Neumann algebras. Preprint Department of Information Science University of Tokyo.
Quoted in sect.4.5.

Hilbert,D. (*1904*): Grundzüge einer allgemeinen Theorie der Integralgleichungen. Nachr.Kgl.Ges.Wiss.Göttingen, Math.-Phys.Kl. 1904, 49-91.
Quoted in sect.3.2.

Hilbert,D. (*1912*): Grundzüge einer allgemeinen Theorie der linearen Integralgleichungen. Teubner, Leipzig.
Quoted in sect.3.2.

Hilbert,D., Neumann,J.von and Nordheim,L. (*1927*): Ueber die Grundlagen der Quantenmechanik. Mathematische Annalen 98, 1-30.
Quoted in sect.3.2.

Hirschfelder,J.O., Curtiss,C.F. and Bird,R.B. (*1954*): Molecular theory of gases and liquids. Wiley, New York.
Quoted in sect.3.3.

Hofstadter,D.R. (*1979*): Gödel, Escher, Bach: an eternal golden braid. The Harvester Press, Sussex.
Quoted in sect.6.2.

Holevo,A.S. (*1972a*): Some statistical problems for quantum fields. Theory Probability Appl.(USSR) 17, 347-351. (Russian original: Teor.Verojatnost.i Primenen. 17, 360-365, 1972).
Quoted in sect.4.5.

Holevo,A.S. (*1972b*): An analog of the theory of statistical decisions in noncommutative probability theory. Trans.Moscow Math.Soc. 26, 133-149. (Russian original: Trudy Moscov.Mat.Obšč. 26, 133-149).
Quoted in sect.4.5.

Holevo,A.S. (*1972c*): On the mathematical theory of quantum communication channels. Problems of Information Transmission 8, 47-54. (Russian original: Problemy Peredači Informacii 8, 62-71).
Quoted in sect.4.5.

Holevo,A.S. (*1973a*): **Statistical problems in quantum physics.** In: "Proceedings of the second Japan-USSR symposium on probability theory". Lecture Notes in Mathematics, Vol.330; Springer, Berlin; pp. 104-119.
Quoted in sect.4.5.

Holevo,A.S. (*1973b*): **Statistical decision theory for quantum systems.** J.Multivariate Analysis 3, 337-394.
Quoted in sect.4.5.

Holevo,A.S. (*1973c*): **Optimal quantum measurements.** Theoretical and Mathematical Physics 17, 1172-1177 (1974). (Russian original: Teoreticheskaya i Matematicheskaya Fiz. 17, 319-326, 1973).
Quoted in sect.4.5.

Holevo,A.S. (*1974*): **The theory of statistical decisions on an operator algebra.** Sov.Math.Dokl. 15, 1276-1281. (Russian original: Dokl.Akad.Nauk SSSR 218, 54-57).
Quoted in sect.4.5.

Holevo,A.S. (*1975a*): **Some statistical problems for quantum Gaussian states.** IEEE Trans.Inform.Theory IT-21, 533-543.
Quoted in sect.4.5.

Holevo,A.S. (*1975b*): **Mathematical problems in the theory of quantum channels.** Proc. Steklov Institute of Mathematics 1978, 231-233. (Russian original: Trudy Mat.Inst. Steklov 135, 1975).
Quoted in sect.4.5.

Holevo,A.S. (*1977a*): **Commutation superoperator of a state and its applications to the noncommutative statistics.** Reports on Mathematical Physics 12, 251-271.
Quoted in sect.4.5.

Holevo,A.S. (*1977b*): **Problems in the mathematical theory of quantum communication channels.** Reports on Mathematical Physics 12, 273-278.
Quoted in sect.4.5.

Holevo,A.S. (*1978*): **Estimation of shift parameters of a quantum state.** Reports on Mathematical Physics 13, 379-399.
Quoted in sect.4.5.

Holland,S.S. (*1964*): **Distributivity and perspectivity in orthomodular lattices.** Trans.Amer.Math.Soc. 112, 330-343.
Quoted in sect.4.6.

Holland,S.S. (*1970*): **The current interest in orthomodular lattices.** In: "Trends in lattice theory"; ed. by J.C.Abbott; van Nostrand-Reinhold, New York; pp. 41-126.
Quoted in sect.4.4.

Hooker,C.A. (*1970*): **Concerning Einstein's, Podolsky's, and Rosen's objection to quantum theory.** Amer.J.Phys. 38, 851-857.
Quoted in sect.3.7.

Hooker,C.A. (*1971a*): **Sharp and the refutation of the Einstein, Podolsky, Rosen paradox.** Philosophy of Science 38, 224-233.
Quoted in sect.3.7.

Hooker,C.A. (*1971b*): **'Against Krips' resolution of two paradoxes in quantum mechanics.** Philosophy of Science 38, 418-428.
Quoted in sect.3.7.

Hooker,C.A. (*1972a*): **Concerning measurements in quantum theory: A critique of a recent proposal.** Int.J.Theor.Phys. 5, 231-250.
Quoted in sect.3.7.

Hooker,C.A. (*1972b*): The nature of quantum mechanical reality: Einstein versus Bohr. In: "Paradigms and paradoxes: The philosophical challenge of quantum domain"; Vol.5 in the University of Pittsburgh Series in the Philosophy of Science; ed. by R.G.Colodny; University of Pittsburgh Press.
Quoted in sects.3.4, 3.7.

Hooker,C.A. (*1975*): The logico-algebraic approach to quantum mechanics. Volume I. Historical evolution. Reidel, Dordrecht-Holland.
Quoted in sect.4.4.

Hooker,C.A. (*1979a*): The logico-algebraic approach to quantum mechanics. Volume 2. Contemporary consolidation. Reidel. Dordrecht-Holland.
Quoted in sect.4.4.

Hooker,C.A. (*1979b*): Physical theory as logico-operational structure. Reidel, Dordrecht-Holland.
Quoted in sect.4.4.

Horwitz,L.P. and Biedenharn,L.C. (*1979*): Exceptional gauge groups and quantum theory. J.Math.Phys.(N.Y.) 20, 269-298.
Quoted in sect.4.2.

Hove,L.van (*1952*): Les difficultés de divergences pour un modèle particulier de champ quantifié. Physica(Utrecht) 18, 145-159.
Quoted in sect.4.2.

Hove,L.van (*1955*): Quantum-mechanical perturbation giving rise to a statistical transport equation. Physica(Utrecht) 21, 517-540.
Quoted in sect.6.3.

Hudson,R.L. (*1979*): The strong Markov property for canonical Wiener processes. J.Functional Analysis 34, 266-281.
Quoted in sect.4.5.

Hugenholtz,N.M. (*1967*): On the factor type of equilibrium states in quantum statistical mechanics. Commun.Math.Phys. 6, 189-193.
Quoted in sect.4.3.

Hugenholtz,N.M. (*1972*): States and representations in statistical mechanics. In: "Mathematics of contemporary physics"; ed. by R.F.Streater, Academic Press, London; pp. 145-182.
Quoted in sect.4.3.

Hugenholtz,N.M. and Wieringa,J.D. (*1969*): On locally normal states in quantum statistical mechanics. Commun.Math.Phys. 11, 183-197.
Quoted in sect.4.3.

Husimi,K. (*1937*): Studies on the foundation of quantum mechanics. Proc.Phys.Math.Soc. Japan 19, 766-789.
Quoted in sects.3.3, 4.4.

Ikebe,T. and Kato,T. (*1962*): Uniqueness of the self-adjoint extension of singular elliptic differential operators. Arch.Ration.Mech.Anal. 9, 77-92.
Quoted in sect.3.3.

Ingleby,M. (*1971*): Some criticism of quantum logic. Helv.Phys.Acta 44, 299-307.
Quoted in sect.4.4.

Inglis,D.R. (*1961*): Completeness of quantum mechanics and charge-conjugation correlation of theta particles. Rev.Mod.Phys. 33, 1-7.
Quoted in sect.3.7.

Ingólfsson,K. (*1976*): Notes on the classical and nonrelativistic limits in quantum mechanics. Letters in Mathematical Physics 1, 351-359.
Quoted in sect.6.4.

Inönü,E. *and Wigner*,E.P. *(1952)*: Representations of the Galilei group. Nuovo Cimento 9, 705-718.
Quoted in sect.3.3.

Itoh,T. *(1965)*: Derivation of nonrelativistic Hamiltonian for electrons from quantum electrodynamics. Rev.Mod.Phys. 37, 159-165.
Quoted in sect.3.3.

Ivert,P.A. *and Sjödin*,T. *(1978)*: On the impossibility of a finite propositional lattice for quantum mechanics. Helv.Phys.Acta 51, 635-636.
Quoted in sect.4.4.

Jacobi,C.G.J. *(1866)*: Vorlesungen über Dynamik. Herausgegeben von A.Clebsch; Reimer Berlin. Second edition as supplement to the collected works, ed. by E.Lottner; Reimer Berlin 1884.
Quoted in sect.3.3.

Jammer,M. *(1954)*: Concepts of space. A history of the theories of space in physics. Harvard University Press, Cambridge, Mass.
Quoted in sect.3.3.

Jammer,M. *(1961)*: Concepts of mass in classical and modern physics. Harvard University Press, Cambridge, Mass.
Quoted in sect.3.3.

Jammer,M. *(1966)*: The conceptual development of quantum mechanics. McGraw-Hill, New York.
Quoted in sects.3.1, 3.4.

Jammer,M. *(1974)*: The philosophy of quantum mechanics. The interpretations of quantum mechanics in historical perspective. Wiley, New York.
Quoted in sects.3.1, 3.4, 4.4, 6.5.

Jánossy,L. *and Nagy*,K. *(1956)*: Ueber eine Form des Einsteinschen Paradoxes der Quantentheorie. Annalen der Physik 17, 115-121.
Quoted in sect.3.7.

Jasselette,P. *and Voisin*,J. *(1970a)*: Sur le problème de l'observation en mécanique quantique. Nuovo Cimento B 66, 229-238.
Quoted in sect.3.5.

Jauch,J.M. *(1960)*: Systems of observables in quantum mechanics. Helv.Phys.Acta 33, 711-720.
Quoted in sects.3.3, 4.6, 5.5.

Jauch,J.M. *(1964a)*: Gauge invariance as a consequence of Galilei-invariance for elementary particles. Helv.Phys.Acta 37, 284-292.
Quoted in sect.3.3.

Jauch,J.M. *(1964b)*: The problem of measurement in quantum mechanics. Helv.Phys.Acta 37, 293-316.
Quoted in sect.3.5.

Jauch,J.M. *(1968a)*: Foundations of quantum mechanics. Addison-Wesley, Reading, Massachusetts.
Quoted in sects.3.5, 3.7, 4.4.

Jauch,J.M. *(1968b)*: Projective representation of the Poincaré group in a quaternionic Hilbert space. In: "Group theory and its applications"; ed. by E.M.Loebl; Academic Press, New York; pp. 131-182.
Quoted in sect.4.4.

Jauch,J.M. *(1970)*: Scattering theory in general quantum mechanics. In: "Analytic methods in mathematical physics; ed. by R.P.Gilbert and R.G.Newton; Gordon and Breach, New York; pp. 185-205.
Quoted in sect.4.4.

Jauch,J.M. (*1971*): Foundations of quantum mechanics. In: "Proceedings of the international school of physics 'Enrico Fermi'. Course 49: Foundations of quantum mechanics; ed. by B.d'Espagnat; Academic Press, New York; pp. 20-55.
Quoted in sects.3.5, 4.2, 4.4.

Jauch,J.M. and Misra,B. (*1961*): Supersymmetries and essential observables. Helv. Phys.Acta 34, 699-709.
Quoted in sect.4.6.

Jauch,J.M. and Piron,C. (*1963*): Can hidden variables be excluded in quantum mechanics? Helv.Phys.Acta 36, 827-837.
Quoted in sects.3.4, 4.4.

Jauch,J.M. and Piron,C. (*1968*): Hidden variables revisited. Rev.Mod.Phys. 40, 228-229.
Quoted in sects.3.4, 4.4.

Jauch,J.M. and Piron,C. (*1969*): On the structure of quantal proposition systems. Helv.Phys.Acta 42, 842-848.
Quoted in sect.4.4.

Jauch,J.M. and Piron,C. (*1970*): What is 'quantum logic'? In: "Quanta. Essays on theoretical physics dedicated to Gregor Wentzel"; ed. by P.O.G.Freund, C.J.Goebel, and Y.Nambu; University of Chicago Press, Chicago; pp. 166-181.
Quoted in sect.4.4.

Jauch,J.M., Wigner,E.P. and Yanase,M.M. (*1967*): Some comments concerning measurements in quantum mechanics. Nuovo Cimento 48, 144-151.
Quoted in sect.3.5.

Jenč,F. (*1972*): Some theorems on atomicity in axiomatic quantum mechanics. J.Math.Phys.(N.Y.) 13, 1675-1680.
Quoted in sect.4.4.

Jenč,F. (*1974*): Atomicity and maximality in axiomatic quantum theory. Reports on Mathematical Physics 6, 253-264.
Quoted in sect.4.4.

Jones,V.F.R. (*1976*): Quantum mechanics over fields of non-zero characteristic. Letters in Mathematical Physics 1, 99-103.
Quoted in sect.4.4.

Jordan,P. (*1926a*): Ueber kanonische Transformationen in der Quantenmechanik. Z.Phys. 37, 383-386.
Quoted in sect.3.2.

Jordan,P. (*1926b*): Ueber kanonische Transformationen in der Quantenmechanik.II. Z.Phys. 38, 513-517.
Quoted in sect.3.2.

Jordan,P. (*1932a*): Ueber eine Klasse nichtassoziativer hyperkomplexer Algebren. Nachr.Ges.Wiss.Göttingen, Math.Phys.Kl. 1932, 569-575.
Quoted in sect.4.2.

Jordan,P. (*1932b*): Die Quantenmechanik und die Grundprobleme der Biologie und Psychologie. Naturwissenschaften 20, 815-821.
Quoted in sect.3.5.

Jordan,P. (*1933a*): Ueber die Multiplikation quantenmechanischer Grössen. Z.Phys. 80, 285-291.
Quoted in sect.4.2.

Jordan,P. (*1933b*): Ueber Verallgemeinerungsmöglichkeiten des Formalismus der Quantenmechanik. Nachr.Ges.Wiss.Göttingen, Math.Phys.Kl. 1933.
Quoted in sect.4.2.

Jordan,P. (*1934*): **Ueber die Multiplikation quantenmechanischer Grössen.II.** Z.Phys. 87, 505-512.
Quoted in sects.4.2, 4.4.

Jordan,P. (*1936*): **Anschauliche Quantentheorie.** Springer, Berlin.
Quoted in sect.3.4.

Jordan,P. (*1938*): **Die Verstärkertheorie der Organismen in ihrem gegenwärtigen Stand.** Naturwissenschaften 26, 537-545.
Quoted in sect.3.5.

Jordan,P. (*1941*): **Die Physik und das Geheimnis des organischen Lebens.** Vieweg, Braunschweig.
Quoted in sect.3.5.

Jordan,P. (*1944*): **Quantenphysik und Biologie.** Naturwissenschaften 32, 309-316.
Quoted in sect.3.5.

Jordan,P. (*1947*): **Verdrängung und Komplementarität.** Stromverlag, Hamburg.
Quoted in sect.6.5.

Jordan,P. (*1949*): **On the process of measurement in quantum mechanics.** Philosophy of Science 16, 269-278.
Quoted in sect.3.5.

Jordan,P., Neumann,J.von and Wigner,E. (*1934*): **On an algebraic generalization of the quantum mechanical formalism.** Annals of Mathematics 35, 29-64.
Quoted in sect.4.2.

Jørgensen,C.K. (*1974*): **Models and paradoxes in quantum chemistry.** Theor.Chim.Acta 34, 189-198.
Quoted in sect.1.4.

Jung,C.G. (*Coll.Works*): **The collected works of C.G.Jung.** Princeton University Press, Princeton, N.J. and Routledge and Kegan Paul, London. Volumes 1-20, 1954-1979.
Quoted in sects.1.1, 2.1, 2.5, 2.6, 6.5.

Jung,C.G. (*1960*): **A letter by C.G.Jung of February 1960.** English translation is published under the title: "A letter on parapsychology and synchronicity", in: Spring, 1961, pp. 50-57, Spring Publications, The Analytical Psychology Club of New York.
Quoted in sect.2.5.

Kadanoff,L.P. and Baym,G. (*1962*): **Quantum statistical mechanics.** Benjamin, New York.
Quoted in sect.4.3.

Kadison,R.V. (*1951*): **Isometries of operator algebras.** Annals of Mathematics 54, 325-338.
Quoted in sects.4.2, 5.4.

Kadison,R.V. (*1962*): **Normalcy in operator algebras.** Duke Math.J. 29, 459-464.
Quoted in sect.5.5.

Kadison,R.V. (*1965*): **Transformations of states in operator theory and dynamics.** Topology 3, Suppl.2, 177-198 (1965).
Quoted in sects.4.2, 4.5, 4.6.

Kakutani,S. and Mackey,G.W. (*1946*): **Rings and lattice characterization of complex Hilbert space.** Bull.Amer.Math.Soc. 52, 727-733.
Quoted in sect.4.4.

Kalman,R.E. (*1968a*): **New developments in systems theory relevant to biology.** In: "Systems theory and biology". Ed. by M.D.Mesarović. Springer, Berlin, pp. 222-232.
Quoted in sects.3.8, 4.5.

Kalman,R.E. (*1968b*): On the mathematics of model building. In: "Neural networks"; ed. by E.R. Caianiello; Springer, Berlin; pp. 170-177.
Quoted in sects.3.8, 4.5.

Kalman,R.E., Falb,P.L. and Arbib,M.A. (*1969*): Topics in mathematical system theory. McGraw-Hill, New York.
Quoted in sects.3.8, 4.5.

Kalmár, I.G. (*1978*): Atomistic orthomodular lattices and a generalized probability theory. Publ.Math.Debrecen 25, 139-153.
Quoted in sect.4.5.

Kalmbach,G. (*1974*): Orthomodular logic. Z.Math.Logik Grundlagen d.Math. 20 395-406.
Quoted in sect.4.4.

Kamber,F. (*1964*): Die Struktur des Aussagenkalküls in einer physikalischen Theorie. Nachr.Ges.Wiss.Göttingen, Math.-Phys.Kl.II 1964, 103-124.
Quoted in sects.3.4, 4.4.

Kamber,F. (*1965*): Zweitwertige Wahrscheinlichkeitsfunktionen auf orthokomplementären Verbänden. Mathematische Annalen 158, 158-196.
Quoted in sect.4.4.

Kankeleit,O. (*1959*): Das Unbewusste als Keimstätte des Schöpferischen. Ernst Reinhardt Verlag, München.
Quoted in sect.2.5.

Kanthack,L. and Wegener,U. (*1976*): Zum Zusammenhang zwischen Projektionsoperatoren und Eigenschaften. Z.allg.Wissenschaftstheorie 7, 249-257.
Quoted in sect.5.2.

Kaplan,I.G. (*1975*): The exclusion principle and indistinguishability of identical particles on quantum mechanics. Usp.Fiz.Nauk 117, 691-704 (1975) (English translation: Sov.Phys.Usp. 18, 988-994, 1976).
Quoted in sect.3.3.

Kaplansky,I. (*1951*): Projections in Banach algebras. Annals of Mathematics 53, 235-249.
Quoted in sects.4.4, 4.6.

Kaplansky,I. (*1955*): Any orthocomplemented complete modular lattice is a continuous geometry. Annals of Mathematics 61, 524-541.
Quoted in sect.4.4.

Kappos,D.A. (*1969*): Probability algebras and stochastic spaces. Academic Press, New York.
Quoted in sect.4.5.

Kappos,D.A. (*1974*): Generalized probability with applications to the quantum mechanics. In: "Proceedings of the C.Carathéodory International Symposium", Greek Math.Soc., Athens; pp. 253-270.
Quoted in sect.4.5.

Kappos,D.A. (*1976*): Measure theory on orthomodular posets and lattices. In: "Measure theory", ed. by A.Bellow, Vol.541; Springer, Berlin.
Quoted in sect.4.5.

Karplus,R. (*1964*): The science curriculum improvement study. J. of Research in Science Teaching 2, 293-303.
Quoted in sect.2.4.

Kasday,L. (*1970*): Thesis. Columbia University, New York.
Quoted in sects.3.4, 3.7.

Kasday, L. (*1971*): **Experimental test of quantum predictions for widely separated photons.** In: "Proceedings of the international school of physics 'Enrico Fermi'. Course 49: Foundations of quantum mechanics; ed. by B.d'Espagnat; Academic Press, New York; pp. 195-210.
Quoted in sects.3.4, 3.7.

Kasday,L., Ullman,J. and Wu,C.S. (*1970*): **The Einstein-Podolsky-Rosen argument: positron annihilation experiment.** Bull.Amer.Phys.Soc. 15, 586.
Quoted in sect.3.7.

Kastler,D. (*1975*): **Stability and equilibrium in quantum statistical mechanics.** In: "non-commutative harmonic analysis", ed. by J.Carmona, J.Dixmier and M.Vergne; Lecture Notes in Mathematics, vol.466, 1975; pp. 86-100.
Quoted in sect.4.3.

Kastler,D. (*1976*): **Equilibrium states of matter and operator algebras.** In: "Symposia Mathematica", vol.20, ed. by S.Doplicher, Academic Press, London; pp. 49-107.
Quoted in sect.4.3.

Kastler,D. and Robinson,D.W. (*1966*): **Invariant states in statistical mechanics.** Commun.Math.Phys. 3, 151-180.
Quoted in sect.4.3.

Kastler,D., Pool,J.C.T. and Poulsen,E.T. (*1969*): **Quasi-unitary algebras attached to temperature states in statistical mechanics. A comment on the work of Haag, Hugenholtz and Winnink.** Commun.Math.Phys. 12, 175-192.
Quoted in sect.4.3.

Kato,T. (*1951*): **Fundamental properties of Hamiltonian operators of Schrödinger type.** Trans.Amer.Math.Soc. 70, 212-219.
Quoted in sect.3.3.

Kato,T. (*1957*): **On the eigenfunctions of many-particle systems in quantum mechanics.** Commun.Pure Appl.Math. 10, 151-177.
Quoted in sect.3.3.

Kato,T. (*1966*): **Perturbation theory for linear operators.** Springer, Berlin.
Quoted in sects.3.3, 6.2.

Kato,T. (*1972*): **Schrödinger operators with singular potentials.** Israel J.Math. 13, 135-148.
Quoted in sect.3.3.

Kato,T. (*1974*): **A second look at the essential selfadjointness of the Schrödinger operators.** In: "Physical reality and mathematical description", ed. by C.P.Enz and J.Mehra; Reidel, Dordrecht-Holland; pp. 193-201.
Quoted in sect.3.3.

Kemble,E.C. (*1937*): **The fundamental principles of quantum mechanics.** McGraw-Hill, New York.
Quoted in sect.3.4.

Kemble,E.C. (*1951*): **Reality, measurement, and the state of the system in quantum mechanics.** Philosophy of Science 18, 237-299.
Quoted in sect.3.4.

Kemeny,J.G. and Oppenheim,P. (*1956*): **On reduction.** Philosophical Studies 7, 6-19.
Quoted in sect.6.1.

Kholevo,A.S see: *Holevo,A.S.*

Kibble,T.W.B. (*1961*): **Lorentz invariance and the gravitational field.** J.Math.Phys. (N.Y.) 2, 212-221.
Quoted in sect.3.3.

Kirsten,C. and Körber,H.-G. (1975): Physiker über Physiker. Wahlvorschläge zur Aufnahme von Physikern in die Berliner Akademie, 1870-1929. Akademie-Verlag, Berlin.
Quoted in sect.3.2.

Klein,M.J. (1962): Max Planck and the beginning of the quantum theory. Arch.History Exact Sci. 1, 459-479.
Quoted in sect.3.1.

Knops,H.J.F. and Verboven,E.J. (1969): On a class of extremal Euclidean invariant states. Physics Letters 29A, 368-387.
Quoted in sect.4.3.

Kochen,S. and Specker,E.P. (1965a): Logical structures arising in quantum theory. In: "The theory of models"; ed. by J.Addison, L.Henkin, and A.Tarski; North-Holland, Amsterdam; pp. 177-189.
Quoted in sects.4.4, 5.3.

Kochen,S. and Specker,E.P. (1965b): The calculus of partial propositional functions. In: "Logic, methodology, and philosophy of science"; ed. by Y.Bar-Hillel; North-Holland, Amsterdam; pp. 45-57.
Quoted in sect.4.4.

Kochen,S. and Specker,E.P. (1967): The problem of hidden variables in quantum mechanics. J.Mathematics and Mechanics 17, 59-88.
Quoted in sects.3.4, 4.4.

Kocher,C.A. and Commins,E.D. (1967): Polarization correlation pf photons emitted in an atomic cascade. Phys.Rev.Lett. 18, 575-577.
Quoted in sects.3.4, 3.7, 6.4.

Koestler,A. (1964): The act of creation. Macmillan, New York.
Quoted in sect.2.5.

Kolmogoroff,A.N. see: *Kolmogorov,A.N.*

Kolmogorov,A.N. (1933): Grundbegriffe der Wahrscheinlichkeitsrechnung. Springer, Berlin. (English translation: "Foundations of the theory of probability"; Chelsea, New York; 1950).
Quoted in sects.3.2, 3.3, 4.4, 4.5.

Kolmogorov,A.N. (1963): On tables of random numbers. Sankhya, Indian J.Statistics, Ser.A, 25, 369-376.
Quoted in sect.4.5.

Kolmogorov,A.N. (1965): Three approaches to the quantitative definition of information. Int.J. of Computer Mathematics 2, 157-168 (1968). (Russian original: Problemy Peredachi Informacii 1, 3-11, 1965).
Quoted in sect.4.5.

Kolos,W. and Roothaan,C.C.J. (1960): Accurate electronic wave functions for the H_2 molecule. Rev.Mod.Phys. 32, 219-232.
Quoted in sect.2.3.

Kolos,W. and Wolniewicz,L. (1968): Improved theoretical ground-state energy of the hydrogen molecule. J.Chem.Phys. 49, 404-410.
Quoted in sect.2.3.

Komar,A. (1962): Indeterminate character of the reduction of the wave packet in quantum theory. Phys.Rev. 126, 365-369.
Quoted in sect.3.4.

König,H. and Meixner,J. (1958): Lineare Systeme und lineare Transformationen. Math. Nachrichten 19, 265-322.
Quoted in sect.4.3.

Koopman,B.O. (*1957*): Quantum theory and the foundation of probability. In: "Proceedings of Symposia in Applied Mathematics. Vol.7"; American Mathematical Society, Providence, Rhode Island; pp. 97-102.
Quoted in sect.4.5.

Körner,S. (*1957*): On philosophical arguments in physics. In: "Observation and interpretation"; ed. by S.Körner; Butterworth, London (reprinted by Dover, New York, 1962); pp. 97-101.
Quoted in sect.3.8.

Kossakowski,A., Frigerio,A., Gorini,V. and Verri,M. (*1977*): Quantum detailed balance and KMS condition. Commun.Math.Phys. 57, 97-110.
Quoted in sect.4.3.

Kotas,J. (*1963*): Axioms for Birkhoff-v.Neumann quantum logic. Bull.Acad.Pol.Sci. Ser.Sci.Math.Astron.Phys. 11, 629-632.
Quoted in sect.4.4.

Kraus,K. (*1971*): General state changes in quantum theory. Annals of Physics 64, 311-335.
Quoted in sect.4.5.

Krause,U. (*1974*): The inner orthogonality of convex sets in axiomatic quantum mechanics. In: "Foundations of quantum mechanics and ordered linear spaces"; ed. by A.Hartkämper and H.Neumann; Lecture Notes in Physics, vol.29, Springer, Berlin; pp. 269-280.
Quoted in sect.4.5.

Krips,H.P. (*1968*): Theory of measurement. Nuovo Cimento Suppl. 6, 1127-1135.
Quoted in sect.3.5.

Krips,H.P. (*1969a*): Fundamentals of measurement theory. Nuovo Cimento B 60, 278-290.
Quoted in sect.3.5.

Krips,H.P. (*1969b*): Axioms of measurement theory. Nuovo Cimento B 61, 12-24.
Quoted in sect.3.5.

Krips,H.P. (*1969c*): Two paradoxes in quantum mechanics. Philosophy of Science 36, 145-152.
Quoted in sect.3.7.

Krips,H.P. (*1971*): Defence of a measurement theory. Nuovo Cimento B 1, 23-33.
Quoted in sect.3.7.

Kristensen,P., Mejlbo,L. and Poulsen,E.T. (*1965*): Tempered distributions in infinitely many dimensions. Commun.Math.Phys. 1, 175-214.
Quoted in sect.3.3.

Kruszyński,P. (*1976*): Automorphisms of quantum logics. Reports on Mathematical Physics 10, 213-217.
Quoted in sect.5.4.

Kubo,R. (*1957*): Statistical-mechanical theory of irreversible processes.I. J.Phys.Soc.Japan 12, 570-586.
Quoted in sect.4.3.

Kuhn,T.S. (*1962*): The structure of scientific revolutions. University of Chicago Press, Chicago.
Quoted in sects.2.3, 2.4, 5.1.

Kuhn,T.S., Heilbron,J.L., Forman,P. and Allen,L. (*1967*): Sources for history of quantum mechanics. An inventory and report. Mem.Amer.Phil.Soc., vol.68. The American Philosophical Society, Philadelphia.
Quoted in sect.3.1.

Kunsemüller,H. (*1964*): Zur Axiomatik der Quantenlogik. Philosophia Naturalis 8, 363-376.
Quoted in sect.4.4.

Küpfmüller,K. (*1924*): Ueber Einschwingvorgänge in Wellenfiltern. Elektrische Nachrichten-Technik 1, 141-152.
Quoted in sect.3.8.

Kutzelnigg,W. (*1978*): Einführung in die theoretische Chemie. Band 2. Die chemische Bindung. Verlag Chemie, Weinheim.
Quoted in sect.1.3.

Lakatos,I. (*1974*): Popper on demarcation and induction. In: "The philosophy of Karl Popper"; ed. by P.A.Schilpp, Open Court, La Salle, Illinois; Book I, pp. 241-273.
Quoted in sect.2.3.

Lamb,W.E. (*1969*): An operational interpretation of nonrelativistic quantum mechanics. Phys.Today 22, No.4, April 1969, 23-28.
Quoted in sect.3.5.

Lamehi-Rachti,M., and Mittig,W. (*1976*): Quantum mechanics and hidden variables: a test of Bell's inequality by the measurement of the spin correlation in low-energy proton-proton scattering. Phys.Rev. D 14, 2543-2555.
Quoted in sects.3.4, 3.7.

Landau,L., and Peierls,R. (*1931*): Erweiterung des Unbestimmtheitsprinzips für die relativistische Quantentheorie. Z.Phys. 69, 56-69.
Quoted in sect.3.5.

Lanford,O.E., and Robinson,D.W. (*1968*): Mean entropy of states in quantum-statistical mechanics. J.Math.Phys.(N.Y.) 9, 1120-1125.
Quoted in sect.4.3.

Lanford,O.E., and Ruelle,D. (*1967*): Integral representations of invariant states on B*-algebras. J.Math.Phys.(N.Y.) 8, 1460-1463.
Quoted in sect.4.3.

Lanford,O.E., and Ruelle,D. (*1969*): Observables at infinity and states with short range correlations in statistical mechanics. Commun.Math.Phys. 13, 194-215.
Quoted in sect.4.3.

Lange,L. (*1885*): Ueber die wissenschaftliche Fassung des Galileischen Beharrungsgesetzes. Ber.kgl.Ges.Wiss.Math.-phys.Kl. 1885, 333-351.
Quoted in sect.3.3.

Langer,S.K. (*1942*): Philosophy in a new key. A study in the symbolism of reason, rite and art. Harvard University Press, Cambridge, Massachusetts.
Quoted in sect.6.5.

Langer,S.K. (*1957*): Problems of art. Charles Scribner's Sons, New York.
Quoted in sect.6.3.

Langhoff,H. (*1960*): Die Linearpolarisation der Vernichtungsstrahlung von Positronen. Z.Phys. 160, 186-193.
Quoted in sect.3.7.

Lanz,L., Prosperi,G.M. and Sabbadini,A. (*1971*): Time scales and the problem of measurements in quantum mechanics. Nuovo Cimento B 2, 184-192.
Quoted in sect.3.5.

Latham,R.E. (*1951*): English translation of Lucretius, De rerum natura, published as "The nature of the universe", Penguin Books, Harmondsworth, Middlesex.
Quoted in sect.3.3.

Lebesgue,H. (*1901*): Sur une généralisation de l'intégrale définie. C.R.Acad.Sci. (Paris) 132, 1025-1028.
Quoted in sect.3.2.

Lebesgue,H. (*1902*): **Intégrale, longueur, aire.** Ann.Mat.Pura Appl. 7, 231-359.
Quoted in sect.3.2.

Lec,S.J. (*1959*): **Mysli nieuczesane.** Wydawnictwo Literackie, Krakow. (English translation by J.Galaska, "Unkempt thoughts", St.Martins, New York, 1962).
Chapter 1.

Lehmann,E.L. (*1959*): **Testing statistical hypotheses.** Wiley, New York.
Quoted in sect.4.5.

Lenk,H. (*1969*): **Philosophische Kritik an Begründungen von Quantenlogiken.** Philosophia Naturalis 11, 413-425.
Quoted in sect.4.4.

Leplae,L., *Umezawa*,H. *and Mancini*,F. (*1974*): **Derivation and application of the boson method in superconductivity.** Physics Reports (Sect.C of Physics Letters) 10, 151-272.
Quoted in sect.6.2.

Leveille,J.P. *and Roman*,P. (*1975*): **Group representations in certain lattices of propositions.** Int.J.Theor.Phys. 14, 73-90.
Quoted in sect.4.4.

Lévy-Leblond,J.M. (*1963*): **Galilei group and non-relativistic quantum mechanics.** J.Math.Phys.(N.Y.) 4, 776-788.
Quoted in sect.3.3.

Lévy-Leblond,J.M. (*1967*): **Nonrelativistic particles and wave equations.** Commun. Math.Phys. 6, 286-311.
Quoted in sect.3.3.

Lévy-Leblond,J.M. (*1971*): **Galilei group and Galilei invariance.** In: "Group theory and its applications. Volume II"; ed. by E.M.Loebl; Academic Press, New York; pp. 221-299.
Quoted in sect.3.3.

Lévy-Leblond,J.M. (*1974*): **The pedagogical role and epistemological significance of group theory in quantum mechanics.** Rivista del Nuovo Cimento 4, 99-143.
Quoted in sect.3.3.

Lévy-Leblond,J.M. (*1976*): **One more derivation of the Lorentz transformation.** Amer.J.Phys. 44, 271-277.
Quoted in sect.3.3.

Lewis,J.T. (*1972*): **The free boson gas.** In: "Mathematics of contemporary physics"; ed. by R.F.Streater; Academic Press, London; pp. 209-226.
Quoted in sect.4.3.

Lewis,J.T. *and Pulè*,J.V. (*1974*): **The equilibrium states of the free boson gas.** Commun.Math.Phys. 36, 1-18.
Quoted in sect.4.3.

Lewis,J.T. *and Thomas*,L.C. (*1975*): **On the existence of a class of stationary stochastic processes.** Ann.Inst.Henri Poincaré, Sect.A 22, 241-248.
Quoted in sect.4.5.

Lieberman,A. (*1978*): **Entropy of states of a gauge space.** Acta Sci.Math. 40, 99-105.
Quoted in sect.4.5.

Lindblad,G. (*1976a*): **On the generators of quantum dynamical semigroups.** Commun. Math.Phys. 48, 119-130.
Quoted in sect.4.5.

Lindblad,G. (*1976b*): **Dissipative operators and cohomology of operator algebras.** Letters in Mathematical Physics 1, 219-224.
Quoted in sect.4.5.

Lindblad,G. (*1979a*): Non-Markovian quantum stochastic processes and their entropy. Commun.Math.Phys. 65, 281-294.
Quoted in sect.4.5.

Lindblad,G. (*1979b*): Gaussian quantum stochastic processes on the CCR algebra. J.Math.Phys.(N.Y.) 20, 2081-2087.
Quoted in sect.4.5.

Lodkin,A.A. (*1974*): Every measure on the projectors of a W*-algebra can be extended to a state. Functional Anal.Appl. 8, 318-321 (1974), Errata: ib. 9, 271 (1975). (Russian original: Funktsional'nyi Analiz i Ego Prilozheniya 8, 54-58, 1974).
Quoted in sect.4.5.

Löfgren,L. (*1977*): Complexity of descriptions of systems: a foundational study. Int.J. General Systems 3, 197-214.
Quoted in sect.6.2.

Loinger,A. (*1962*): Galilei group and Liouville equation. Ann.Phys.(New York) 20, 132-144.
Quoted in sects.3.2, 3.3, 5.6.

Loinger,A. (*1968*): Comments on a recent paper concerning the quantum theory of measurement. Nucl.Phys.A 108, 245-249.
Quoted in sect.3.5.

London,F. (*1926*): Winkelvariable und kanonische Transformationen in der Undulationsmechanik. Z.Phys. 40, 193-210.
Quoted in sect.3.2.

London,F. (*1927*): Quantenmechanische Deutung der Theorie von Weyl. Z.Phys. 42, 375-389.
Quoted in sect.3.3.

London,F. (*1938a*): The λ-phenomenon of liquid Helium and the Bose-Einstein degeneracy. Nature(London) 141, 643-644.
Quoted in sect.4.3.

London,F. (*1938b*): On the Bose-Einstein condensation. Phys.Rev. 54, 947-954.
Quoted in sect.4.3.

London,F. *and Bauer*,E. (*1939*): La théorie de l'observation en mécanique quantique. Hermann, Paris.
Quoted in sect.1.5, 3.5.

Loomis,L.H. (*1947*): On the representation of σ-complete Boolean algebras. Bull.Amer.Math.Soc. 53, 757-760.
Quoted in sect.6.1

Loomis,L.H. (*1955*): The lattice theoretical background of the dimension theory of operator algebras. Memoirs of the American Mathematical Society, No.18.
Quoted in sect.4.6.

Lorentz,H.A. (*1892*): La théorie électromagnétique de Maxwell et son application aux corps mouvants. Archives Néerlandaises des Sciences Exactes et Naturelles 25, 363-552.
Quoted in sect.3.3.

Lorentz,H.A. (*1895*): Versuch einer Theorie der elektrischen und optischen Erscheinungen in bewegten Körpern. Leyden.
Quoted in sect.3.3.

Lorentz,H.A. (*1904*): Electromagnetic phenomena in a system moving with any velocity less than that of light. Nederlandse Koninklijeke Akademie van Wetenschappen, Proceedings of the Section of Sciences 6, 809-831.
Quoted in sect.3.3.

Lorenzen,P. (*1955*): Einführung in die operative Logik und Mathematik. Springer, Berlin; 1st ed. 1955, 2nd ed. 1969.
Quoted in sect.4.4.

Lorenzen,P. (*1962*): Metamathematik. Bibliographisches Institut, Mannheim.
Quoted in sect.4.4.

Loupias,G. and Mebkhout,M. (*1979*): Extension of KMS states and angular momentum. J.Math.Phys.(N.Y.) 20, 984-998.
Quoted in sect.4.3.

Lowdenslager,D.B. (*1957*): On postulates for general quantum mechanics. Proc.Amer. Math.Soc. 8, 88-91.
Quoted in sect.4.2.

Lüders,G. (*1951*): Ueber die Zustandsänderung durch den Messprozess. Annalen der Physik 8, 322-328.
Quoted in sects.3.3, 3.5.

Lüders,G. (*1966*): Zum Symmetrisierungs-Postulat in der Quantenmechanik identischer Teilchen. Z.Phys. 192, 449-461.
Quoted in sect.3.5.

Ludwig,G. (*1953*): Der Messprozess. Z.Phys. 135, 483-511.
Quoted in sect.3.5.

Ludwig,G. (*1954*): Die Grundlagen der Quantenmechanik. Springer, Berlin.
Quoted in sects.3.1, 3.5.

Ludwig,G. (*1955*): Zur Deutung der Beobachtung in der Quantenmechanik. Physikalische Blätter 11, 489-494.
Quoted in sects.1.4, 3.4, 3.5.

Ludwig,G. (*1957*): Zum Ergodensatz und zum Begriff der makroskopischen Observablen. Z.Naturforsch. A 12, 662-663.
Quoted in sect.3.5.

Ludwig,G. (*1958a*): Zum Ergodensatz und zum Begriff der makroskopischen Observablen.I. Z.Phys. 150, 346-374.
Quoted in sect.3.5.

Ludwig,G. (*1958b*): Zum Ergodensatz und zum Begriff der makroskopischen Observablen.II. Z.Phys. 152, 98-115.
Quoted in sect.3.5.

Ludwig,G. (*1961a*): Gelöste und ungelöste Probleme des Messprozesses in der Quantenmechanik. In: "Werner Heisenberg und die Physik unserer Zeit"; hg. von F.Bopp; Vieweg, Braunschweig; pp. 150-181.
Quoted in sects.1.4, 3.4, 3.5.

Ludwig,G. (*1961b*): Axiomatic quantum statistics of macroscopic systems (ergodic theory). In: "Proceedings of the International School of Physics 'Enrico Fermi'. Course 14. Ergodic theories"; ed. by P.Caldirola; Academic Press, New York; pp. 57-132.
Quoted in sect.3.5.

Ludwig,G. (*1963a*): Zur Begründung der Thermodynamik auf Grund der Quantenmechanik. Z.Phys. 171, 476-486.
Quoted in sect.3.5.

Ludwig,G. (*1963b*): Zur Begründung der Thermodynamik auf Grund der Quantenmechanik. Teil 2. Masterequation. Z.Phys. 173, 232-240.
Quoted in sect.3.5.

Ludwig,G. (*1964*): Versuch einer axiomatischen Grundlegung der Quantenmechanik und allgemeiner physikalischer Theorien. Z.Phys. 181, 233-260.
Quoted in sects.4.5, 4.6.

Ludwig,G. (1967a): Attempt of an axiomatic foundation of quantum mechanics and more general theories II. Commun.Math.Phys. 4, 331-348.
Quoted in sect.4.5.

Ludwig,G. (1967b): Hauptsätze über das Messen als Grundlage der Hilbert-Raumstruktur der Quantenmechanik. Z.Naturforsch. A 22, 1303-1323.
Quoted in sect.4.5.

Ludwig,G. (1967c): Ein weiterer Hauptsatz über das Messen als Grundlage der Hilbert-Raumstruktur der Quantenmechanik. Z.Naturforsch. A 22, 1324-1327.
Quoted in sect.4.5.

Ludwig,G. (1967d): An axiomatic foundation of quantum mechanics on a nonsubjective basis. In: "Quantum theory and reality"; ed. by M.Bunge; Springer, Berlin; pp. 98-104.
Quoted in sect.4.5.

Ludwig,G. (1968): Attempt of an axiomatic foundation of quantum mechanics and more general theories III. Commun.Math.Phys. 9, 1-12.
Quoted in sect.4.5.

Ludwig,G. (1970): Deutung des Begriffs 'physikalische Theorie' und axiomatische Grundlegung der Hilbertraumstruktur der Quantenmechanik durch Hauptsätze des Messens. Springer, Berlin.
Quoted in sect.4.5.

Ludwig,G. (1972): An improved formulation of some theorems and axioms in the axiomatic foundation of the Hilbert space structure of quantum mechanics. Commun.Math.Phys. 26, 78-86.
Quoted in sect.4.5.

Ludwig,G. (1976): Einführung in die Grundlagen der theoretischen Physik. Band 3: Quantenmechanik. Vieweg, Braunschweig.
Quoted in sect.4.5.

Łukasiewicz,J. (1951): Aristotle's syllogistic. Oxford, 2nd ed. 1957.
Quoted in sect.4.4.

Mackey,G.W. (1945): On infinite-dimensional spaces. Trans.Amer.Math.Soc. 57, 155-207.
Quoted in sect.4.4.

Mackey,G.W. (1949): Imprimitivity for representations of locally compact groups I. Proc.Nat.Acad.Sci.(USA) 35, 537-545.
Quoted in sect.3.3.

Mackey,G.W. (1957): Quantum mechanics and Hilbert space. Amer.Math.Monthly 64, 45-57.
Quoted in sect.4.4.

Mackey,G.W. (1960): Lecture notes on the mathematical foundations of quantum mechanics. Harvard University, Cambridge, Mass., mimeographed.
Quoted in sect.4.4.

Mackey,G.W. (1963a): The mathematical foundations of quantum mechanics. Benjamin, New York.
Quoted in sects.3.3, 4.4.

Mackey,G.W. (1963b): Infinite-dimensional group representations. Bull.Amer.Math.Soc. 69, 628-686.
Quoted in sect.4.2.

Mackey,G.W. (1968): Induced representations of groups and quantum mechanics. Benjamin, New York.
Quoted in sect.3.3.

Mackey,G.W. (*1976*): The theory of unitary group representations. University of Chicago Press, Chicago.
Quoted in sect.3.3.

MacLaren,M.D. (*1964*): Atomic orthocomplemented lattices. Pacific J.Math. 14, 597-612.
Quoted in sect.4.4.

Mączyński,M.J. (*1967*): A remark on Mackey's axiom system for quantum mechanics. Bull.Acad.Pol.Sci.Ser.Sci.Math.Astron.Phys. 15, 583-587.
Quoted in sect.4.4.

Mączyński,M.J. (*1970*): Quantum families of Boolean algebras. Bull.Acad.Pol.Sci.Ser. Sci.Math.Astron.Phys. 18, 93-96.
Quoted in sect.4.4.

Mączyński,M.J. (*1971*): Boolean properties of observables in axiomatic quantum mechanics. Reports on Mathematical Physics 2, 135-150.
Quoted in sect.4.4.

Mączyński,M.J. (*1972*): Hilbert space formalism of quantum mechanics without the Hilbert space axiom. Reports on Mathematical Physics 3, 209-219.
Quoted in sect.4.4.

Mączyński,M.J. (*1973a*): The field of real numbers in axiomatic quantum mechanics. J.Math.Phys.(N.Y.) 14, 1469-1471.
Quoted in sect.4.4.

Mączyński,M.J. (*1973b*): The orthogonality postulate in axiomatic quantum mechanics. Int.J.Theor.Phys. 8, 353-360.
Quoted in sect.4.4.

Mączyński,M.J. (*1974*): Functional properties of quantum logics. Int.J.Theor.Phys. 11, 149-156.
Quoted in sect.4.4.

Mączyński,M.J. (*1976*): Orthomodularity and lattice characterization of Hilbert spaces. Bull.Acad.Polon.Sci.Ser.sci.math., astr. et phys. 24, 481-484.
Quoted in sect.4.4.

Maeda,F. (*1958*): Kontinuierliche Geometrien. Springer, Berlin.
Quoted in sect.4.4.

Maeda,S. (*1977*): On the distance between two projections in a C*-algebra. Mathematica Japanica (Kobe) 22, 61-65.
Quoted in sect.5.4.

Maeda,F. and Maeda,S. (*1970*): Theory of symmetric lattices. Springer, Berlin.
Quoted in sects.4.4, 4.6.

Maksimov,V.M. (*1975*): Nonautomorphic dynamics in algebraic statistical mechanics. Theoretical and Mathematical Physics 25, 944-950, 1976. (Russian original: Teoreticheskaya i Matematicheskaya Fiz. 25, 10-19, 1975).
Quoted in sect.4.2.

Maksimov,V.M. (*1976a*): Dynamics in the state space and Heisenberg equations. Theoretical and Mathematical Physics 26, 259-262, 1976. (Russian original: Teoreticheskaya i Matematicheskaya Fiz. 26, 382-286, 1976).
Quoted in sect.4.2.

Maksimov,V.M. (*1976b*): Heisenberg dynamics. Covariant representations. Theoretical and Mathematical Physics 28, 715-720, 1976. (Russian original: Teoreticheskaya i Matematicheskaya Fiz. 28, 180-188, 1976).
Quoted in sect.4.2.

Mandel,L. and Wolf,E. (*1973*): Coherence and quantum optics. Plenum Press, New York.
Quoted in sect.6.4.

Manin,Yu.I. (*1977*): **A course in mathematical logic.** Springer, New York.
Quoted in sect.4.4.

Mann,T. (*1936*): **Freud und die Zukunft.** Bermann-Fischer, Wien. (English translation in: T.Mann, "Freud, Goethe, Wagner", Knopf, New York, 1937).
Chapter 2.

Manuceau,J. (*1968*): **C*-algèbre de relations de commutation.** Ann.Inst. Henri Poincaré, A, 8, 139-161.
Quoted in sect.5.4.

Marchand,J.P. (*1977*): **Relative coarse graining.** Foundations of Physics 7, 35-49.
Quoted in sect.4.5.

Marchand,J.P. and Wyss,W. (*1977*): **Statistical inference and entropy.** J.Statistical Physics 16, 349-355.
Quoted in sect.4.5.

Margenau,H. (*1936*): **Quantum-mechanical description.** Phys.Rev. 49, 240-242.
Quoted in sects.3.5, 3.7.

Margenau,H. (*1949*): **Einstein's conception of reality.** In: "Albert Einstein: Philosopher-Scientist"; ed. by P.A.Schilpp; Library of Living Philosphers, Evanstone, Illinois; pp. 243-268.
Quoted in sects.3.5, 3.7.

Margenau,H. (*1950*): **The nature of physical reality.** McGraw Hill, New York.
Quoted in sect.2.5.

Margenau,H. (*1958*): **Philosophical problems concerning the meaning of measurement in physics.** Philosophy of Science 25, 23-33.
Quoted in sect.3.5.

Margenau,H. (*1963a*): **Measurements and quantum states.** Philosophy of Science 30, 1-16, 138-157.
Quoted in sects.3.4, 3.5.

Margenau,H. (*1963b*): **Measurements in quantum mechanics.** Annals of Physics 23, 469-485.
Quoted in sects.3.4, 3.5.

Margenau,H. (*1966*): **Exclusion principle and measurement theory.** In: "Quantum theory of atoms, molecules, and the solid state. A tribute to John C.Slater"; ed. by P.O.Löwdin; Academic Press, New York; pp. 81-91.
Quoted in sect.1.4.

Margenau,H. and Hill,R.N. (*1961*): **Correlation between measurements in quantum theory.** Progr.Theor.Phys. 26, 722-738.
Quoted in sect.3.5.

Margenau,H. and Park,J.L. (*1967*): **Objectivity in quantum mechanics.** In: "Delaware Seminar in the Foundation of Science"; ed. by M.Bunge; Springer, Berlin; pp. 161-187.
Quoted in sect.3.5.

Marshall,T.W. (*1963*): **Random electrodynamics.** Proc.Roy.Soc. London, A, 276, 475-491.
Quoted in sect.6.4.

Martin,P.C. and Schwinger,J. (*1959*): **Theory of many-particles systems.I.** Phys.Rev. 115, 1342-1373.
Quoted in sect.4.3.

Matsumoto,H. and Umezawa,H. (*1976*): **A rigorous formulation of the boson method in superconductivity.** Fortschr.Phys. 24, 357-404.
Quoted in sect.6.2.

Maurin,K. (*1966*): Allgemeine Eigenfunktionsentwicklungen, unitäre Darstellungen lokalkompakter Gruppen und automorphe Funktionen. Mathematische Annalen 165, 204-222.
Quoted in sect.3.3.

Maurin,K. (*1967*): Methods of Hilbert spaces. Pańѕtwowe Wydawnictwo Naukowe, Warsaw, 1967; 2nd ed. 1972. (Polish original: Państ.Wydawn.Nauk, Warsaw, 1959).
Quoted in sect.3.3.

Maxwell,J.C. (*1877*): Matter and motion. Reprinted by Dover, New York.
Quoted in sect.3.3.

McGuire,J.H. and Fry,E.S. (*1973*): Restrictions on nonlocal hidden-variable theory. Phys.Rev.D 2, 555-557.
Quoted in sect.3.4.

McKinsey,J.C.C. and Suppes,P. (*1954*): Review of: Paulette Destouches-Février, La Structure des théories physiques. J.Symbolic Logic 19, 52-55.
Quoted in sect.4.4.

McKnight,J.L. (*1957*): The quantum theoretical concept of measurement. Philosophy of Science 24, 321-330.
Quoted in sect.3.5.

McKnight,J.L. (*1958*): An extended latency interpretation of quantum mechanical measurement. Philosophy of Science 25, 209-222.
Quoted in sect.3.5.

Mehra,J. (*1972*): 'The golden age of theoretical physics': P.A.M. Dirac's scientific work from 1924 to 1933. In: "Aspects of quantum theory", ed. by A.Salam and E.P.Wigner, Cambridge University Press, London; pp. 17-59.
Quoted in sect.3.2.

Meisel,W.S. (*1972*): Computer-oriented approaches to pattern recognition. Academic Press, New York.
Quoted in sect.6.3.

Meixner,J. (*1954*): Thermodynamische Erweiterung der Nachwirkungstheorie. Z.Physik 139, 30-43.
Quoted in sect.4.3.

Meixner,J. (*1969a*): Thermodynamik der Vorgänge in einfachen fluiden Medien und die Charakterisierung der Thermodynamik irreversibler Prozesse. Z.Phys. 219, 79-104.
Quoted in sect.4.3.

Meixner,J. (*1969b*): Processes in simple thermodynamic materials. Arch.Ration.Mech. Anal. 33, 33-53.
Quoted in sect.4.3.

Mercereau,J.E. (*1969*): Macroscopic quantum phenomena. In: "Superconductivity"; ed. by R.D.Parks; Dekker, New York; Vol.1, pp. 393-421.
Quoted in sect.3.7.

Meyer-Abich,K.M. (*1965*): Korrespondenz, Individualität und Komplementarität. Eine Studie zur Geistesgeschichte der Quantentheorie in den Beiträgen Niels Bohrs. Steiner, Wiesbaden.
Quoted in sect.6.5.

Mielnik,B. (*1968*): Geometry of quantum states. Commun.Math.Phys. 9, 55-80.
Quoted in sects.4.4, 4.5.

Mielnik,B. (*1969*): Theory of filters. Commun.Math.Phys. 15, 1-46.
Quoted in sects.4.4, 4.5.

Mielnik,B. (*1974*): Generalized quantum mechanics. Commun.Math.Phys. 37, 221-256.
Quoted in sect.4.5.

Mielnik,B. (1976): **Quantum logic: is it necessarily orthocomplemented?** In: "Quantum mechanics, determinism, causality, and particles", ed. by M.Flato, Z.Mavic, A.Milojevic, D.Sternheimer and J.P.Vigier; Reidel, Dordrecht; pp. 117-135.
Quoted in sect.4.4.

Miller,W.H. (1978): **A classical/semiclassical theory for the interaction of infrared radiation with molecular systems.** J.Chem.Phys. 69, 2188-2195.
Quoted in sect.6.4.

Milonni,P.W. (1976): **Semiclassical and quantum-electrodynamical approaches in nonrelativistic radiation theory.** Physics Reports (Section C of Phys.Lett.) 25, 1-81.
Quoted in sect.6.4.

Mirman,R. (1973): **The Einstein-Podolsky-Rosen paradox.** Nuovo Cimento B 16, 398-403.
Quoted in sect.3.7.

Mises,R.von (1919): **Grundlagen der Wahrscheinlichkeitsrechnung.** Math.Z. 5, 52-99.
Quoted in sect.4.5.

Mises,R.von (1931): **Wahrscheinlichkeitsrechnung.** Deuticke, Leipzig. (New English version: "Mathematical theory of probability and statistics"; edited and complemented by H.Geiringer; Academic Press, New York; 1964).
Quoted in sect.4.5.

Misra,B. (1967): **When can hidden variables be excluded in quantum mechanics?** Nuovo Cimento A 47, 841-859.
Quoted in sect.3.4.

Mittelstaedt,P. (1959): **Untersuchungen zur Quantenlogik.** Sitzungsberichte der Bayrischen Akad.d.Wiss., Math.-Naturwiss.Kl. 1959, 321-386. (Essentially the same paper: "Quantenlogik". Fortschr.Phys. 9, 106-147, 1961).
Quoted in sect.4.4.

Mittelstaedt,P. (1960): **Ueber die Gültigkeit der Logik in der Natur.** Naturwissenschaften 47, 385-391.
Quoted in sect.4.4.

Mittelstaedt,P. (1962): **Logik und Quantenlogik.** Physikalische Blätter 18, 23-29.
Quoted in sect.4.4.

Mittelstaedt,P. (1963): **Philosophische Probleme der modernen Physik.** Bibliographisches Institut, Mannheim.
Quoted in sect.4.4.

Mittelstaedt,P. (1970): **Quantenlogische Interpretation orthokomplementärer quasimodularer Verbände.** Z.Naturforsch.A 25, 1773-1778.
Quoted in sect.4.4.

Mittelstaedt,P. (1972): **On the interpretation of the lattice of subspaces of the Hilbert space as a propositional calculus.** Z.Naturforsch.A 27, 1358-1362.
Quoted in sect.4.4.

Mittelstaedt,P. (1973): **Objektivierbarkeit, Quantenlogik und Wahrscheinlichkeit.** In: "Einheit und Vielheit. Festschrift für Carl Friedrich v. Weizsäcker zum 60. Geburtstag"; ed. by E.Scheibe and G.Süssmann; Vandenhoeck and Ruprecht, Göttingen; pp. 119-139.
Quoted in sect.4.4.

Mittelstaedt,P. (1974): **Ueber das Einstein-Podoslky-Rosen Paradoxon.** Z.Naturforsch. A 29, 539-548.
Quoted in sect.3.7.

Mittelstaedt,P. (1977): **Time dependent propositions and quantum logic.** J.Philosophical Logic 6, 463-472.
Quoted in sect.4.4.

Mittelstaedt,P. (1978): Quantum logic. Reidel, Dordrecht.
Quoted in sect.4.4.

Mittelstaedt,P. (1979): Quantum logic. In: "Problems in the foundations of physics"; ed. by G.Taraldo di Francia, Proceedings of the International School of Physics 'Enrico Fermi', course 72; North Holland, Amsterdam; pp. 264-299.
Quoted in sect.4.4.

Mittelstaedt,P. and Stachow,E.W. (1974): Operational foundation of quantum logic. Foundations of Physics 4, 355-365.
Quoted in sect.4.4.

Miyatake,O. (1952a): On the non-existence of solution of field equations in quantum mechanics. J.Inst.Polytechnics, Osaka City Univ. Ser.A 2, 89-99.
Quoted in sect.4.2.

Miyatake,O. (1952b): On the singularity of the perturbation-term in the field quantum mechanics. J.Inst.Polytechnics, Osaka City Univ. Ser.A 3, 145-155.
Quoted in sect.4.2.

Moldauer,P.A. (1972a): A reinterpretation of von Neumann's theory of measurement. Foundations of Physics 2, 41-47.
Quoted in sect.3.5.

Moldauer,P.A. (1972b): Is there a quantum measurement problem? Phys.Rev. D 5, 1028-1032.
Quoted in sect.3.5.

Moldauer,P.A. (1974): Reexamination of the arguments of Einstein,Podolsky, and Rosen. Foundations of Physics 4, 195-205.
Quoted in sect.3.7.

Morgan,C.L. (1923): Emergent evolution. London.
Quoted in sect.6.1.

Moroz,B.Z. (1971): Formal systems arising in analysis of physical theories. Sov.Math.Dokl. 12, 934-937. (Russian original: Dokl.Akad.Nauk SSSR 198, 1018-1020).
Quoted in sect.4.4.

Morris,C.W. (1938): Foundations of the theory of signs. In: "International Encyclopaedia of unified Science. Vol.1,: ed. by O.Neurath, R.Carnap and C.Morris; University of Chicago Press, Chicago; pp. 77-137.
Quoted in sect.2.1.

Moss,R.E. (1973): Advanced molecular quantum mechanics. Chapman and Hall, London.
Quoted in sect.3.3.

Mould,R.A. (1962): Quantum theory of measurement. Annals of Physics 17, 404-417.
Quoted in sect.3.5.

Moy,S.C. (1954): Characterizations of conditional expectations as a transformation on function spaces. Pacific J.Math. 4, 47-64.
Quoted in sect.4.5.

Moya,R.P. (1975): Equilibrium states for the infinite ideal Bose gas. J.Math.Phys. (N.Y.) 16, 147-149.
Quoted in sect.4.3.

Murray,F.J. and Neumann,J.von (1936): On rings of operators. Annals of Mathematics 37, 116-229.
Quoted in sects.4.4, 5.5, 5.6.

Murray,F,J. and Neumann,J.von (1943): On rings of operators IV. Annals of Mathematics 44, 716-808.
Quoted in sect.5.5.

Nagel,E. *(1961)*: The structure of science. Harcourt, Brace and World, New York.
Quoted in sects.5.5, 6.1.

Nakamura,M. *and Turumaru*,T. *(1954)*: Expectations in an operator algebra. Tôhoku Math.J. 6, 182-188.
Quoted in sect.4.5.

Nakamura,M. *and Umegaki*,H. *(1961a)*: A note on the entropy for operator algebras. Proc.Japan Acad. 37, 149-154.
Quoted in sect.4.5.

Nakamura,M. *and Umegaki*,H. *(1961b)*: On the Blackwell theorem in operator algebra. Proc.Japan Acad. 37, 312-315.
Quoted in sect.4.5.

Nakamura,M. *and Umegaki*,H. *(1962)*: On von Neumann's theory of measurements in quantum statistics. Mathematica Japonicae(Kobe) 7, 151-157.
Quoted in sect.4.5.

Nakamura,M., *Takesaki*,M. *and Umegaki*,H. *(1960)*: A remark on the expectations of operator algebras. Kōdai Math.Sem.Reports 12, 82-90.
Quoted in sect.4.5.

Narnhofer,H. *(1978)*: Time evolution in equilibrium states without time automorphism. Acta Phys.Austriaca 49, 207-217.
Quoted in sect.4.2.

Narnhofer,H. *and Robinson*,D.W. *(1975)*: Dynamical stability and pure thermodynamic phases. Commun.Math.Phys. 41, 89-97.
Quoted in sect.4.3.

Naudts,J. *(1970)*: Observables at infinity. Ann.Soc.Sci.Bruxelles 84, 257-266.
Quoted in sect.4.3.

Naville,A. *(1920)*: Classification des sciences. Paris.
Quoted in sect.3.8.

Nayfeh,A.H. *(1973)*: Perturbation methods. Wiley, New York.
Quoted in sect.6.2.

Nelson,E. *(1959)*: Analytic vectors. Annals of Mathematics 70, 572-615.
Quoted in sect.3.3.

Nelson,E. *(1967)*: Dynamical theories of Brownian motion. Princeton University Press, Princeton.
Quoted in sects.3.3, 4.5.

Nelson,E. *(1974)*: Notes on non-commutative integration. J.Functional Analysis 15, 103-116.
Quoted in sect.4.5.

Nelson,E. *and Stinespring*,W.F. *(1959)*: Representation of elliptic operators in an enveloping algebra. Amer.Math.J. 81, 547-560.
Quoted in sect.3.3.

Nerode,A. *(1958)*: Linear automaton transformations. Proc.Amer.Math.Soc. 9, 541-544.
Quoted in sect.3.3.

Neumann,E. *(1954)*: The origins and history of consciousness. Princeton University Press, Princeton. (German original: "Ursprungsgeschichte des Bewusstseins", Rascher, Zürich, 1949).
Quoted in sect.2.5.

Neumann,H. *(1971)*: Classical systems and observables in quantum mechanics. Commun.Math.Phys. 23, 100-116.
Quoted in sect.4.5.

Neumann,H. (*1972*): **Transformation properties of observables.** Helv.Phys.Acta 45, 811-819.
Quoted in sect.4.5.

Neumann,H. (*1974*): **A new physical characterization of classical systems in quantum mechanics.** Int.J.Theor.Phys. 9, 225-228.
Quoted in sects.3.3, 4.5.

Neumann,J.von (*1927a*): **Mathematische Begründung der Quantenmechanik.** Nachr.Ges. Wiss.Göttingen, Math.Phys.Kl. 1927, 1-57.
Quoted in sect.3.2.

Neumann,J.von (*1927b*): **Wahrscheinlichkeitstheoretischer Aufbau der Quantenmechanik.** Nachr.Ges.Wiss.Göttingen, Math.Phys.Kl. 1927, 245-272.
Quoted in sect.3.2.

Neumann,J.von (*1927c*): **Thermodynamik quantenmechanischer Gesamtheiten.** Nachr.Ges. Wiss.Göttingen, Math.Phys.Kl. 1927, 273-291.
Quoted in sect.3.2.

Neumann,J.von (*1929*): **Allgemeine Eigenwerttheorie Hermitescher Funktionaloperatoren.** Mathematische Annalen 102, 49-131.
Quoted in sect.3.2.

Neumann,J.von (*1931*): **Die Eindeutigkeit der Schrödingerschen Operatoren.** Mathematische Annalen 104, 570-578.
Quoted in sects.3.2, 4.2.

Neumann,J.von (*1932*): **Mathematische Grundlagen der Quantenmechanik.** Springer, Berlin. (English translation: "Mathematical foundations of quantum mechanics". Princeton University Press, Princeton, New Jersey, 1955).
Quoted in sects.1.5, 3.1, 3.2, 3.3, 3.4, 3.5, 3.8, 4.4, 4.5.

Neumann,J.von (*1935*): **Representations and ray-representations in quantum mechanics.** Bull.Amer.Math.Soc. 41, 305.
Quoted in sect.3.3.

Neumann,J.von (*1936*): **On an algebraic generalization of the quantum mechanical formalism (Part I).** Mat.Sbornik 1, 415-484. (A part II was never published).
Quoted in sect.4.2.

Neumann,J.von (*1937a*): **Continuous geometries with a transition probability.** Unpublished manuscript, reviewed by I.Halperin in: "John von Neumann. Collected Works", Vol.4, pp. 191-194; ed. by A.H.Taub; Pergamon Press, Oxford, 1962.
Quoted in sect.4.4.

Neumann,J.von (*1937b*): **Quantum logics (strict and probability logic).** Unfinished manuscript, reviewed by A.H.Taub. In: "John von Neumann. Collected Works. Vol.4"; Pergamon Press, Oxford, 1962; pp. 195-197.
Quoted in sect.4.4.

Neumann,J.von (*1938*): **On infinite direct products.** Compos.Math. 6, 1-77.
Quoted in sect.4.2.

Neumann,J.von (*1942*): **Approximative properties of matrices of high finite order.** Portugaliae Math. 3, 1-62. (Reprinted in: "John von Neumann. Collected Works"; Vol.4, pp. 270-331; ed. by A.H.Taub; Pergamon Press, Oxford, 1962).
Quoted in sect.4.4.

Neumann,J.von (*1946*): **Statement of John von Neumann before the special senate committee on atomic energy.** Reprinted in: "John von Neumann. Collècted Works", ed. by A.H.Taub, Vol.6, Pergamon Press, Oxford, 1963; pp. 499-502.
Quoted in sect.2.6.

Neumann,J.von (1955a): **Can we survive technology?** Fortune, June 1955. (Reprinted in: "John von Neumann. Collected Works", ed. by A.H.Taub, Vol.6, Pergamon Press, Oxford, 1963; pp. 504-519).
Quoted in sect.2.6.

Neumann,J.von (1955b): **Method in the physical sciences.** In: "The unity of knowledge", ed. by L.Leary, Doubleday, New York, pp. 157-164. (Reprinted in: "John von Neumann. Collected Works", ed. by A.H.Taub, Vol.6, Pergamon Press, Oxford, 1963; pp. 491-498).
Chapter 4.

Neumann,J.von (1960): **Continuous geometry.** Princeton University Press, Princeton.
Quoted in sect.4.4.

Neumann,J.von (1961): Undated and untitled manuscript; reviewed under the title **Von Neumann's characterization of factors of type II_1,** by I.Kaplansky. In: "John von Neumann. Collected Works. Vol.III"; Pergamon Press, Oxford, 1961; pp. 562-563.
Quoted in sect.4.4.

Neumann,J.von (1966): **Theory of self-reproducing automata.** Edited and completed by A.W.Burks. University of Illinois Press. Urbana.
Quoted in sect.6.2.

Neumann,J.von and Morgenstern,O. (1944): **Theory of games and economic behavior.** Princeton University Press, Princeton, New Jersey.
Quoted in sect.4.5.

Newton,J.S. (1687): **Philosophiae naturalis principia mathematica.** London. (Translation into English by Andrew Motte in 1729. The translation revised, and supplied with an historical and explanatory appendix, by Florian Cajori, University of California Press, Berkeley, 1960).
Quoted in sect.3.3.

Newton,T.D. and Wigner,E.P. (1949): **Localized states for elementary systems.** Rev.Mod.Phys. 21, 400-406.
Quoted in sect.3.3.

Nikodým,O.M. (1966): **The mathematical apparatus for quantum-theories. Based on the theory of Boolean lattices.** Springer, Berlin.
Quoted in sect. 4.4.

Nilson,D.R. (1976): **Bibliography on the history and philosophy of quantum mechanics.** In: "Logic and probability in quantum mechanics", ed. by S.Suppes; Reidel, Dordrecht-Holland; pp. 457-520.
Quoted in sect.3.4.

Noether,E. (1918): **Invariante Variationsprobleme.** Nachr.Ges.Wiss.Göttingen, Math.-Phys.Kl. 1918, 235-257.
Quoted in sect.3.3.

Ochs,W. (1971): **Can quantum theory be presented as a classical ensemble theory?** Z.Naturforsch. A 26, 1740-1753.
Quoted in sect.3.4.

Ochs,W. (1972a): **On the covering law in quantal proposition systems.** Commun.Math. Phys. 25, 245-252.
Quoted in sect.4.4.

Ochs,W. (1972b): **On Gudder's hidden-variable theorems.** Nuovo Cimento B 10, 172-184.
Quoted in sect.3.4.

Ochs,W. (1972c): **On the foundation of quantal proposition systems.** Z.Naturforsch. A 27, 893-900.
Quoted in sect.4.4.

Ochs,W. and Spohn,H. (1978): A characterization of the Segal entropy. Reports on Mathematical Physics 14, 75-87.
Quoted in sect.4.5.

Oppenheim,P. and Putnam,H. (1958): Unity of science as a working hypothesis. In: "Concepts, theories, and the mind-body problem", ed. by H.Feigl, G.Maxwell and M.Scriven, Minnesota Studies in the Philosophy of Science, vol.2; pp. 3-36.
Quoted in sect.6.1.

Packard,A.S. and Cope,E.D. (1883): Editors' table. American Naturalist, 17, 175.
Quoted in sect.2.6.

Padmanabhan,A.R. (1979): Probabilistic aspects of von Neunmann algebras. J.Functional Analysis 31, 139-149.
Quoted in sect.4.5.

Pais,A. (1979): Einstein and the quantum theory. Rev.Mod.Phys. 51, 863-914.
Quoted in sect.4.3.

Papaliolios,C. (1967): Experimental test of a hidden-variable quantum theory. Phys.Rev.Lett. 18, 622-625.
Quoted in sect.3.4.

Park,J.L. (1968a): Nature of quantum states. Amer.J.Phys. 36, 211-226.
Quoted in sects.3.4, 3.5.

Park,J.L. (1968b): Quantum theoretical concepts of measurement. Philosophy of Science 35, 205-231, 389-411.
Quoted in sects.3.4, 3.5.

Park,J.L. (1970): The concept of transition in quantum mechanics. Foundations of Physics 1, 23-33.
Quoted in sect.3.5.

Park,J.L. and Margenau,H. (1968): Simultaneous measurability in quantum theory. Int.J.Theor.Phys. 1, 211-283.
Quoted in sect.3.5.

Parker,W.H., Langenberg,D.N., Denenstein,A. and Taylor,B.N. (1969): Determination of e/ħ, using macroscopic quantum phase coherence in superconductors.I. Experiment. Phys.Rev. 177, 639-664.
Quoted in sects.3.7, 4.2.

Pauli,W. (1933): Die allgemeinen Prinzipien der Wellenmechanik. In: "Handbuch der Physik", Bd. 24/1, herausgegeben von H.Geiger und K.Scheel; Springer-Verlag, Berlin, 2.Auflage. (An almost identical version appeared in: "Handbuch der Physik"; herausgegeben von S.Flügge, Band V, Teil 1; Springer-Verlag, Berlin, 1958).
Quoted in sects.3.1, 3.2, 3.3, 3.4, 3.5, 6.5.

Pauli,W. (1948): The concept of complementarity. Editorial. Dialectica 2, 307-311.
Quoted in sects.3.4, 6.5.

Pauli,W. (1950): Die philosophische Bedeutung der Idee der Komplementarität. Experientia 6, 72-81.
Quoted in sect.3.4.

Pauli,W. (1953): Remarques sur le problème des paramètres cachés dans la mécanique quantique et sur la théorie de l'onde pilote. In: "L.de Broglie, Physicien et Penseur"; A.Michel, Paris; pp. 33-42.
Quoted in sect.3.4

Pauli,W. (1954): Wahrscheinlichkeit und Physik. Dialectica 8, 112-124.
Quoted in sect.3.4.

Pauli,W. *(1955)*: **Matter.** In: "Man's right to knowledge", Columbia University Press, New York; pp. 10-18.
Quoted in sect.3.4.

Pauli,W. *(1957)*: **Phänomenen und physikalische Realität.** Dialectica 11, 36-48.
Quoted in sect.3.4.

Pauli,W. *(1958)*: **Albert Einstein in der Entwicklung der Physik.** Neue Zürcher Zeitung No.89, 12.Januar 1958. (Reprinted in: Universitas 13, 593-598, 1958; Physikalische Blätter 15, 241-245, 1959).
Quoted in sect.3.4.

Pauling,L. *(1958)*: **The significance of chemistry.** In: "Frontiers in science", ed. by E.Hutchings; New York.
Quoted in sect.1.4.

Pauling,L. *(1959)*: **Quantum theory and chemistry.** In: "Max Planck Fetschrift", ed. by B.Kockel, W.Macke, and A.Papapetrou, VEB Deutscher Verlag der Wissenschaften, Berlin; pp. 385-388.
Quoted in sect.1.3.

Pauri,M. and Prosperi,G.M. *(1968)*: **Canonical realizations of the Galilei group.** J.Math.Phys.(N.Y.) 9, 1146-1162.
Quoted in sect.3.4.

Pearle,P.M. *(1970)*: **Hidden-variable example based upon data rejection.** Phys.Rev. D 2, 1418-1425.
Quoted in sect.3.4.

Pedersen,G.K. *(1972)*: **Operator algebras with weakly closed abelian subalgebras.** Bull.London Math.Soc. 4, 171-175.
Quoted in sect.4.6.

Pedersen,G.K. *(1979)*: **C*-algebras and their automorphism groups.** Academic Press, London.
Quoted in sects.4.2, 4.6.

Pedersen,G.K. and Takesaki,M. *(1972)*: **The operator equation THT=K.** Proc.Amer.Math. Soc. 36, 311-312.
Quoted in sect.4.5.

Pedersen,G.K. and Takesaki,M. *(1973)*: **The Radon-Nikodym theorem for von Neumann algebras.** Acta Math. 130, 53-87.
Quoted in sect.4.5.

Peirce,C.S. *(Collected Papers)*: **The collected papers of Charles Sanders Peirce.** Vol.1-6, ed. by C.Hartshorne and P.Weiss, Harvard University Press, Cambridge, Mass., 1931-1935; Vol.7-8, ed. by A.Burks, Harvard University Press, Mass., 1958.
Quoted in sect.2.1.

Peña-Auerbach,L.de la, and Cetto,A.M. *(1977)*: **Why Schrödinger's equations?** Int.J. Quantum Chem. 12, Suppl. 1, 23-37.
Quoted in sect.3.4.

Peña-Auerbach,L.de la, and Cetto,A.M. *(1979)*: **The quantum harmonic oscillator revisited: A new look from stochastic electrodynamics.** J.Math.Phys.(N.Y.) 20, 469-483.
Quoted in sect.6.4.

Peres,A. and Rosen,N. *(1964a)*: **Macroscopic bodies in quantum theory.** Phys.Rev. 135, B1486-B1488.
Quoted in sect.3.5.

Peres,A. and Rosen,N. *(1964b)*: **Measurement of a quantum ensemble by a classical apparatus.** Ann.Phys.(New York) 29, 366-377.
Quoted in sect.3.5.

Peres,A. *and Singer*,P. (*1960*): On possible experimental tests for the paradox of Einstein, Podolsky and Rosen. Nuovo Cimento 15, 907-915.
Quoted in sect.3.7.

Personick,S.D. (*1971*): An image-band interpretation of optical heterodyne noise. Bell Syst.Tech.J. 50, 213-216.
Quoted in sect.4.5.

Petersen,A. (*1963*): The philosophy of Niels Bohr. Bull.Atomic Scientists 19, No.7, 8-11.
Quoted in sects.3.1, 3.4.

Pfeifer,P. (*1980*): Chiral molecules - a superselection rule induced by the radiation field. Dissertation ETH Zürich, No.6551. ok Gotthard S+D A.G., Zürich.
Quoted in sect.5.6.

Pfeifer,P. (*1981*): A nonlinear Schrödinger equation yielding the shape of molecules by spontaneous symmetry breaking. In: "Classical, semiclassical and quantum mechanical problems in mathematics, chemistry and physics", ed. by K.Gustafson and W.P.Reinhardt, Plenum Press, New York, in press.
Quoted in sect.5.6.

Piaget,J. (*1970*): Le structuralisme. Presses Universitaires de France, Paris, 4e éd. (English translation: "Structuralism", Basic Books, 1970).
Quoted in sect.2.4.

Picht,G. (*1969*): Mut zur Utopie. Piper, München.
Quoted in sect.6.5.

Picht,G. (*1976*): Philosophie oder vom Wesen und rechten Gebrauch der Vernunft. In: "Meyers Enzyklopädisches Lexikon", Band 18, Bibliographisches Institut, Mannheim; pp. 587-591.
Quoted in sect.2.6.

Piron,C. (*1964*): Axiomatique quantique. Helv.Phys.Acta 37, 439-468.
Quoted in sects.4.4, 5.3.

Piron,C. (*1969*): Les règles de supersélection continues. Helv.Phys.Acta 42, 330-338.
Quoted in sect.4.4.

Piron,C. (*1971*): Observables in general quantum theory. In: "Proceedings of the international school of physics 'Enrico Fermi'. Course 49: Foundations of quantum mechanics"; ed. by B.d'Espagnat; Academic Press, New York; pp. 274-286.
Quoted in sect.4.4.

Piron,C. (*1972*): Survey of general quantum physics. Foundations of Physics 2, 287-314.
Quoted in sect.4.4.

Piron,C. (*1976*): Foundations of quantum physics. Benjamin, Reading, Massachusetts.
Quoted in sect.4.4.

Piziak,R. (*1974*): Orthomodular lattices as implication algebras. J.Philosophical Logic 3, 413-418.
Quoted in sect,4,4.

Planck,M. (*1900*): Zur Theorie des Gesetzes der Energieverteilung im Normalspectrum. Verhandlungen der Deutschen Physikalischen Gesellschaft 2, 237-345.
Quoted in sect.3.2.

Platt,J.R. (*1961*): The chemical bond and the distribution of electrons in molecules. In: "Handbuch der Physik", ed. by S.Flügge, vol.37/2. Springer, Berlin; pp. 173-281.
Quoted in sect.1.3.

Platt,J.R. (*1964*): Strong inference. Science 146, 347-353.
Quoted in sect.2.3.

Plymen,R.J. (*1968a*): C*-algebras and Mackey's axioms. Commun.Math.Phys. 8, 132-146.
Quoted in sect.4.6.

Plymen,R.J. (*1968b*): A modification of Piron's axioms. Helv.Phys.Acta 41, 69-74.
Quoted in sect.4.4, 4.6.

Plymen,R.J. (*1968c*): Dispersion-free normal states. Nuovo Cimento A 54, 862-870.
Quoted in sect.3.4.

Poincaré,H. (*1902*): La science et l'hypothèse. Flammarion, Paris. (English translation: "Science and hypothesis", London, 1905; reprinted by Dover, New York, 1952).
Quoted in sect.3.3.

Poincaré,H. (*1904*): L'état actuel et l'avenir de la physique mathématique. Bull.Sci. Math.(Paris) 28, 302-324. (English translation: "The principles of mathematical physics"; The Monist 15, 1-24 (1905).
Quoted in sect.3.3.

Poincaré,H. (*1905*): Sur la dynamique de l'électron. C.R.Acad.Sci.(Paris) 140, 1504-1508.
Quoted in sect.3.3.

Poincaré,H. (*1913*): The foundations of Science. Translated by G.B.Halsted, Science Press, New York.
Quoted in sect.2.5.

Polanyi,M. (*1962*): Tacit knowing: its bearing on some problems of philosophy. Rev.Mod.Phys. 34, 601-616.
Quoted in sect.2.3.

Polanyi,M. (*1963*): The potential theory of absorption. Science 141, 1010-1013.
Quoted in sect.2.4.

Polanyi,M. (*1967*): Life transcending physics and chemistry. Chemical Engineering News 45, 54-66.
Quoted in sect.6.1.

Polanyi,M. (*1968*): Life's irreducible structure. Science 160, 1308-1312.
Quoted in sect.6.1.

Pool,J.C.T. (*1968a*): Baer*-semigroups and the logic of quantum mechanics. Commun. Math.Phys. 9, 118-141.
Quoted in sects.4.4, 4.5, 4.6.

Pool,J.C.T. (*1968b*): Semimodularity and the logic of quantum mechanics. Commun. Math.Phys. 9, 212-228.
Quoted in sects.4.4, 4.5, 4.6.

Popper,K.R. (*1935*): Logik der Forschung. Springer, Wien. (English translation: "The logic of scientific discovery", Hutchinson, London, 1959).
Quoted in sects.2.3, 3.5.

Popper,K.R. (*1967*): Quantum mechanics without the observer. In: "Quantum theory and reality"; ed. by M.Bunge; Springer, Berlin; pp. 7-44.
Quoted in sect.3.5.

Popper,K.R. (*1968*): Birkhoff and von Neumann's interpretation of quantum mechanics. Nature(London) 219, 682-685.
Quoted in sect.4.4.

Prigogine,I., George,C., Henin,F. and Rosenfeld,L. (*1973*): A unified formulation of dynamics and thermodynamics. Chemica Scripta 4, 5-32.
Quoted in sect.3.5.

Primack,J.R. and Brodsky,S.J. (*1969*): The electromagnetic interaction of composite systems. Annals of Physics 52, 315-365.
Quoted in sect.3.3.

Primas,H. (*1961*): Ueber quantenmechanische Systeme mit einem stochastischen Hamiltonoperator. Helv.Phys.Acta 34, 36-57.
Quoted in sect.6.4.

Primas,H. (*1975*): Pattern recognition in molecular quantum mechanics. I. Background dependence of molecular states. Theor.Chim.Acta 39, 127-148.
Quoted in sect.5.4.

Primas,H. (*1977*): Theory reduction and non-Boolean theories. J.Math.Biology 4, 281-301.
Quoted in sects.5.5, 6.1.

Primas,H. (*1978*): Kinematical symmetries in molecular quantum mechanics. In: "Group theoretical methods in physics", ed. by P.Kramer and A.Rieckers, "Lecture Notes in Physics", vol.79, Springer, Berlin; pp. 72-91.
Quoted in sect.6.3.

Primas,H. (*1980*): Foundations of theoretical chemistry. In: "Quantum dynamics of molecules: The new experimental challenge to theorists". NATO Advanced Study Institutes Series, vol.57, ed. by R.G.Woolley; Plenum New York; pp. 39-113.
Quoted in sects.4.4, 4.5, 4.6, 5.3, 5.4.

Primas,H. and Gans,W. (*1979*): Quantenmechanik, Biologie und Theoriereduktion. In: "Materie-Leben-Geist. Zum Problem der Reduktion der Wissenschaften", ed. by B.Kanitscheider, Duncker & Humblot, Berlin; pp. 15-42.
Quoted in sects.5.5, 6.1.

Primas,H. and Müller-Herold,U. (*1978*): Quantum mechanical system theory. Adv.Chem. Phys. 38, 1-107.
Quoted in sect.3.5, 4.6, 5.4.

Prosperi,G.M. (*1971a*): Macroscopic physics, quantum mechanics and quantum theory of measurement. In: "Quantum theory and beyond"; ed. by T.Bastin; Cambridge University Press, London; pp. 55-64.
Quoted in sect.3.5.

Prosperi,G.M. (*1971b*): Macroscopic physics and the problem of measurement in quantum mechanics. In: "Proceedings of the international school of physics 'Enrico Fermi'. Course 49: Foundations of quantum mechanics"; ed. by B.d'Espagnat; Academic Press, New York; pp. 97-126.
Quoted in sect.3.5.

Prosperi,G.M. (*1974*): Models of the measuring process and of macro-theories. In: "Foundations of quantum mechanics and ordered linear spaces", ed. by A.Hartkämper and H.Neumann, "Lecture Notes in Physics", vol.29; Springer, Berlin, pp. 163-198.
Quoted in sect.3.5.

Przibram,K. (*editor*) (*1963*): Briefe zur Wellenmechanik. Springer, Wien.
Quoted in sect.3.1.

Pulmannová,S. (*1975*): Note on the structure of quantal proposition system. Acta Phys.Slovaca 25, 234-240.
Quoted in sect.4.4.

Pulmannová,S. (*1976*): A superposition principle in quantum logics. Commun.Math.Phys. 49, 47-51.
Quoted in sect.4.4.

Pulmannová,S. (*1977*): Symmetries in quantum logics. Int.J.Theor.Phys. 16, 681-688.
Quoted in sect.4.4.

Pusz,W. and Woronowicz,S.L. (*1978*): **Passive states and KMS states for general quantum systems.** Commun.Math.Phys. 58, 273-290.
Quoted in sect.4.3.

Putnam,H. (*1957*): **Three-valued logic.** Philosophical Studies (University of Minnesota) 8, 73-80.
Quoted in sect.4.4.

Putnam,H. (*1962*): **The analytic and the synthetic.** In: "Minnesota Studies in the Philosophy of Science", Vol.III, ed. by H.Feigl and G.Maxwell; University of Minnesota Press, Minneapolis, pp. 358-397.
Quoted in sect.2.3.

Putnam,H. (*1969*): **Is logic empirical?** In: "Boston studies in the philosophy of science. Vol.5"; ed. by R.S.Cohen and M.W.Wartofsky; Reidel, Dordrecht-Holland; pp. 216-241.
Quoted in sect.4.4.

Putnam,H. (*1975*): **Mathematics, matter and method. Philosophical Papers, Volume 1.** Cambridge University Press, London.
Quoted in sect.4.4.

Pylyshyn,Z.W. (*1973*): **What the mind's eye tells the mind's brain: a critique of mental imagery.** Psychological Bulletin 80, 1-24 (Reprinted in: "Images, perception, and knowledge", ed. by J.M.Nicholas; Reidel, Dordrecht, 1977; pp. 1-36.
Quoted in sect.2.5.

Quine,W.V. (*1953*): **Two dogmas of empiricism.** In: W.V.Quine, "From a logical point of view", Harvard University Press, Cambridge, Mass. 1953; pp. 20-46.
Quoted in sect.2.3, 4.4.

Radin,C. (*1973*): **Dynamics of limit models.** Commun.Math.Phys. 33, 283-292.
Quoted in sect.4.2.

Raggio,G.A. (*1978*): **Dispersionfree states on the center of C*- and W*-algebras.** Internal progress report, Lab.Phys.Chem. ETH-Zürich, October 1978. Unpublished.
Quoted in sect.5.6.

Raggio,G.A. (*1981*): **States and composite systems in W*-algebraic quantum mechanics.** Thesis ETH Zürich.
Quoted in sects.5.3, 5.4, 5.6.

Ramsay,A. (*1966*) : **A theorem on two commuting observables.** J.Mathematics and Mechanics 15, 227-234.
Quoted in sect.4.4.

Randall,C.H. (*1966*): **A mathematical foundation for empirical science - with special reference to quantum theory. Part I. A calculus of experimental propositions.** Thesis, Rensselaer Polytechnic Institute, Troy, N.Y. (reissued as Knolls Atomic Power Laboratory Report KAPL-3147, June, 1966).
Quoted in sect.4.4.

Randall,C.H. and Foulis,D.J. (*1970*): **An approach to empirical logic.** Amer.Math. Monthly 77, 363-374.
Quoted in sect.4.4.

Randall,C.H. and Foulis,D.J. (*1973*): **Operational statistics.II. Manuals of operations and their logics.** J.Math.Phys.(N.Y.) 14, 1472-1480.
Quoted in sect.4.4.

Randall,C.H. and Foulis,D.J. (*1976*): **A mathematical setting for inductive reasoning.** In: "Foundations of probability theory, statistical inference, and statistical theories of science", volume III: "Foundations and philosophy of statistical theories in the physical sciences", ed. by W.L.Harper and C.A.Hooker, Reidel, Dordrecht; pp. 169-205.
Quoted in sect.4.4.

Randall,C.H. and Foulis,D.J. (*1979*): **The operational approach to quantum mechanics.** In: "Physical theories as logico-operational structure", ed. by C.A.Hooker, Reidel, Dordrecht; pp. 167-201.
Quoted in sect.4.4.

Reece,G. (*1973*): **The theory of measurement in quantum mechanics.** Int.J.Theor.Phys. 7, 81-117.
Quoted in sect.3.5.

Reed,M. and Simon,B. (*1975*): **Methods of modern mathematical physics. II Fourier analysis, self-adjointness.** Academic Press, New York.
Quoted in sect.3.3.

Reichenbach,H. (*1928*): **Philosophie der Raum-Zeit-Lehre.** Gruyter, Berlin. (English translation: "The philosophy of space and time"; Dover, New York; 1958).
Quoted in sect.5.2.

Reichenbach,H. (*1944*): **Philosophic foundations of quantum mechanics.** University of California Press, Berkeley. (German translation: "Philosophische Grundlagen der Quantenmechanik"; Birkhäuser, Basel, 1949).
Quoted in sects.3.7, 4.4.

Reichenbach,H. (*1951*): **Ueber die erkenntnistheoretische Problemlage und den Gebrauch einer dreiwertigen Logik in der Quantenmechanik.** Z.Naturforsch. A 6, 569-575.
Quoted in sect.4.4.

Reichenbach,H. (*1956*): **The direction of time.** University of California Press, Berkeley.
Quoted in sect.5.2.

Reid,C. (*1970*): **Hilbert.** Springer, Berlin.
Quoted in sect.3.2.

Richards,W.G. (*1979*): **Ab initio methods and the study of molecular hydration.** In: "Water. A comprehensive treatise. Volume 6. Recent advances", ed. by F.Franks, Plenum Press, New York; p. 123.
Quoted in sect.6.2.

Richter,E. (*1964*): **Bemerkungen zur 'Quantenlogik'.** Philosophia Naturalis 8, 225-231.
Quoted in sect.4.4.

Riesz,F. (*1907*): **Ueber orthogonale Funktionssysteme.** Nachr.Kgl.Ges.Wiss. Göttingen, Math.Phys.Kl. 1907, 116-122.
Quoted in sect.3.2.

Roberts,J.E. and Roepstorff,G. (*1969*): **Some basic concepts of algebraic quantum theory.** Commun.Math.Phys. 11, 321-338.
Quoted in sect.4.2.

Robinson,D.W. (*1965*): **The ground state of the Bose gas.** Commun.Math.Phys. 1, 159-174.
Quoted in sect.4.3.

Robinson,D.W. (*1970*): **Normal and locally normal states.** Commun.Math.Phys. 19, 219-234.
Quoted in sect.4.3.

Robinson,D.W. (*1973*): **Return to equilibrium.** Commun.Math.Phys. 31, 171-189.
Quoted in sect.4.3.

Robinson,D.W. (*1975*): A characterization of clustering states. Commun.Math.Phys. 41, 79-88.
Quoted in sect.4.3.

Robinson,D.W. and Ruelle,D. (*1967*): Extremal invariant states. Ann.Inst.Henri Poincaré, A, 6, 299-310.
Quoted in sect.4.3.

Rose,G. (*1964*): Zur Orthomodularität von Wahrscheinlichkeitsfeldern. Z.Physik 181, 331-332.
Quoted in sect.4.4.

Rosen,R. (*1977*): Complexity as a system property. Int.J. General Systems 3, 227-232.
Quoted in sect.6.2.

Rosenberg,A. (*1953*): The number of irreducible representations of simple rings with no minimal ideals. Amer.J.Math. 75, 523-530.
Quoted in sect.5.4.

Rosenfeld,L. (*1953*): Strife about complementarity. Science Progress 41, 393-410. (Revised version of: "L'évidence de la complementarité", in: "Louis de Broglie, physician et penseur"; Albin Michel, Paris, 1953, pp. 43-65).
Quoted in sect.3.4.

Rosenfeld,L. (*1957*): Misunderstandings about the foundation of quantum theory. In: "Observation and interpretation"; ed. by S.Körner, Butterworth, London. (Reprinted by Dover Publications, New York, 1962); pp. 41-45.
Quoted in sects.3.1, 3.4.

Rosenfeld,L. (*1960*): Heisenberg, physics and philosophy. Book review. Nature(London) 186, 830-832.
Quoted in sect.3.4.

Rosenfeld,L. (*1961*): Foundations of quantum theory and complementarity. Nature (London) 190, 384-388.
Quoted in sect.3.4.

Rosenfeld,L. (*1965*): The measuring process in quantum mechanics. Progr.Theor.Phys. Suppl. (commemoration issue for H.Yukawa), 222-231.
Quoted in sect.3.5.

Rosenfeld,L. (*1968*): Questions of method in the consistency problem of quantum mechanics. Nucl.Phys.A 108, 241-244.
Quoted in sect.3.5.

Ruedenberg,K. (*1962*): The physical nature of the chemical bond. Rev.Mod.Phys. 34, 326-376.
Quoted in sect.1.3.

Ruedenberg,K. (*1975*): The nature of the chemical bond, an energetic view. In: "Localization and delocalization in quantum chemistry", vol.1; ed. by O.Chalvet, R.Doudel, S.Diner, and J.P.Malrien. Reidel, Dordrecht; pp. 224-245.
Quoted in sect.1.3.

Ruelle,D. (*1966*): States of physical systems. Commun.Math.Phys. 3, 133-150.
Quoted in sect.4.3.

Ruelle,D. (*1969*): Statistical mechanics. Benjamin, New York.
Quoted in sect.4.3.

Ruelle,D. (*1970*): Integral representation of states on a C*-algebra. J.Functional Analysis 6, 116-151.
Quoted in sect.4.3.

Runtz,G.R., Bader,R.F.W., and Messer,R.R. (*1977*): Definition of bond paths and bond directions in terms of the molecular charge distribution. Can.J.Chem. 55, 3040-3045.
Quoted in sect.1.3.

Ruskai,M.B. (*1971*): Time development of quantum lattice systems. Commun.Math.Phys. 20, 193-204.
Quoted in sect.4.2.

Ruskai,M.B. (*1973*): A generalization of entropy using traces on von Neumann algebras. Ann.Inst. Henri Poincaré, A, 19, 357-373.
Quoted in sect.4.5.

Russell,B. (*1948*): Human knowledge. Its scope and limits. Allen and Unwin, London.
Quoted in sect.6.1.

Rüttimann,G.T. (*1970*): On the logical structure of quantum mechanics. Foundations of Physics 1, 173-182.
Quoted in sects.4., 4.6.

Rüttimann,G.T. (*1977a*): Jauch-Piron states. J.Math.Phys.(N.Y.) 18, 189-193.
Quoted in sect.4.4.

Rüttimann,G.T. (*1977b*): Logikkalküle der Quantenphysik. Duncker und Humblot, Berlin.
Quoted in sect.4.4.

Ryle,G. (*1949*): The concept of mind. Hutchinson, London.
Quoted in sect.3.8.

Saint-Exupéry,A.de (*1939*): Wind, sand and stars. Translated by Lewis Galantière; Harcourt, Brace and World Inc.. (French original: "Terre des hommes", 1939).
Chapter 5.

Saint-Exupéry,A.de (*1948*): The wisdom of the sands. Translated by Stuart Gilbert; Harcourt, Brace and World Inc., 1950).
Quoted in sect.4.6.

Sakai,S. (*1956*): A characterization of W*-algebras. Pacific J.Math. 6, 763-773.
Quoted in sect.4.6.

Sakai,S. (*1957*): On topological properties of W*-algebras. Proc.Japan. Acad. 33, 439-444.
Quoted in sect.4.3.

Sakai,S. (*1965*): A Radon-Nikodym theorem in W*-algebras. Bull.Amer.Math.Soc. 71, 149-151.
Quoted in sect.4.5.

Sakai,S. (*1971*): C*-algebras and W*-algebras. Springer, Berlin.
Quoted in sects.4.2, 4.5, 5.4, 5.5.

Salecker,H. and Wigner,E.P. (*1958*): Quantum limitation of the measurement of space-time distances. Phys.Rev. 109, 571-577.
Quoted in sect.5.2.

Schaffner,K.F. (*1967*): Antireductionism and molecular biology. Science 157, 644-647.
Quoted in sect.6.1.

Schatten,R. (*1960*): Norm ideals of completely continuous operators. Springer, Berlin.
Quoted in sects.3.3., 4.5.

Scheibe,E. (*1964*): Die kontingenten Aussagen in der Physik. Axiomatische Untersuchungen zur Ontologie der klassischen Physik und der Quantentheorie. Athenäum, Frankfurt.
Quoted in sect.3.4.

Scheibe,E. (*1967*): Bibliographie zu Grundlagenfragen der Quantenmechanik. Philosophia Naturalis 10, 249-290.
Quoted in sect.3.4.

Scheibe,E. (*1973*): The logical analysis of quantum mechanics. Pergamon Press, Oxford.
Quoted in sect.3.7.

Scheibe,E. (*1974*): Popper and quantum logic. Brit.J.Phil.Sci. 25, 319-328.
Quoted in sect.4.4.

Schlieder,S. (*1968*): Einige Bemerkungen zur Zustandsänderung von relativistischen quantenmechanischen Systemen durch Messungen und zur Lokalitätsforderung. Commun.Math.Phys. 7, 305-331.
Quoted in sects.3.5, 3.7.

Schmidt,E. (*1907a*): Zur Theorie der linearen und nichtlinearen Integralgleichungen.I. Mathematische Annalen 63, 433-476.
Quoted in sects.3.2, 3.7.

Schmidt,E. (*1907b*): Zur Theorie der linearen und nichtlinearen Integralgleichungen.II. Mathematische Annalen 64, 161-174.
Quoted in sect.3.2.

Schmidt,E. (*1908*): Ueber die Auflösung linearer Gleichungen mit unendlich vielen Unbekannten. Rend.Circ.Mat. Palermo 25, 53-77.
Quoted in sect.3.2.

Schnorr,C.P. (*1969*): Eine Bemerkung zum Begriff der zufälligen Folge. Z.Wahrscheinlichkeitstheorie verw.Geb. 14, 27-35.
Quoted in sect.4.5.

Schnorr,C.P. (*1970a*): Ueber die Definition von effektiven Zufallstests.I. Z.Wahrscheinlichkeitstheorie verw.Geb. 15, 297-312.
Quoted in sect.4.5.

Schnorr,C.P. (*1970b*): Ueber die Definition von effektiven Zufallstests.II. Z.Wahrscheinlichkeitstheorie verw.Geb. 15, 313-328.
Quoted in sect.4.5.

Schnorr,C.P. (*1970c*): Klassifikation der Zufallgesetze nach Komplexität und Ordnung. Z.Wahrscheinlichkeitstheorie verw.Geb. 16, 1-21.
Quoted in sect.4.5.

Schnorr,C.P. (*1971a*): A unified approach to the definition of random sequences. Math.Systems Theory 5, 246-258.
Quoted in sect.4.5.

Schnorr,C.P. (*1971b*): Zufälligkeit und Wahrscheinlichkeit. Eine algorithmische Begründung der Wahrscheinlichkeitstheorie. Lecture Notes in Mathematics, Vol.218, Springer, Berlin.
Quoted in sects.3.4, 4.5.

Schnorr,C.P. (*1973*): Process complexity and effective random tests. J.Computer and Systems Sciences 7, 376-388.
Quoted in sect.4.5.

Schrödinger,E. (*1926a*): Quantisierung als Eigenwertproblem. (Erste Mitteilung). Annalen der Physik 79, 361-376.
Quoted in sect.3.2.

Schrödinger,E. (*1926b*): Quantisierung als Eigenwertproblem. (Zweite Mitteilung). Annalen der Physik 79, 489-527.
Quoted in sect.3.2.

Schrödinger,E. (1926c): Ueber das Verhältnis der Heisenberg-Born-Jordanschen Quantenmechanik zu der meinen. Annalen der Physik 79, 734-756.
Quoted in sect.3.2.

Schrödinger,E. (1926d): Quantisierung als Eigenwertproblem. (Vierte Mitteilung). Annalen der Physik 81, 109-139.
Quoted in sect.3.2.

Schrödinger,E. (1927): Energieaustausch nach der Wellenmechanik. Annalen der Physik 83, 955-968.
Quoted in sect.3.7.

Schrödinger,E. (1935a): Discussion of probability relations between separated systems. Proc.Cambridge Phil.Soc. 31, 555-563.
Quoted in sects.1.4, 3.7, 5.6.

Schrödinger,E. (1935b): Die gegenwärtige Situation in der Quantenmechanik. Naturwissenschaften 23, 807-812, 823-828, 844-849.
Quoted in sects.1.4, 3.5, 3.7, 5.6.

Schrödinger,E. (1936): Probability relations between separated systems. Proc.Cambridge Phil.Soc. 32, 446-452.
Quoted in sects.1.4, 3.7, 5.6.

Schwinger,J. (1962): Gauge invariance and mass. Phys.Rev. 125, 397-398.
Quoted in sect.3.3.

Scott,W.T. (1967): Erwin Schrödinger. An introduction to his writings. University of Massachusetts Press, Amherst, Massachusetts.
Quoted in sect.3.1.

Segal,I.E. (1947): Postulates for general quantum mechanics. Annals of Mathematics 48, 930-948.
Quoted in sect.4.2.

Segal,I.E. (1953): A non-commutative extension of abstract integration. Annals of Mathematics 57, 401-457. Correction: ib 58, 595-596 (1953).
Quoted in sect.4.5.

Segal,I.E. (1954): Abstract probability spaces and a theorem of Kolmogoroff. Amer.J.Math. 76, 721-732.
Quoted in sect.4.5.

Segal,I.E. (1958): Distributions in Hilbert space and canonical systems of operators. Trans.Amer.Math.Soc. 88, 12-41.
Quoted in sect.4.2.

Segal,I.E. (1959): The mathematical meaning of operationalism in quantum mechanics. In: "The axiomatic method"; ed. by L.Henkin, P.Suppes, and A.Tarski; North-Holland, Amsterdam; pp. 341-352.
Quoted in sect.4.2.

Segal,I.E. (1960): A note on the concept of entropy. J.Mathematics and Mechanics 9, 623-629.
Quoted in sect.4.5.

Sen,R.N. and Zahavi,D. (1972): Galilei invariance and superfluidity. Physica 59, 379-384.
Quoted in sect.3.3.

Sewell,G.L. (1973): States and dynamics of infinitely extended physical systems. Commun.Math.Phys. 33, 43-51.
Quoted in sect.4.2.

Shapiro,D. (1965): Neurotic styles. Basic Books, New York.
Quoted in the preface.

Sharp,D.H. (1961): The Einstein-Podolsky-Rosen paradox re-examined. Philosophy of Science 28, 225-233.
Quoted in sect.3.7.

She,C.Y. and Heffner,H. (1966): Simultaneous measurement of noncommuting observables. Phys.Rev. 152, 1103-1110.
Quoted in sect.4.5.

Sherman,S. (1951): Non-negative observables are squares. Proc.Amer.Math.Soc. 2, 31-33.
Quoted in sect.4.2.

Sherman,S. (1956): On Segal's postulates for general quantum mechanics. Annals of Mathematics 64, 593-601.
Quoted in sect.4.2.

Shimony,A. (1963): Role of the observer in quantum theory. Amer.J.Phys. 31, 755-773.
Quoted in sects.3.4, 3.5, 3.6.

Shimony,A. (1971): Experimental test of local hidden-variable theories. In: "Proceedings of the international school of physics 'Enrico Fermi'. Course 49: "Foundations of quantum mechanics"; ed. by B.d'Espagnat; Academic Press, New York; pp. 182-194.
Quoted in sect.3.4.

Shimony,A. (1974): Approximate measurement in quantum mechanics.II. Phys.Rev. D 9, 2321-2323.
Quoted in sect.3.5.

Shultz,F.W. (1974): A characterization of state spaces of orthomodular lattices. J. of Combinatorial Theory 17, 317-328.
Quoted in sect.4.5.

Shultz,F.W. (1977): Events and observables in axiomatic quantum mechanics. Int.J.Theor.Phys. 16, 259-272.
Quoted in sect.4.6.

Shultz,F.W. (1979): On normed Jordan algebras which are Banach dual spaces. J.Functional Analysis 31, 361-376.
Quoted in sect.4.2.

Sikorski,R. (1960): Boolean algebras. Springer, Berlin; second revised edition, 1962; third edition (corrected reprint), 1968.
Quoted in sects.4.4, 4.5.

Simon,B. (1976): Quantum dynamics: from automorphism to Hamiltonian. In: "Studies in Mathematical Physics. Essays in Honor of Valentine Bargmann", ed. by E.H.Lieb, B.Simon and A.S.Wightman; Princeton University Press, Princeton, New Jersey; pp. 327-349.
Quoted in sect.4.2.

Simon,H.A. (1962): The architecture of complexity. Proc.Amer. Philosophical Soc. 106, 467-482.
Quoted in sect.6.2.

Sirugue,M. and Testard,D. (1971): Some connections between ground states and temperature states of thermodynamical systems. Commun.Math.Phys. 22, 223-237.
Quoted in sect.4.3.

Slansky,J. (1973): Pattern recognition: introduction and foundations. Dowden, Hutchinson and Ross, Stroudsburg, Pennsylvania.
Quoted in sect.6.3.

Sklar, L. *(1976)*: **Thermodynamics, statistical mechanics and the complexity of reductions.** In: "Proceedings of the 1974 Biennial Meeting, Philosophy of Science, vol.32, ed. by R.S.Cohen, C.A.Hooker, A.C.Michalos and J.W.van Evra; Reidel, Dordrecht; pp. 15-32.
Quoted in sect.2.4.

Skolem, T. *(1934)*: **Ueber die Nichtcharakterisierbarkeit der Zahlenreihe mittels endlich oder abzählbar unendlich vieler Aussagen mit ausschliesslich Zahlenvariablen.** Fund.Math. 23, 150-161.
Quoted in sect.4.1.

Slater, J.C. *(1960)*: **Quantum theory of atomic structure. Vol.III.** McGraw Hill, New York.
Quoted in sect.3.3.

Slawny, J. *(1972)*: **On factor representations and the C*-algebra of canonical commutation relations.** Commun.Math.Phys. 24, 151-170.
Quoted in sect.5.4.

Smith, D.R. *(1975)*: **The multivariable method in singular perturbation analysis.** SIAM Review 17, 221-273.
Quoted in sect.6.2.

Sneed, J.D. *(1966)*: **Von Neumann's argument for the projection postulate.** Philosophy of Science 33, 22-39.
Quoted in sect.3.5.

Sneed, J.D. *(1970)*: **Quantum mechanics and classical probability theory.** Synthese 21, 34-64.
Quoted in sect.4.5.

Sneed, J.D. *(1971)*: **The logical structure of mathematical physics.** Reidel, Dordrecht.
Quoted in sect.2.3.

Solvay *(1928)*: **Electrons et photons.** Rapports et discussions du cinquième conseil de physique tenu à Bruxelles du 24 au 29 octobre 1927 sous les auspices de l'institut international de physique Solvay. Gauthier-Villars, Paris.
Quoted in sect.3.4.

Sommerfeld, A. *(1914)*: **Ueber die Fortpflanzung des Lichtes in dispergierenden Medien.** Annalen der Physik 44, 177-202.
Quoted in sect.3.8.

Specker, E. *(1960)*: **Die Logik nicht gleichzeitig entscheidbarer Aussagen.** Dialectica 14, 239-246.
Quoted in sect.4.4.

Spohn, H. *(1979)*: **Die Herleitung von kinetischen Gleichungen aus Hamiltonscher Dynamik: Markoffsche Limites.** Habilitationsschrift, Fakultät für Physik der Ludwig-Maximilians-Universität, Münschen. Published as: "Kinetic equations from Hamiltonian dynamics: Markovian limits. Rev.Mod.Phys. 52, 569-615 (1980).
Quoted in sect.6.3.

Spohn, H. *and Lebowitz*, J.L. *(1978)*: **Irreversible thermodynamics for quantum systems weakly coupled to thermal reservoirs.** Adv.Chem.Phys. 38, 109-142.
Quoted in sect.6.3.

Stachow, E.W. *(1976)*: **Completeness of quantum logic.** J.Philosophical Logic 5, 237-280.
Quoted in sect.4.4.

Stachow, E.W. *(1978)*: **Quantum logical calculi and lattice structures.** J.Philosophical Logic 7, 347-386.
Quoted in sect.4.4.

Stapp,H.P. (*1971*): S-matrix interpretation of quantum theory. Phys.Rev. D 3, 1303-1320.
Quoted in sect.3.4.

Stapp,H.P. (*1972*): The Copenhagen interpretation. Amer.J.Phys. 40, 1098-1116.
Quoted in sect.3.4.

Stegmüller,W. (*1974*): Theorie und Erfahrung. Erster Halbband: Begriffsformen, Wissenschaftssprache, empirische Signifikanz und theoretische Begriffe. Springer, Berlin.
Quoted in sect.4.5.

Stegmüller,W. (*1975*): Structures and dynamics of theories. Erkenntnis 9, 75-100.
Quoted in sect.2.3.

Stein,H. *and Shimony*,A. (*1971*): Limitations of measurement. In: "Proceedings of the international school of physics 'Enrico Fermi'. Course 49: Foundations of quantum mechanics"; ed. by B.d'Espagnat; Academic Press; pp. 56-76.
Quoted in sect.3.5.

Stinespring,W.F. (*1955*): Positive functions of C*-algebras. Proc.Amer.Math.Soc. 6, 211-216.
Quoted in sect.4.5.

Stolz,P. (*1969*): Attempt of an axiomatic foundation of quantum mechanics and more general theories.V. Commun.Math.Phys. 11, 303-313.
Quoted in sect.4.5.

Stolz,P. (*1971*): Attempt of an axiomatic foundation of quantum mechanics and more general theories.VI. Commun.Math.Phys. 23, 117-126.
Quoted in sect.4.5.

Stone,M.H. (*1930*): Linear transformations in Hilbert space. III. Operational methods and group theory. Proc.Nat.Acad.Sci.U.S. 16, 172-175.
Quoted in sects.3.3, 4.2.

Stone,M.H. (*1932*): On one-parameter unitary groups in Hilbert space. Annals of Mathematics 33, 643-648.
Quoted in sect.3.3.

Stone,M.H. (*1949*): Postulates for the barycentric calculus. Ann.Math. Pura Appl. 29, 25-30.
Quoted in sect.4.5.

Stone,M.H. (*1966*): Some current trends in mathematical research. In: "Symposia on theoretical physics", Vol.2; pp. 159-164. Ed. by A.Ramakrishnan. Plenum Press, New York.
Quoted in the preface.

Størmer,E. (*1974*): Positive linear maps of C*-algebras. In: "Foundations of quantum mechanics and ordered linear spaces", ed. by A.Hartkämper and H.Neumann; Lecture Notes in Physics, vol.29; Springer, Berlin; pp. 85-106.
Quoted in sect.4.5.

Størmer,E. (*1976*): Jordan algebras versus C*-algebras. In: "Quantum dynamics: models and mathematics", ed. by L.Streit; Springer, Wien; pp. 1-13.
Quoted in sect.4.2.

Strătilă,S. *and Zsidó*,L. (*1979*): Lectures on von Neumann algebras. Editura Academiei, Bucharest, and Abacus Press, Wells.
Quoted in sect.4.5.

Strauss,M. (*1936*): **Zur Begründung der statistischen Transformationstheorie der Quantenphysik.** Sitzungsberichte der Preussischen Akad.d.Wiss., Phys.-Math. Kl. 27, 382-398. (English translation with a postscript 1971: "The logic of complementarity and the foundation of quantum theory". In: M.Strauss, "Modern physics and its philosophy"; Reidel, Dordrecht-Holland, 1972; pp. 186-203).
Quoted in sect.4.4.

Strauss,M. (*1938*): **Mathematics as logical syntax - a method to formalize the language of a physical theory.** J.Unified Science (Erkenntnis) 7, 147-153. (Revised reprint. In: M.Strauss, "Modern physics and its philosophy"; Reidel, Dordrecht-Holland, 1972; pp. 71-76).
Quoted in sect.4.4.

Strauss,M. (*1967*): **Grundlagen der modernen Physik.** In: "Mikrokosmos-Makrokosmos. Vol.2"; ed. by H.Ley and R.Löther; Akademie-Verlag, Berlin; pp. 55-92. (English translation of part III: "Foundations of quantum mechanics". In: M.Strauss, "Modern physics and its philosophy"; Reidel, Dordrecht-Holland, 1972; pp. 226-238).
Quoted in sect.4.4.

Strauss,M. (*1970*): **Intertheory relations.** In: "Induction, physics, and ethics"; ed. by P.Weingartner and G.Zecha; Reidel, Dordrecht-Holland; pp. 220-248.
Quoted in sects.3.1, 3.3, 4.4.

Strauss,M. (*1972*): **Modern physics and its philosophy. Selected papers in the logic, history, and philosophy of science.** Reidel, Dordrecht.
Quoted in sect.4.4.

Strauss,M. (*1973*): **Logics for quantum mechanics.** Foundations of Physics 3, 265-276.
Quoted in sect.4.4.

Stueckelberg,E.C.G. (*1960*): **Quantum theory in real Hilbert space.** Helv.Phys.Acta 33, 727-752.
Quoted in sect.4.4.

Stueckelberg,E.C.G. *and Guenin*,M. (*1961*): **Quantum theory in real Hilbert space.II.** Helv.Phys.Acta 34, 621-628.
Quoted in sect.4.4.

Stueckelberg,E.C.G. *and Guenin*,M. (*1962*): **Théorie des quanta dans l'espace de Hilbert réel. IV.** Helv.Phys.Acta 35, 673-695.
Quoted in sect.4.4.

Stueckelberg.E.C.G., *Guenin*,M., *Piron*,C. *and Ruegg*,H. (*1961*): **Quantum theory in real Hilbert space.III.** Helv.Phys.Acta 34, 675-698.
Quoted in sect.4.4.

Suppe,F. (*ed.*) (*1974*): **The structure of scientific theories.** University of Illinois Press, Urbana, Ill.
Quoted in sect.2.3.

Suppes,P. (*1965*): **Logics appropriate to empirical theories.** In: "Theories of models"; ed. by J.W.Addison, L.Henkin, and A.Tarski; North-Holland, Amsterdam; pp. 364-375.
Quoted in sect.4.4.

Süssmann,G. (*1957*): **An analysis of measurement.** In: "Observation and interpretation"; ed. by S.Körner; Butterworth, London, 1957. (Reprinted by Dover, New York, 1962); pp. 131-136.
Quoted in sect.3.4.

Süssmann,G. (*1958*): **Ueber den Messvorgang.** Bayrische Akademie der Wissenschaften, Math.-Naturw.Kl., Abhandlungen, Neue Folge, Heft 88, 41 Seiten.
Quoted in sects.3.4, 3.5.

Takesaki,M. *(1970a)*: Tomita's theory of modular Hilbert algebras and its applications. Lecture Notes in Mathematics, vol.128; Springer, Berlin.
Quoted in sects.4.3, 4.5.

Takesaki,M. *(1970b)*: Disjointness of the KMS-states of different temperatures. Commun.Math.Phys. 17, 33-41.
Quoted in sect.4.3.

Takesaki,M. *(1972)*: Conditional expectations in von Neumann algebras. J.Functional Analysis 9, 306-321.
Quoted in sect.4.5.

Takesaki,M. *(1973)*: States and automorphisms of operator algebras, standard representations, and the Kubo-Martin-Schwinger boundary condition. In: "Statistical mechanics and mathematical problems", ed. by A.Lenard. Lecture Notes in Physics, vol.20, pp. 205-246; Springer, Berlin.
Quoted in sect.4.5.

Takesaki,M. *(1979)*: Theory of operator algebras.I. Springer, New York.
Quoted in sects.4.2, 5.6.

Takesaki,M. *and Winnink*,M. *(1973)*: Local normality in quantum statistical mechanics. Commun.Math.Phys. 30, 129-152.
Quoted in sect.4.3.

Tausk,K.S. *(1966)*: Relation of measurement with ergodicity, macroscopic systems, information and conservation laws. International Atomic Energy Agency, Internal Report 14/1966 (Trieste, August 1966).
Quoted in sect.3.5.

Teller,E. *(1961)*: Der quantenmechanische Messprozess und die Entropie. In: "Werner Heisenberg und die Physik unserer Zeit"; hg. von F.Bopp; Vieweg, Braunschweig; pp. 90-92.
Quoted in sect.3.4.

Theimer,O. *(1971)*: Derivation of the blackbody radiation spectrum by classical statistical mechanics. Phys.Rev. D 4, 1597-1601.
Quoted in sect.3.2.

Thirring,W. *(1980)*: Lehrbuch der mathematischen Physik.4. Quantenmechanik grosser Systeme. Springer, Wien.
Quoted in sect.4.3.

Thomas,L. *(1978)*: Hubris in science? Science 200, 1459-1462.
Quoted in sect.2.6.

Tischer,J. *(1979)*: A note on the Radon-Nikodym theorem by Pedersen and Takesaki. Acta Sci.Math. 41, 411-418.
Quoted in sect.4.5.

Tischer,J. *(1980)*: Gleason's theorem for type I von Neumann algebras. Preprint Mathematisches Institut, Universität Erlangen.
Quoted in sect.4.5.

Tjøstheim,D. *(1975)*: Multiplicity theory for multivariate wide sense stationary generalized processes. J.Multivariate Analysis 5, 314-321.
Quoted in sect.3.8.

Tjøstheim,D. *(1976)*: A commutation relation for wide sense stationary processes. SIAM J.Appl.Math. 30, 115-122.
Quoted in sect.3.8.

Tomita,M. *(1967)*: Standard forms of von Neumann algebras. Reported at the Vth functional analysis symposium of the Mathematical Society of Japan, Sendai. (Unpublished, quoted from Takesaki, 1970a).
Quoted in sect.4.3.

Tomijama,J. (*1957*): On the projection of norm one in W*-algebras. Proc.Japan Acad. 33, 608-612.
Quoted in sect.4.5.

Tomiyama,J. (*1969*): On the tensor products of von Neumann algebras. Pacific J.Math. 30, 263-270.
Quoted in sect.5.6.

Topping,D.M. (*1967*): Asymptoticity and semimodularity in projection lattices. Pacific J.Math. 20, 317-325.
Quoted in sects.4.4, 4.6.

Tou,J.T. and Gonzalez,R.C. (*1974*): Pattern recognition principles. Addison-Wesley, Reading, Mass.
Quoted in sect.6.3.

Toyoda,T. (*1973*): Necessity of complex Hilbert space for quantum mechanics. Progr.Theor.Phys. 49, 707-713.
Quoted in sect.4.4.

Truesdell,C. (*1976*): History of classical mechanics. Naturwissenschaften 63, 119-130.
Quoted in sect.3.3.

Turing,A.M. (*1936*): On computable numbers, with an application to the Entscheidungsproblem. Proc.London Math.Soc. 42, 230-265. Correction ib. 43, 544-546(1937).
Quoted in sect.4.5.

Turner,J.E. (*1968*): Violation of the quantum ordering of propositions in hidden variable theories. J.Math.Phys.(N.Y.) 9, 1411-1415.
Quoted in sect.3.4.

Uhlhorn,U. (*1962*): Representation of symmetry transformations in quantum mechanics. Ark.Fys. 23, 307-340.
Quoted in sects.3.3, 4.2.

Ulam,S.M. (*1976*): Adventures of a mathematician. Scribner's Sons, New York.
Quoted in sect.1.1.

Umegaki,H. (*1954*): Conditional expectation in an operator algebra. Tôhoku Math.J. 6, 177-181.
Quoted in sect.4.5.

Umegaki,H. (*1956*): Conditional expectations in an operator algebra.II. Tôhoku Math.J. 8, 86-100.
Quoted in sect.4.5.

Umegaki,H. (*1959*): Conditional expectation in an operator algebra.III. Kōdai Math. Sem.Reports 11, 51-64.
Quoted in sect.4.5.

Umegaki,H. (*1962*): Conditional expectation in an operator algebra.IV. Kōdai Math. Sem.Reports 14, 59-85.
Quoted in sect.4.5.

Utiyama,R. (*1956*): Invariant theoretical interpretation of interaction. Phys.Rev. 101, 1597-1607.
Quoted in sect.3.3.

Van der Waals,J.D., see: *Waals,J.D. van der*

Van der Waerden,B.L., see: *Waerden,B.L. van der*

Van Hove,L., see: *Hove,L. van*

Varadarajan,V.S. (*1962*): Probability in physics and a theorem on simultaneous observability. Commun.Pure Appl.Math. 15, 189-217. (Erratum: ib. 18, 757 (1963).
Quoted in sect.4.4.

Varadarajan,V.S. (1968): Geometry of quantum theory. Vol.1. Van Nostrand, Princeton, New Jersey.
Quoted in sect.4.4.

Varadarajan,V.S. (1970): Geometry of quantum theory. Vol.2. Van Nostrand-Reinhold, New York.
Quoted in sect.4.4.

Varadarajan,V.S. (1977): Book review of "Foundations of quantum physics" by C.Piron". Bull.Amer.Math.Soc. 83, 226-231.
Quoted in sect.4.4.

Verhagen,C.J.D.M. (1975): Some general remarks about pattern recognition; its definition; its relation with other disciplines; a literature survey. Pattern recognition 7, 109-116.
Quoted in sect.6.3.

Voisin,J. (1965): On some unitary representations of the Galilei group.I. Irreducible representations. J.Math.Phys.(N.Y.) 6, 1519-1529.
Quoted in sect.3.3.

Von Neumann,J., see: *Neumann,J. von*

Von Weizsäcker,C.F., see: *Weizsäcker,C.F. von*

Vowden,B.J. (1967): On the Gelfand-Neumark theorem. J.London Math.Soc. 42, 725-731.
Quoted in sect.4.2.

Waals,J.D. van der (1927): Lehrbuch der Thermostatik. Erster Teil. Allgemeine Thermostatik. Bearbeitet von Ph.Kohnstamm. Ambrosius Barth, Leipzig.
Quoted in sect.6.2.

Waerden,B.L. van der (1966): On the measurements in quantum mechanics. Z.Phys. 190, 99-109.
Quoted in sect.3.5.

Waerden,B.L. van der (1967): Sources of quantum mechanics. North-Holland, Amsterdam.
Quoted in sect.3.1.

Waerden,B.L. van der (1973): From matrix mechanics and wave mechanics to unified quantum mechanics. In: "The physicist's conception of nature", ed. by J.Mehra; Reidel, Dordrecht; pp. 276-293.
Quoted in sect.3.2.

Wakita,H. (1960) : Measurement in quantum mechanics. Progr.Theor.Phys. 23, 32-40.
Quoted in sect.3.5.

Wakita,H. (1962a): Measurement in quantum mechanics.II. Reduction of a wave packet. Progr.Theor.Phys. 27, 139-144.
Quoted in sect.3.5.

Wakita,H. (1962b): Measurement in quantum mechanics.III. Macroscopic measurement and statistical operators. Progr.Theor.Phys. 27, 1156-1164.
Quoted in sect.3.5.

Wald,A. (1937): Die Widerspruchsfreiheit des Kollektivbegriffes der Wahrscheinlichkeitsrechnung. Ergeb. eines mathematischen Kolloquiums 8, 38-72.
Quoted in sect.4.5.

Wallace,J. (1974): Planck distribution in a classical nonlinear coupled harmonic-oscillator system. Nuovo Cimento 22B, 22-42.
Quoted in sect.3.2.

Watanabe,S. (1961): A model of mind-body relation in terms of modular logic. Synthese 13, 261-302.
Quoted in sect.4.4.

Watanabe,S. (*1965*): Conditional probability in physics. Progr.Theor.Phys.Suppl., Extra Number 1965, 135-160.
Quoted in sect.4.4.

Watanabe,S. (*1966*): Algebra of observation. Progr.Theor.Phys.Suppl. 37/38, 350-367.
Quoted in sect.4.4.

Watanabe,S. (*1969a*): Knowing and guessing. A quantitative study of inference and information. Wiley, New York.
Quoted in sects.4.4, 6.3.

Watanabe,S. (*1969b*): Modified concepts of logic, probability, and information based on generalized continuous characteristic function. Information and Control 15, 1-21.
Quoted in sect.4.4.

Watanabe,S. (*1969c*): Pattern recognition as an inductive process. In: "Methodologies of pattern recognition", ed. by S.Watanabe, Academic Press, New York, pp. 521-534.
Quoted in sect.6.3.

Watanabe,S. (*editor*) (*1972*): Frontiers of pattern recognition. Academic Press, New York.
Quoted in sect.6.3.

Webster,N. (*1961*): Webster's third new international dictionary of the English language, unabridged. Ed. by P.B.Gove, Merriam Company, Springfield, Massachusetts.
Quoted in sect.5.6.

Weidlich,W. (*1960*): Zur Interpretation der Quantenmechanik. Z.Naturforsch. A 15, 651-654.
Quoted in sect.3.4.

Weidlich,W. (*1967*): Problems of the quantum theory of measurement. Z.Phys. 205, 199-220.
Quoted in sect.3.5.

Weidlich,W. and Haake,F. (*1969*): On the quantum statistical theory of the measuring process. J.Phys.Soc.Jap., Suppl. 26, 231-232.
Quoted in sect.3.5.

Weidmann,J. (*1976*): Lineare Operatoren in Hilberträumen. Teubner, Stuttgart.
Quoted in sect.3.3.

Weinberg,J. (*1968*): Abstraction in the formation of concepts. In: "Dictionary of the history of ideas", ed. by P.P.Wiener, Charles Scribner's Sons, New York; vol.I, pp. 1-9.
Quoted in sect.6.3.

Weinberg,S. (*1965*): Photons and gravitons in perturbation theory: derivation of Maxwell's and Einstein's equations. Phys.Rev. 138, B988-B1002.
Quoted in sect.5.6.

Weinberg,S. (*1974*): Recent progress in gauge theories of the weak, electromagnetic and strong interactions. Rev.Mod.Phys. 46, 255-277.
Quoted in sect.3.3.

Weizsäcker,C.F. von (*1941*): Zur Deutung der Quantenmechanik. Z.Phys. 118, 489-509.
Quoted in sect.3.4.

Weizsäcker,C.F. von (*1955*): Komplementarität und Logik. Naturwissenschaften 42, 521-529, 545-555.
Quoted in sect.4.4.

Weizsäcker,C.F. *von* (*1958*): **Die Quantentheorie der einfachen Alternative.** Z.Naturforsch. A 13, 245-253.
Quoted in sect.4.4.

Weizsäcker,C.F. *von* (*1961*): **Die Einheit der Physik.** In: "Werner Heisenberg und die Physik unserer Zeit"; hg. von F.Bopp; Vieweg, Braunschweig; pp. 23-46.
Quoted in sect.3.4.

Weizsäcker,C.F. *von* (*1971a*): **The unity of physics.** In: "Quantum theory and beyond"; ed. by T.Bastin; Cambridge University Press, London; pp. 229-262.
Quoted in sect.4.4.

Weizsäcker,C.F. *von* (*1971b*): **Die Einheit der Natur.** Hauser Verlag, München.
Quoted in sect.4.4.

Weizsäcker,C.F. (*1973*): **Classical and quantum descriptions.** In: "The physicist's conception of nature", ed. by J.Mehra, Reidel, Dordrecht; pp. 635-667.
Quoted in sect.4.4.

Weizsäcker,C.F. *von*, *Scheibe*,E. *and Süssmann*,G. (*1958*): **Komplementarität und Logik. III. Mehrfache Quantelung.** Z.Naturforsch. A 13, 705-721.
Quoted in sect.4.4.

Welton,T.A. (*1948*): **Some observable effects of the quantum-mechanical fluctuations of the electromagnetic field.** Phys.Rev. 74, 1157-1157.
Quoted in sect.6.4.

Weyl,H. (*1918*): **Gravitation und Elektrizität.** Sitzungsberichte der Königlich Preussischen Akademie der Wissenschaften zu Berlin, 1918, 465-480.
Quoted in sect.3.3.

Weyl,H. (*1922*): **Space, time, matter.** Methuen, London.
Quoted in sect.5.2.

Weyl,H. (*1923*): **Raum, Zeit, Materie.** 5.Auflage, Springer, Berlin.
Quoted in sect.3.3.

Weyl,H. (*1927*): **Quantenmechanik und Gruppentheorie.** Z.Phys. 46, 1-46.
Quoted in sects.3.3, 6.3.

Weyl,H. (*1928*): **Gruppentheorie und Quantenmechanik.** Hirzel, Leipzig. (English translation of the second German edition: "The theory of groups and quantum mechanics", Methuen, London, 1931; reprinted by Dover, New York, 1950).
Quoted in sect.3.3.

Weyl,H. (*1929a*): **Gravitation and the electron.** Proc.Nat.Acad.Sci.(USA) 15, 323-334.
Quoted in sect.3.3.

Weyl,H. (*1929b*): **Gravitation and the electron.** The Rice Institute Pamphlet 16, 280-295.
Quoted in sect.3.3.

Weyl,H. (*1929c*): **Elektron und Gravitation.** Z.Phys. 56, 330-352.
Quoted in sect.3.3.

Weyl,H. (*1949a*): **Wissenschaft als symbolische Konstruktion des Menschen.** In: Eranos-Jahrbuch 1948, Band 16, hg. von O.Fröbe-Kapteyn; Rhein-Verlag, Zürich; pp. 375-431.
Quoted in sect.4.1.

Weyl,H. (*1949b*): **Philosophy of mathematics and natural science.** Princeton University Press, Princeton.
Quoted in sect.3.7.

Wheeler,J.A. (*1946*): **Polyelectrons.** Annals of the New York Academy of Science 48, 219-238.
Quoted in sect.3.7.

Wheeler,J.A. (*1957*): **Assessment of Everett's 'relative state' formulation of quantum theory.** Rev.Mod.Phys. 29, 463-465.
Quoted in sect.3.6.

Whittaker,E. (*1951/53*): **A history of the theories of aether and electricity.** Volume one: "The classical theories"; Volume two: "The modern theories 1900-1926". Nelson, London 1951 and 1953; reprinted by Harper, New York, 1960.
Quoted in sects.3.1, 3.3.

Whitten-Wolfe,B. *and Emch*,G.G. (*1976*): **A mechanical quantum measuring process.** Helv.Phys.Acta 49, 45-55.
Quoted in sect.3.5.

Wick,G.C., *Wightman*,A.S., *and Wigner*,E.P. (*1952*): **The intrinsic parity of elementary particles.** Phys.Rev. 88, 101-105.
Quoted in sects.3.3, 4.4.

Wiener,N. (*1933*): **The Fourier integral and certain of its applications.** Cambridge University Press, New York.
Quoted in sect.3.2.

Wiener,N. (*1948*): **Cybernetics.** M.I.T. Press, Cambridge, Massachusetts; second edition 1961.
Quoted in sect.3.2.

Wightman,A.S. (*1948*): **Note on polarization effects on Compton scattering.** Phys.Rev. 74, 1813-1817.
Quoted in sect.3.7.

Wightman,A.S. (*1962*): **On the localizability of quantum mechanical systems.** Rev. Mod.Phys. 34, 845-872.
Quoted in sect.3.3.

Wightman,A.S. (*1969*): **What is the point of so-called 'axiomatic field theory'?** Phys.Today 22, 53-58(No.9, Sept. 1969).
Quoted in sect.4.1.

Wightman,A.S. *and Schweber*,S.S. (*1955*): **Configuration space methods in relativistic quantum field theory.I.** Phys.Rev. 98, 812-837.
Quoted in sect.4.2.

Wigner,E.P. (*1931*): **Gruppentheorie und ihre Anwendung auf die Quantenmechanik der Atomspektren.** Vieweg, Braunschweig. (English translation: "Group theory"; Academic Press, New York; 1959).
Quoted in sects.3.3, 4.2.

Wigner,E.P. (*1939*): **On unitary representations of the homogeneous Lorentz group.** Annals of Mathematics 40, 149-204.
Quoted in sect.3.3.

Wigner,E.P. (*1952*): **Die Messung quantenmechanischer Operatoren.** Z.Phys. 133, 101-108.
Quoted in sect.3.5.

Wigner,E.P. (*1960*): **The unreasonable effectiveness of mathematics in the natural sciences.** Commun. Pure Appl.Math. 13, 1-14.
Quoted in sect.2.2.

Wigner,E.P. (*1961*): **The probability of the existence of a self-reproducing unit.** In: "The logic of personal knowledge. Essays presented to Michael Polanyi". Routledge and Kegan Paul, London; pp. 231-238. (Reprinted in: E.P.Wigner, "Symmetries and reflections", Indiana University Press, Bloomington, 1967, pp. 200-208).
Quoted in sect.1.4.

Wigner,E.P. (*1962a*): **Remarks on the mind-body question.** In: "The scientist speculates"; ed. by I.J.Good; Heinemann, London; pp. 284-302. (Reprinted in: E.P.Wigner, "Symmetries and reflections"; Indiana University Press, Bloomington, 1967; pp. 171-184).
Quoted in sects.1.4, 3.5.

Wigner,E.P. (*1962b*): **Theorie der quantenmechanischen Messung.** In: "Physikertagung Wien 1961"; hg. von E.Brüche; Physik Verlag, Mosbach, Baden; pp. 1-8.
Quoted in sect.3.5.

Wigner,E.P. (*1963*): **The problem of measurement.** Amer.J.Phys. 31, 6-15.
Quoted in sects.1.5, 3.5.

Wigner,E.P. (*1964a*): **Symmetry and conservation laws.** Proc.Nat.Acad.Sci.(USA) 51, 956-965.
Quoted in sect.3.3.

Wigner,E.P. (*1964b*): **Two kinds of reality.** The Monist 48, 248-264. (Reprinted in: E.P. Wigner, "Symmetries and reflections"; Indiana University Press, Bloomington, 1967; pp. 185-199).
Quoted in sect.3.5.

Wigner,E.P. (*1969*): **Epistemology of quantum mechanics.** In: "Contemporary Physics. Trieste Symposium 1968", Vol.2; International Atomic Energy Agency, Vienna; pp. 431-437.
Quoted in sect.3.5.

Wigner,E.P. (*1970a*): **On hidden variables and quantum mechanical probabilities.** Amer.J.Phys. 38, 1005-1009.
Quoted in sect.3.4.

Wigner,E.P. (*1970b*): **Physics and the explanation of life.** Foundations of Physics 1, 35-45.
Quoted in sect.3.5.

Wilbur,W.J. (*1977*): **On characterizing the standard quantum logics.** Trans.Amer.Math. Soc. 233, 265-282.
Quoted in sect.4.4.

Willems,J.C. (*1972a*): **Dissipative dynamical systems. Part I: General Theory.** Arch. Rat.Mech.Anal. 45, 321-351.
Quoted in sect.4.3.

Willems,J.C. (*1972b*): **Dissipative dynamical systems. Part II: Linear systems with quadratic supply rates.** Arch.Rat.Mech.Anal. 45, 352-393.
Quoted in sect.4.3.

Willems,J.C. (*1974*): **Consequences of a dissipation inequality in the theory of dynamical systems.** In: "Physical structure in system theory", ed. by JJ.van Dixhoorn and F.J.Evans; Academic Press, London; pp. 193-218.
Quoted in sect.4.3.

Williams,C. (*1931*): **Many dimensions.** Faber, London; new edition 1963.
Quoted in sect.3.6.

Willig,P. (*1974*): **Continuous W*-algebras are non-normal.** Tôhoku Math.J. 26, 483-486.
Quoted in sects.5.5, 5.6.

Winnink,M. (*1968*): **An application of C*-algebras to quantum statistical mechanics of systems in equilibrium.** Proefschrift Rijksuniversiteit te Groningen. V.R.B.-Offsetdrukkerij, Groningen.
Quoted in sect.4.3.

Witmer,E.E. (*1967*): **Interpretation of quantum mechanics and the future of physics.** Amer.J.Phys. 35, 40-52(1967), 36, 277(1968).
Quoted in sect.3.4.

Wittgenstein,L. (*1922*): Tractatus logico-philosophicus. Routledge and Kegan Paul, London.
Quoted in sect.2.4.

Wojciechowski,S. (*1975*): Symmetric transition probabilities in convex model of quantum mechanics. Reports on Mathematical Physics 8, 387-390.
Quoted in sect.4.5.

Woolley,R.G. (*1976a*): Quantum theory and molecular structure. Advan.Phys. 25, 27-52.
Quoted in sect.1.3.

Woolley,R.G. (*1976b*): On the description of high resolution experiments in molecular physics. Chem.Phys.Lett. 44, 73-75.
Quoted in sect.6.4.

Wright,J.D. (*1976*): Wild AW*-factors and Kaplansky-Rickart algebras. J.London Math. Soc. 13, 83-89.
Quoted in sect.4.6.

Wright,R. (*1978a*): Spin manuals. In: "Mathematical foundations of quantum theory"; ed. by A.R.Marlow, Academic Press, New York; pp. 177-254.
Quoted in sect.4.4.

Wright,R. (*1978b*): The state of the pentagon. A nonclassical example. In: "Mathematical foundations of quantum theory"; ed. by A.R.Marlow, Academic Press, New York; pp. 255-274.
Quoted in sect.4.4.

Wu,C.S., and Shakanov,I. (*1950*): The angular correlation of scattered annihilation radiation. Phys.Rev. 77, 136.
Quoted in sect.3.7.

Yanase,M.M. (*1961*): Optimal measuring apparatus. Phys.Rev. 123, 666-668(1961).
Quoted in sect.3.5.

Yanase,M.M. (*1964*): Remarks on the theory of measurement. Amer.J.Phys. 32, 208-211.
Quoted in sect.3.5.

Yanase,M.M. (*1971*): Optimal measuring apparatus. In: Proceedings of the international school of physics 'Enrico Fermi'. Course 49: Foundations of quantum mechanics"; ed. by B.d'Espagnat! Academic Press, New York; pp. 77-83.
Quoted in sect.3.5.

Yang,C.N. (*1950*): Selection rules for the dematerialization of a particle into two photons. Phys.Rev. 77, 242-245.
Quoted in sect.3.7.

Yang,C.N. and Mills,R.L. (*1954*): Conservation of isotopic spin and isotopic gauge invariance. Phys.Rev. 96, 191-195.
Quoted in sect.3.3.

Young,T.Y. and Calvert,T.W. (*1974*): Classification, estimation and pattern recognition. American Elsevier, New York.
Quoted in sect.6.3.

Yukawa,H. (*1973*): Creativity and intuition. Kodansha, Tokyo.
Quoted in sect.2.5.

Zabey,P.C. (*1975*): Reconstruction theorems in quantum mechanics. Foundations of Physics 5, 323-342.
Quoted in sect.4.4.

Zadeh,L.A. (*1965*): Fuzzy sets. Information and Control 8, 338-353.
Quoted in scet.5.4.

Zeh,H.D. (*1970*): On the interpretation of measurement in quantum theory. Foundations of physics 1, 69-76.
Quoted in sects.3.5, 3.6, 3.8, 5.6.

Zeh,H.D. (*1971*): On the irreversibility of time and observation in quantum theory. In: "Proceedings of the international school of physics 'Enrico Fermi'. Course 49: Foundations of quantum mechanics"; ed. by B.d'Espagnat; Academic Press, New York; pp. 263-273.
Quoted in sects.3.6, 3.8, 5.6.

Zeh,H.D. (*1973*): Toward a quantum theory of observation. Foundations of Physics 3, 109-116.
Quoted in sects.3.6, 3.8.

Zeman,J. (*1971*): Time in science and philosophy. Elsevier, Amsterdam.
Quoted in sect.5.2.

Zierler,N. (*1961*): Axioms for non-relativistic quantum mechanics. Pacific J.Math. 11, 1151-1169.
Quoted in sect.4.4.

Zierler,N. (*1966*): On the lattice of closed subspaces of Hilber space. Pacific J.Math. 19, 583-586.
Quoted in sect.4.4.

Zierler,N. and Schlessinger,M. (*1965*): Boolean embeddings of orthomodular sets and quantum logic. Duke Math.J. 32, 251-262.
Quoted in sect.3.4.

INDEX